PHYSICS FROM PLANET EARTH
An Introduction to Mechanics

PHYSICS FROM PLANET EARTH
An Introduction to Mechanics

Joseph C. Amato
Enrique J. Galvez

CRC Press is an imprint of the
Taylor & Francis Group, an **informa** business

CRC Press
Taylor & Francis Group
6000 Broken Sound Parkway NW, Suite 300
Boca Raton, FL 33487-2742

Printed on acid-free paper
Version Date: 20150325

International Standard Book Number-13: 978-1-4398-6783-9 (Pack - Book and Ebook)

Library of Congress Cataloging-in-Publication Data

Amato, Joseph C., author.
Physics from planet Earth - an introduction to mechanics / Joseph C. Amato and Enrique J. Galvez.
pages cm
Includes bibliographical references and index.
ISBN 978-1-4398-6783-9 (hardcover : alk. paper) 1. Mechanics. I. Galvez, Enrique Jose, 1956- author. II. Title.

QA805.A48 2015
531--dc23 2015001675

Visit the Taylor & Francis Web site at
http://www.taylorandfrancis.com

and the CRC Press Web site at
http://www.crcpress.com

To our former colleagues, who helped shape our careers while laying the groundwork for a dynamic, intellectually vibrant department at Colgate University:

Jack Dodd

Hugh Helm

Charles Holbrow

James Lloyd

Shimon Malin

Victor Mansfield

and

Donald Holcomb (Cornell)

CONTENTS

PREFACE FOR INSTRUCTORS

Physics from Planet Earth (*PPE*) is a one-semester calculus-based introduction to classical mechanics intended for first-year college students studying physics, chemistry, astronomy, or engineering. It is based on a course that has been taught successfully to students at Colgate University for over 15 years, using materials that were developed and refined during the past two decades and updated continually to include present-day research developments. The course differs from the traditional introductory course in several significant ways: (1) many topics are motivated or illustrated by contemporary applications in astronomy, planetary science, and space travel; (2) the three conservation laws (momentum, energy, and angular momentum) are introduced as fundamental laws of nature from which secondary concepts such as force and torque are derived; and (3) topics are arranged around the conservation laws, avoiding the "math overload" confronting students at the beginning of the standard course.

PPE was motivated by the authors' earlier experience teaching the traditional introductory course. Far too often, students were bewildered and dismayed by the rapid-fire presentation of physical concepts, mathematics, and problem-solving strategies. For too many good students, physics was a survival rite rather than an intellectual pleasure. In our opinion, textbooks bear some of the blame. By attempting to survey the full spectrum of physics, traditional textbooks sacrifice clarity and cogency in favor of coverage. Yet aside from their more sophisticated formatting and bolder use of color, today's texts are nearly interchangeable with the classic texts first published 70 years ago by the venerable authors Francis Sears, Mark Zemansky, David Halliday, and Robert Resnick.

In writing *PPE*, we endeavored to offer a clear and concise exposition of classical mechanics, one that students can read and *want* to read, and one that instructors will enjoy using. We acknowledge that the textbook is only one source of instruction—perhaps not the primary one—that students will access during the semester. Whatever its role, we want ours to amplify, enrich, and facilitate the instructor's unique way of exposing students to the elegant world of physics. All topics in *PPE* other than the core concepts (conservation, force, etc.) should be regarded as optional, and we encourage you to select, neglect, or append material to accommodate your personal teaching style and match your course to the background, ability, and aspirations of your students.

In *PPE*, the laws of physics are assigned the task of exploring the heavens—the Moon, Sun, solar system, galaxy, and universe—the same task shouldered by Newton 330 years ago at the birth of classical mechanics. *PPE* describes how humans, stranded in space on planet Earth, learned the laws of mechanics and used them to discover the structure and workings of the universe and identify their place within it. In our experience, this motif consistently captures student interest and provides numerous opportunities to describe the often haphazard, serendipitous, *human* aspect of scientific discovery. The traditional core concepts are addressed, but the conservation laws are given more prominence than usual, which seems appropriate given their predictive power and broad applicability throughout the diverse fields of science and engineering. Moreover, the conservation laws afford a cleaner, less cluttered platform for the exposition of classical mechanics.

Physics from Planet Earth is divided into five distinct sections. Section I is a review of the basic mathematical tools needed to study mechanics: geometry, trigonometry, and vectors. (Calculus is less of a concern; the simple derivatives and integrals encountered in the early chapters should be familiar to most students.) Rather than simply presenting the mathematics in a barebones fashion, we show how Greek astronomers used geometry and trigonometry to determine the scale of the solar system, and then we apply vectors plus elementary wave principles to discuss velocity, relative velocity, and the expansion of the universe (Hubble's law). We depart from the standard sequence of topics, wherein acceleration follows on the heels of velocity, reserving the former for later, when forces are introduced. There's plenty of good physics to discuss before then.

Section II deals with the conservation of momentum. Generalizing from simple 1-D collisions, we define mass and momentum, discover the conservation law, and use it to study 3-D collisions, rocket propulsion, and the "slingshot" effect. Then we dissect a collision to introduce force and Newton's laws. A broad range of lab-scale applications follow: projectile motion, friction, terminal velocity, rocket propulsion, circular motion, and oscillation. Next, we study Kepler's laws, which lead to Newton's discovery of universal gravitation and an array of astronomical applications: planetary orbits, weighing the Earth and stars, exoplanets, the shape of astronomical bodies, tidal effects, and the Roche limit.

Section III presents evidence for a second conservation principle—energy—which allows us to describe motion as a function of position rather than time. Applications include oscillators (linear and nonlinear), escape velocity, planetary atmospheres, orbits, and the physics of interplanetary travel.

Section IV is about angular momentum and its conservation. While it is mathematically the most challenging part of the course (employing vector cross products), it is also the most fun. We discuss a wide variety of applications, including pulsars, orbit eccentricity, baseball bats, asteroid collisions, pendula, gyroscopes, the changing length of a solar day, and the precession of the equinoxes.

In Section V (Chapter 14), we wrap things up with a discussion of dark matter, dark energy, and the ultimate fate of the universe.

Throughout the text, we have injected novel—but optional—topics that delve into the simplest aspects of special and general relativity. Our goal is exceedingly modest: to demonstrate that the relativistic expressions for momentum and energy, mass-energy equivalence, time dilation, and the gravitational red shift, are not radical departures from classical physics, but arise from straightforward (albeit bold and brilliant) extensions of Newtonian physics and Galilean relativity. Hopefully, our (oversimplified) treatment of Noether's theorem will convince students that symmetry can reliably guide theory, even at the frontiers of contemporary physics. Finally, the text is peppered with stories of scientific discovery and the human dramas behind them. We believe it is important for students to know that the path to discovery is not smooth, that scientists are fallible and far from perfect beings. We want them to know that, with passion and hard work, they too can make important and lasting contributions to their fields of study.

Chapter 14 deserves some explanation. In the authors' experience, the final week of a semester is a poor time to introduce new "testable" material. Instead, we sought to end the semester on a high note: (1) to show students that they could now appreciate the thrust of contemporary cosmology and (2) to inspire them to continue in their study of physics. The material of Chapter 14 is not meant "to be on the test."

A semester at Colgate University is 14 weeks long, and *PPE* is appropriately organized into 14 chapters. Guided by physics education research, we eliminated much lecture time in favor of active-learning modules. *PPE* is presently taught using 28 lectures interleaved with an equal number of "recitation" periods. The latter are conducted as brief reviews of lecture and reading followed by guided problem-solving. The easier "end-of-chapter" problems included in *PPE* originated as recitation questions. The short exercises sprinkled throughout the text first appeared as questions to be answered collaboratively during lectures. The more challenging end-of-chapter problems are either traditional problems gleaned from other texts or problems concocted by *PPE* instructors using data from recent space missions or astronomy research. If our students can solve the standard "textbook" problems of introductory mechanics, they can just as easily answer questions that reflect the interests of contemporary physicists and astronomers.

Throughout the long period of course development, we benefited from the suggestions of department colleagues teaching the course, who also contributed numerous exam questions that have been converted into end-of-chapter problems. We especially thank physicists Pat Crotty, Jonathan Levine, Beth Parks, and Ken Segall, plus astronomers Tony Aveni, Tom Balonek, and Jeff Bary, for their comments and contributions. The finished manuscript is vastly improved over the original thanks to the close reading of five anonymous reviewers who spotted errors, suggested clarifications, and made numerous insightful recommendations for improvement. Almost without exception, we adopted their recommendations wholeheartedly, even when they necessitated substantial modification of the manuscript. We accept full responsibility for any remaining errors.

We sincerely hope that you will find *Physics from Planet Earth* to be a refreshing and easy-to-use text for your introductory mechanics course, and look forward to receiving your feedback.

Joseph C. Amato
Enrique J. Galvez
Colgate University
Hamilton, New York

PREFACE FOR STUDENTS

Dear Students:

How we envy you! In your lifetime you will learn the answers to cosmic questions that have haunted thoughtful people throughout human history. *Are we alone in the universe*? Twenty-five years ago, we did not know of a single planet beyond those of our own solar system. Now we suspect that one out of every five suns (40 billion stars just within our own galaxy) hosts an Earth-like planet capable of supporting life. Before your career is over you'll likely be able to point to a star with a planet that harbors an intelligent civilization. *What is the universe made of*? We now know that only 5% of the universe is the ordinary stuff that forms atoms and molecules. The rest is *dark* matter and *dark* energy, and we have absolutely no idea what it is. Before your career is over, *you* will know. *How did the universe begin*? We've made great progress on this question, but the spark that triggered the cosmos still lies beyond the reach of current physical theories. During your lifetime, eagerly awaited new data and novel theories will bring the world closer to understanding the moment of creation. In the twenty-first century there are just as many unsolved mysteries as there were in the pioneering days of Isaac Newton. It is an exciting time to be studying physics.

Physics from Planet Earth (*PPE*) is a theme-based introduction to classical mechanics. Once you can solve the standard problems (ball and block, projectile, and inclined plane) found in traditional textbooks, you are well equipped to contemplate grander mysteries: What is the size and mass of Earth? Why is it round? How do the planets move, and how far away are celestial bodies? Above all, *how do we learn these things*? In physics, *how* we know is as important as *what* we know. Classical mechanics was developed over 300 years ago to answer questions like these. *PPE* follows a parallel story line. The laws of mechanics are deduced from familiar lab-scale experiments and are then used to construe the structure and behavior of the universe. To achieve this goal, we make a bold assumption: *the laws are the same everywhere in the universe*. All experimental evidence supports this assumption. Our Earth-derived physical laws have universal validity.

Physics is *imagination*—grown-up imagination tempered by exacting observations and tight logic. But imagination challenges the status quo and is hardly ever welcomed with immediate enthusiasm. If no one had told you, would you have imagined that you are on a spinning ball that is orbiting the Sun at a mind-numbing speed of 67,500 miles per hour? If you *had* been told, would you have believed it? It took over 2000 years of imagination, insight, and experimental skill to establish these "simple" facts and convince people of their truth. Envisioning Earth as a *planet*—an isolated outpost in the vastness of space—was one of the greatest achievements of sixteenth-century science. *PPE* is studded with stories of scientific breakthroughs and the human dramas surrounding them. We sincerely hope you will enjoy reading them. Like all of us, physicists are imperfect beings, and the path of science is often erratic and messy. Scientific progress depends as much on passion, luck, and hard work as it does on flashes of brilliance.

Finally, we want you to appreciate the skill and ingenuity of our ancestors. During the semester, take time to study the night sky—it is stunningly beautiful. Look at the crescent Moon. Its visible shape and position (relative to the Sun) prove that it is much closer to us than the Sun. Two thousand years ago, long before telescopes were invented, Greek astronomers understood this. Look for Jupiter, and with binoculars find its largest moons. From night to night, the moons change their positions relative to the planet as the planet drifts across the starry backdrop. In the sixteenth and seventeenth centuries these observations allowed our forebears to discern the geometry of the solar system and infer the ubiquitous presence of gravity. Now locate the constellation Cassiopeia, the one shaped like a giant "W." Using the right hand "V" as a pointer, follow a straight line with your eye until you find a barely visible smudge. One hundred years ago, astronomers determined that this fuzzy patch was too far away to be part of the Milky Way. This is the Andromeda galaxy, the only naked-eye object in the sky that is not within our own galaxy. It holds a trillion stars and is *en route* to a head-on collision with us in about 2 billion years. A century ago, it held the first clue to the true size of the universe. Long before computers, the Internet, or smartphones, our ancestors deciphered these clues and discovered their place in the cosmos. And so can you. Wherever you look, the heavens offer up vast troves of information for discovery and interpretation by educated, inquisitive minds.

While teaching this course over a span of 15 years, the authors frequently added examples drawn from the news of the day, reported in newspapers and numerous science websites. Mission updates issued by NASA, ESA, JAXA, and other space agencies, Hubble Space Telescope images, and data from the Kepler Observatory were routinely added to enrich the content of the course. We challenge you to do the same: download timely articles appearing in the news and try to understand them using the physics you are learning. More often than not, you will discover that they are within your grasp. In an informed way, you will appreciate the imagination and cleverness of present-day scientists and marvel at the magnificence of the cosmos.

Fr. Georges Lemaître, a Catholic priest, distinguished astronomer, and professor of physics in Belgium, was one of the originators of the concept of an expanding universe and the Big Bang theory of creation. In an essay written in 1931, he wrote as follows:

> If I had a question of the infallible oracle..., I think I would choose this: "Has the universe ever been at rest, or did the expansion start at the beginning?" But, I think, I would ask the oracle not to give the answer, in order that a subsequent generation would not be deprived of the pleasure of searching for and of finding the solution.*

So we really do envy you!

Professor Joseph C. Amato
Professor Enrique J. Galvez
Colgate University
Hamilton, New York

* Georges Lemaître, "The beginning of the world from the point of view of quantum theory," *Nature* **128**, 704–706 (1931).

AUTHORS

Joseph C. Amato earned a PhD in experimental solid state physics at Rutgers University, New Jersey, in 1975. He began teaching at Colgate University, Hamilton, New York, in 1977, worked in accelerator physics at Cornell University, Ithaca, New York, in 1981–1984, and retired from Colgate University in 2009 as the William R. Kenan Jr. Professor of Physics. At Colgate University, Dr. Amato conducted research in low temperature physics (funded by the National Science Foundation and Research Corporation), physics education (funded by NSF), and most recently engaged in an experimental study of granular materials. He has been active in physics curriculum development and especially in the design of novel laboratory apparatus and exercises for introductory physics courses. He is a recipient of the Professor of the Year award from the Colgate chapter of American Association of University Professors.

Enrique J. Galvez earned a PhD in physics at the University of Notre Dame, Indiana, in 1986. He has been member of the faculty at Colgate University since 1988 and is currently the Charles A. Dana Professor physics and astronomy. His research interests include atomic and optical physics and physics education. He has received several external grants from Research Corporation and the National Science Foundation. Dr. Galvez's recent research projects include experimental atomic physics with Rydberg atoms, geometric phases in optics, and photon entanglement. His educational projects include modernizing the introductory physics curriculum and the developing new quantum mechanics laboratories.

Section I

MATHEMATICAL TOOLBOX

1

SURVEYING THE SKIES

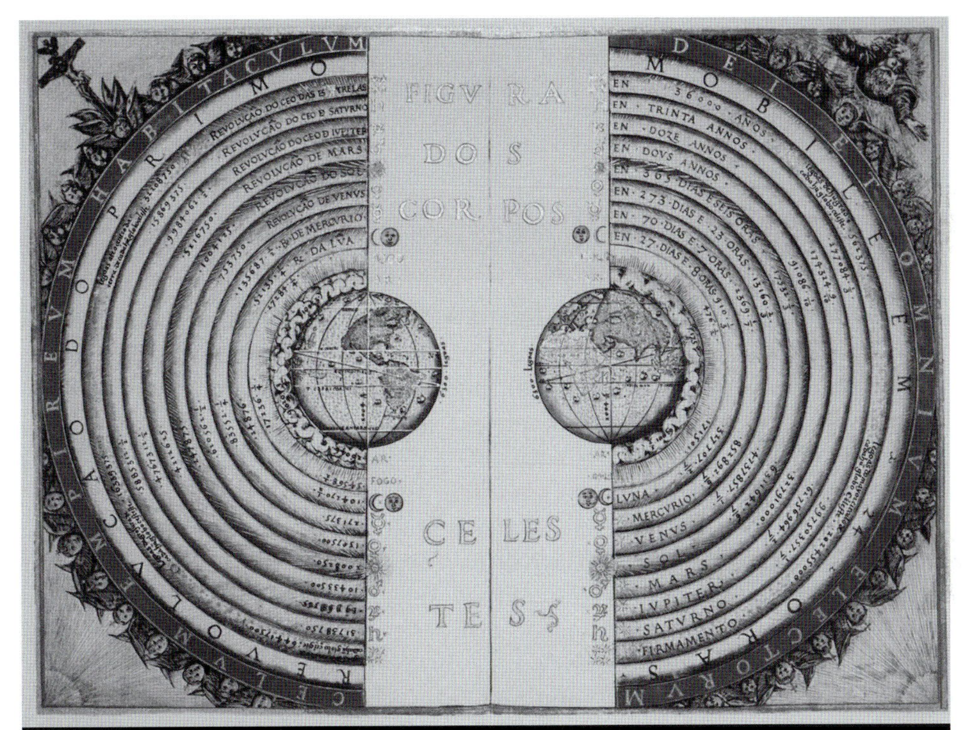

Ptolemy's universe, as depicted by the Portuguese cartographer Bartolomeu Velho in 1568. The Moon, the five known planets, and the Sun each occupy unique spheres, as do the fixed stars (Firmamento). The *synodic periods* of the planets are also indicated. This beautiful drawing is located in the Bibliotheque Nationale in Paris. (From the public domain.)

Why study physics? The original motivation—which inspires us even today—was to make sense of what we saw happening in the sky, to understand the universe and our place within it. How big are the Moon and Sun? How far away are they? How do the planets move? These questions were first addressed nearly 2500 years ago by Greek scholars who applied the newly developed methods of geometry to construct a mathematical model of the solar system.

Throughout the history of physics, geometric analysis has played a key role in understanding terrestrial (Earth-related) as well as astronomical phenomena. Perhaps the great Italian scientist Galileo Galilei said it best, in his book Il Saggiatore (The Assayer), published in 1623:

> *Philosophy [i.e., physics] is written in this grand book—I mean the universe—which stands continually open to our gaze, but it cannot be understood unless one first learns to comprehend the language and interpret the characters in which it is written. It is written in the language of mathematics, and its characters are triangles, circles, and other geometrical figures, without which it is humanly impossible to understand a single word of it; without these, one is wandering around in a dark labyrinth.*

The purpose of this chapter is not merely to erect a historical backdrop for the study of physics, but also to illustrate the central importance of geometry to scientific discovery. We hope it will inspire you to refresh your mastery of this beautiful field of mathematics.

1.1 EARLY ASTRONOMY

In the ancient world, before the widespread use of artificial light, nighttime must have seemed much darker, and the night sky must have appeared much brighter, than it does to us today. Early civilizations worldwide studied the skies with intense and enduring interest. The apparent motion of the Sun, Moon, planets, and stars captured the imagination of ancient sky watchers, shaped their religious beliefs and set the yearly rhythm of their lives. Before 2000 BCE, monuments were erected in northern Europe (e.g., Stonehenge) and Egypt (the pyramids), and later in the New World (e.g., Machu Picchu, Peru, 1450 CE) in precise alignment with the rising Sun at the summer solstice* or other key seasonal events. Egyptians noted that the annual flooding of the Nile River valley began when the star Sirius made its first appearance at sunrise on the eastern horizon. Accordingly, they devised a yearly calendar that began with this celestial signal. Careful observations of Sirius over many years helped to establish the length of the year as $365\frac{1}{4}$ days. By 750 BCE, Babylonian observers were keeping detailed records of planetary observations as well as the dates of lunar and solar eclipses. Based on tens to hundreds of years of record keeping, they discovered, for example, that solar and lunar eclipses occurring 18 years apart (more precisely, 18 years, $11\frac{1}{3}$ days, called a *Saros* cycle) looked *exactly* the same. These observations were used primarily for astrological purposes rather than scientific ones; nevertheless, they indicate a profound curiosity about the heavens, plus a growing willingness to perform careful measurements and record them faithfully over many years to uncover recurring patterns of nature. These are the seeds of experimental science.

* The summer solstice is the longest day of the year, presently June 20 or 21 in the northern hemisphere.

Given the rudimentary state of mathematics and technology in the ancient world, the wealth of astronomical knowledge accumulated by early civilizations is amazing. Even more remarkable, though, is the flowering of mathematics and science that occurred in ancient Greek society from the sixth century BCE to the second century CE. Prior to this period, geometry was primarily concerned with practical matters such as land surveying and architecture.* Starting around 500 BCE, Greek mathematicians (among others, Pythagoras and his followers) transformed geometry from a practical tool into the pure mathematical discipline we are familiar with today, in which relationships between abstractions such as perfect circles, perfectly straight lines, and infinitesimal points are constructed from pure logic rather than by measurement. Euclid's *Elements*, a treatise on geometry and early number theory, was completed in 300 BCE and is widely considered to be one of the most influential books in the history of the world.† Based on "pure" irrefutable logic and dealing with "perfect" shapes, geometry was, in the eyes of Greek astronomers, ideally suited for studying the heavens.

* In fact, the word *geometry*, from the Greek *geometrien*, literally means *land measurement*.

† Euclid's *Elements* was required reading for university students up to the early twentieth century.

1.2 ERATOSTHENES AND THE EARTH'S RADIUS

By the middle of the fifth century BCE, it was widely accepted that the Earth was anything but flat. Travelers reported that as they ventured north, new stars rose above the northern horizon, while stars on the southern horizon sank from view, just as one would expect if the Earth's surface were convex. The Greek astronomer Parmenides correctly deduced that the Moon reflected light from the Sun and argued that a lunar eclipse occurred when the Moon passed through Earth's shadow. By studying the monthly progression of the Moon's appearance,‡ astronomers concluded that the Moon must be a sphere (Figure 1.1). But if the Moon is spherical, perhaps Earth is, too. Indeed, careful observation of lunar eclipses revealed that Earth's shadow is circular (Figure 1.2), proving beyond a doubt that Earth is a sphere.

In about 250 BCE, Eratosthenes, living in Alexandria (Egypt) on the coast of the Mediterranean Sea, conducted an elegant experiment to measure the Earth's radius. He knew that at noon on the summer solstice, the Sun's image could be seen reflected from water at the bottom of deep wells in the town of Syene,§ 800 km south of Alexandria. In other words, the Sun was directly overhead of Syene at that time (see Figure 1.3). In Alexandria on the same date, he erected a vertical rod (called a *gnomon*) of height h on a flat, horizontal surface and measured the length of its shadow ℓ_{sh} cast by the Sun. From the lengths of the rod and its shadow, Eratosthenes calculated the angle $\varphi = \tan^{-1}(\ell_{sh}/h)$ between the

‡ That is, the *phases* of the Moon: crescent, first quarter (or half full), gibbous (more than half full), full, etc.

§ The modern name for Syene is Aswan, the location of one of the world's largest dams.

FIGURE 1.1 Images of the Moon taken from September 29 to October 6, 2003. The curvature of the *terminator* separating the bright and dark regions strongly suggests that the Moon is a sphere illuminated by the Sun. (Photo: Chad Purser.)

rod and the direction of the Sun to be 7.2° or 0.13 rad. (See Box 1.1 for a review of radians.) Referring to Figure 1.3, angles θ and φ are identical because they are opposite interior angles. Therefore,

$$\varphi = \theta = \angle AOS = \frac{d}{R_E} = 0.13\ \text{rad}, \tag{1.1}$$

where R_E is the radius of the planet. Given the distance d = 800 km, R_E is easily found to be 6400 km, in good agreement with modern measurements.

Equation 1.1 is so simple that it is easy to overlook its significance. Geometry, a practical tool long used by surveyors to lay out tracts of land, was now being used to measure distances of astronomical scale. The lines drawn by the surveyor now stretched to the heavens.

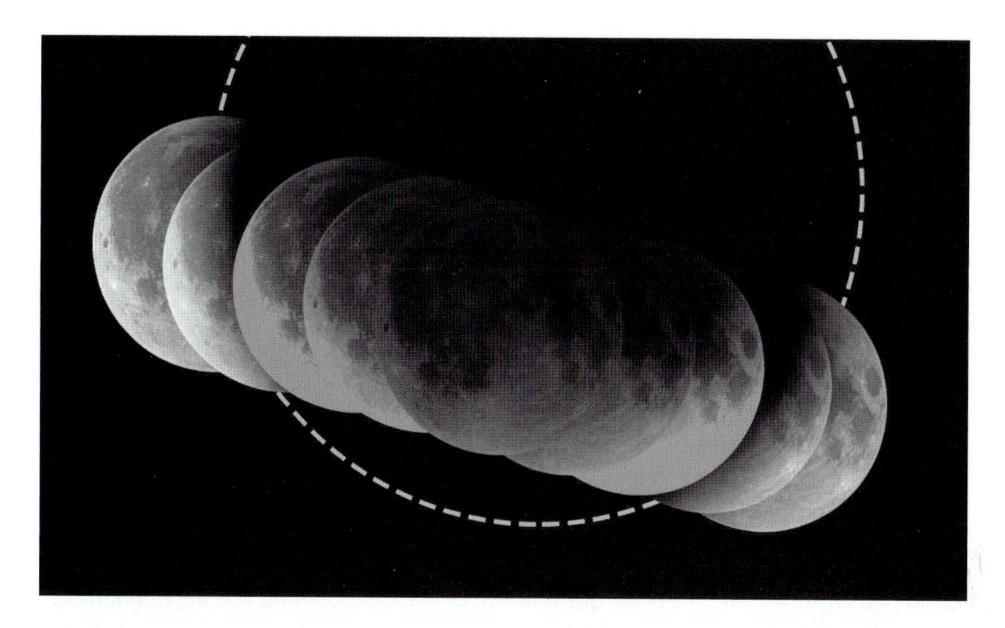

FIGURE 1.2 Total lunar eclipse of April 15, 2014. Nine images, taken at equal time intervals, are shown as the Moon moves through the shadow of the Earth. The dotted circle has a radius equal to 8/3 of the apparent lunar radius. (Image courtesy of Adam Block, Mount Lemmon Sky Center, University of Arizona, Phoenix, AZ, processing with the assistance of Samantha Galvez.)

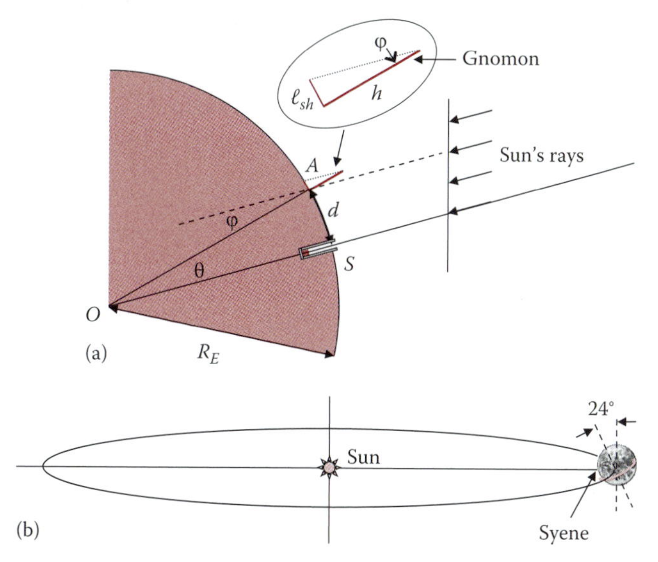

FIGURE 1.3 Eratosthenes' experiment: (a) The locations of Alexandria and Syene are indicated by A and S, respectively. At noon on the summer solstice, the Sun was directly overhead in Syene, but cast shadows in Alexandria. At A, a gnomon of length h cast a shadow of length ℓ_{sh}. (b) The position of Earth at summer solstice. The latitude of Syene is 24°5′. In 250 BCE, the Earth's axis was tilted by almost exactly the same amount.

BOX 1.1 RADIANS AND SMALL ANGLE APPROXIMATIONS

You are no doubt familiar with measuring angles in terms of degrees, a unit of measurement that we inherited from the Babylonians living 5000 years ago. As you recall, a full circle encompasses an angle of 360°; each degree is divided into 60 arc minutes (′), while each minute is further divided into 60 arc seconds (″), or $1'' = \frac{1}{60}{}' = \frac{1}{3600}$ deg. The *radian* is an alternate measure of angle, and is defined as follows. Imagine a thin rod of length r that is free to pivot about one end. Rotate this rod through an angle θ, as shown in Figure B1.1a. The angle θ, expressed in radians, is defined as the arc length s swept out by the free end of the bar, divided by its length r:

$$\theta\,(\text{rad}) = \frac{s}{r}. \quad \text{(B1.1)}$$

Because s and r have units of length, their quotient θ is dimensionless. Radians are a "natural" way to describe angles, totally independent of our units of length (cm, inches, etc.). If we ever encounter intelligent extraterrestrials (who have never heard of Babylonia), they would know nothing about degrees, minutes, and seconds, but would surely be quite familiar with expressing angles in terms of radians.

It's easy to convert between radians and degrees. Look back at Figure B1.1a, and increase θ until the rotating rod sweeps out a full circle. Then, the arc length is the circle's circumference, $s = 2\pi r$, and the angle θ in degrees is 360°. In terms of radians,

$$\theta\,(\text{rad}) = \frac{2\pi r}{r} = 2\pi = 360°,$$

or 2π (rad) = 360°. Therefore, to convert from degrees to radians,

$$\theta\,(\text{rad}) = \frac{2\pi}{360} \times \theta\,(\text{deg}). \quad \text{(B1.2)}$$

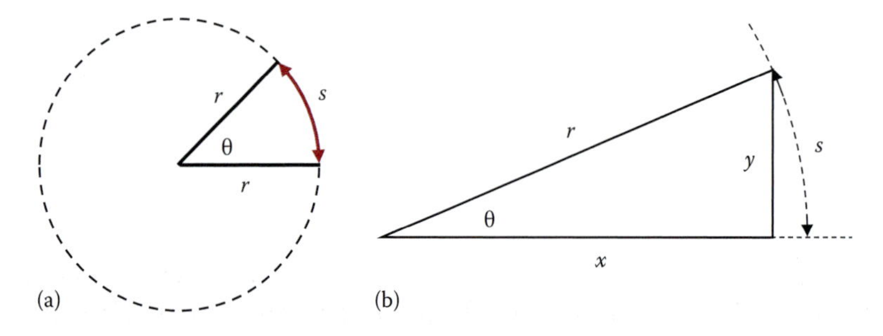

FIGURE B1.1 (a) The angle θ, measured in radians, is defined as the ratio of the arc length s to the radius r, $\theta = s/r$. (b) As $\theta \to 0$, the arc length s approaches the altitude y, and the hypotenuse r approaches the base x.

Exercise B1.1: (a) Express the following angles in terms of radians: 30°, 45°, 60°, 90°, and 180°. It is common to express each of these angles as a rational fraction of π. (b) How many degrees are equal to 1 rad?

Answer (b) 57.3°

In physics, radians are the preferred way to express angles. To see why, consider the triangle shown in Figure B1.1b. Imagine that the triangle is inscribed within a circle of radius r, so r is its hypotenuse, and $\sin\theta = y/r$. Note that the arc length s is not very different in length from y; in fact, as θ gets smaller, s and y become indistinguishable. Therefore, in the limit as $\theta \to 0$, $s \to y$ and $y/r \to s/r$, or

$$\sin\theta \simeq \theta \quad \text{for } \theta\,(\text{rad}) \ll 1. \tag{B1.3}$$

This is called the *small angle approximation* for $\sin\theta$.

Similarly, for small angles such that $\theta \ll 1$ (rad), $x \to r$ so $y/x \to y/r \simeq s/r$, or

$$\tan\theta \simeq \sin\theta \simeq \theta \quad \text{for } \theta\,(\text{rad}) \ll 1. \tag{B1.4}$$

Exercise B1.2: Use the small angle approximations of Equations B1.3 and B1.4 to estimate the sine and tangent of the following angles: θ (rad) = 0.2, 0.1, 0.05, and 0.01. Then use your calculator to find exact values for the sine and tangent of these angles. Compare your exact results with your approximations. Do you see a trend?

You will encounter radians throughout this course, and in all advanced physics courses as well. Be sure that you know how to use your calculator to convert between radians and degrees; and when calculating trigonometric functions, always be sure that you know what mode (radian or degree) your calculator is currently in!

EXERCISE 1.1

A 2 m gnomon is erected in Syene. At noon on the summer solstice, the rod does not cast a shadow. At a different time on the same day, a shadow of length 1.16 m is cast to the east of the gnomon. What is the time?

Answer 2:00 PM

1.3 ARISTARCHUS, HIPPARCHUS, AND THE DISTANCE TO THE MOON AND SUN

When we look at an object, its *apparent* size is proportional to the angle it *subtends* at our location. (See Figure 1.4.) From Earth, the angle subtended by the Sun, 0.53°, is almost exactly equal to the angle subtended by the Moon, a fact that was well known to Greek astronomers in 300 BCE. The Sun and the Moon therefore appear to be the same size. So how did early astronomers deduce that the Sun was much larger, and much farther away, than the Moon? The answer can be found in Figure 1.5a and b, in which the Earth, Moon, and Sun are located at the vertices of triangle *EMS*. θ_1 is the angle between the Sun and Moon, measured from Earth, whereas θ_2 is unknown. The line $\overline{tt'}$ is the *terminator* separating the bright and dark halves of the Moon, and the line $\overline{vv'}$ separates the near side of the Moon (visible from Earth) from the far side that is hidden from our view. Note that $\overline{tt'} \perp \overline{SM}$ and $\overline{vv'} \perp \overline{EM}$. The area of the Moon that is illuminated by the Sun *and* visible to us on Earth lies between lines $\overline{tt'}$ and $\overline{vv'}$ (i.e., between points t and v) and is proportional to angle φ. Following the construction shown in the inset to Figure 1.5a, $180° = (90° - \theta_1) + (90° - \theta_2) + \varphi$, or $\varphi = \theta_1 + \theta_2$. For example, if $\theta_1 + \theta_2 = 90°$, we will see a half moon from our observation point on Earth; if $\theta_1 + \theta_2 = 45°$, we will see a thin crescent moon, etc. Knowing θ_1, and estimating $\theta_1 + \theta_2$ from the visible fraction of the Moon's surface, we can find θ_2 and construct triangle *EMS* like the one shown in Figure 1.5a. Then, using the law of sines, we can estimate the ratio between the Earth–Sun distance and the Earth–Moon distance.

EXERCISE 1.2

Typically, the Moon and the Sun are not simultaneously visible in the sky. In this case, how might you measure θ_1? (*Hint:* Think about *time*.)

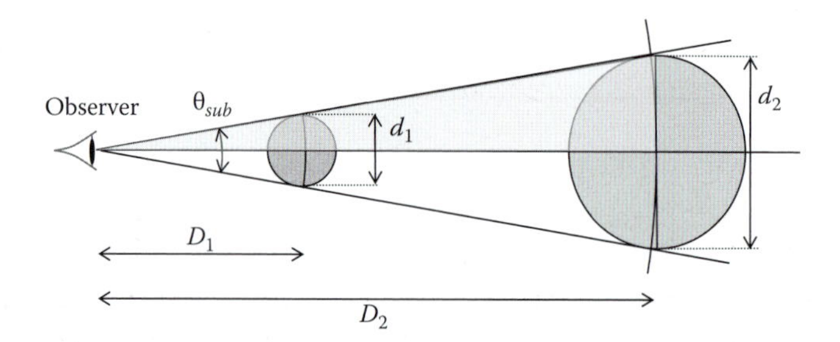

FIGURE 1.4 The angle subtended by a spherical body is defined as the angle between two lines drawn from the observer to the outer extremities of the body. For astronomical bodies, $\theta_{sub} \ll 1$ rad, so that θ_{sub} = diameter/distance. The two bodies shown in the figure have different diameters but the same subtended angle, $\theta_{sub} = d_1/D_1 = d_2/D_2$. This is the situation for the Moon and Sun when viewed from Earth.

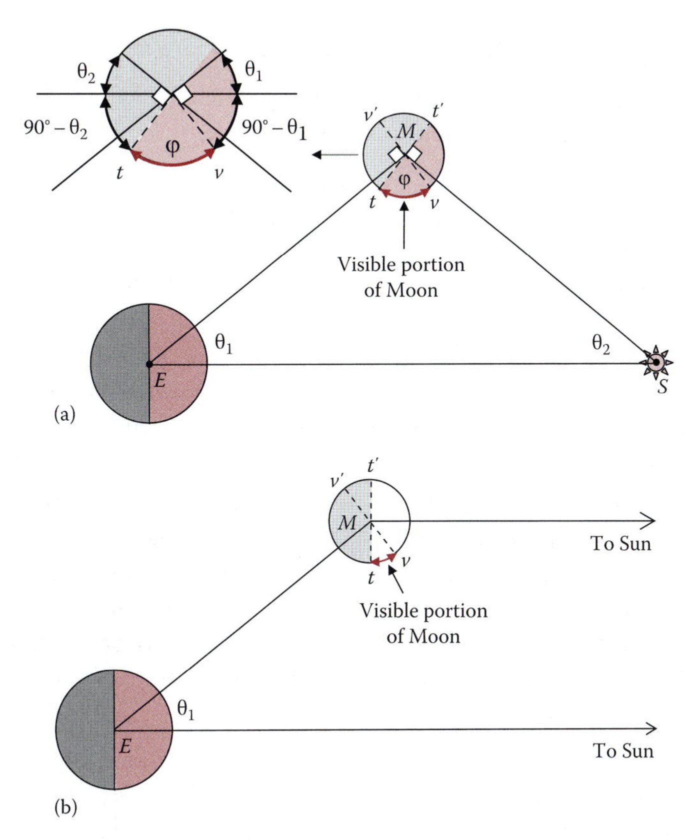

FIGURE 1.5 The visible portion of the Moon, when viewed from Earth, lies between points t and v. In (a), the distances from Earth to the Moon and Sun are assumed to be comparable, that is, $R_{ES} \sim R_{EM}$, whereas in (b), it is assumed that $R_{ES} \gg R_{EM}$. The visible slice of the Moon is much smaller in the latter case.

Figure 1.5a is drawn with $R_{EM} \approx R_{ES}$, whereas Figure 1.5b is drawn with $R_{ES} \gg R_{EM}$, so that $\theta_2 \approx 0$. If $\theta_1 = 45°$, then in the first case the Moon will appear half full, whereas in the second it will appear as a thin crescent. A few casual observations of the (real) Moon will convince you (as they convinced astronomers 2200 years ago) that the second case is correct: the Sun is many times more distant than the Moon, so it is vastly larger as well.

In the early third century BCE, the astronomer Aristarchus attempted to measure the ratio of distances R_{ES}/R_{EM} using the clever method illustrated in Figure 1.6. In the figure, M and M' are the locations of the Moon when it appears half full (or *in dichotomy*) when viewed from Earth. Using the reasoning described above, $\angle EMS = \angle EM'S = 90°$, so the two triangles in the figure are congruent right triangles. Aristarchus measured the total time $t_1 + t_2$ for a complete (360°) orbit of the Moon to be 27.3 days, and the time t_1 between dichotomies (the first and third quarter Moons) to be 13.2 days, allowing him to determine θ_1 $(=\theta_2)$. Using $\cos\theta_1 = R_{EM}/R_{ES}$, he concluded that $R_{ES} \approx 20\,R_{EM}$, which, although exceedingly inaccurate (the correct ratio is ≈ 400!), remained unchallenged for over 1800 years! Such was the unquestioned authority of the early Greek astronomers.

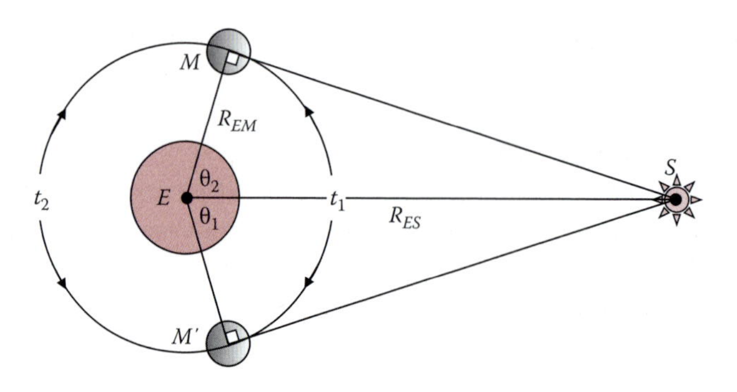

FIGURE 1.6 Aristarchus' measurement of R_{ES}/R_{EM}. Points M and M' indicate the positions of the first and third quarter Moon, when it is half full as viewed from the Earth. The fact that θ_1 $(= \theta_2) \approx 90°$ proved that $R_{ES} \gg R_{EM}$.

EXERCISE 1.3

(a) Use Aristarchus' numbers for the orbital period of the Moon and the time t_1 to calculate his value of θ_1. (b) If the Sun were *infinitely* far from Earth, what would be the value of θ_1? (c) What would t_1 have been? Why was Aristarchus' result so poor? (How accurately can *you* estimate the exact times of dichotomy?)

Answer (a) 87° (b) 90° (c) 13.65 days

In spite of the error, Aristarchus had correctly concluded that $R_{ES} \gg R_{EM}$. Moreover, by *imagining* the Earth, Moon, and Sun as points at the vertices of a giant celestial triangle, and using geometry to solve for the side lengths of that triangle, he had achieved the first calculation of an astronomical distance from experimental data. In many respects, his work marked the birth of the science of astronomy.

In the second century BCE, the astronomer Hipparchus combined this result with lunar eclipse observations to determine the Earth–Moon distance with uncanny accuracy. His ingenious analysis is outlined in Figure 1.7. The figure (which is *not* drawn to scale) depicts the Sun, Earth, and Moon at the midpoint of a total lunar eclipse. The lines $\overline{SEO}$ and $\overline{S'E'O}$ drawn tangent to the Sun and Earth coincide with the edges of the Earth's shadow cast by the Sun. R_{ES} and R_{EM} represent the center-to-center distances between the Earth and Sun and the Earth and Moon, whereas r_S, r_E, and r_M are the radii of the Sun, Earth, and Moon.

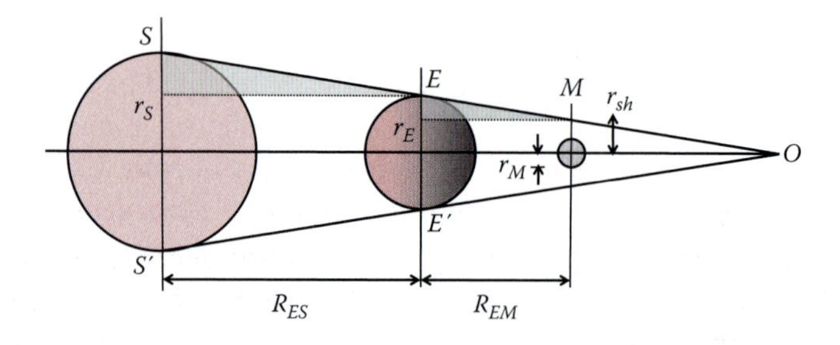

FIGURE 1.7 Hipparchus' calculation of R_{EM}, showing the Sun, Earth, and Moon at the midpoint of a full lunar eclipse. r_{sh} is the radius of the Earth's shadow at a distance equal to R_{EM}.

In addition, r_{sh} is the radius of the Earth's shadow projected onto the Moon. Over 2200 years ago, long before the time lapse photography of Figure 1.2 was possible, Greek astronomers had determined that $r_{sh} \simeq (8/3) r_M$, in excellent agreement with modern measurements.

Following Aristarchus, Hipparchus assumed that $R_{ES} = nR_{EM}$, where $n \gg 1$. (As you will see, the exact value of n is not important to the analysis.) Since the angles subtended by the Sun and Moon are equal, $r_S = nr_M$. To simplify Hipparchus' analysis, it is convenient to draw two (dotted) lines parallel to the central line in Figure 1.7. All of the triangles in the figure are similar, so the ratios of their corresponding side lengths are equal. In particular, for the two *shaded* triangles,

$$\frac{r_E - r_{sh}}{R_{EM}} = \frac{r_S - r_E}{R_{ES}},$$

or

$$\frac{r_E - r_{sh}}{r_S - r_E} = \frac{R_{EM}}{R_{ES}} = \frac{1}{n}. \tag{1.2}$$

Rearranging Equation 1.2 and expressing r_S and r_{sh} in terms of r_M, we obtain

$$n\left(r_E - \frac{8}{3} r_M\right) = nr_M - r_E,$$

or

$$(n+1) r_E = n\left(1 + \frac{8}{3}\right) r_M.$$

But if $n \gg 1$, then $(n + 1) \simeq n$, so $r_M \simeq (3/11) r_E = 0.27 r_E$. This is Hipparchus' first result. Let's see how accurate it is. The currently accepted values of the Earth and Moon radii are $r_E = 6371$ km and $r_M = 1740$ km, so $r_M = 0.273 r_E$. Hipparchus' result (from about 250 BCE) is astonishingly accurate!

To continue, recall that the angle φ subtended by the Moon, as viewed from Earth, is 0.53°, or 9.25×10^{-3} rad, that is, $\varphi = 2r_M/R_{EM} = 9.25 \times 10^{-3}$ (see Figure 1.4). Substituting Hipparchus' result for r_M,

$$R_{EM} = \frac{(6/11) r_E}{9.25 \times 10^{-3}} = 59 r_E.$$

Since φ is the angle subtended for an observer at the Earth's *surface*, our result is the distance from the Earth's surface to the center of the Moon. To find the *center-to-center* Earth–Moon distance, we need to add one extra Earth radius, yielding $R_{EM} = 60 r_E$. The currently accepted mean distance to the Moon is 3.82×10^5 km, yielding $R_{EM} = 60.0 r_E$. Once again, Hipparchus' result is in spectacular agreement with our modern value! Its significance is that it was the first *accurate* determination of an astronomical scale distance. No wonder Hipparchus is widely considered to be the greatest of the early Greek astronomers.

1.4 PTOLEMY AND THE EARTH-CENTERED UNIVERSE

Very little is known about the progress of astronomy during the two centuries after Hipparchus. Greece and Egypt were annexed by the Roman Empire, but Alexandria remained a thriving cultural and scientific center. In 140 CE, the Alexandrian astronomer and mathematician Ptolemy (or Claudius Ptolemaeus—he was a Roman citizen) published one of the most important and enduring books of the ancient world. Part star catalogue, part compendium of planetary and eclipse data, Ptolemy's *Almagest** is the most valuable single source of historical information regarding early Greek astronomy. More importantly, the *Almagest* contains a detailed mathematical model describing the apparent motion of the planets, Sun, and stars, an Earth-centered (or *geocentric*) model of the universe that remained unchallenged for over 1400 years.

Ptolemy's universe (Figure 1.8) is a set of concentric spheres with Earth at its center. The Moon, Sun, and planets occupy spheres of increasing radius, with the stars assigned to the largest sphere, farthest from Earth and closest to the domain of the gods. Among these bodies, the stars appear to move in the

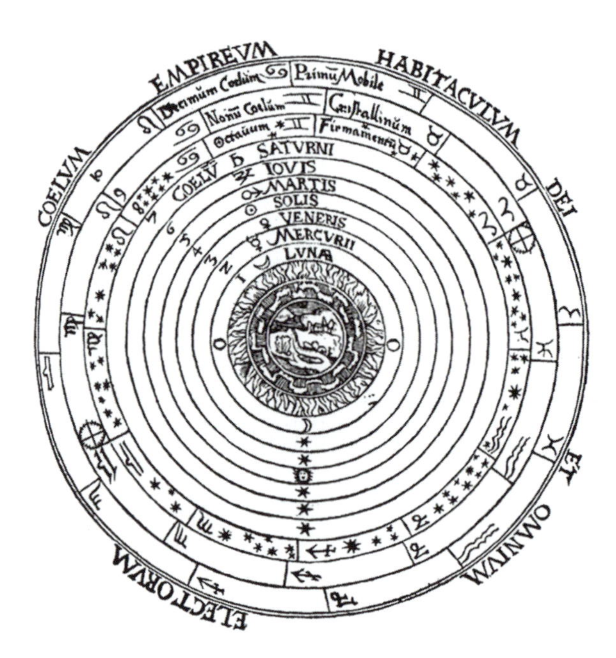

FIGURE 1.8 Ptolemy's Earth-centric universe, drawn by Peter Apian, in his *Cosmographia* (1539). Note the concentric circles occupied by the Moon, Mercury, Venus, Sun, Mars, Saturn, Jupiter, and the stars, surrounded by the realm of the gods. (From *Cosmographia*, by Peter Apian [1539], public domain.)

* The title *Almagest* is historically interesting. Ptolemy referred to his work as "The Great Compilation." During the Middle (or Dark) Ages, the book was preserved and translated by Arab scholars, who truncated the name to "The Greatest," or *Al Magesti*. In the twelfth century CE, the book was translated (from Arabic) into Latin as the *Almagesti*, and later the title was anglicized.

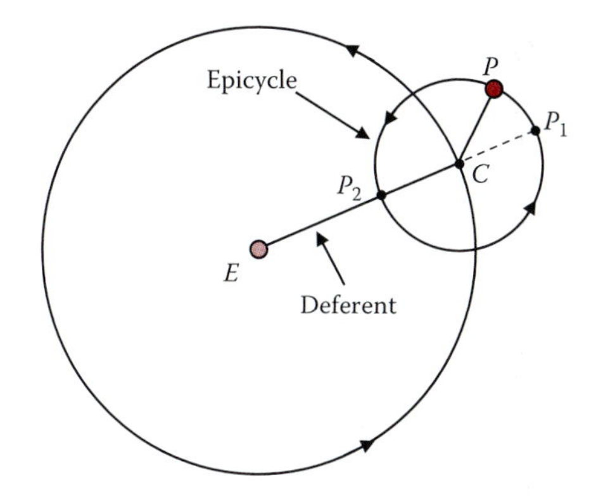

FIGURE 1.9 Ptolemy's model for planetary motion, showing the *deferent* $\overline{EC}$ and the epicycle centered on point *C*. The planet *P* rotates uniformly about point *C* as *C* itself rotates about the Earth. Together, the two motions account for the puzzling retrograde motion of the planets.

simplest manner: their positions relative to one another are fixed and their sphere makes one full revolution per day about Earth. (As we now know, it is the Earth that turns once per day). The motion of the planets is more complicated. Their positions relative to one other, and relative to the stellar background, change profoundly during the course of a year.* Each day, the planets follow the Sun across the sky, rising in the east and setting in the west. Over longer periods of time, they appear to drift slowly from west to east among the stars (called *prograde* motion), but during certain intervals, they reverse their drift direction and move east to west relative to the starry backdrop (called *retrograde* motion). The challenge to Greek astronomers was to "save the appearances," that is, to model this puzzling behavior geometrically using perfect circles, the only type of motion deemed appropriate for celestial bodies. Building on the work of earlier astronomers (especially Hipparchus), Ptolemy solved the puzzle by describing each planet's path as a combination of two interlocking circles, shown in Figure 1.9. The first circle (called the *deferent*) is centered on Earth *E*, and it defines the uniform rotation of a point *C* that is the center of the second circle, called the *epicycle*. The planet *P* rotates around its epicycle as its center *C* rotates about Earth. The planet appears to have its fastest prograde motion at point P_1, when it is farthest from Earth. At point P_2, when it is closest to Earth, it has its fastest retrograde rate, and at other points on the epicycle it briefly appears to be stationary (i.e., its apparent direction of travel changes from prograde to retrograde and *vice versa*) relative to the background of distant stars.†

The introduction of epicycles (and other small corrections) satisfactorily described the wealth of planetary data collected over centuries by earlier astronomers, while it preserved the Greeks' philosophical preference for perfect circular motion. As a physical model, however, it was somewhat suspect because the planets were not all treated in the same fashion. Planetary observations indicated that the angle between Venus and the Sun (called the angle of *elongation*) never exceeded 46°, and the angle between Mercury and the Sun never exceeded 23°, whereas the elongation of Mars, Jupiter, and Saturn passed through all possible values (up to 180°) over the course of a few years. To account for the anomalous behavior of Mercury and Venus, Ptolemy placed the centers of their epicycles in line with the Sun (see Figure 1.10), and forced C_V, C_M, and *S* to circle Earth *together* once per year. The epicycle centers for Mars, Jupiter, and Saturn, on the other hand, were free to circle Earth at different rates, permitting those planets to drift far from alignment with the Sun. No justification was offered for the two distinctly different behaviors, perhaps because Ptolemy's goal was to *describe* planetary motions geometrically rather than to search for *causes*. From a scientific point of view, the more complicated a model needs to be, the less convincing it is.‡ Ptolemy's

* The word *planet* is derived from the Greek word *wanderer*.

† Ptolemy's model was somewhat more sophisticated than this, but the added complications are of historical interest only.

‡ This statement is often referred to as *Occam's Razor*.

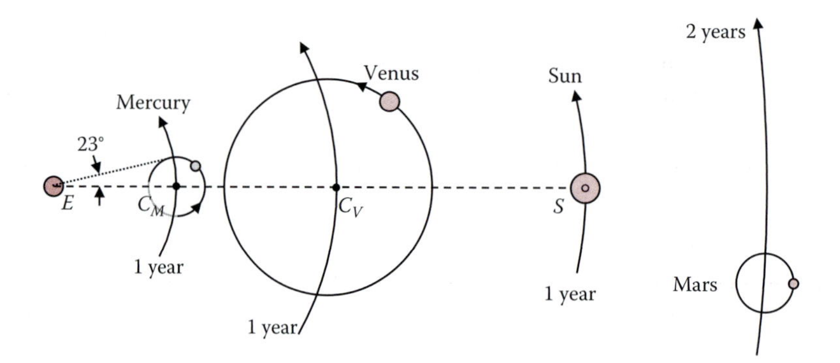

FIGURE 1.10 In the Ptolemaic system, C_M, C_V, and S rotate together about Earth with the same 1 year period, so that the elongation of either planet remains small. The epicycle centers of the superior planets, such as Mars (shown above), rotate with different periods, allowing them to drift out of alignment with the Sun. For example, if Mars is initially aligned with the Sun, 1 year later it will be opposed to the Sun, that is, it will be on the opposite side of Earth from the Sun.

special treatment of Mercury and Venus must surely have bothered his contemporaries and successors; nevertheless, no one would propose a better model of the universe until the Renaissance, 1400 years after the appearance of the *Almagest*.

1.5 COPERNICUS AND GALILEO

In the sixteenth century CE, the Polish scholar Nicholas Copernicus adopted the near-forgotten speculations of Aristarchus, Pythagoras, and other classical-era Greek astronomers to construct a simpler—but highly counterintuitive—model of the solar system. Copernicus understood that the apparent daily rotation of the celestial sphere of the stars could be explained equally well by a counter rotation of the Earth. Moreover, the erratic motion of the planets, requiring the introduction of epicycles, could be described much more simply by assuming that all planetary orbits are Sun-centered (*heliocentric*) rather than geocentric and that Earth itself orbits the Sun as do Mercury, Venus, Mars, Jupiter, and Saturn. In his monumental work called *De Revolutionibus Orbium Coelestium* (*On the Revolutions of the Heavenly Spheres*, published in 1543), Copernicus showed that a heliocentric planetary system is consistent with all planetary observations—including retrograde motion—while it preserves the preference for circular motion. Moreover, unlike the geocentric system of the Greeks, all planets can be treated in the same manner.* Because the Copernican system contradicted the classical world view of Ptolemy, and displaced Earth from its unique position at the center of the universe, it was scorned by the scientific and religious communities.

* Invoking Occam's Razor, Copernicus argued in favor of his simpler model of the solar system: "We thus rather follow Nature, who producing nothing vain or superfluous often prefers to endow one cause with many effects."

Scholars sought to dismiss it by posing "common sense" questions, such as: How can something so heavy as the Earth move? If the Earth spins, what holds it together, and why don't we then fly off? If the Earth is in orbit about the Sun, shouldn't the apparent positions of the stars change throughout the year? The ensuing debate, which occupied scholars for nearly a century after Copernicus, was settled by the Italian astronomer, physicist, and mathematician Galileo Galilei using a revolutionary new invention—the telescope.

The telescope was invented in the Netherlands in the early seventeenth century. Soon after acquiring one, Galileo was grinding his own lenses in Italy, eventually constructing an improved instrument with an overall magnification of about 30×. Training his telescope on the sky, he promptly observed the craters on the Moon (and calculated their height), sunspots, the four brightest moons of Jupiter revolving about that planet, and—most importantly—the phases of Venus. He reported his observations in 1610 in a slender publication called *Siderius Nuncius* (*The Starry Messenger*). According to Galileo, the latter two discoveries offered incontrovertible experimental support for the Copernican model, sparking the infamously bitter dispute between Galileo and the Roman Catholic Church.*

The discovery of the moons of Jupiter was significant because it provided a counterexample to the geocentric model of Ptolemy. Here was a system of astronomical bodies, analogous to the planets, which orbited a body other than Earth. All motion in the sky was *not* centered on Earth! But Galileo's observations of Venus dealt the decisive blow to the Ptolemaic model, in which Venus' epicycle is positioned somewhere between Earth and the Sun. Study Figure 1.10 carefully and compare it to the earlier discussion of the phases of the Moon (Figure 1.5). If Figure 1.10 is drawn roughly to scale and $\theta_1 \leq 46°$ (the observed angle of maximum elongation of Venus), then $\theta_1 + \theta_2 \leq 90°$, so according to Ptolemy, no matter when we observe the planet, we will never see it more than half full. Conversely, in the Copernican picture, when Earth and Venus are in opposition (on opposite sides of the Sun), Venus will appear to us as a nearly full (*gibbous*) disk, which is just what Galileo observed. Figure 1.11 is a composite of many images of Venus, taken on a succession of nights using a modern telescope. These images offer irrefutable evidence against the Ptolemaic model. The debate between Galileo and the Church reached an ugly head in 1643, when Galileo published a comparison of the Copernican and Ptolemaic models, called *Dialogue Concerning the Two Chief World Systems*. The book was written for the general public and was written in the form of a play, a script with three characters. Simplicio, the aptly-named character espousing the Ptolemaic system (the official view of the Church), is unable to defend his views against the reasoning of Sagredo and Salviati, who champion the Copernican system. Not surprisingly, the book was construed as a mockery of the Church's position and Galileo was subsequently summoned to Rome to face trial for heresy, a crime punishable by death. Due to his advanced age (70 years), his life was spared but he was committed to house arrest until his death 7 years later. During that time, he wrote his final—and perhaps greatest—book, *Discourses Concerning Two New Sciences*, an account of his extensive experimental and theoretical research on the laws of motion. In *Discourses*, Galileo presented experimental data refuting long-accepted theories of motion (attributed to Aristotle) and laid the foundation for modern scientific

* It was a dangerous business to contradict the Church. In 1600, the Dominican monk Giordano Bruno was burned at the stake for heresy. His open support for the Copernican model did not help his case.

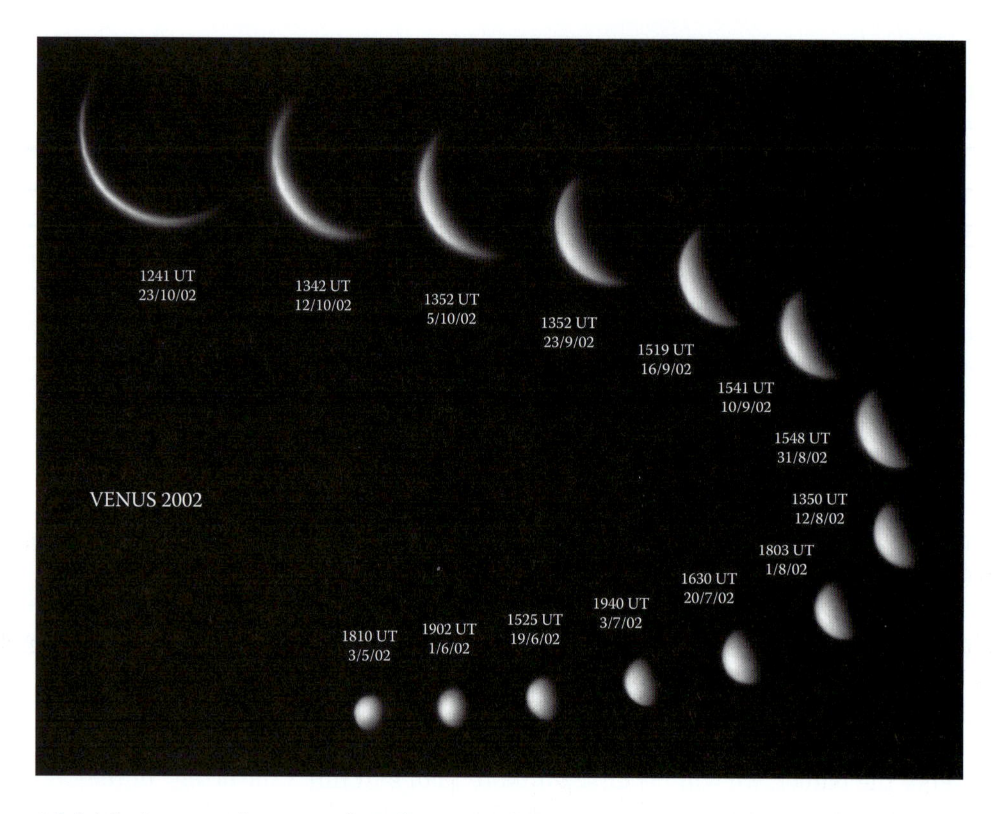

FIGURE 1.11 Multiple images of Venus taken from March to October 2002. The nearly full (gibbous) appearance of Venus in March is strong evidence against the Ptolemaic model. During this time period, Venus moved from its farthest point from Earth to its closest point to Earth. Using this composite image, you can estimate the radius of the planet's orbit in terms of AU. Can you see how? (Photo: Chris Proctor, Torbay Astronomical Society, Torquay, U.K.)

method, wherein theoretical models are accepted subject to confirmation by experimental evidence. *Discourses* signaled the birth of *classical mechanics*, and, in many respects, is the forerunner of your present textbook.

1.6 MAPPING THE SOLAR SYSTEM

Once the Copernican model was accepted, it was straightforward to determine the radii of each planet's orbit about the Sun, relative to the radius of Earth's orbit. The procedure is illustrated in Figure 1.12 for *inferior* planets (those closer than Earth to the Sun). Let θ be the *elongation* angle, measured from Earth (E), between the direction to the Sun (S) and the direction to the planet (P). The maximum elongation of the planet θ_{max} occurs when line $\overline{EP}$ is tangent to the planet's orbit, which is assumed to be

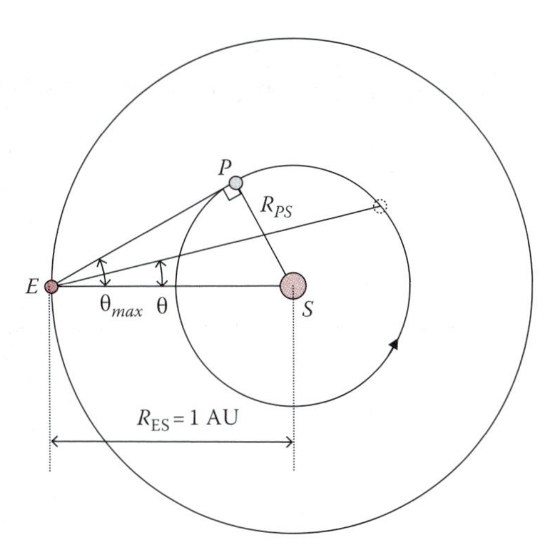

FIGURE 1.12 Measuring the orbital radius of an inferior planet. The maximum elongation θ_{max} occurs when line *EP* is tangent to the circular trajectory of the planet, so that $\angle EPS$ is a right angle.

circular. Since the tangent to a circle is perpendicular the circle radius, triangle *ESP* is a right triangle with $\overline{ES}$ as its hypotenuse, so

$$\frac{R_{PS}}{R_{ES}} = \sin\theta_{max}. \tag{1.3}$$

The Earth–Sun distance R_{ES} is a convenient and important unit of length in astronomy. Accordingly, it is given its own name, the *astronomical unit*, or AU: $R_{ES} = 1$ AU. For Venus, whose maximum elongation is 46°, the radius of its orbit about the Sun is therefore $R_{VS} = 1\ \text{AU} \times \sin 46° = 0.72\ \text{AU}$.

EXERCISE 1.4

The maximum elongation of Mercury is 23°. What is its orbital radius in AU?

Answer 0.39 AU

The procedure for finding the orbital radii of the *superior* planets (those that are farther than Earth from the Sun) is similar and is left as an end-of-chapter problem.

But how many meters equal 1 AU? This was not known in Galileo's time. In fact, the best estimate of the Earth–Sun distance was not significantly different from the egregiously incorrect value proposed by Aristarchus in 300 BCE! Without an accurate measured value for the AU, astronomers had no way to determine the true size of the solar system; measuring the Earth–Sun distance was therefore a high priority in the seventeenth century. In fact, it was a task that occupied astronomers well into the mid-twentieth century.

1.7 MEASURING THE EARTH–SUN DISTANCE

The telescopes used by Galileo (*Galilean* telescopes) were noninverting instruments that were better suited for terrestrial applications such as navigation or military use. As optical technology improved, astronomers had access to bigger and better lenses as well as improved telescope designs.* *Keplerian* telescopes (introduced by the great astronomer Johannes Kepler) offered higher magnification but produced an inverted image, acceptable for astronomical use but not for terrestrial use. More importantly, the Keplerian design allowed astronomers to insert finely adjustable crosshairs within the telescope to allow them to measure very small angles.† As you will see, the ability to measure small angles was the key to determining the distance to the Sun.

One of the first credible measurements of the Earth–Sun distance was performed by Gian Domenico Cassini and Jean Richer during the years 1671–1673. The idea was to observe Mars simultaneously from two widely separated points on Earth. Each observer saw Mars at a slightly different position relative to the background of distant stars. This apparent shift in position, called *parallax*, is explained in Box 1.2. If the distance to Mars (measured in AU) and the Earth radius (measured in meters) are known, R_{ES} can be calculated. Cassini, a distinguished astronomer, observed Mars from the gardens of the Royal Observatory in Paris. His assistant Jean Richer was dispatched to the muggy island of Cayenne, just off the northeast coast of South America in present day French Guiana,‡ to carry out simultaneous observations of Mars. Their experiment was one of the first intercontinental scientific expeditions ever attempted, and scientists throughout Europe eagerly awaited the results.

The two astronomers observed Mars when it was in opposition to the Sun and therefore at its closest distance to Earth (Figure 1.13). Since the radius of Mars' orbit is 1.52 AU, the Earth–Mars distance R_{EM} at this time was 0.52 AU. As explained in Box 1.2, astronomers use their parallax measurements to calculate the *horizontal parallax angle* θ_P, the angle that would be measured if observations were taken when Mars is on each observer's horizon and triangle *OEM* in Figure 1.13b is a right triangle. In this case,

$$\tan\theta_P \simeq \theta_P\,(\text{rad}) = \frac{r_E}{R_{EM}}.$$

Cassini concluded that $\theta_P = 25\hat{''}$, where the symbol $\hat{''}$ indicates seconds of arc ($1\hat{''} = \frac{1}{3600}$ deg).

---

* Simple lenses cannot focus light perfectly. In particular, they focus light of different colors at different places, an effect known as *chromatic aberration*. The greater the lens diameter, the worse the aberration (error). (Achromatic lenses were not introduced until the late eighteenth century.) Reflecting telescopes, in which the large objective lens is replaced by a parabolic mirror (free from chromatic aberration), were introduced in 1660–1670 by James Gregory (*Gregorian* telescope), Isaac Newton (*Newtonian* telescope), and Laurent Cassegrain. The latter design is the basis of the Hubble space telescope and its successor, the James Webb space telescope.

† These crosshairs are magnified by the telescope's eyepiece lens, and appear in focus along with the image of the astronomical object being observed.

‡ Cayenne, once called Devil's Island, was the site of the notorious penal colony featured in the movie *Papillon*. Today, French Guiana harbors the launch site of the European Space Agency. Because of its location close to the equator, it is ideally suited for launching satellites into geosynchronous orbit.

BOX 1.2 PARALLAX AND THE MEASUREMENT OF ASTRONOMICAL DISTANCES

The best way to learn about parallax is to see it for yourself. Look at a distant vertical object such as a tree trunk or a telephone pole. Hold up one finger vertically, a few inches from your eyes and in line with your nose and the distant object. Close your left eye, but keep your right eye open. Your finger will appear to have shifted to the left of the background object. Close your right eye and open your left. Your finger will now appear to have moved to the right of the distant object. The apparent shift in position of a nearby object (your finger) relative to a distant object (the tree or pole) is called *parallax*.

Now let's see how to use parallax to measure astronomical distances. Suppose you wish to find the distance from Earth to a nearby celestial body such as a planet in our solar system. (The planet replaces your finger.) For the background object, choose a star in nearly the same direction as the planet, but much farther away. (The star replaces the tree of our earlier example.) Instead of your two eyes that are only a few cm apart, imagine two observers that are located at widely separated spots on the Earth (see Figure B1.2a). Observer A measures an angle θ_A between the planet P and the background star S, and observer B measures θ_B at the same time in the same way. Since the star is *much* farther away than the

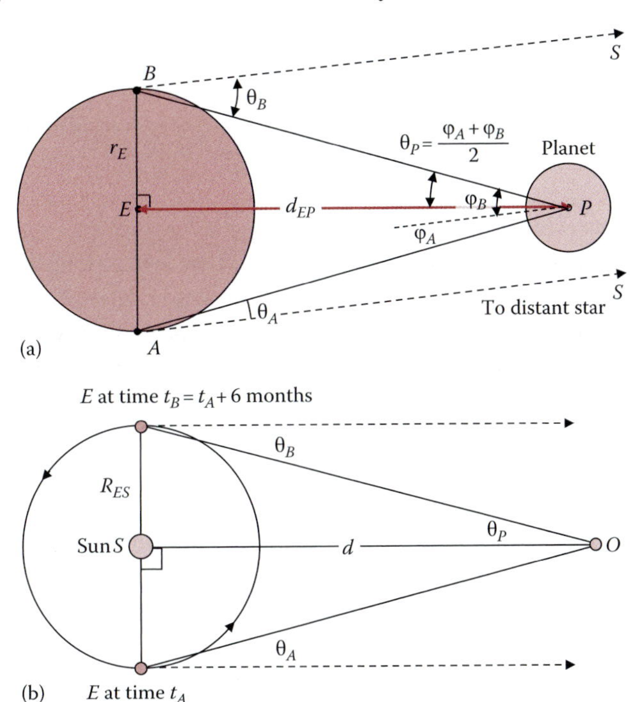

FIGURE B1.2 (a) Observers A and B, at different locations on Earth, see planet P in slightly different positions relative to the background of distant stars. This apparent shift in position, or *parallax*, allows the observers to calculate the distance d to the planet. Angles θ_A, θ_B, φ_A, and φ_B are *greatly* exaggerated in the drawing. (b) To measure the distance to more remote bodies, such as stars, the two observation points must be separated by a distance much greater than $2r_E$. This can be accomplished by observing the body at two times 6 months apart, when the Earth has moved halfway around its solar orbit. The two observation points are then separated by 2 AU.

planet, the dotted lines $\overline{AS}$ and $\overline{BS}$ are almost exactly parallel. Therefore, $\theta_A + \theta_B = \varphi_A + \varphi_B$, where $\varphi_A + \varphi_B$ is the angle subtended by the observers' baseline $\overline{AB}$, as viewed from the planet. Knowing the longitude and latitude of each observer, it is possible to calculate the angles and side lengths of triangles *EAP* and *EBP*, and so determine the distance d_{EP} from Earth to the planet.

The subtended angle increases with the length of baseline $\overline{AB}$, and the maximum possible angle occurs when the baseline length is $2r_E$. No matter where on Earth the two observers are actually located, they use their measurements to calculate this largest possible angle, and report the *horizontal parallax angle* θ_P, defined as $\theta_P = (1/2)(\theta_A + \theta_B)_{max}$ in their papers.

Keep in mind that the figure is not drawn to scale: for example, the closest distance to Mars is 12,000 times greater than the radius of Earth, so the subtended angle $\varphi_A + \varphi_B$ is *very* small. Therefore,

$$d_{EM} = \frac{r_E}{\tan\theta_P} \simeq \frac{r_E}{\theta_P}, \tag{B1.5}$$

where we have employed the small angle approximation for $\tan\theta_P$. Note that as the distance d gets larger, θ_P gets smaller and consequently it is more difficult to measure accurately. To measure the Sun's distance from Earth by parallax, it is necessary to measure accurately a parallax angle of

$$\theta_P = \frac{r_E}{R_{ES}} = \frac{6.4\times10^3 \text{ km}}{1.5\times10^8 \text{ km}} \approx 4\times10^{-5} \text{ (rad)} \approx 8''.$$

Even with modern telescopes, this is a daunting task.

To measure longer distances, a longer baseline is needed. One way to lengthen $\overline{AB}$ is to take observations from the same location on Earth, but at times 6 months apart, when the Earth has moved halfway around its solar orbit. Parallax measured in this way is called *annual parallax* (see Figure B1.2b). The new baseline is now 2 AU, and r_E is replaced by 1 AU in Equation B1.5, an increase by a factor of about 24,000. This allows us to determine distances to objects quite far from our solar system. In fact, this is such a common procedure that astronomers have defined a unit of length called the *parsec* (pc) in terms of the annual parallax angle. The object *O* in Figure B1.2b is *defined* to be at a distance $d = 1$ pc when the parallax angle θ_P $(= \theta_A = \theta_B)$ is equal to $1''$. Let's express the parsec in terms of more familiar units. First, recall that $1'' = \frac{1}{3600}\text{ deg} = 4.848\times10^{-6}$ (rad), so, using Equation B1.5,

$$d \equiv 1 \text{ pc} = \frac{1 \text{ AU}}{4.848\times10^{-6}} = 2.063\times10^5 \text{ AU}.$$

This is about twice the diameter of our Milky Way Galaxy. To express the parsec in terms of meters, recall that 1 AU = 1.5×10^{11} m, so 1 pc = 3.09×10^{16} m.

Exercise B1.3: The *light year* (ly), defined as the distance traveled by a light wave in 1 year. Express the parsec in terms of light years. (The speed of light in vacuum is $c = 3\times10^8$ m/s.)

Answer 1 pc = 3.26 ly

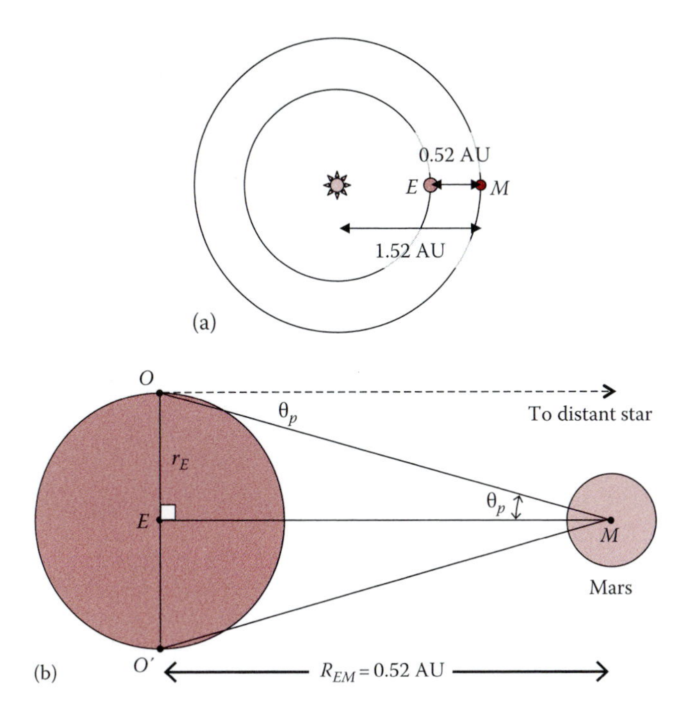

FIGURE 1.13 (a) When Mars is in opposition to the Sun, its distance from Earth is 0.52 AU. (b) Finding the distance to Mars by parallax. The horizontal parallax angle θ_P is the parallax angle an observer at O would measure if Mars were on O's horizon. Since $R_{EM} \gg r_E$, and $\theta_P \ll 1$ rad, triangle OEM is essentially a right triangle.

EXERCISE 1.5

(a) Convert θ_P to radians. (b) Use the small angle approximation to determine Cassini's estimate for R_{EM}. (c) Then find his estimate for the AU. Compare this to today's accepted value of 1.496×10^8 km. Also, compare this to Aristarchus' estimate.

Answer (a) 1.21×10^{-4} (b) 5.26×10^7 km (c) 1.01×10^8 km

The accuracy of these parallax measurements was limited by the instrumentation available to astronomers in the seventeenth century, and the value of θ_P quoted by Cassini is better interpreted as an upper limit, that is, $\theta_P \leq 25''$. A century later, in 1761 and 1769, astronomers scattered to locations around the globe to observe the so called *transits* of Venus (see Figure 1.14), when the planet passes directly across the face of the Sun (as viewed from Earth).* Since Venus has an orbit of radius 0.72 AU, it is only 0.28 AU

* Among these observers was the famous British Captain James Cook, who sailed to the South Pacific island of Tahiti to witness the Venus transit of 1769.

from Earth when it transits the Sun, so its parallax is about twice the angle measured for Mars by Cassini and Richer. An analysis of the transit technique is given as an end-of-chapter problem.

Parallax measurements were performed throughout the nineteenth century, and well into the twentieth century, to determine the value of the AU with increasing accuracy. In 1961, the method was supplanted by *radar ranging* experiments, in which the distance to Venus was determined to high accuracy by measuring the round-trip travel time of radar pulses beamed from Earth and reflected from the surface of Venus. The quest to determine the Earth–Sun distance was finally over—3 years after the launch of the *Sputnik* satellite and the birth of the space age. Subsequent radar ranging experiments led to the modern definition (adopted in 2012) of the astronomical unit as 149,597,870,700 m.

FIGURE 1.14 The 2004 transit of Venus (the small dot) across the face of the Sun, viewed from the authors' home institution. To determine the parallax of Venus, astronomers record the exact time when the planet's rim coincides with the Sun's rim. Observers at different longitudes on Earth will record different times, allowing θ_P to be calculated. (Photo courtesy of Professor Roger Rowlett.)

EXERCISE 1.6

When Earth and Venus are at their minimum separation, a radar pulse is emitted from Earth toward Venus and the reflected pulse is detected 279.3 s later.

(a) Using this data, show that 1 AU = 1.496×10^5 km. (The speed of light is 3×10^5 km/s.)
(b) How long would it take a radar pulse to reach Venus and return to Earth when the two planets are at their *maximum* separation?

Answer (b) 28.6 min

1.8 MEASURING INTERGALACTIC DISTANCES

The measurement of the Earth–Sun distance was one of the most important—and challenging—achievements in the history of astronomy. Once the astronomical unit was known, astronomers could find the actual sizes of planetary orbits and, using *annual parallax* measurements (see Box 1.2), determine the distance to a few nearby stars.* For more distant stars, annual parallax angles are too small to measure. How do we reach beyond these nearby stars (which are all in our Milky Way Galaxy) to measure the vastly larger distances to other galaxies? How do we go from measuring the scale of our solar system to measuring the scale of the universe?

One way to proceed is to use a star's *apparent brightness* to gauge its distance: the farther a star is from us, the dimmer it appears. Let L be the star's *luminosity*, the total power emitted by the star in all directions. At a distance d from the star, this power is spread over a spherical surface of area $4\pi d^2$, so the power per unit area is

$$b = \frac{L}{4\pi d^2}. \tag{1.4}$$

The amount of energy detected by our eyes, or by our telescopes, is proportional to b, the apparent brightness of the star. Two identical stars, one twice as far away as the other, will differ in brightness by a factor of 4. To determine the distance d, we need to know both b and L. We can easily measure b, but how can we find L?

In 1912, Henrietta Leavitt, a research assistant at the Harvard Observatory, made a key discovery. While working for the prominent astronomer E.M. Pickering, she inspected thousands of photographs of the Small Magellenic Cloud (*SMC*), a relatively compact cluster of stars that hovers near our Milky Way Galaxy. A tiny fraction of these stars are *Cepheid variables*, stars that oscillate in brightness in a precisely repeatable manner. Leavitt found that the peak brightness b_{max} of a Cepheid variable was mathematically expressible as a function of the star's oscillation period P, $b_{max} = b(P)$. Since the stars of the *SMC* are all about the same distance d from Earth, Equation 1.4 shows that differences in brightness among the Cepheids must be due to differences in luminosity; hence, $L = L(P)$. This suggests that *any* two Cepheids, in the same or separate galaxies *anywhere* in space, which oscillate with the same period P, have the same luminosity. So if two Cepheids from separate regions of space are found with the same period and if one appears dimmer than the other by a factor of 4, then the dimmer star must be twice as far away as the brighter one.

To use Cepheids to measure *intergalactic* distances, we first need to *calibrate* Leavitt's period–luminosity relationship. We need to find Cepheids near enough to Earth† that their distances d can be found by parallax, and then measure the brightness b_{max} and period P of these variable stars. Together with Leavitt's measurements, this will allow us to find the function $L(P)$. Then, when a Cepheid is located in another galaxy, its period can be measured to determine its luminosity, and its brightness measured to calculate the distance to that galaxy by using Equation 1.4.

In 1913, the year following Leavitt's discovery, the Danish astronomer Ejner Hertzsprung used parallax to determine the distance to nearby Cepheids, providing the necessary calibration of $L(P)$.

* First done by the German astronomer Friedrich Wilhelm Bessel in 1838.

† Polaris, the North Star, is a Cepheid variable relatively close (430 ly) to Earth.

A better measurement was provided in 1918 by the American astronomer Harlow Shapley, who used his result to estimate the diameter of the Milky Way galaxy (roughly 100,000 light-years [ly]*). This set the stage for the "great debate" of 1920 between astronomers Shapley and Heber Curtis regarding the large scale structure of the universe. Shapley argued that the Milky Way encompassed the entire universe, including the "nebulae" that we now know to be separate galaxies. Curtis argued that those nebulae were "island universes" resembling the Milky Way, and, as such, were at nearly inconceivable distances from us. The debate was settled in 1924 by Edwin Hubble, who used the newly constructed 100 in. (diameter) telescope at Mt. Wilson Observatory in California to detect Cepheid variables in the galaxy Andromeda. By measuring their periods and brightness, Hubble estimated the distance to Andromeda as 930,000 ly, far beyond the boundary of the Milky Way. While his estimate was too small by nearly a factor of 3, Hubble nevertheless provided strong evidence for our modern conception of the universe: that Earth is a tiny outpost circling one of about 10^{11} stars in the Milky Way galaxy, which is one of at least as many similar galaxies, or "island universes," populating an inconceivably large volume of space.

* A light year (ly) is the distance traveled by light in 1 year: 1 ly = 3×10^8 m/s $\times$ 3.15×10^7 s = 9.46×10^{15} m.

1.9 SUMMARY

Over the past 2500 years, our estimate of the size of the universe has expanded dramatically. Who among the Greeks would have guessed the true distance to the stars or imagined the uncountable number of them that are visible using even a small telescope? In the seventeenth century, who would have imagined—or accepted—the existence of other "island" galaxies, each containing billions of suns and planets too distant to be seen by the naked eye, and too remote to have any conceivable connection to human life on Earth? Throughout this period of discovery, our conception of cosmic distance has grown as our ability to measure tiny angles has improved. Astronomers today routinely measure parallax angles as small as a few milliarcseconds, and these are converted into distances using simple geometry.† To accomplish this, two fundamental lengths had to be determined: the Earth–Sun distance, or AU, and the radius of the Earth r_E. We needed the AU to convert annual parallax measurements into stellar distances, to calibrate Henrietta Leavitt's period–luminosity law, so we could use brightness to calculate intergalactic distance. But to determine the AU, we first needed to find the radius of the Earth, to convert parallax measurements of Mars and Venus into distances, to discover the true size of the solar system. r_E is the first rung on our cosmic distance ladder, the fundamental distance upon which all other cosmic distance measurements depend. As a striking example of the humble workings of science, it is worth remembering that r_E was originally determined by measuring the shadow of a stick erected in Alexandria, Egypt, 2500 years ago.

† An annual parallax of 1 milliarcsecond corresponds to a distance of 1000 pc, or 3160 ly.

PROBLEMS

1.1 Copernicus measured the distance from the Sun to the outer (superior) planets in the following way. Begin at $t = 0$ with the Earth E and the outer planet P in *opposition* (Sun–Earth–Planet aligned). At a later time t they will be in *quadrature* (Sun–Earth–Planet forming a right triangle), as shown in the figure. If we know the angle $P'SE'$ then we can determine R_{PS} in terms of R_{ES} (=1 AU) by simple trigonometry.

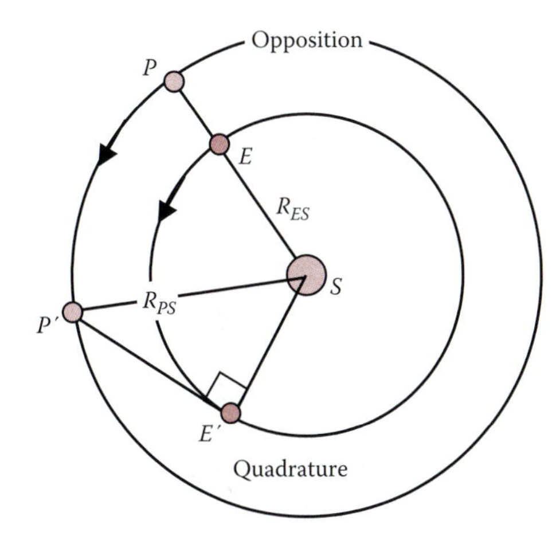

Mars has a revolution period that is 1.881 times Earth's period (365.26 days). Mars and Earth are in quadrature 106 days after being in opposition.

(a) What is the angle PSP'?

(b) What is the angle ESE'?

(c) What is the angle $P'SE'$?

(d) From $P'SE'$, find R_{PS} for Mars in units of AU (1 AU = R_{ES}).

1.2 In January, 2001, astronomers announced the discovery of two new "solar systems" about stars other than our own. One of them, centered on the star Gleise 876 (15 ly away from Earth), has two planets whose periods have been measured accurately. Assuming circular orbits, and applying the laws of classical mechanics, the orbital radii of the two planets are found to be 0.132 and 0.210 AU. (You'll find out how to do this later.)

(a) Make a careful sketch of this "solar system."

(b) Suppose you are on the second (outer) planet. What is the maximum elongation of the inner planet?

(c) Suppose now that you are on the inner planet, and that this planet has a moon in circular orbit with a period equal to 30 days. Following Aristarchus, you measure the times of dichotomy to be 13.6 days apart. Find the radius of the moon's orbit in terms of AU.

1.3 One of the many groundbreaking achievements of Galileo, during his initial experiments using his telescopes, was to measure the height of crater walls on the Moon. The figure below illustrates the technique when the Moon is in dichotomy (half full). D is the distance of the crater wall from the *terminator* (the edge of the illuminated region), L_{sh} is the length of the wall's shadow, and H is the height of the crater wall.

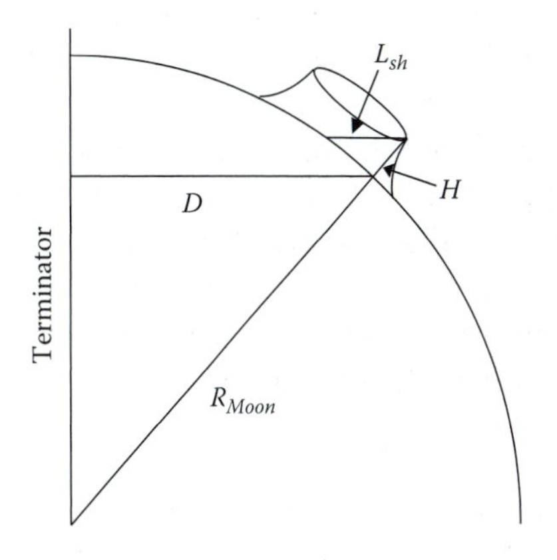

(a) From the geometry, complete the following equation: $H/L_{sh} =$.

(b) Using a ruler to measure the figure below, estimate the value of H/L_{sh}.

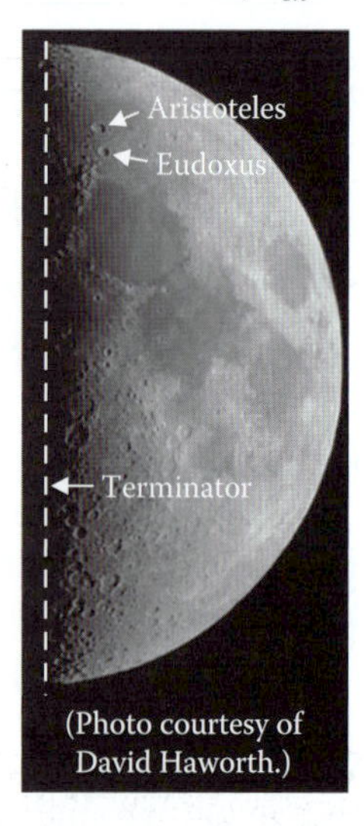

(Photo courtesy of David Haworth.)

(c) The figure below is a blowup of the one above. The diameter of Aristoteles is 87 km. Use this plus the results of parts (a) and (b) to estimate the height of the crater wall.

(Photo courtesy of David Haworth.)

1.4 The apparent size of an object is proportional to the angle subtended by the object. Figure 1.11 shows images of the planet Venus as it moves from its farthest distance from us to its closest approach to us. When it is farthest away, Venus subtends an angle 9.8″ and at its closest approach, 59.2″. Let the radius of the Earth's orbit about the Sun equal 1 AU and the radius of Venus' orbit equal x AU, where $x < 1$.

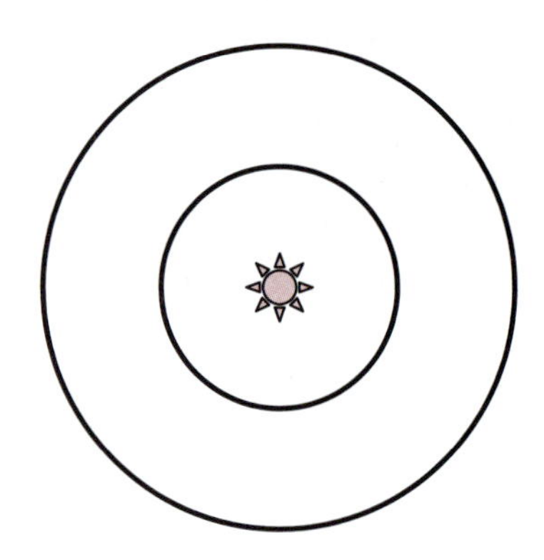

(a) Copy the figure above and sketch the positions of the two planets when they are closest to and farthest from each other.

(b) Use the information given above to estimate x.

1.5 The "parsec" was introduced by Bessel in 1840. One parsec (pc) is the distance to a star whose apparent position, relative to background stars that are much farther away, shifts by 2 arcseconds (2″ or simply 2″) as the Earth moves halfway around its orbit (2 AU). See Box 1.2 for a discussion of parallax. The star Alpha Centauri shifts by 1.49″ in half a year (see the figure).

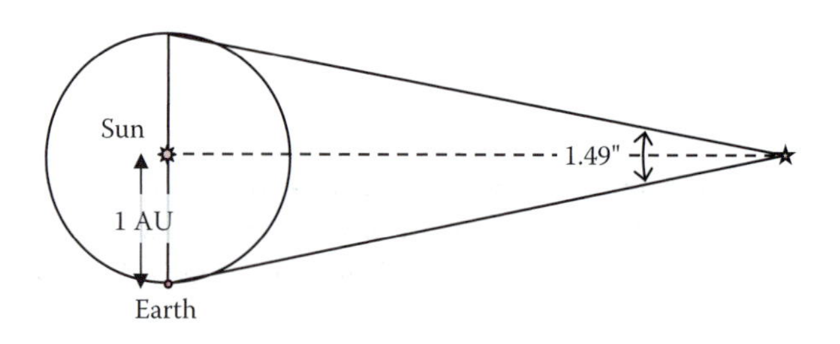

(a) How far away is Alpha Centauri in AU?

(b) How far away is the star in parsecs?

(c) How far away is the star in light-years?

1.6 In 1671–1673, Cassini and Richer determined the Earth–Sun distance by measuring the parallax of Mars. A better scheme was proposed by Edmund Halley in 1693, and carried out by astronomers throughout Europe, Asia, and even the United States (the "colonies") in 1761 and 1769, long after Halley's death. On rare occasions (which occur in pairs 8 years apart), Venus passes across the Sun's disc as viewed from the Earth. During such a *transit*, the planet appears to observers on Earth as a tiny dark spot on the surface of the Sun (see Figure 1.14). Imagine two observers separated by a baseline equal to one Earth radius R_E. At the instant when one observer sees Venus start across the face of the Sun, the other observer sees Venus at an angle of 31.25" relative to the outer rim of the star. This is shown in the figure. (Note that the figure is not drawn to scale! In a properly scaled drawing, the Sun, Earth, and Venus would appear as small dots along a single straight line. *Hint*: 31″ is a very small angle.)

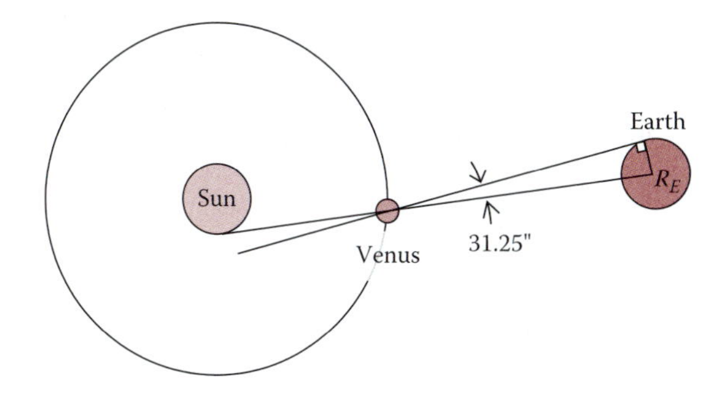

(a) The *maximum elongation* of Venus is 46°. Show that this implies that at its closest approach to Earth, Venus is 0.281 AU from Earth.

(b) From the information given above (summarized in the figure), find the value of the astronomical unit (AU), that is, the distance from Earth to the Sun, in meters.

1.7 The figure below is a composite image showing two views of the Moon, against the background of stars, taken simultaneously at two locations on Earth 5220 km apart. Recall that the angle subtended by the Moon from Earth is 0.52°.

(a) Using a ruler, take measurements directly from the figure to find the parallax angle between the two images.

(b) From the parallax angle, find the distance to the Moon. (Do not expect better than 10% accuracy.)

(Photo courtesy of Pete Lawrence [U.K.] and Pete Cleary [Canada].)

1.8 The January 5, 2006 issue of *Nature* had two articles describing the *occultation* of a distant star by Charon, the largest of Pluto's moons. The figure shows the intensity of starlight as Charon passed in front of the star, measured from two different locations on Earth. Starlight was blocked by the moon for 55.05 ± 0.21 s. The shadow of the moon cast by the star on Earth was measured to be moving at a speed of 20.93 km/s. Assuming that the light beam from the star crossed Charon's diameter, find the moon's radius (with uncertainty).

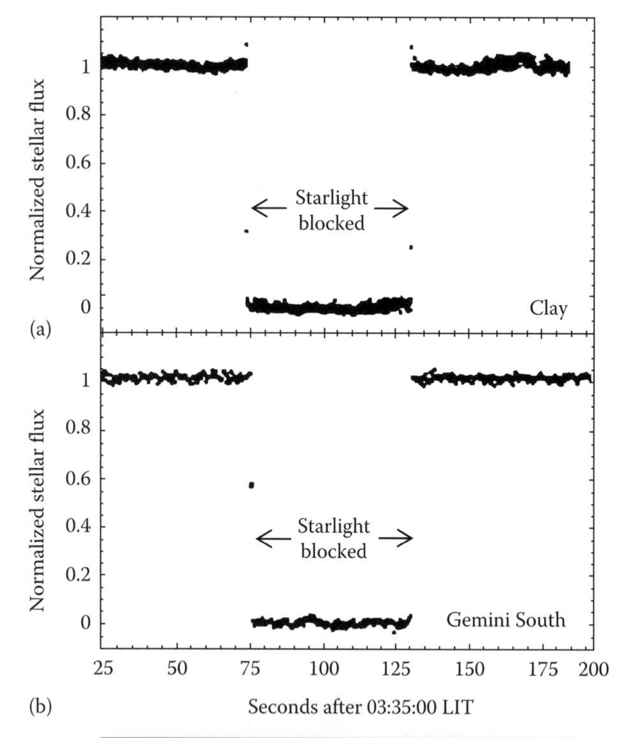

(Figures from Gulbis, A.A.S. et al., *Nature*, 439, 48, 2006.)

1.9 On August 27th, 2003, Mars was at its closest point to Earth in 50,000 years. Below is a picture of Mars and the Moon taken in July 2003. Very visible on the Moon is crater Anaxagoras with a diameter 52 km. The distance between the Moon and Earth is 384,400 km, and the diameter of Mars is 6,790 km.

(a) Estimate the distance between Earth and Mars in the photo.

(b) Compare your answer to the *closest possible* approach between the two planets.

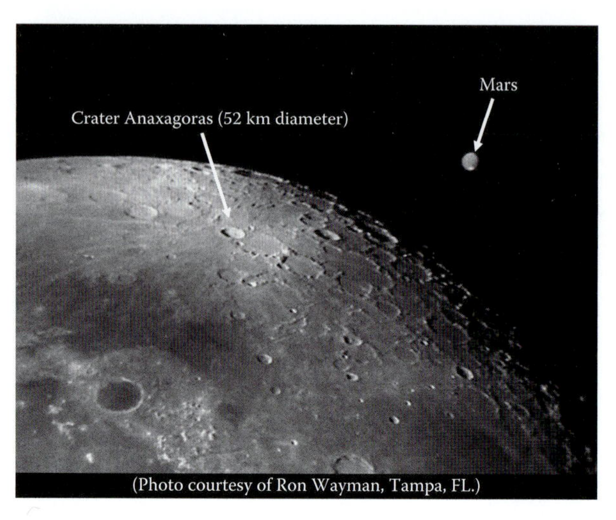

(Photo courtesy of Ron Wayman, Tampa, FL.)

1.10 Algol, in the constellation Perseus, is an eclipsing binary star system; that is, it is a pair of stars that circulate around a fixed point (called the center of mass—to be discussed soon) such that the smaller—and dimmer—star periodically passes in front of the larger, brighter star when viewed from Earth. When this happens—once every 2.867 days—the light reaching Earth from the star system dips by a factor of about 3, a phenomenon that was first recorded in 1667. During this time period t_0, Earth moves at a rate of 30 km/s in its orbit about the Sun. Because light travels with finite speed, the time between eclipses, measured by an observer on Earth, is variable. As the Earth travels from point A to point C, the measured interval between eclipses will be longer than t_0; from C to A, it will be shorter. Because of Earth's motion, the eclipse at C occurs *16.7 min later* than the time expected from the period t_0 given above.

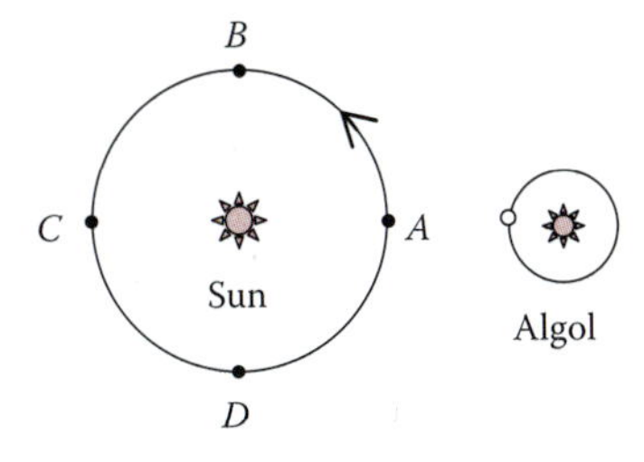

(a) How many eclipses of Algol occur while Earth is moving from *A* to *C*?

(b) Use the information given above to find the speed of light (1 AU = 1.5×10^{11} m).

(c) At what location(s) of Earth (*A*, *B*, *C*, and *D* in the figure) would you measure the period of Algol to be equal to t_0? At what location(s) would the measured period be longest? Shortest?

(d) At point *B*, how much would Algol's measured period differ from t_0?

2

VECTORS

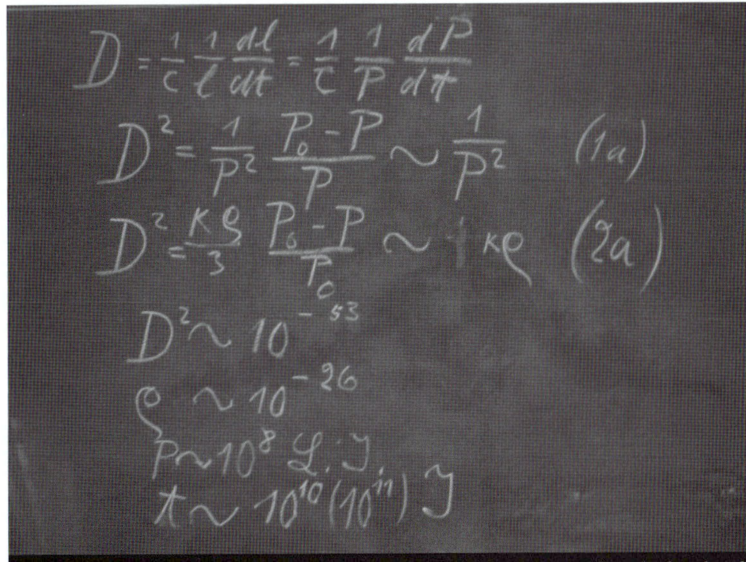

Mathematics allows the most sweeping properties of the universe to be calculated and expressed concisely. This blackboard was used by Albert Einstein during a lecture given at Oxford University in 1931. The last three lines contain estimates of the density ($\rho = 10^{-26}$ kg/m^3), size ($P = 10^8$ ly), and age ($t = 10^{10}$–10^{11} years) of the universe. The blackboard is on permanent display at the Museum of the History of Science in Oxford, England. (From the public domain.)

Physics and mathematics are inextricably entwined. How lucky we are, and how amazing it is, that the laws of physics can be expressed so precisely and elegantly in mathematical form. In contrast to conventional wisdom, mathematics simplifies *the analysis of physical phenomena, enabling us to study increasingly complicated—hence more realistic—physical situations. Whenever new (higher?) mathematics is introduced, it is done to ease and empower analysis and to promote deeper understanding. Imagine how difficult it would be to describe motion without the use of negative numbers. Plane geometry, used throughout Chapter 1, is restricted to two dimensions. To describe motion in the real three-dimensional world, we need to introduce new, more versatile, mathematical tools. These powerful tools, called* vectors, *are the subject of this chapter.*

2.1 INTRODUCTION

Blame it on René Descartes. As the story goes, the seventeenth century physicist, mathematician, and philosopher* was lying in bed one morning lazily watching a fly wander about on the ceiling above him. Descartes wondered how he could describe the fly's position precisely and unambiguously. He soon realized he could do this by defining the fly's position in terms of its perpendicular distance from each of the two adjacent walls of the room (Figure 2.1a). Nowadays, we would call these two distances the fly's *coordinates*. We would define the line joining the ceiling and one wall as a *coordinate axis*, the x-axis, say. The line where the ceiling and the second wall intersect might similarly be called the y-axis. Finally, if the bug were to leave the ceiling (to pester us, perhaps), we would need a third coordinate, the perpendicular distance of the fly from the ceiling, to describe its position completely. The third, or z-axis, would lie along the line of intersection between the two walls. Assuming that the room is a rectangular box, our three axes would be mutually perpendicular. This strategy for defining position and, more generally, for describing three-dimensional space is called a *Cartesian coordinate system*† (Figure 2.1b). A body's x-coordinate is its perpendicular distance from the plane formed by the y- and z-axes (the y–z plane) and likewise for its y- and z-coordinates. Positive and negative x-values are assigned to positions on the opposite sides of the y–z plane; positive and negative y- and z-values are defined in the same way. Descartes' concept of coordinates was the first step in the development

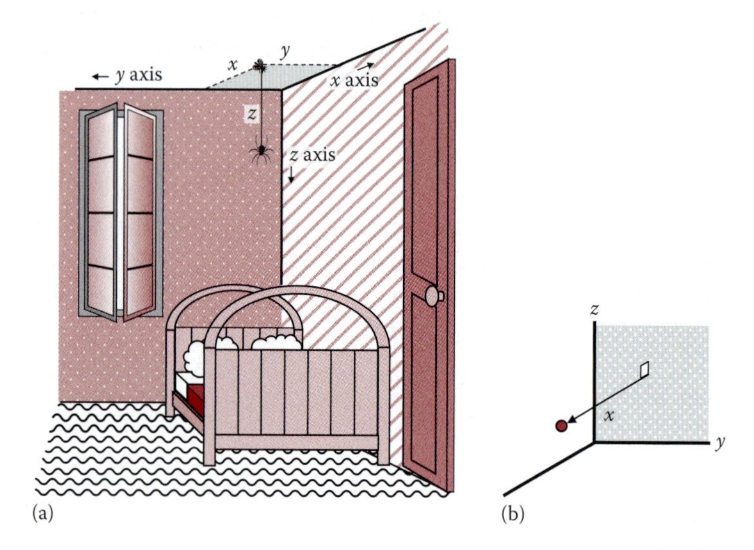

FIGURE 2.1 (a) A fly rests on the ceiling near one corner of a room. The perpendicular distance from each wall defines the fly's x- and y-coordinates. The z-coordinate of a spider hanging below the ceiling is defined as its perpendicular distance from the ceiling. The x-, y- and z-axes lie along the intersections of the ceiling with each wall, and along the intersection of the two walls. (b) A Cartesian coordinate system. The x-coordinate of a body is its perpendicular distance from the y–z plane, and likewise for the y- and z-coordinates.

* Who said, famously, "I think, therefore I am."

† The proper modifier "Cartesian" refers to René Descartes.

of *analytic geometry*, wherein the theorems of classical two-dimensional (plane) geometry are superseded by algebraic *vector* equations that make it easy to analyze motion in the real three-dimensional world. Strange that Descartes was inspired by the meandering motion of a pesky insect (or so the story goes).

2.2 POSITION AND DISPLACEMENT VECTORS

Imagine you are approached by a stranger, who hands you a note that says, "The girl/boy of your dreams is waiting for you. To find her/him, proceed from your present location a meters east; then go b meters north, where you will find a vertical ladder; climb the ladder to a height of c meters." In the interests of science (your *first* love!), you decide to reexpress these instructions mathematically before starting off. Begin by setting up a Cartesian coordinate system with its *origin* (where the three axes cross and $(x, y, z) = (0, 0, 0)$) at your present location. Let the positive x-axis, or $\hat{i}$ *direction*, point east and the positive y-axis, or $\hat{j}$ *direction*, point north. Similarly, let the positive z-axis, or $\hat{k}$ *direction*, point vertically upward. (The symbols $\hat{i}$, $\hat{j}$, and $\hat{k}$ are used worldwide to indicate direction; we will adhere to them throughout this text.) Figure 2.2 is an illustration of your path. You start at (0, 0, 0) and end up at (a, b, c). In physics, your overall change of position, or *displacement*, is represented by a straight line drawn from the initial to the final location. An arrowhead is attached to this line at the final position to indicate your direction of travel. The resulting arrow is labeled $\vec{s}$ in Figure 2.2.

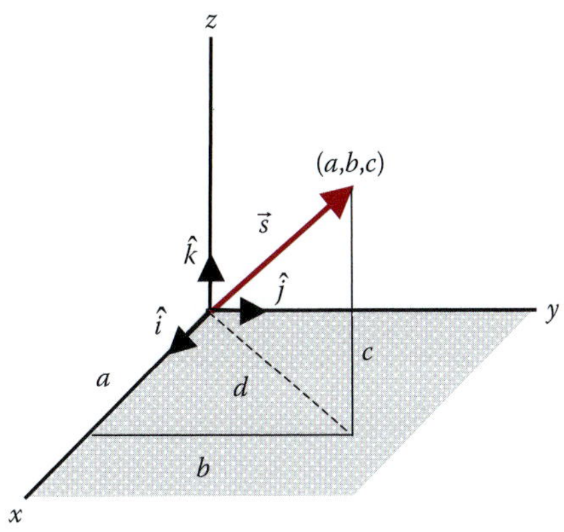

FIGURE 2.2 Starting at the origin (0, 0, 0) you travel a meters in the east or $\hat{i}$ direction, then b meters in the north or $\hat{j}$ direction, and then ascend the stairs c meters in the vertical $\hat{k}$ direction. Your overall displacement from the origin $\vec{s} = a\hat{i} + b\hat{j} + c\hat{k}$ is depicted by the arrow drawn from your initial position to your final position.

Alternatively, all of the graphic information shown in Figure 2.2 can be expressed equally well by the succinct algebraic equation:

$$\vec{s} = a\hat{i} + b\hat{j} + c\hat{k}\ \text{m}. \tag{2.1}$$

Equation 2.1 is the algebraic equivalent of the displacement arrow $\vec{s}$ drawn in Figure 2.2. The displacement $\vec{s}$ has a magnitude (length) and a direction and is called a *vector*.* Its direction is obvious, but what is its magnitude? Applying the Pythagorean theorem in the x–y plane of Figure 2.2, $d = \sqrt{a^2 + b^2}$. Using the Pythagorean theorem a second time to include the vertical path, $s = |\vec{s}| = \sqrt{d^2 + c^2}$, where s or $|\vec{s}|$ designates the magnitude of $\vec{s}$. Hence, the magnitude of $\vec{s} = a\hat{i} + b\hat{j} + c\hat{k}$ is

$$s = |\vec{s}| = \left|a\hat{i} + b\hat{j} + c\hat{k}\right| = \sqrt{a^2 + b^2 + c^2}. \tag{2.2}$$

Equation 2.2 is the general expression for the magnitude of a vector written in Cartesian coordinates. In this case, since you started out at the origin of coordinates, your displacement $\vec{s}$ is the same as your final position.

* A quantity that has a magnitude but no direction, e.g., temperature, is called a *scalar*.

2.3 UNIT VECTORS

In the preceding section, we regarded $\hat{i}$, $\hat{j}$, and $\hat{k}$ as symbols, synonymous with the x-, y-, and z-directions. In fact, they are themselves vectors. To see this, start with the general expression for a vector: $\vec{A} = a\hat{i} + b\hat{j} + c\hat{k}$. How would you write a vector that is parallel to the x- or $\hat{i}$ direction? Referring to Figure 2.2, $\vec{A}$ points in the x-direction if $b = c = 0$, so $\vec{A} = a\hat{i}$. Now, how would you write a vector that points in the x-direction *and has a magnitude of* 1 *unit*? Clearly, $a = 1$, so $\vec{A} = \hat{i}$. Hence, $\hat{i}$ is itself a vector with a magnitude of one unit pointing in the positive x-direction; it is appropriately called a *unit vector*. Likewise, $\hat{j}$ and $\hat{k}$ are unit vectors pointing in the positive y- and z-directions.

Imagine that you are at the location specified by the position vector $\vec{r} = x\hat{i} + y\hat{j} + z\hat{k}$ and you set out from there in a direction parallel to $\hat{i}$. As you do so, only the value of x changes; the values of y and z remain the same. Similarly, if you move parallel to $\hat{j}$, only the value of y changes. In this sense, movements in the $\hat{i}$, $\hat{j}$, and $\hat{k}$ directions are independent of, or *orthogonal* to, one other. Orthogonality is a very useful feature of coordinate systems, and it will often be helpful to think of an arbitrary displacement as a combination of independent orthogonal displacements in the $\hat{i}$, $\hat{j}$, and $\hat{k}$ directions.

EXERCISE 2.1

Cartesian coordinates are not the only way to specify position. The global positioning system (GPS) uses radio signals from orbiting satellites to calculate your position near the surface of the Earth. GPS coordinates are latitude, longitude, and altitude relative to the surface of the Earth. Are these coordinates orthogonal? That is, can you change each coordinate independently, without affecting the other two?

2.4 VECTOR ARITHMETIC

Now let's return to the story begun earlier (Section 2.2). Suppose you carry out the instructions given to you by the stranger, but when you arrive at your destination there is no one waiting for you, only a second set of instructions. Although bitterly disappointed, you decide to continue this *treasure hunt*, and using these new instructions you set out once again in search of your beloved. How can we describe your trip mathematically? Figure 2.3 illustrates your two-segment trip. To simplify the drawing, each segment of the trip has been restricted to the x–y plane (i.e., no ladder), but the analysis is easily generalized to three dimensions. In Figure 2.3, your first displacement is represented by the vector $\vec{s}_1 = a_1\hat{i} + b_1\hat{j}$ and your second displacement by $\vec{s}_2 = a_2\hat{i} + b_2\hat{j}$. As shown, the coordinates of your final position are $(a_1 + a_2, b_1 + b_2)$; you could have reached this same final position by first going $(a_1 + a_2)$ units in the $\hat{i}$ direction and then going $(b_1 + b_2)$ units in the $\hat{j}$ direction, so your *total* vector displacement from the origin is $\vec{s}_{tot} = (a_1 + a_2)\hat{i} + (b_1 + b_2)\hat{j}$. But the total displacement is just the *sum* of the individual displacements, so we may write (now generalizing to three dimensions),

$$\vec{s}_{tot} = \vec{s}_1 + \vec{s}_2 = (a_1 + a_2)\hat{i} + (b_1 + b_2)\hat{j} + (c_1 + c_2)\hat{k}. \tag{2.3}$$

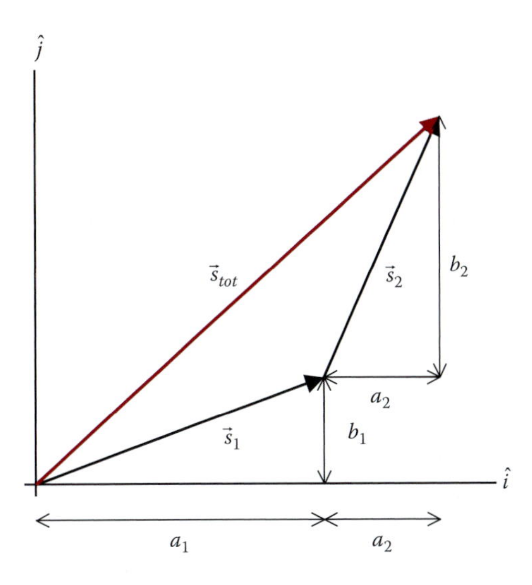

FIGURE 2.3 Vector additions of two successive displacements. For simplicity, the two displacements are confined to the x–y plane. The final position could have been reached by first traveling $(a_1 + a_2)$ meters in the $\hat{i}$ direction and then proceeding $(b_1 + b_2)$ meters in the $\hat{j}$ direction.

Equation 2.3 provides the general rule for adding vectors: the total $\hat{i}$-component is the sum of the individual $\hat{i}$-components, etc. Each of the new $\hat{i}$, $\hat{j}$, and $\hat{k}$ components is computed *separately*. Note, from Equation 2.3, that vector addition is commutative: $\vec{s}_1 + \vec{s}_2 = \vec{s}_2 + \vec{s}_1$.

EXERCISE 2.2

If you wanted to add *three* vectors together, how would you rewrite Equation 2.3?

Vector subtraction is defined similarly:

$$\vec{s}_2 - \vec{s}_1 = (a_2 - a_1)\hat{i} + (b_2 - b_1)\hat{j} + (c_2 - c_1)\hat{k}. \tag{2.4}$$

This operation is shown graphically in Figure 2.4. As an example of vector subtraction, suppose you are at position $\vec{s}_1 = a_1\hat{i} + b_1\hat{j}$ and you want to meet a friend who is at position $\vec{s}_2 = a_2\hat{i} + b_2\hat{j}$. The displacement vector taking you from $\vec{s}_1$ to $\vec{s}_2$ is $\Delta\vec{s} = \vec{s}_2 - \vec{s}_1 = (a_2 - a_1)\hat{i} + (b_2 - b_1)\hat{j}$ as indicated in the figure. The new $\hat{i}$ component is just the difference in the individual $\hat{i}$ components, and likewise for the $\hat{j}$ and $\hat{k}$ components.

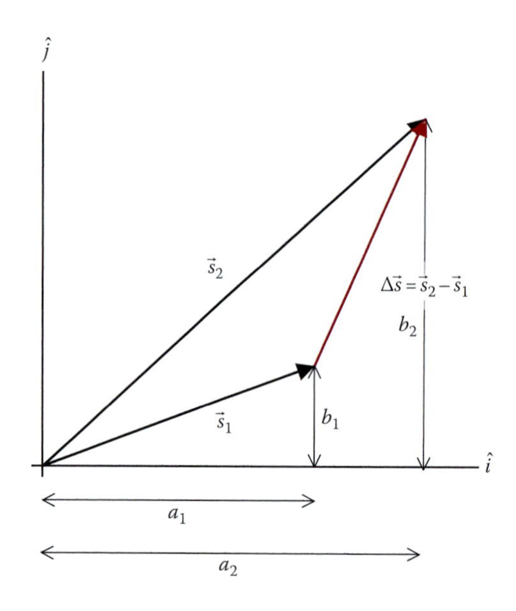

FIGURE 2.4 Vector subtraction: $\Delta\vec{s} = \vec{s}_2 - \vec{s}_1$. Note the direction of $\Delta\vec{s}$, which conforms to the rules of vector addition, $\vec{s}_1 + \Delta\vec{s} = \vec{s}_2$.

EXERCISE 2.3

A cube of side length a has its sides aligned with the x, y, and z-axes of a Cartesian coordinate system, as shown. Write the displacement vector describing a change in position from point A to point B, the two points shown in the following figure.

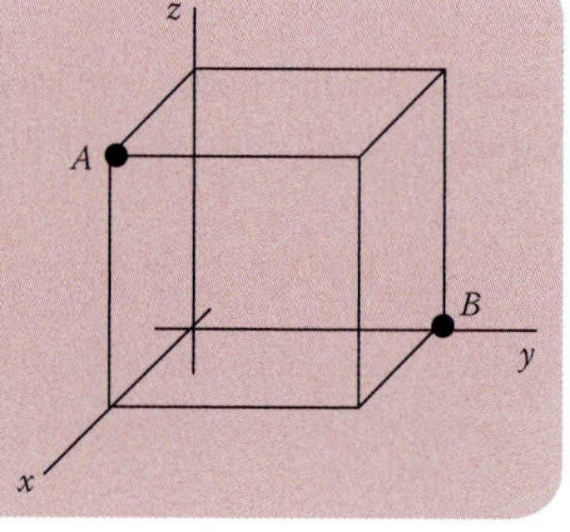

Answer: $\Delta\vec{s} = -a\hat{i} + a\hat{j} - a\hat{k}$

It is also useful to define the multiplication of a vector by a scalar. If $\vec{s} = a\hat{i} + b\hat{j} + c\hat{k}$, the product of the vector $\vec{s}$ and a scalar n is

$$n\vec{s} = na\hat{i} + nb\hat{j} + nc\hat{k}. \tag{2.5}$$

Each component of $\vec{s}$ is separately multiplied by the factor n. In agreement with Equation 2.2, this changes the magnitude of $\vec{s}$ by a factor n. If $n > 1$, the vector is stretched; if $n < 1$, it is shrunk. If $n > 0$, its direction does not change; but if $n < 0$, its direction is reversed. All of these features are shown in Figure 2.5.

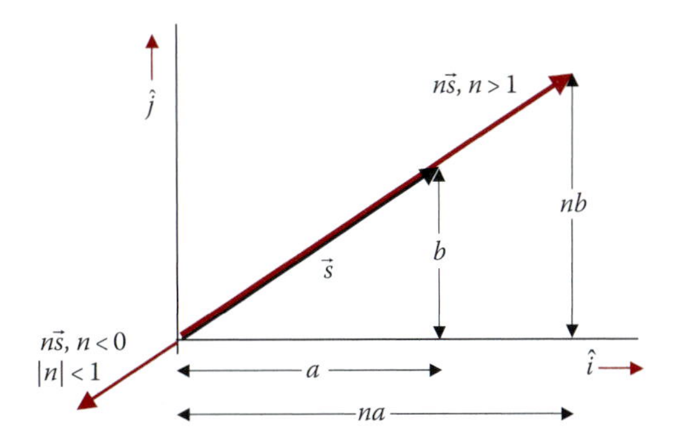

FIGURE 2.5 Multiplication of vector $\vec{s}$ by a scalar n. Depending on n, the original vector is stretched ($n > 1$), shrunk ($0 < n < 1$), or has its direction reversed ($n < 0$).

Note that $-\vec{s} = (-1)\vec{s} = -a\hat{i} - b\hat{j} - c\hat{k}$ has the same magnitude as $\vec{s}$ but points in the opposite direction. Vector subtraction (Equation 2.4) is thus equivalent to the addition of the negative of a vector: $\vec{A} - \vec{B} = \vec{A} + (-\vec{B})$.

The three unit vectors $\hat{i}$, $\hat{j}$, and $\hat{k}$ allow us to interpret the general expression for a vector literally: the vector $\vec{A} = a\hat{i} + b\hat{j} + c\hat{k}$ is the sum of three *component* vectors $a\hat{i}$, $b\hat{j}$, and $c\hat{k}$, each of which is a stretched version of one of the unit vectors $\hat{i}$, $\hat{j}$, or $\hat{k}$. It will often be useful to think of vectors in terms of their components.

EXERCISE 2.4

Let $\vec{A} = a\hat{i} + b\hat{j} + c\hat{k}$ and $\vec{B}$ be a unit vector parallel to $\vec{A}$. Find $\vec{B}$ in terms of a, b, and c. (*Hint:* Let $\vec{B} = n\vec{A} = na\hat{i} + nb\hat{j} + nc\hat{k}$. According to the earlier discussion, this has the same direction as $\vec{A}$ if $n > 0$. Find n so that $|\vec{B}| = 1$.)

Answer $\vec{B} = n\vec{A} = \dfrac{a\hat{i} + b\hat{j} + c\hat{k}}{\sqrt{a^2 + b^2 + c^2}}$

EXERCISE 2.5

Find the unit vector parallel to $\vec{V} = \hat{i} - 2\hat{j} + 4\hat{k}$. Also find the unit vector *anti*parallel to $\vec{V}$.

Answer $\hat{V} = \dfrac{1}{\sqrt{21}}(\hat{i} - 2\hat{j} + 4\hat{k})$, $\hat{V}' = -\hat{V} = \dfrac{1}{\sqrt{21}}(-\hat{i} + 2\hat{j} - 4\hat{k})$

2.5 ANGLES BETWEEN VECTORS

Equation 2.2 tells us that if we know the components of a vector, we can easily determine its magnitude. But these components also tell us the *direction* of the vector relative to the axes of our chosen coordinate system. This is easy to see in two dimensions and not much harder to understand in three dimensions.

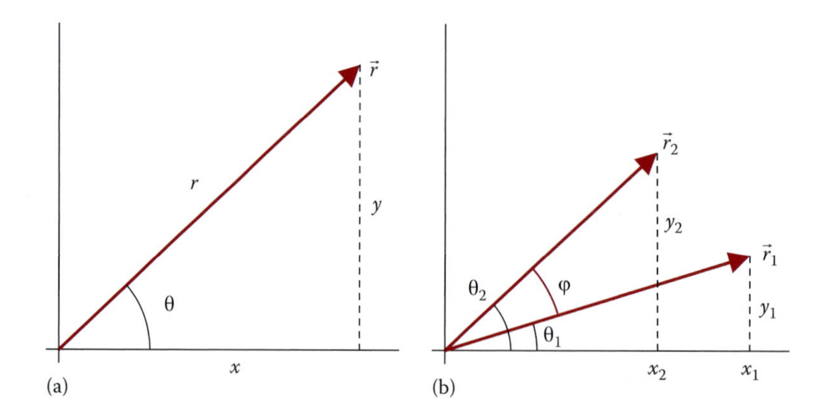

FIGURE 2.6 (a) Vector $\vec{r}$ makes an angle $\theta = \cos^{-1}(x/r)$ with the x- or $\hat{i}$ axis. (b) Vectors $\vec{r}_1 = x_1\hat{i} + y_1\hat{j}$, $\vec{r}_2 = x_2\hat{i} + y_2\hat{j}$ make angles θ_1 and θ_2 with the x-axis. The angle between the vectors is θ.

In Figure 2.6a, the vector $\vec{r} = x\hat{i} + y\hat{j}$ makes an angle θ with respect to the $\hat{i}$ axis. Knowing x and y, it is easy to find θ by simple trigonometry: $\cos\theta = x/r$ (and $\sin\theta = y/r$). Now consider Figure 2.6b, which shows two vectors $\vec{r}_1 = x_1\hat{i} + y_1\hat{j}$ and $\vec{r}_2 = x_2\hat{i} + y_2\hat{j}$ that make angles θ_1 and θ_2 with the $\hat{i}$ direction. Let's derive an expression for the angle φ between the two vectors, $\varphi = \theta_2 - \theta_1$, directly in terms of the components x_1, x_2, y_1, and y_2. First recall the trigonometric identity:

$$\cos(\theta_2 - \theta_1) = \cos\theta_1 \cos\theta_2 + \sin\theta_1 \sin\theta_2.$$

Substituting for sines and cosines ($\cos\theta_1 = x_1/r_1$, etc.), we find

$$\cos\varphi = \cos(\theta_2 - \theta_1) = \frac{x_1}{r_1}\frac{x_2}{r_2} + \frac{y_1}{r_1}\frac{y_2}{r_2} = \frac{x_1x_2 + y_1y_2}{r_1r_2}.$$

In three dimensions, the corresponding equation for the cosine of the angle φ between two vectors $\vec{r}_1 = x_1\hat{i} + y_1\hat{j} + z_1\hat{k}$ and $\vec{r}_2 = x_2\hat{i} + y_2\hat{j} + z_2\hat{k}$ is

$$\cos\varphi = \frac{x_1x_2 + y_1y_2 + z_1z_2}{r_1r_2}. \tag{2.6}$$

Finding the angle between two vectors is such an important task that we introduce a new mathematical symbol to express Equation 2.6 in a more compact way. We define the *dot product* of two vectors as

$$\vec{r}_1 \cdot \vec{r}_2 = x_1x_2 + y_1y_2 + z_1z_2, \tag{2.7}$$

so that Equation 2.6 can be expressed as

$$\cos\varphi = \frac{\vec{r}_1 \cdot \vec{r}_2}{r_1 r_2}. \tag{2.8}$$

From Equation 2.8, $\vec{r}_1 \cdot \vec{r}_2 = r_1 r_2 \cos\varphi$, and from Equation 2.7, $\vec{r}_1 \cdot \vec{r}_2 = \vec{r}_2 \cdot \vec{r}_1$, that is, the dot product is commutative.

Example

Let $\vec{u} = \hat{i} + 2\hat{j} + 2\hat{k}$ and $\vec{v} = 2\hat{i} + 4\hat{j} + 4\hat{k}$. Find the angle between $\vec{u}$ and $\vec{v}$. Solution: First find the magnitude of each vector: $u = \sqrt{1^2 + 2^2 + 2^2} = 3$. Similarly, $v = 6$. Then, using Equation 2.6 $\cos\varphi = (1\cdot 2 + 2\cdot 4 + 2\cdot 4)/3\cdot 6 = 18/18 = 1$, so $\varphi = \cos^{-1}(1) = 0$, that is, the two vectors are parallel. (In fact, this should have been obvious, since $\vec{v} = 2\vec{u}$.)

EXERCISE 2.6

Find the angle φ between the vectors $\vec{r}_1 = \hat{i} + 2\hat{j} + 2\hat{k}$ and $\vec{r}_2 = 2\hat{i} - 2\hat{j} + \hat{k}$. Is $\vec{r}_2$ the *only* vector making this angle φ with $\vec{r}_1$? If not, find another one.

Answer 90°

EXERCISE 2.7

Complete the following statement: two vectors are perpendicular to each other if their dot product equals ... Evaluate the following dot products: $\hat{i}\cdot\hat{j} = \cdots$, $\hat{i}\cdot\hat{k} = \cdots$, $\hat{j}\cdot\hat{k} = \cdots$; also $\hat{i}\cdot\hat{i} = \cdots$, $\hat{j}\cdot\hat{j} = \cdots$, $\hat{k}\cdot\hat{k} = \cdots$

We are finally in a position to determine the direction of a vector relative to our coordinate axes in three dimensions. Let $\vec{r} = a\hat{i} + b\hat{j} + c\hat{k}$. Taking the dot product of $\vec{r}$ with the unit vector $\hat{i}$ and using the results of Exercise 2.7 we find

$$\vec{r}\cdot\hat{i} = |\vec{r}||\hat{i}|\cos\theta_x = r\cos\theta_x = a\hat{i}\cdot\hat{i} + b\hat{j}\cdot\hat{i} + c\hat{k}\cdot\hat{i} = a\hat{i}\cdot\hat{i} = a,$$

where θ_x is the angle between $\vec{r}$ and $\hat{i}$, that is, between $\vec{r}$ and the x-axis. Solving for θ_x,

$$\cos\theta_x = \frac{\vec{r}\cdot\hat{i}}{r} = \frac{a}{r}. \tag{2.9}$$

This result looks the same as the one we obtained earlier for the two-dimensional case using trigonometry. But in this case, we are working in three dimensions, and Equation 2.9 holds for $\vec{r}$ pointing in *any* direction relative to our coordinate axes. Because $\cos\theta_x$ helps us determine the direction of a vector, it is called a *direction cosine*.

To completely determine the vector's direction relative to our coordinate axes, we also need to find the other direction cosines $\cos\theta_y$ and $\cos\theta_z$. Problem 2.11 shows that the three direction cosines are not independent.

EXERCISE 2.8

You'll need your calculator to solve this one. The figure shows a cube of side length a aligned with its edges parallel to the x-, y-, and z-axes of a Cartesian coordinate system. Vector $\vec{d}$ is the cube diagonal extending from the corner at the origin to the opposite corner of the cube. What is the angle θ_x between $\vec{d}$ and the x-axis? What is the angle between $\vec{d}$ and either of the other two axes?

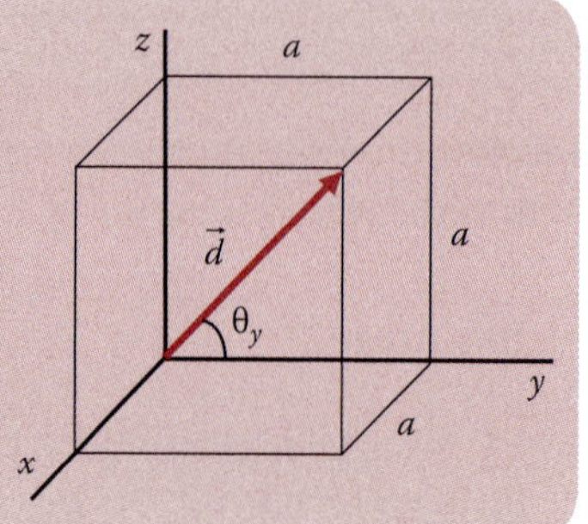

Answer $\theta_x = \theta_y = \theta_z = \cos^{-1}(1/\sqrt{3}) = 55°$

2.6 VECTORS WITHOUT COORDINATE AXES

All of the aforementioned mathematical operations have been defined using the $\hat{i}$, $\hat{j}$, and $\hat{k}$ components of the vectors involved. The magnitude of each component depends on the particular coordinate system we have chosen, that is, the orientation of its axes in space. But vector equations are independent of the coordinate system we are using: if $\vec{C} = \vec{A} + \vec{B}$ in one coordinate system, then $\vec{C} = \vec{A} + \vec{B}$ in any other coordinate system. To see this, study the following example.

Example

In the x–y coordinate system shown, $\vec{A} = \hat{i} + \hat{j}$, $\vec{B} = -\hat{i} + \hat{j}$, so $\vec{C} = \vec{A} + \vec{B} = 2\hat{j}$. $\vec{C}$ has a magnitude 2 and points in the vertical direction. Now let's write these vectors in the primed coordinate system wherein the x- and y-axes are rotated by 45°. Note that $|\vec{A}| = |\vec{B}| = \sqrt{2}$. $\vec{A}$ lies along the x'-axis, so $\vec{A} = \sqrt{2}\hat{i}'$. Similarly, $\vec{B} = \sqrt{2}\hat{j}'$. In the primed system, $\vec{A} + \vec{B} = \sqrt{2}\hat{i}' + \sqrt{2}\hat{j}'$, which once again has a magnitude of 2 units and points in the vertical (original $\hat{j}$) direction. Whether you add the vectors using their primed or unprimed components, the sum $\vec{A} + \vec{B}$ results in the same vector $\vec{C}$.

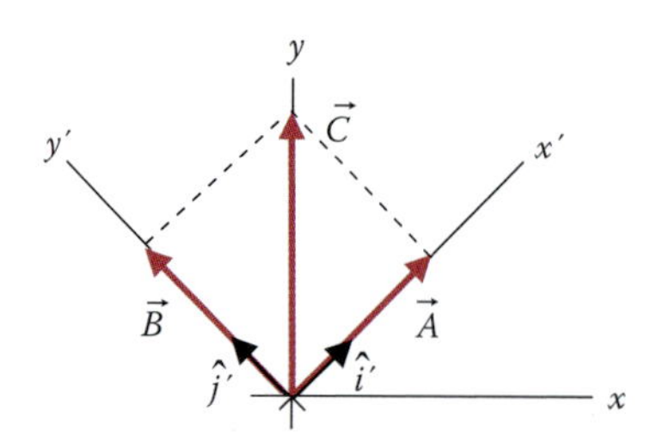

Changing the coordinate system changes the values of C_x, C_y, C_z, but the components of $\vec{A}$ and $\vec{B}$ change as well, so that $C_{x'} = A_{x'} + B_{x'}$, etc., no matter how we orient the axes. Apart from ease of calculation, the coordinate system we choose really doesn't matter. In fact, we can define and carry out all of the operations discussed (addition, subtraction, scalar multiplication, and the dot product) by graphical means, without reference to the vectors' $\hat{i}$, $\hat{j}$, and $\hat{k}$ components. To unleash the full power of vectors, *we will erase the coordinate axes*!

This is illustrated in Figure 2.7 for the case of vector addition. Figure 2.7a is a copy of Figure 2.3. Figure 2.7b is the same drawing without the x- and y-axes. In the latter drawing, there are no a_i's and b_i's to sum; instead, vector addition $\vec{s}_{tot} = \vec{s}_1 + \vec{s}_2$ is shown as a tip-to-tail concatenation of the two vectors $\vec{s}_1$ and $\vec{s}_2$. The tail of $\vec{s}_2$ is attached to the tip of $\vec{s}_1$, and $\vec{s}_{tot}$ is the vector drawn from the tail of the first vector to the tip of the second. Figure 2.7a and b shows that the two descriptions of vector addition—with and without axes—are equivalent.

With or without an explicit coordinate system a vector is still defined by its magnitude and direction. Without a coordinate system we cannot define a vector's direction relative to $\hat{i}$, $\hat{j}$, and $\hat{k}$ directions, but we can still specify the vector's direction relative to another vector and that will be sufficient to analyze physical phenomena. Having freed a vector from a particular coordinate system, we are also free to move it around. As long as we do not change its magnitude or direction, we can move the vector without altering it. Figure 2.8a illustrates this, using vector addition as an example. By shifting the positions of $\vec{s}_1$ and $\vec{s}_2$, *without changing their directions or magnitudes*, we can arrange them so that the tail of $\vec{s}_1$ is attached to the tip of $\vec{s}_2$ rather than the other way around. The vectors $\vec{s}_1$, $\vec{s}_2$, and their shifted copies form a parallelogram, showing that $\vec{s}_{tot} = \vec{s}_1 + \vec{s}_2 = \vec{s}_2 + \vec{s}_1$: vector addition is commutative, just as we learned earlier using the coordinate representation (Equation 2.3).

Vector subtraction is illustrated in Figure 2.8b. The difference vector $\Delta\vec{s} = \vec{s}_2 - \vec{s}_1$ can be found by first drawing $\vec{s}_1$ and $\vec{s}_2$ with their tails attached; then, $\Delta\vec{s} = \vec{s}_2 - \vec{s}_1$ fits between the tips of $\vec{s}_1$ and $\vec{s}_2$ and points

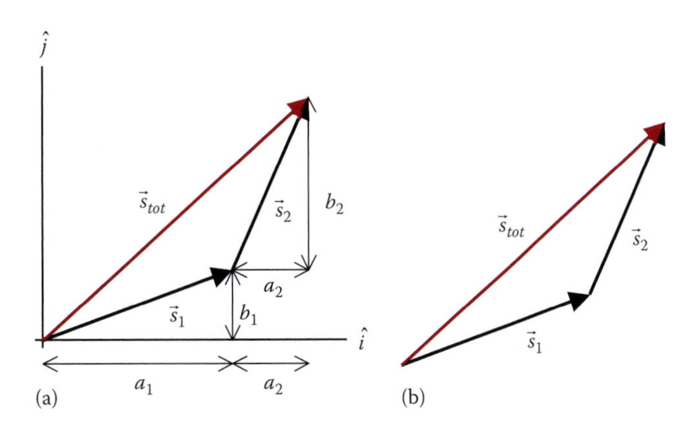

FIGURE 2.7 Vector addition: (a) with coordinate axes shown and (b) without coordinate axes.

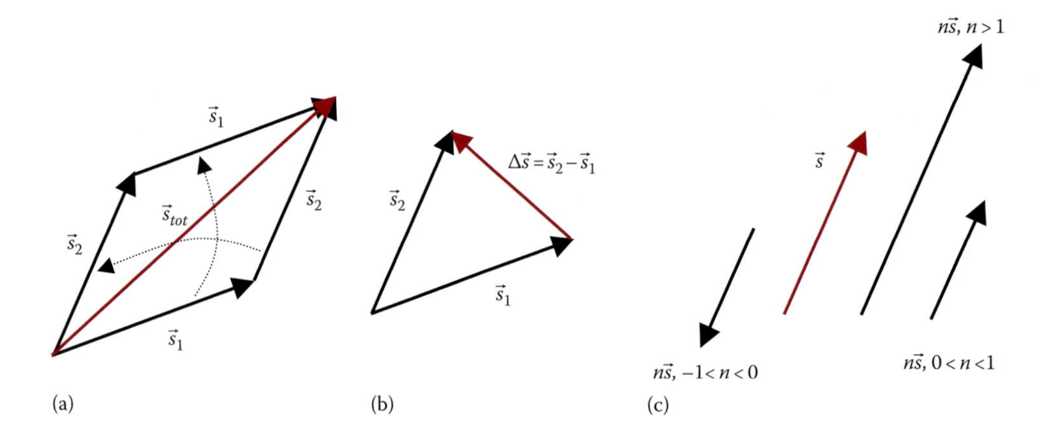

FIGURE 2.8 (a) $\vec{s}_{tot} = \vec{s}_1 + \vec{s}_2$. Transposing $\vec{s}_1$ and $\vec{s}_2$, $\vec{s}_{tot} = \vec{s}_2 + \vec{s}_1$ showing that vector addition is commutative. The original two vectors plus their transposed copies form a parallelogram. (b) Vector subtraction $\Delta\vec{s} = \vec{s}_2 - \vec{s}_1$ without coordinate axes. The vectors $\vec{s}_1$ and $\vec{s}_2$ are arranged with their tails attached and $\Delta\vec{s}$ points from the tip of the $\vec{s}_1$ to the tip of $\vec{s}_2$. (c) Vector multiplication by a scalar n.

from the tip of $\vec{s}_1$ to the tip of $\vec{s}_2$. (In other words, $\Delta\vec{s} = \vec{s}_2 - \vec{s}_1$ starts from the tip of the vector with the minus sign in front of it and ends at the tip of the vector with the plus sign.) Be careful! It is very easy to assign the wrong direction to $\Delta\vec{s}$. One way to ensure that you have chosen the correct direction is to note that vector subtraction must be consistent with vector addition: the direction of $\Delta\vec{s}$ must be chosen so that $\vec{s}_1 + \Delta\vec{s} = \vec{s}_1 + (\vec{s}_2 - \vec{s}_1) = \vec{s}_2$, as it is in Figure 2.8b.

Scalar multiplication of a vector is illustrated in Figure 2.8c. For clarity, we have purposely shifted the vectors laterally; remember, shifting a vector without changing its direction does not change its identity.

EXERCISE 2.9

Vectors $\vec{A}$ and $\vec{B}$ define two sides of a parallelogram with diagonals $\vec{C}$ and $\vec{D}$. In terms of $\vec{A}$ and $\vec{B}$, find expressions for the following:

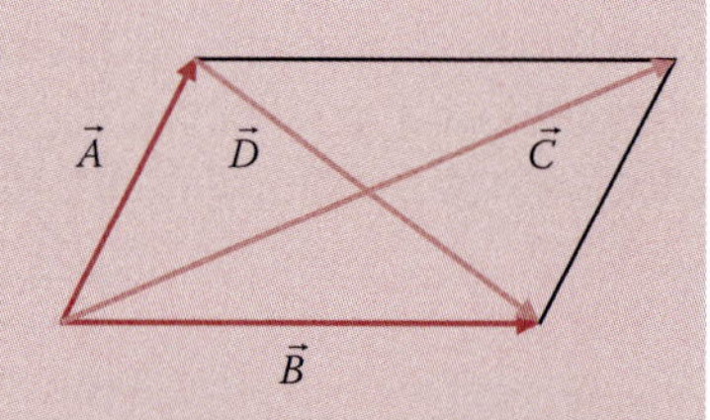

(a) $\vec{C} =$
(b) $\vec{D} =$
(c) $\vec{C} - \vec{D} =$

Answer (a) $\vec{A} + \vec{B}$ (b) $\vec{B} - \vec{A}$ (c) $2\vec{A}$

EXERCISE 2.10

Find an expression for the vector $\vec{V}$ extending from the intersection of $\vec{A}$ and $\vec{B}$ to the midpoint of the line joining the tips of $\vec{A}$ and $\vec{B}$.

Answer $\vec{V} = \frac{1}{2}(\vec{B}+\vec{A}) = \frac{1}{2}(\vec{A}+\vec{B})$

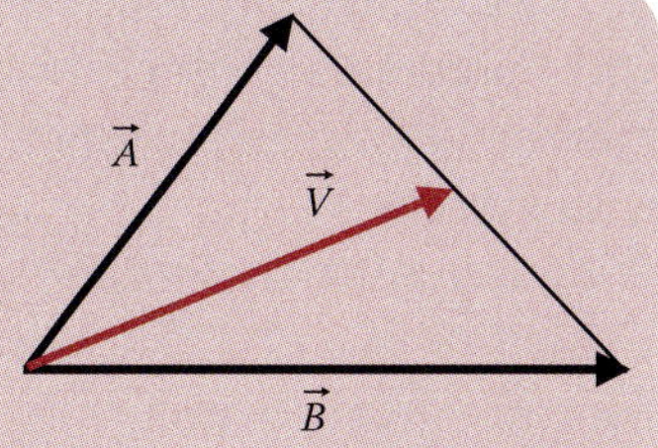

EXERCISE 2.11

The vectors $\vec{A}$, $\vec{B}$, and $\vec{C}$ are of equal length and are parallel to three sides of a regular hexagon of side length a. Sketch the following vectors:

(a) $\vec{U} = \vec{A} + \vec{B} - \vec{C}$.
(b) $\vec{V} = \vec{A} + \vec{B} + \vec{C}$.

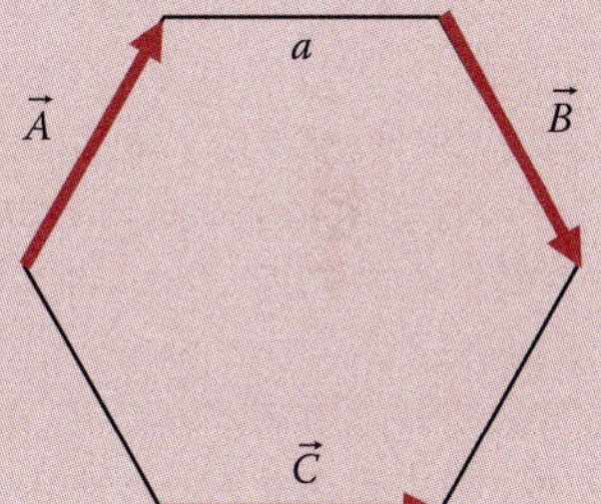

2.7 COMPARING THE COPERNICAN AND PTOLEMAIC WORLDS: VECTORIALLY

As an application of vector algebra, let's once again compare the Copernican and Ptolemaic models of the universe. Recall that epicycles were conceived to explain the phenomenon of retrograde motion, the puzzling reversal in the apparent direction of planetary motion as observed from Earth (see Figure 2.9). In the Ptolemaic model, Earth sits motionless at the center of the universe,* and retrograde motion is a consequence of a planet's epicycle. In Copernicus' heliocentric model, retrograde motion occurs because all of the planets, including Earth, are in nearly circular orbits centered on

* The epicycle model was originally proposed by the Greek astronomer Appolonius in the third century BCE. Ptolemy refined the theory to bring it into better agreement with planetary observations.

the Sun. Since both models agree equally well with planetary observations, they must be mathematically equivalent. Let's prove this with vectors.

Figure 2.10a shows the Earth E, Sun S, and a superior planet P (such as Mars, Jupiter, or Saturn) configured according to the Copernican model. $\vec{R}_{SE}$ and $\vec{R}_{SP}$ are vectors from the Sun to Earth and planet P, respectively. These two vectors rotate at different rates, and their tips trace out the circular planetary orbits of E and P. Using the rule for vector subtraction, the vector $\vec{R}_{EP}$ from Earth to planet P obeys the equation $\vec{R}_{EP} = \vec{R}_{SP} - \vec{R}_{SE}$. Let $\vec{R}_{ES}$ be the vector pointing from the Earth to the Sun. Since $\vec{R}_{ES} = -\vec{R}_{SE}$, then $\vec{R}_{EP} = \vec{R}_{SP} + \vec{R}_{ES}$. We are free to rearrange these vectors as long as we do not change their magnitudes or directions. This has been done in Figure 2.10b. $\vec{R}_{ES}$ and $\vec{R}_{SP}$ still rotate, and their tips still trace out circles, precisely as they did before we moved them around. But Figure 1.10b exactly depicts the Ptolemaic model for superior planets. (Compare to Figure 1.9.) For each superior planet, the deferent traces out its Copernican orbit and the epicycle mirrors the Earth's orbit about the Sun.

FIGURE 2.9 Retrograde motion of Mars from June to December, 2003. Each of the superimposed images was taken about 1 week apart, with the earliest image to the right (west). The brightest image occurred when the Earth and Mars were closest to each other. The faint dotted line in the background is the planet Uranus. (Photo: Tunç Tezel, Bursa, Turkey.)

EXERCISE 2.12

Mars orbits the Sun with a period of 1.88 years and an orbital radius of 1.52 AU. In the Ptolemaic model, what are (a) the radius of its epicycle? (b) the radius of its deferent? (c) the period of its epicycle? and (d) the period of its deferent?

The same vector equation holds for the inferior planets Mercury and Venus. Figure 2.10c and d compares the two models for these bodies. The roles of deferent and epicycle are now switched. Each inferior planet has a deferent of length 1 AU and a period of 1 year, and the planet's epicycle is a replica of its heliocentric orbit. Note that the epicycle centers for Mercury and Venus are not just *in line* with the Sun, they are *coincident* with the Sun. This refinement of the Ptolemaic model makes it mathematically equivalent to the Copernican system for *all* of the planets, and was advanced by the famed Danish astronomer Tycho Brahe in 1588. Even though it solved the problem of the phases of Venus (see Section 1.5), the Tychonic

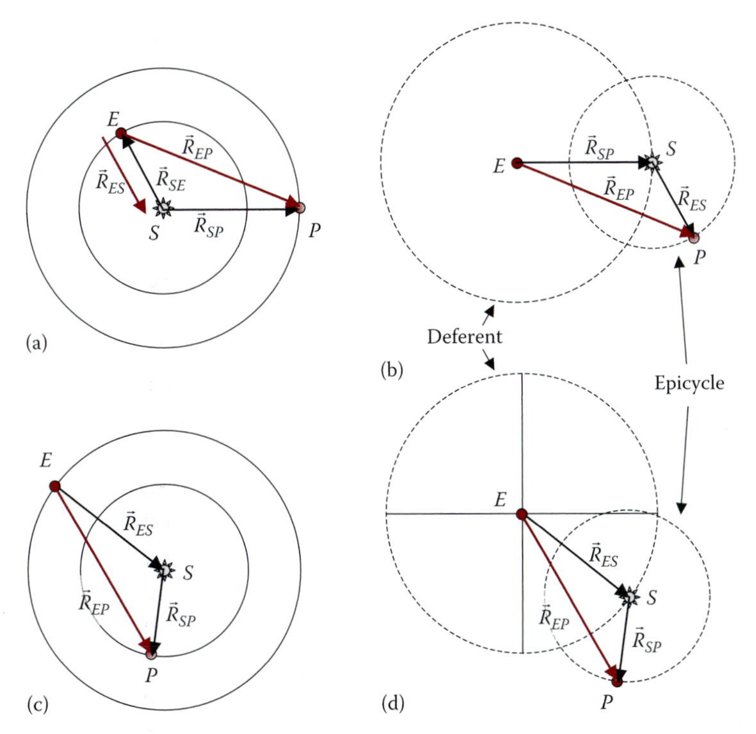

FIGURE 2.10 (a) The Earth (E), Sun (S), and superior planet (P) according to the Copernican model. (b) The same configuration as in (a), but $\vec{R}_{SP}$ and $\vec{R}_{ES}$ $(= -\vec{R}_{SE})$ have been translated to new positions to show the equivalence with the Ptolemaic model. The dotted lines show the deferent and epicycle of planet P. The orbit of an inferior planet (P) is shown in the (c) Copernican and (d) Ptolemaic models. In this case, the deferent has a length 1 AU and an epicycle with the same radius as the planet's heliocentric orbit.

system—wherein Mercury and Venus orbited the Sun while the other planets orbited Earth—was rejected as an unscientific hodgepodge by Galileo and was not once mentioned in his *Dialogues*.

There is a valuable lesson to learn here. Ptolemy was surely baffled by the behavior of Mercury and Venus, and he must have wondered why the epicycle periods of Mars, Jupiter, and Saturn were exactly the same as the observed orbital period of the Sun. Nevertheless, he spent most of his life refining the epicycle theory to *save the appearances*, that is, to account for planetary observations using a *geocentric* model in which planetary orbits were restricted to circles (or combinations of circles). By adding successive layers of complexity to the basic epicycle model, he dramatically improved the agreement between theory and observation. But in spite of his ingenuity, or perhaps because of it, he missed the bigger prize. Blinded by the dominant scientific paradigm of his day (due to Aristotle), he overlooked the evidence for a heliocentric planetary system.* Stories like this are common throughout the history of science. It is a lot easier to tinker with an existing theory than it is to recognize the need for a new one. Recalling the trial of Galileo, it is also a lot less risky!

* In fact, the heliocentric system had been proposed by Aristarchus in the third century BCE. But bound by the teachings of Aristotle, it was rejected by later Greek astronomers.

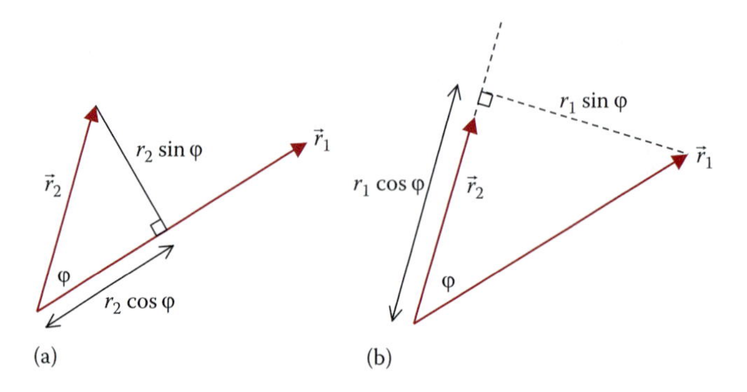

FIGURE 2.11 The dot product $\vec{r}_1 \cdot \vec{r}_2 = r_1 r_2 \cos\varphi$ without coordinate axes. (a) $r_2 \cos\varphi$ is the component of $\vec{r}_2$ parallel to $\vec{r}_1$ and (b) $r_1 \cos\varphi$ is the component of $\vec{r}_1$ parallel to $\vec{r}_2$.

2.8 DOT PRODUCT REVISITED

Recall the definition of the dot product from Section 2.4: the dot product of two vectors $\vec{r}_1 = a_1\hat{i} + b_1\hat{j} + c_1\hat{k}$ and $\vec{r}_2 = a_2\hat{i} + b_2\hat{j} + c_2\hat{k}$ is defined as $\vec{r}_1 \cdot \vec{r}_2 = a_1 a_2 + b_1 b_2 + c_1 c_2 = r_1 r_2 \cos\varphi$, where φ is the angle between the two vectors. Without a specified coordinate system, the coefficients a_i, b_i, and c_i are undetermined, but the dot product may still be defined in terms of the vectors' magnitudes and the angle between them. Consider two vectors $\vec{r}_1$ and $\vec{r}_2$ whose directions differ by an angle φ. Vector $\vec{r}_2$ can be thought of as the sum of two vectors, one which is parallel to $\vec{r}_1$ and has a magnitude $r_2 \cos\varphi$, and the other which is perpendicular to $\vec{r}_1$ with a magnitude $r_2 \sin\varphi$. These vectors are shown in Figure 2.11a. Then,

$$\vec{r}_1 \cdot \vec{r}_2 = r_1 r_2 \cos\varphi = r_1 (r_2 \cos\varphi) = r_1 \times (\text{the component of } \vec{r}_2 \text{ parallel to } \vec{r}_1). \tag{2.10}$$

Alternatively, for the same two vectors $\vec{r}_1$ and $\vec{r}_2$ (see Figure 2.11b),

$$\vec{r}_1 \cdot \vec{r}_2 = r_1 r_2 \cos\varphi = r_2 (r_1 \cos\varphi) = r_2 \times (\text{the component of } \vec{r}_1 \text{ parallel to } \vec{r}_2). \tag{2.11}$$

Equations 2.8, 2.10, and 2.11 are equivalent expressions for the dot product. Lastly, note that

$$\vec{r}_1 \cdot \vec{r}_1 = r_1 r_1 \cos(0) = r_1^2, \tag{2.12}$$

that is, *the dot product of any vector with itself equals its magnitude squared.* This will often come in handy as you will see in the next section.

2.9 APPLICATION: THE LAW OF COSINES

Vectors make it easy to solve many geometry problems. To appreciate this, let's use vector algebra to derive a result that should be familiar from earlier coursework in mathematics. Figure 2.12 illustrates the addition of two vectors: $\vec{c} = \vec{a} + \vec{b}$, where $\vec{a}$ and $\vec{b}$ point in directions that differ by an angle θ. The three

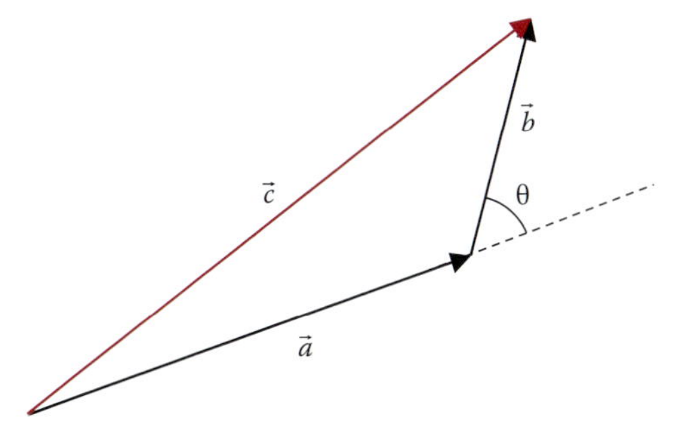

FIGURE 2.12 Illustrating the law of cosines. θ is the angle between the direction of $\vec{a}$ and $\vec{b}$. Note that this is the exterior angle of the triangle.

vectors $\vec{a}$, $\vec{b}$, and $\vec{c}$ form a triangle.* (Note that θ is the *exterior* angle of the triangle.) Taking the dot product of $\vec{c}$ with itself we get, according to the preceding discussion,

$$\vec{c}\cdot\vec{c} = c^2 = \left(\vec{a}+\vec{b}\right)\cdot\left(\vec{a}+\vec{b}\right)$$
$$= a^2 + b^2 + 2\vec{a}\cdot\vec{b} = a^2 + b^2 + 2ab\cos\theta.$$

This is the familiar law of cosines, derived in two easy steps!†

EXERCISE 2.13

Lines $\overline{AC}$ and $\overline{BD}$ are any two diagonals of a circle. Prove that lines $\overline{AD}$ and $\overline{BC}$ are parallel and of equal length. (*Hint:* Define *vector* $\overrightarrow{AD} = \vec{r}_1 + \vec{r}_3$ and vector $\overrightarrow{BC} = \vec{r}_2 + \vec{r}_4$, where each $\vec{r}_i$ has a magnitude equal to the radius of the circle. Show that $\overrightarrow{AD} = \overrightarrow{BC}$.)

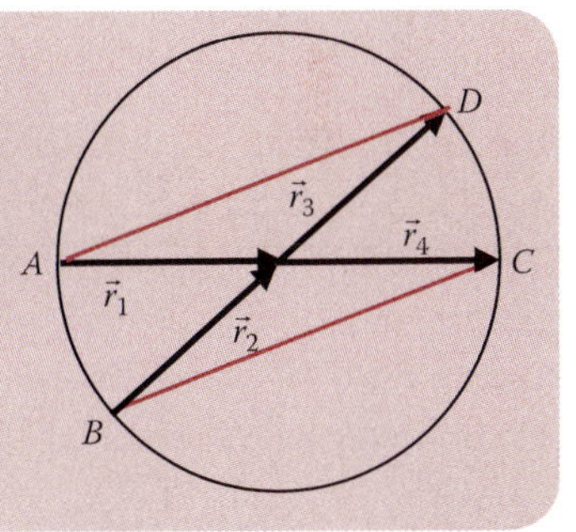

* Vectors $\vec{a}$ and $\vec{b}$ lie in a single plane, with no component of either vector perpendicular to that plane. Therefore, vector $\vec{c}$ must also lie in the same plane, and the three vectors can be arranged to form a triangle.

† You might remember the law of cosines as $c^2 = a^2 + b^2 - 2ab\cos\varphi$. In this equation, φ is the *interior* angle of the triangle, that is, $\varphi = 180°-\theta$. Since $\cos(180 - \theta) = -\cos\theta$, the two expressions are equivalent.

2.10 APPLICATION: CALCULATING THE NUMBER OF HOURS OF DAYLIGHT

The Earth's axis of rotation, passing through its north and south poles, is tilted by an angle $\alpha = 23.5°$ from the plane of its orbit about the Sun (see Figure 2.13). This tilt causes the daily dose of sunlight reaching a spot *O* on Earth to vary with *O*'s season and latitude. In the northern hemisphere, the shortest *day* of the year occurs on December 21 or 22 called the December (or winter) *solstice*. The longest day falls on June 20 or 21 called the June (or summer) solstice. From December to June in the northern hemisphere, the span of daylight increases continually and on the *equinoxes* (March 20 and September 22–23) there are 12 h of daylight everywhere on Earth.‡ During winter (summer) in the northern hemisphere, there are fewer (more)

‡ This discussion ignores refraction of light by the Earth's atmosphere and assumes that the Earth's orbit is circular. Of course, the months of winter and summer are reversed in the southern hemisphere.

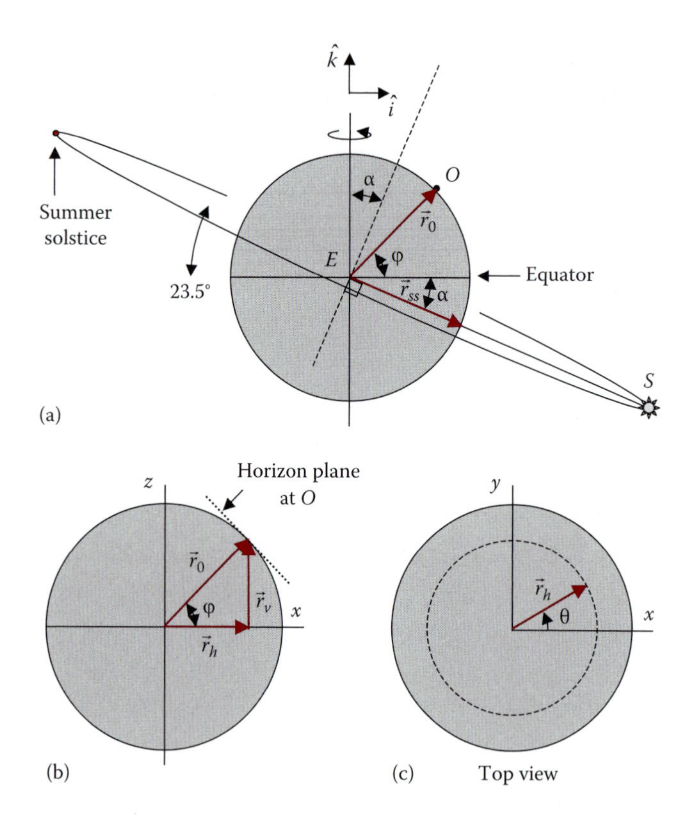

FIGURE 2.13 (a) Earth and Sun on the December solstice. In the Ptolemaic picture, the Sun revolves about Earth once each year. (The Sun and its orbit are not shown to scale.) Angle α is the planet's axial tilt, angle φ is the latitude of O, and the dotted line is perpendicular to the plane of the Sun's orbit. Angle φ is the latitude of the observation point O. The subsolar point is located at $\vec{r}_{ss}$ on the line between the Earth and Sun. As the year passes, $\vec{r}_{ss}$ rotates in the plane of the orbit. (b) Vector $\vec{r}_O = \vec{r}_v + \vec{r}_h$, where $r_h = R_E \cos\varphi$. (c) Top view of (b). Due to the Earth's daily rotation, the horizontal component $\vec{r}_h$ sweeps out a full circle once per day.

hours of daylight at higher latitudes than at lower latitudes. For example, on the December solstice there is zero daylight above the Arctic Circle located at latitude $\varphi = 66.5° = 90° - \alpha$. Conversely, on the June solstice the Sun is visible for 24 hours above the Arctic Circle. This complex dependence of daylight on time and latitude is difficult to analyze using ordinary trigonometry, but relatively easy to understand using vectors.

Figure 2.13a depicts the Earth and Sun on the day of the December solstice. For an observer stationed at point *O*, the Sun reaches its highest altitude above *O*'s horizon at noon (local time), when the points *O*, *S*, and *E* (the Earth's center) lie in the plane of the figure. Let $\vec{r}_O$ be the vector from *E* to *O* and $\vec{r}_{ss}$ be the vector from *E* to the *subsolar* point where line $\overline{ES}$ intersects the planet's surface. For simplicity, we'll use a pseudo-Ptolemaic description wherein a stationary Earth rotates on its axis once per day and the Sun circles Earth once per year.* Define a fixed coordinate system with origin at *E* and *z*-axis pointed

* From Section 2.7, we know that this is mathematically equivalent to the Copernican model.

along the Earth's N–S axis so that the Equator lies in the x–y plane. Let the y-axis point into the page in Figure 2.13a. Vector $\vec{r}_{ss}$ makes an angle δ = −α with respect to the equatorial plane (δ is negative when $\vec{r}_{ss}$ lies south of the Equator), so

$$\vec{r}_{ss} = R_E(\cos\delta\hat{i} + \sin\delta\hat{k}), \tag{2.13}$$

where R_E is the Earth's radius. Angle δ is called the Sun's *declination*. For observer O at latitude φ at noon,

$$\vec{r}_O = R_E(\cos\varphi\hat{i} + \sin\varphi\hat{k}). \tag{2.14}$$

Note that $\vec{r}_O$ has a horizontal component $r_h = R_E\cos\varphi$ and a vertical component $r_v = R_E\sin\varphi$ (see Figure 2.13b). As the Earth turns, $\vec{r}_v$ remains constant but $\vec{r}_h$ rotates along with point O, tracing out a full circle once per day. Let θ be the angle of rotation since noon (θ = 0 at noon, θ < 0 in the morning), so $\vec{r}_h(\theta) = r_h(\cos\theta\hat{i} + \sin\theta\hat{j})$ (Figure 2.13c). For example, at 6:00 pm local time the planet has rotated 90° since noon and $\vec{r}_h(\theta = 90°) = r_h(\cos 90°\hat{i} + \sin 90°\hat{j}) = r_h\hat{j}$. Combining this result with Equation 2.14, the expression for $\vec{r}_O$ at any time of the day is

$$\vec{r}_O(\theta) = R_E\cos\varphi\left(\cos\theta\hat{i} + \sin\theta\hat{j}\right) + R_E\sin\varphi\hat{k}. \tag{2.15}$$

Now we need to figure out the times, or angles θ, of sunrise and sunset at location O. First note that $\vec{r}_{ss}$ points directly from Earth to the Sun. Because the Sun is so far from Earth ($R_E \ll 1$ AU), $\vec{r}_{ss}$ is antiparallel to the direction of sunlight reaching any point on Earth. The *horizon* at O is defined by the plane perpendicular to $\vec{r}_O$ and sunrise and sunset occur when $\vec{r}_{ss}$ lies on the horizon, when $\vec{r}_{ss} \perp \vec{r}_O(\theta)$ or $\vec{r}_{ss}\cdot\vec{r}_O(\theta) = 0$. From Equations 2.13 and 2.15,

$$\vec{r}_{ss}\cdot\vec{r}_O(\theta) = R_E^2(\cos\delta\cos\varphi\cos\theta + \sin\delta\sin\varphi) = 0$$

or

$$\cos\theta = -\tan\delta\tan\varphi. \tag{2.16}$$

Solving this equation for θ allows us to determine the times of sunrise (θ < 0) and sunset (θ > 0) on the day of the December solstice. For example, on the Equator φ = 0, so cos θ = 0 or θ = ±π/2. The Sun rises when θ = −π/2, is at its highest altitude when θ = 0 (noon), and sets when θ = +π/2. Daylight occurs over the range Δθ = π (rad) = 180°. Since a full rotation (360°) takes 24 h, there will be 12 h of daylight at all points on the Equator on the winter solstice (or, in fact, on any day of the year).*

* The authors thank their colleague Professor J. Levine for providing this derivation.

EXERCISE 2.14

(a) Many of the world's major cities lie within ±5° of latitude 45°N. These include Boston, New York, Chicago, Toronto, Montreal, Paris, Munich, Geneva, and Rome. Estimate the amount of daylight in these cities on the December solstice. (b) Prove that for all points on or above the Arctic Circle (latitude 66.5°N) there are zero hours of daylight on the same day.

Answer (a) $\theta = \pm 64.2°$, or 8.6 h of daylight

What about other days of the year? The Sun's declination δ is the angle the subsolar vector $\vec{r}_{ss}$ makes with the equatorial plane. As suggested by Figure 2.13a, δ varies from $-\alpha$ at the December solstice to $+\alpha$ at the June solstice and passes through zero (the equatorial plane) on the two equinoxes midway between the solstices. To a fair approximation,

$$\delta(t) = -\alpha \cos\left(\frac{2\pi t}{365.25}\right), \tag{2.17}$$

where t is the number of days since the December solstice.* To find the number of daylight hours on day t after the winter solstice, simply replace δ with $\delta(t)$ in Equation 2.16. For example, on the equinoxes $\delta(t) = 0$, so, $\cos\theta = 0$ and $\theta = \pm\pi/2$ regardless of latitude φ. On these 2 days, all locations on Earth enjoy 12 h of daylight.

EXERCISE 2.15

How many hours of daylight will there be in Montreal on June 21?

Answer $\theta = \pm 116°$, about 15.4 h

* The correct expression is found from $\sin\delta(t) = -\sin\alpha\cos(2\pi t/365.25)$. For $\alpha = 23.5°$, Equation 2.17 agrees with the correct result to within 0.25°. We are ignoring the eccentricity of Earth's orbit.

2.11 SUMMARY

Vectors are a powerful extension of geometry. Using them, the elegant but fussy strategies of classical geometry are buttressed—or replaced altogether—by simple algebraic expressions. In this chapter we introduced vectors, first in terms of their $\hat{i}$, $\hat{j}$, and $\hat{k}$ components in a particular coordinate system and later without reference to a coordinate system. Vector addition, subtraction, and scalar multiplication were defined with and without coordinate axes, and the two descriptions were shown to be equivalent. The direction of a vector, relative to coordinate axes or relative to another vector, was determined using a new mathematical operation: the dot product. These operations make geometric analysis in three dimensions easy. As an example, we derived a set of equations that allow us to calculate the duration of daylight at any location on Earth as a function of latitude and date. In subsequent chapters, we will use vectors to study motion in three dimensions, for Earth-bound and celestial bodies.

Advances in physics are often accompanied by advances in mathematics. Where would observational astronomy be without geometry and trigonometry? Where would Newtonian mechanics (coming soon!) be without calculus? In this chapter, the mathematics of vectors allowed us to prove—almost effortlessly—that the Ptolemaic and the Copernican descriptions of planetary motion are mathematically equivalent. What new paradigm shifts are in store as we extend our understanding of the universe? Will new mathematics be required to change our present paradigms? At the present time, we know almost nothing about 95% of the contents of the universe. (This is true—really!) Are we any better prepared than Ptolemy to accept a new paradigm or to champion a modern Copernicus, Galileo, Newton, or Einstein?

PROBLEMS

2.1 Two vectors $\vec{A}$ and $\vec{B}$ have equal magnitudes of 10 units. They are oriented as shown in the figure below and their vector sum is $\vec{R}$.

(a) Draw $\vec{R}$.

(b) Find the $\hat{i}$ and $\hat{j}$ components of $\vec{A}$ and $\vec{B}$.

(c) Find the magnitude of $\vec{R}$.

(d) Find a unit vector parallel to $\vec{R}$.

(e) Find the angle that $\vec{R}$ makes with the positive x-axis.

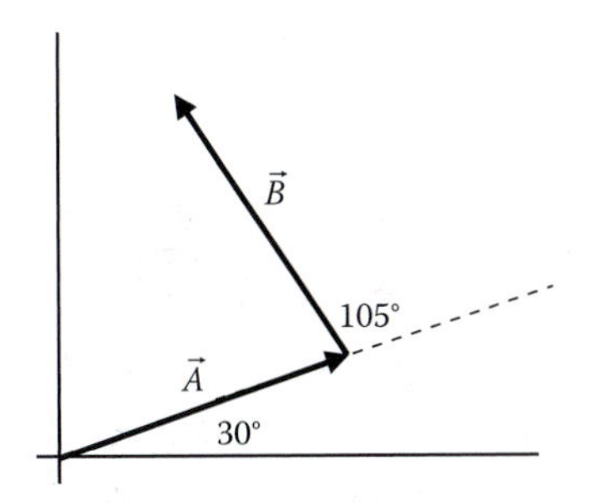

2.2 Let the vectors $\vec{A}$ and $\vec{B}$ be given by $\vec{A} = 3\hat{i} + 4\hat{j}$, and $\vec{B} = -4\hat{i} + 3\hat{j}$.

(a) Using the dot product prove that $\vec{A}$ and $\vec{B}$ are perpendicular.

(b) If $\vec{C} = \vec{A} + \vec{B}$ and $\vec{D} = \vec{A} - \vec{B}$, show that $\vec{C}$ and $\vec{D}$ are also perpendicular.

(c) Two intersecting lines have slopes b/a and $-a/b$. Show that the two lines are perpendicular. (*Hint:* Construct vectors parallel to each line.)

2.3 If $\vec{U} + \vec{V} \perp \vec{U} - \vec{V}$ prove that $U = V$.

2.4 If $\vec{A} = 2\hat{i} + 5\hat{j} - 4\hat{k}$ and $\vec{B} = 3\hat{i} - 6\hat{j} - 6\hat{k}$, show that $\vec{A}$ and $\vec{B}$ are perpendicular.

2.5 Suppose it is known that $\vec{A} \cdot \vec{B} = \vec{A} \cdot \vec{C}$ and $\vec{A}$ is not zero. Is it permissible to cancel $\vec{A}$ from both sides of the equation? Explain your answer with a specific example.

2.6 Find the expression for a vector that is perpendicular to the y-axis, makes an angle of 30° with the x-axis, has a length equal to 10, and has a positive z-component.

2.7 Two vectors $\vec{R}$ and $\vec{S}$ lie in x–y plane. Their magnitudes are 4.5 and 7.3 units, respectively, and their directions are 320° and 85°

measured counterclockwise from the x-axis. Draw the vectors. Evaluate $\vec{R} \cdot \vec{S}$.

2.8 Vector $\vec{A}$ has a magnitude of 10 and makes an angle $\pi/4$ (45°) with respect to the positive x-axis. Write a nonzero vector $\vec{B}$ such that $\vec{A}+\vec{B}$ has the same magnitude as $\vec{A}-\vec{B}$.

2.9 It is known that $\vec{A}+\vec{B}+\vec{C}=0$ and that $A = B = C$. What is the angle between $\vec{A}$ and $\vec{B}$?

2.10 The following figure shows a box of dimensions 3 × 4 × 2 units with one corner at the origin.

(a) Draw the vectors given by $\vec{A}=3\hat{i}+4\hat{j}+2\hat{k}$ and $\vec{B}=4\hat{j}+2\hat{k}$.

(b) What are the lengths of $\vec{A}$ and $\vec{B}$?

(c) Calculate the dot product $\vec{A} \cdot \vec{B}$.

(d) Calculate the angle between $\vec{A}$ and $\vec{B}$.

(e) Find the vector $\vec{C}$ such that $\vec{A}+\vec{B}+\vec{C}=0$.

(f) Find a unit vector parallel to $\vec{A}$.

(g) Find a unit vector perpendicular to $\vec{B}$. Is your answer the only possible answer?

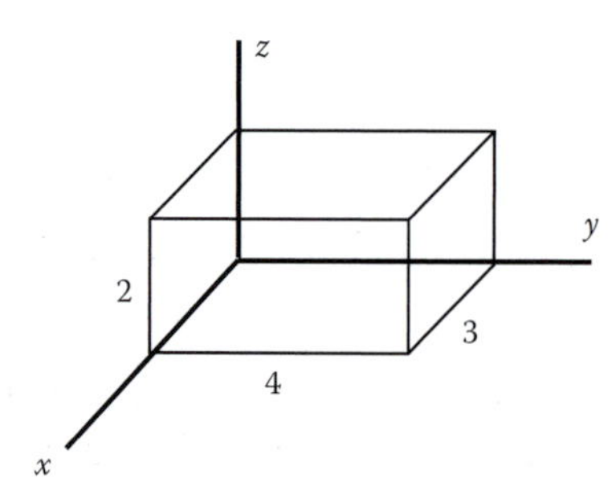

2.11 Let θ_x be the angle between vector $\vec{R}=a\hat{i}+b\hat{j}+c\hat{k}$ and the x-axis, θ_y be the angle between $\vec{R}$ and the y-axis, and similarly for θ_z. (See the figure below.) Show $\cos^2\theta_x + \cos^2\theta_y + \cos^2\theta_z = 1$.

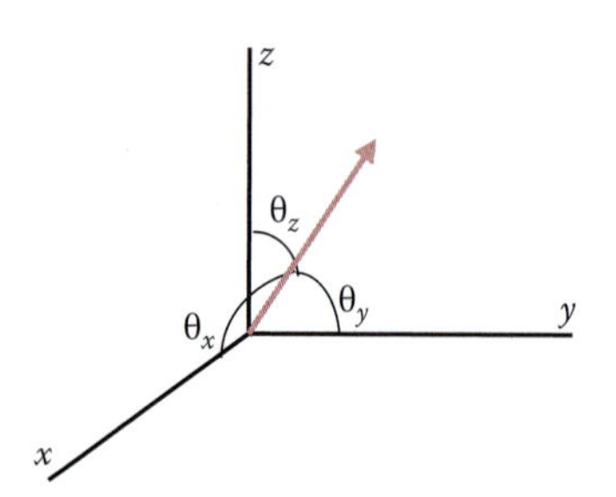

2.12 The angles θ and φ of the following figure are inscribed within a circle. Point O is at the center of the circle. Using vectors, prove that $\theta = (1/2)\varphi$ (*Hint:* Draw a line from O to the midpoint of line B. Show that this line is parallel to A.)

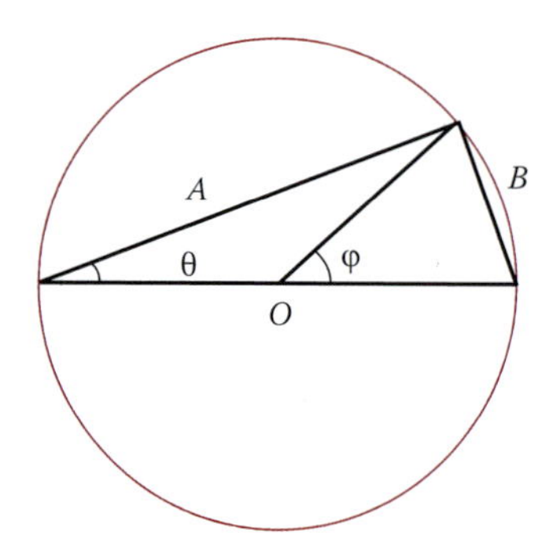

2.13 Consider a cube positioned along a coordinate axis system shown in the following figure. The length of each side of the cube is 4 cm.

(a) Write a vector (in unit-vector notation) for the body diagonal indicated by $\vec{A}$.

(b) Similarly, write a vector for the face diagonal $\vec{B}$.

(c) Find the angle between $\vec{A}$ and $\vec{B}$.

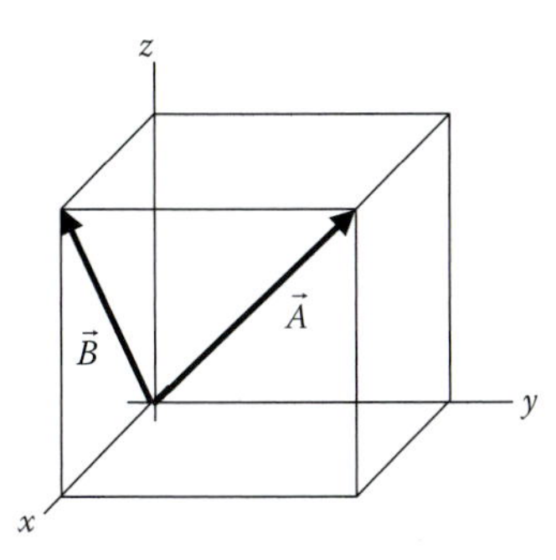

2.14 Consider the diagram of vectors shown in the following figure.

(a) Write expressions for $\vec{A}$ and $\vec{B}$ in terms of $\vec{C}$ and $\vec{D}$.

(b) Using the distributive property of the dot product with respect to addition $\{\vec{a}\cdot(\vec{b}+\vec{c})=\vec{a}\cdot\vec{b}+\vec{a}\cdot\vec{c}\}$, show that $\vec{A}$ and $\vec{B}$ are perpendicular.

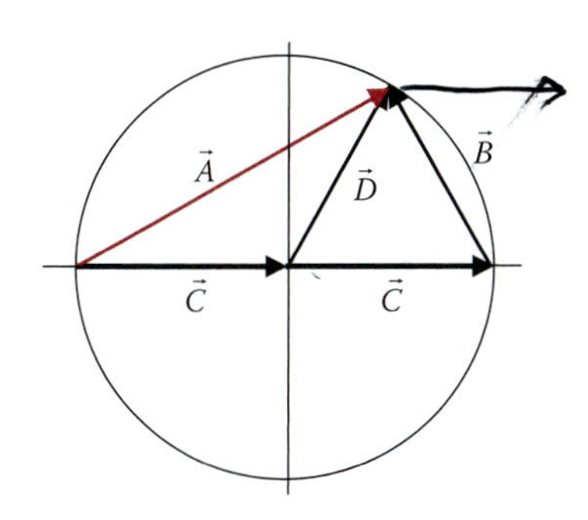

2.15 The *center of mass* of two point masses m_1 and m_2 located at positions $\vec{r}_1$ and $\vec{r}_2$, respectively, is defined as

$$\vec{r}_{CM} = \frac{m_1\vec{r}_1 + m_2\vec{r}_2}{m_1 + m_2}.$$

(a) Prove that $\vec{r}_{CM}$ can be written in the following form:

$$\vec{r}_{CM} = \vec{r}_1 + \alpha\left(\vec{r}_2 - \vec{r}_1\right),$$

where $0 \le \alpha \le 1$.

(b) Show from (a) that $\vec{r}_{CM}$ extends from the origin to a point along the line joining $\vec{r}_1$ and $\vec{r}_2$ as shown in the following figure.

(c) For $m_2 = 3m_1$, what is the value of α?

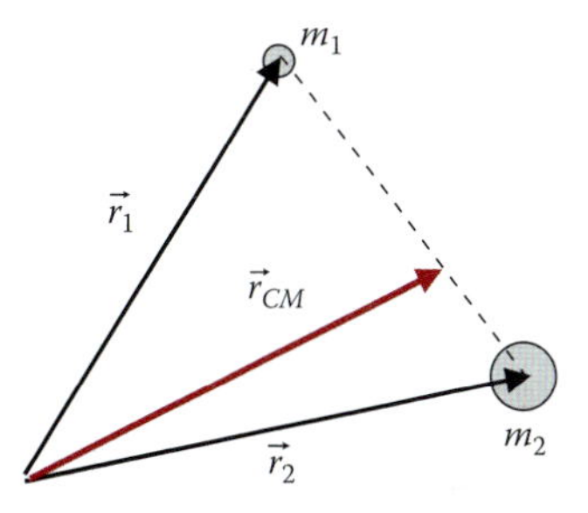

2.16 Mars orbits the Sun with an orbital radius of 1.52 AU. See the following figure. Suppose you are on Mars looking back towards Earth.

(a) What is the maximum elongation φ_{max} of Earth as viewed from Mars?

(b) When Earth is in the position E' the angle SME' is 30°. From your vantage point on Mars, what *phase* of Earth do you see at this time? Make a careful sketch to accompany your calculations. (*Hint:* Use the law of sines and don't forget that $\sin\theta = \sin(180°-\theta)$.)

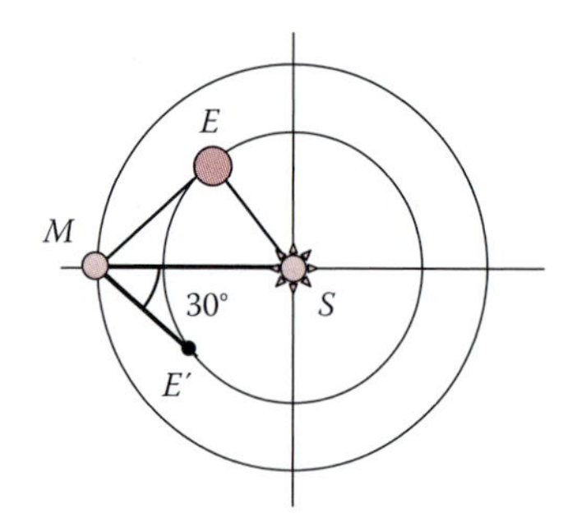

2.17 The radius of Mars' orbit around the Sun is 1.52 times that of the Earth's, while its period of revolution is approximately two times that of the Earth (it is actually 1.88 year).

(a) Copy the figure below, which shows the Sun, Earth and Mars when the two planets are closest to each other. Indicate on your drawing the positions of both planets 1, 2 and 3 months before and after the time shown in the figure.

(b) Show that Mars' motion exhibits *retrograde motion* to a viewer on Earth.

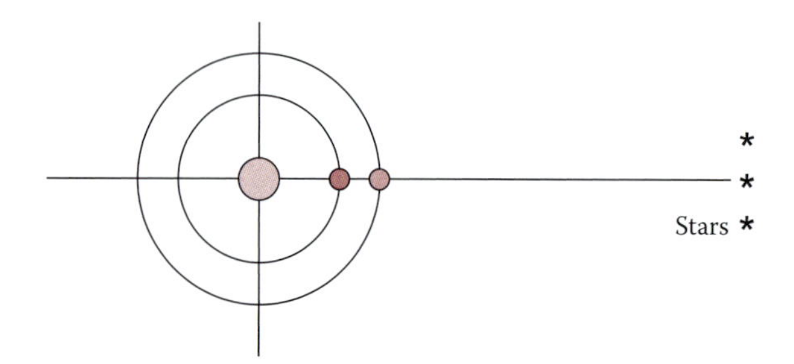

(c) Match the different positions of Earth with the images of Mars shown in Figure 2.9. Which image of Mars is seen when
(1) the planets are as shown above,
(2) 3 months earlier than this, and
(3) 3 months later than this?

(d) In the figure, Mars is said to be in opposition. How many years will pass before Mars will once again be in opposition?

(e) Would Jupiter also exhibit retrograde motion when viewed from Earth? Jupiter has a period of about 12 years and lies about 5 AU from the Sun.

2.18 In a rectangular x–y coordinate system, the vector $\vec{R} = x\hat{i} + y\hat{j}$. In a second rectangular x'–y' coordinate system, rotated by an angle θ, the same vector is $\vec{R} = x'\hat{i}' + y'\hat{j}'$, where $\hat{i}'$ and $\hat{j}'$ are unit vectors parallel to the rotated axes (see the following figure).

(a) Show that the coordinates in the rotated system are related to the original coordinates by

$$x' = x\cos\theta + y\sin\theta,$$
$$y' = y\cos\theta - x\sin\theta.$$

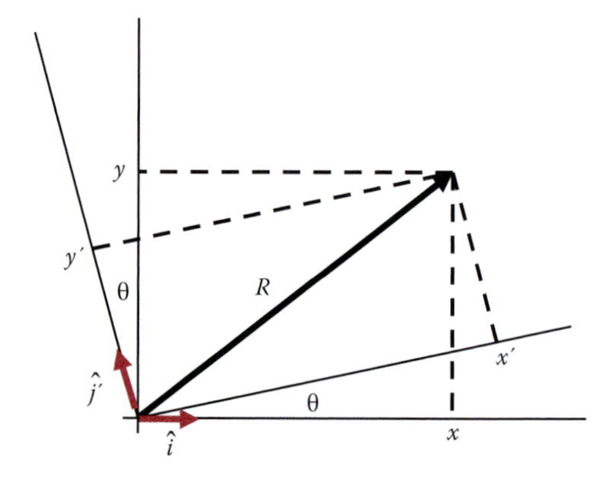

Hint: Use the dot product: $x' = \vec{R}\cdot\hat{i}' = \left(x'\hat{i}' + y'\hat{j}'\right)\cdot\hat{i}'$, etc.

(b) Prove that the magnitude of $\vec{R}$ is independent of the coordinate system by showing that

$$x^2 + y^2 = x'^2 + y'^2.$$

2.19 Recall the experiment of Eratosthenes discussed in Chapter 1. At noon on the summer solstice the Sun was directly overhead in Syene, Egypt.

(a) What is the latitude of Syene?

(b) How many daylight hours were there on the day of the experiment? How many hours was the Sun visible in Syene on a clear day 6 months later?

3

USING VECTORS TO DESCRIBE MOTION

The Hubble Ultra Deep Field 2014: A tiny portion of the sky in the constellation Formax, contains 10,000 visible galaxies. The angular width of the image is about 1/10 the subtended angle of the Moon. Light from the most distant galaxies originated only a few 100 million years after the Big Bang. (Photo courtesy of NASA, ESA, H. Teplitz and M. Rafelski [IPAC/Caltech], A. Koekemoer [STScI], R. Windhorst [ASU], and Z. Levay [STScI].)

We are, by definition, in the very center of the observable region. We know our immediate neighborhood rather intimately. With increasing distance, our knowledge fades, and fades rapidly. Eventually, we reach the dim boundary—the utmost limits of our telescopes. There, we measure shadows, and we search among ghostly errors of measurement for landmarks that are scarcely more substantial. The search will continue. Not until the empirical resources are exhausted, need we pass on to the dreamy realms of speculation.

Edwin Hubble
The Realm of the Nebulae (1936)

3.1 INTRODUCTION

Voyager 1 was launched from Cape Canaveral on September 5, 1977. It is now outside the solar system, 120 AU from the Sun, the farthest any man-made object has ever ventured from Earth. At this distance, the gravitational influences of the Sun and the planets are negligible, and *Voyager* is moving at a constant speed of 17 km/s along a straight line path. On November 20, 1998, the first module of the *International Space Station* (ISS) was placed in orbit. Since then, the ISS has hosted over 200 astronauts from 15 nations, men and women who have conducted research in the physical and biological sciences while coping with the harsh conditions of living in space. The ISS is largest artificial satellite to orbit Earth; it is moving at constant speed in a nearly perfect circular orbit 350 km above the planet's surface.* Motion at constant speed along a straight line and motion at constant speed along a circular path are two of the simplest but most important types of motion encountered in the physical world.† From subatomic particles to stars and galaxies, nature provides us with countless examples of both. In this chapter, we will carefully develop the fundamental concepts of *displacement*, *speed*, and *velocity*, and derive precise mathematical descriptions of straight line and circular motion. We will then introduce the concept of *relative motion* and argue that all motion must be understood as relative. Finally, we will use these ideas to interpret a century's worth of astronomical research to probe the origins of the universe and estimate its age!

* The 32 satellites that presently comprise the global positioning system are also in near-perfect circular orbits, at a higher altitude of 20,000 km.

† A third, equally important, type of motion is called simple harmonic motion. It will be discussed in Chapter 7.

3.2 DISPLACEMENT AND VELOCITY

One weekend just after midterm exams, you and your classmate decide to leave campus and drive to a nearby city for a well deserved study break. You want to take the thruway, a straight four-lane road with a 65 mile/h speed limit. Your classmate wants to take the scenic route, a winding two-lane country road with a 55 mile/h limit. Since this road is rarely patrolled by police (or so he believes), he argues that it will be the faster route. Unable to agree on which road to take, you each decide to take your own car and meet up in the city. The two routes are illustrated in Figure 3.1. The starting (initial) point for either of you is $\vec{r}_i$ and your common (final) destination is $\vec{r}_f$. You both depart at the same time t_i, but you arrive at $\vec{r}_f$ just a few minutes earlier than your classmate, even though he has ignored the speed limit and driven faster than you throughout the trip. Being a serious physics student, you politely point out that, in spite of his

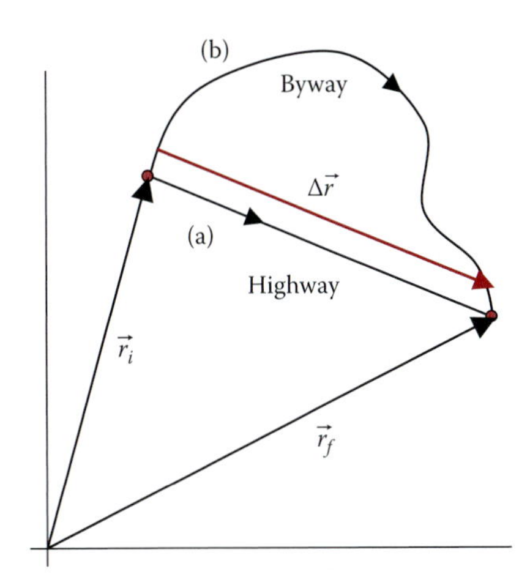

FIGURE 3.1 (a) The straight highway route between campus ($\vec{r}_i$) and the city ($\vec{r}_f$) and (b) the winding country road between the same two locations. The displacement vector $\Delta\vec{r}$ is the same for both routes.

recklessness, your *displacement* was exactly the same as his. Better yet, your *average velocity* was greater! How is this possible?

The answer lies in what we mean by displacement and velocity. The displacement $\Delta\vec{r}$ is defined as the vector difference between the initial and final position vectors, $\Delta\vec{r} = \vec{r}_f - \vec{r}_i$, as shown in Figure 3.1. $\Delta\vec{r}$ is the vector extending from the initial position $\vec{r}_i$ to the final position $\vec{r}_f$. All trips (including the two shown in the figure) that originate at $\vec{r}_i$ and end at $\vec{r}_f$ have the same displacement. The *average velocity* is defined in terms of the displacement and the time taken for the trip:

$$\vec{v}_{avg} = \frac{\text{displacement}}{\text{change in time}} = \frac{\Delta\vec{r}}{t_f - t_i}, \tag{3.1}$$

where t_f is the arrival time. Note that $\vec{v}_{avg}$ is the product of a vector $\Delta\vec{r}$ and a scalar $(t_f - t_i)^{-1}$, so it is itself a vector pointing in the same direction as the displacement $\Delta\vec{r}$. Returning to our example, since you arrived at $\vec{r}_f$ earlier than your roommate, and since you both had equal displacements, your average velocity is greater than his, even though his speedometer reading may have been higher than yours throughout the trip.

The average *speed* is distinctly different from the average velocity: it is just the distance traveled (i.e., the length of the path traveled) divided by the change in time. For our example, it is the change in the car's odometer reading divided by the elapsed time. The exercise below illustrates the difference between average speed and average velocity.

EXERCISE 3.1

The Indianapolis 500 is an auto race consisting of 200 laps around a 2.5 mile oval track. In 2010, Dario Franchitti won the event in a time of 3 h 6 min.

(a) What was his average speed?
(b) What was his overall displacement?
(c) What was his average velocity?

Answer (a) 161 mph (b) zero (c) zero

The *instantaneous velocity* is defined in terms of the average velocity. Consider the wiggly path shown as a dotted line in Figure 3.2a. At time t_1, a body is at $\vec{r}(t_1)$ and its position at other times is given by the vector $\vec{r}(t)$, the tip of which traces out the curved (dotted) path shown. At time t_2, the body is at $\vec{r}(t_2)$, and its average velocity (which is parallel to the displacement) during the time interval $\Delta t = t_2 - t_1$ is found from Equation 3.1:

$$\vec{v}_{avg} = \frac{\vec{r}(t_2) - \vec{r}(t_1)}{t_2 - t_1} = \frac{\Delta\vec{r}}{\Delta t}.$$

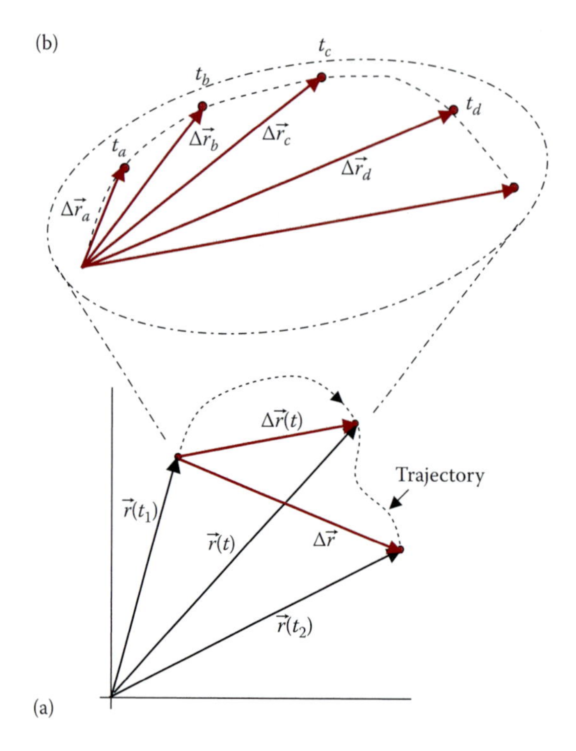

FIGURE 3.2 (a) A body moves along the dotted path shown, starting at $\vec{r}(t_1)$ and arriving Δt s later at $\vec{r}(t_2)$. (b) As Δt is made smaller ($t_a < t_b < t_c < t_d$), the displacement vector $\Delta\vec{r}$ changes in both length and direction and becomes tangent to the trajectory at $\vec{r}(t_1)$ as $\Delta t \to 0$.

For smaller intervals of time, the body will be at intermediate points on the trajectory, such as those shown at points a, b, c, and d in Figure 3.2b, and $\vec{v}_{avg}$ changes with the interval. The instantaneous velocity at time t_1 is defined as the average velocity over a tiny time interval $(t_1, t_1 + \Delta t)$, in the limit as Δt shrinks to zero. Mathematically, the instantaneous velocity at any time t is expressed as

$$\vec{v}(t) = \lim_{\Delta t \to 0} \vec{v}_{avg} = \lim_{\Delta t \to 0} \frac{\vec{r}(t+\Delta t) - \vec{r}(t)}{\Delta t}. \tag{3.2}$$

Figure 3.2b shows that in the limit $\Delta t \to 0$, $\Delta\vec{r}$ is tangent to the trajectory at the point $\vec{r}(t_1)$, so the instantaneous velocity $\vec{v}(t)$ is also parallel to the trajectory at $\vec{r}(t)$. Equation 3.2 should look suspiciously familiar; it is simply the defining equation of a *derivative*. In this case, the function we are differentiating is the vector $\vec{r}(t)$:

$$\vec{v}(t) = \lim_{\Delta t \to 0} \frac{\Delta\vec{r}}{\Delta t} = \frac{d\vec{r}(t)}{dt}. \tag{3.3}$$

Using coordinate notation, $\vec{r}(t) = x(t)\hat{i} + y(t)\hat{j} + z(t)\hat{k}$. Then, since the unit vectors $\hat{i}$, $\hat{j}$, and $\hat{k}$ are constants,

$$\vec{v}(t) = \frac{d\vec{r}}{dt} = \frac{dx}{dt}\hat{i} + \frac{dy}{dt}\hat{j} + \frac{dz}{dt}\hat{k} = v_x\hat{i} + v_y\hat{j} + v_z\hat{k},$$

where v_x, v_y, and v_z are the components of the instantaneous velocity vector $\vec{v}$. The instantaneous *speed* at time t is simply the magnitude of $\vec{v}(t)$. For our automotive example, the instantaneous speed $v(t)$ is simply the speedometer reading at time t.

EXERCISE 3.2

The position of a body is given by $\vec{r}(t) = (3+t)\hat{i} + (4-t^2)\hat{j}$ m.

(a) What is the body's position at $t = 0$?
(b) Derive an expression for the body's instantaneous velocity $\vec{v}(t)$.
(c) In what direction is the body moving at $t = 0$? In what direction is it moving at $t = 10$ s?

Answer (a) $\vec{r}(0) = 3\hat{i} + 4\hat{j}$ m (b) $\vec{v}(t) = \frac{d\vec{r}}{dt} = \hat{i} - 2t\hat{j}$ m/s (c) $\vec{v}(0)$ is →, $\vec{v}(10)$ is ↓

EXERCISE 3.3

The position of a body is given by $\vec{r}(t) = (3t^2 - 12t)\hat{i} + (4-2t)\hat{j} + 3\hat{k}$ m. At what time, if ever, is the velocity parallel to the y-axis?

3.3 CONSTANT VELOCITY MOTION

Let's continue with the example of the previous section. As you recall, you took the straight highway from campus to the city. Suppose you also activated the cruise control feature of your car so that your speed and direction of travel were both constant throughout the trip. Since speed and direction define velocity, then $\vec{v}(t)$ was constant during your trip: $\vec{v}(t) = \vec{v}_{avg} \equiv \vec{v}$ (a constant) for all times $t_i < t < t_f$. Using Equation 3.1 and setting your departure time $t_i = 0$,

$$\vec{v}_{avg} = \vec{v} = \frac{\vec{r}(t) - \vec{r}_i}{t} \quad \text{or} \quad \vec{r}(t) = \vec{r}_i + \vec{v}t. \tag{3.4}$$

Equation 3.4 describes *constant velocity motion*, straight line motion with constant speed. The same equation describes the present motion of the *Voyager 1* spacecraft, or a ball rolling without friction on a horizontal track, or a gas molecule in flight between collisions. Constant velocity motion occurs on all length scales throughout the universe and is the simplest and most fundamental type of motion.

EXERCISE 3.4

A body is moving with constant velocity. At time $t = 0$, it is at position $\vec{r}(0) = 3\hat{i} + 4\hat{j}$ m. At $t = 2$ s, the body is at $\vec{r}(2) = 7\hat{i}$ m.

(a) What is the body's velocity?
(b) Where will it be at $t = 3$ s?

Answer (a) $\vec{v} = 2\hat{i} - 2\hat{j}$ m/s (b) $\vec{r}(3) = 10\hat{i} - 2\hat{j}$ m

3.4 UNIFORM CIRCULAR MOTION

Now imagine a body that is constrained to move in a circular path about the origin of an x–y coordinate system, as shown in Figure 3.3a. If the radius of the path is R, the position of the body is given by

$$\vec{r}(t) = R\cos\theta(t)\hat{i} + R\sin\theta(t)\hat{j}, \tag{3.5}$$

where $\theta(t)$ is the time-dependent angle between vector $\vec{r}(t)$ and unit vector $\hat{i}$. When the body is moving *counter*clockwise, $\theta(t)$ is *increasing* with time. Let $\theta(t)$ and θ_0 be the values of θ (expressed in rad) at times t and t_0, respectively. The length of the circular arc (Figure 3.3b) traversed by the body during the time interval $\Delta t = t - t_0$ is $\Delta s = R(\theta(t) - \theta_0) = R\Delta\theta$. The average speed (*not* the average velocity!) during this time interval is $\Delta s/\Delta t = R\Delta\theta/\Delta t$. In the limit $\Delta t \to 0$, the instantaneous speed is $v(t) = ds/dt = R\, d\theta/dt$. If $v(t)$ is constant, the *angular speed* $\omega \equiv d\theta/dt$ must also be constant,* so

$$\frac{d\theta}{dt} \equiv \omega = \frac{\theta(t) - \theta_0}{t - t_0},$$

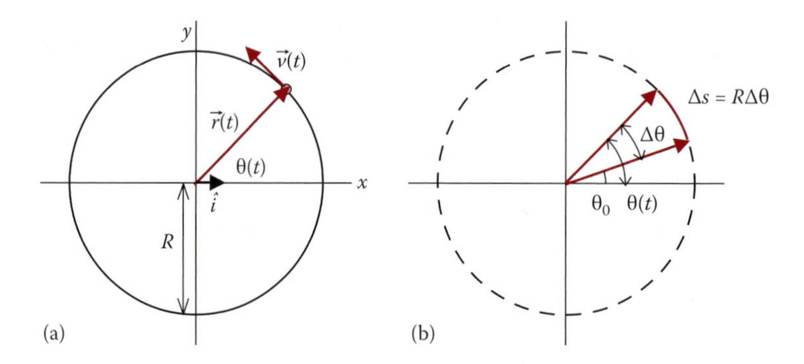

FIGURE 3.3 (a) A body in circular motion about the origin of coordinates. The radius of its trajectory is R, and at time t its position vector $\vec{r}(t)$ makes an angle $\theta(t)$ with the unit vector $\hat{i}$ pointing along the positive x-axis. (b) When $\vec{r}(t)$ sweeps out an angle $\Delta\theta$, the body traces out a circular arc of length $\Delta s = R\Delta\theta$.

* In Chapter 12, we will assign a direction to ω, which specifies the rotation axis and sense of rotation (clockwise or counterclockwise). This will change ω from a scalar to a vector.

or, setting $t_0 = 0$,

$$\theta(t) = \omega t + \theta_0.$$

Because $\theta(t)$ increases at a uniform rate, the position vector $\vec{r}(t)$ rotates uniformly with time, and the body is said to undergo *uniform circular motion*. Although it circles the origin with constant speed, its direction of motion is continually changing; therefore, its *vector* velocity $\vec{v}(t)$ is *not* constant.

In Equation 3.5, $\theta(t)$ is the argument of the sine and cosine functions. It is customary in physics to express the argument of a trigonometric function as a dimensionless number. As you recall, the *radian* is a dimensionless measure of angle. Unless explicitly noted (by the degree symbol [°]), you should assume that all arguments of trigonometric functions, such as $\theta(t)$ in Equation 3.5, are expressed in terms of rad. Consequently, the units of ω are rad/s, or simply s^{-1} (since radians are dimensionless). Suppose θ increases by 30° every second. In terms of radians, $30° = \pi/6$, so $\omega = d\theta/dt = \pi/6\ s^{-1}$. More generally, let T be the rotation period, the time needed for a full rotation ($\Delta\theta = 360°$ or 2π rad). Then,

$$\omega = \frac{2\pi}{T}. \tag{3.6}$$

This relation between angular velocity ω and rotation period T is ubiquitous in physics, so be absolutely sure to remember it.

EXERCISE 3.5

A body undergoes uniform circular motion. At $t = 0$ s and $t = 2$ s, the body is at the positions shown in the figure.

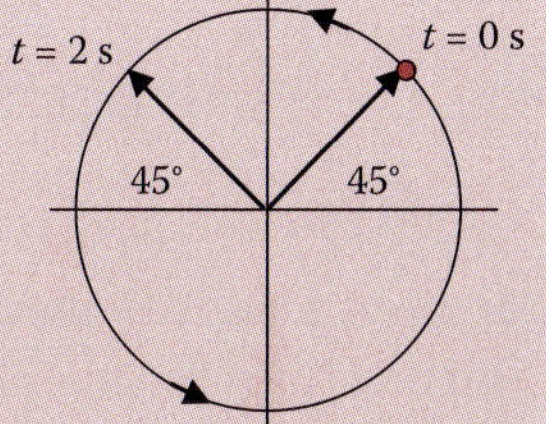

(a) What is the value of θ_0 (in radians)?
(b) What is the value of ω?
(c) What is the direction of the velocity vector at $t = 8$ s?
(d) What is the average velocity from $t = 0$ s to $t = 8$ s?

Answer: (a) $\pi/4$ (b) $\pi/4\ s^{-1}$ (c) ↘ (d) zero

In Section 3.2, we saw that a body's instantaneous velocity $\vec{v}(t)$ is tangent to its trajectory at position $\vec{r}(t)$. Let's verify this for a body in uniform circular motion. First we'll find an expression for $\vec{v}(t)$ by taking the time derivative of $\vec{r}(t)$, using Equation 3.5 with $\theta(t) = \omega t + \theta_0$:

$$\vec{v}(t) = \frac{d\vec{r}(t)}{dt} = \frac{d}{dt}\left[R\cos(\omega t + \theta_0)\hat{i} + R\sin(\omega t + \theta_0)\hat{j}\right].$$

Although this looks complicated, it is easy to evaluate if we first define $u(t) = \omega t + \theta_0$ (so $du/dt = \omega$) and then use the chain rule for derivatives. For the cosine term,

$$\frac{d}{dt}\cos(\omega t + \theta_0) = \frac{d}{dt}\cos u(t) = \frac{d}{du}\cos u\frac{du}{dt} = -\sin u\frac{du}{dt} = -\omega\sin(\omega t + \theta_0),$$

and likewise for the derivative of the sine term. The resulting expression for the velocity of a body in uniform circular motion is

$$\vec{v}(t) = \frac{d\vec{r}}{dt} = -\omega R \sin(\omega t + \theta_0)\hat{i} + \omega R \cos(\omega t + \theta_0)\hat{j}. \tag{3.7}$$

EXERCISE 3.6

In Figure 3.3, vector $\vec{r}$ makes an angle of $\pi/4$ (45°) with respect to the $\hat{i}$ direction at the moment depicted, that is, $\omega t + \theta_0 = \pi/4$. Show, from Equation 3.7, that at this time $\vec{v}$ is in the direction indicated in the figure.

From elementary geometry, you know that a line that is tangent to a circle at point P is perpendicular to a line $\overline{OP}$ drawn from the center O of the circle to P. Replace $\overline{OP}$ with vector $\vec{r}$. If $\vec{v}(t)$ is tangent to the circular trajectory, it must be perpendicular to $\vec{r}(t)$. Let's show this by evaluating the dot product of $\vec{r}(t)$ and $\vec{v}(t)$. Recalling that $\hat{i}\cdot\hat{i} = \hat{j}\cdot\hat{j} = 1$ and $\hat{i}\cdot\hat{j} = 0$,

$$\begin{aligned}\vec{r}\cdot\vec{v} &= \left(R\cos\theta\hat{i} + R\sin\theta\hat{j}\right)\cdot\left(-\omega R\sin\theta\hat{i} + \omega R\cos\theta\hat{j}\right)\\ &= -\omega R^2 \cos\theta\sin\theta + \omega R^2 \sin\theta\cos\theta = 0.\end{aligned}$$

Since $\vec{r}\cdot\vec{v} = 0$, $\vec{v}$ is perpendicular to $\vec{r}$, and so it is tangent to the circle as expected.

EXERCISE 3.7

Suppose a body is moving in a circular path, but its speed is not constant; that is, $d\theta/dt \neq$ constant. Starting from Equation 3.5, with $\theta = \theta(t)$, a general function of time, show that $\vec{v}(t)$ is always perpendicular to $\vec{r}(t)$, no matter how $\theta(t)$ varies with time.

How is the body's *speed* v related to ω and R? From Equation 2.12, $v^2 = \vec{v}\cdot\vec{v}$, so

$$\begin{aligned}v^2 &= \left(-\omega R\sin\theta\hat{i} + \omega R\cos\theta\hat{j}\right)\cdot\left(-\omega R\sin\theta\hat{i} + \omega R\cos\theta\hat{j}\right)\\ &= \omega^2 R^2\left(\sin^2\theta + \cos^2\theta\right) = \omega^2 R^2,\end{aligned}$$

where we have used the trigonometric identity $\sin^2\theta + \cos^2\theta = 1$. Hence, the speed v of a body in uniform circular motion is a constant given by

$$v = \omega R = \frac{2\pi R}{T}, \tag{3.8}$$

where T is the period of rotation. The second equality in Equation 3.8 is easy to remember: the body travels once around the circle's circumference $2\pi R$ in time T, so its speed = distance/time = $2\pi R/T$.

EXERCISE 3.8

(a) You are standing at a point on the Earth's equator. What is your speed due to the Earth's rotation? Express this in km/h or mile/h (1 mile ≈ 1.6 km). Are you surprised? (b) What is the Earth's orbital speed about the Sun? Once again, express this in familiar units such as km/h or mile/h.

Answer (a) 1,680 km/h ≈ 1,000 mile/h (b) 1.08×10^5 km/h ≈ 67,500 mile/h

3.5 RELATIVE VELOCITY

Imagine you are on a jet plane, flying coast to coast high above an unbroken expanse of cloud cover. The flight is very smooth, and since you cannot see the ground, you have no sensation that you are moving at several hundred miles per hour. Looking out your window, you see a small prop plane below you, pointed in the same direction as your jet. But there is something odd about it; it appears to be traveling *backward*—tail first—which you know is not how planes are designed to fly! Since you are a physics student, you immediately grasp what's going on. If the speed and direction of the two planes were the same, the prop plane would appear to be motionless; if the prop plane were traveling faster (slower) than the jet, you would see it moving forward (backward) with an apparent speed equal to the difference between the speeds of the two planes: $v_{app} = v_{prop} - v_{jet}$. Although pleased by your deduction, you (wisely) refrain from discussing it with the passenger seated next to you. Instead, you quietly ponder: What would you see if the two planes were not traveling in the same direction?

Let's answer this question using vectors. Consider two bodies, 1 and 2, each moving at a constant velocity with positions described by the following expressions (see Figure 3.4):

$$\vec{r}_1(t) = \vec{r}_{10} + \vec{v}_1 t,$$
$$\vec{r}_2(t) = \vec{r}_{20} + \vec{v}_2 t.$$

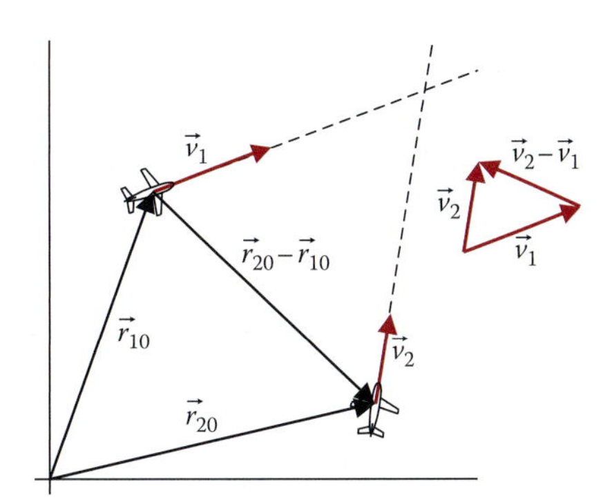

FIGURE 3.4 Two bodies, initially at positions $\vec{r}_{10}$ and $\vec{r}_{20}$, are moving with constant velocities $\vec{v}_1$ and $\vec{v}_2$. The initial position and velocity of body 2 *relative* to body 1 are $\vec{r}_{20} - \vec{r}_{10}$ and $\vec{v}_2 - \vec{v}_1$.

We define the position of body 2, *relative* to body 1, as

$$\vec{r}_{21}(t) \equiv \vec{r}_2(t) - \vec{r}_1(t) = (\vec{r}_{20} - \vec{r}_{10}) + (\vec{v}_2 - \vec{v}_1)t. \tag{3.9}$$

The velocity of body 2, *relative* to body 1, is similarly defined:

$$\vec{v}_{21} \equiv \frac{d\vec{r}_{21}}{dt} = \vec{v}_2 - \vec{v}_1. \tag{3.10}$$

The relative velocity $\vec{v}_{21}$ is the apparent velocity of body 2 as seen by an observer moving with body 1. In the above example, it is the apparent velocity of the prop plane (body 2) as viewed by you in the jet (body 1).

EXERCISE 3.9

One Saturday afternoon, you are preparing for an upcoming trip by shopping in a large department store. As you step onto a *down* escalator you see your roommate directly below you on the neighboring *up* escalator.

Each of you is moving at a speed of 1.41 m/s (hint!) at an angle of 45° relative to the horizontal.

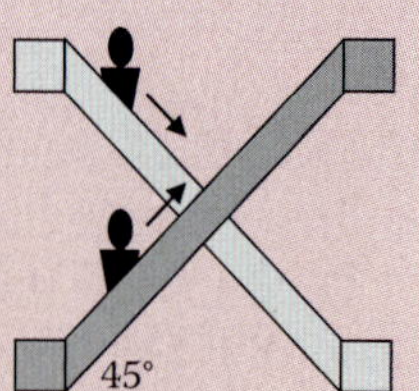

(a) What is the direction of the your roommate's velocity relative to you?
(b) What is the direction of your velocity relative to your roommate?
(c) What is the magnitude of the relative velocity in either case?
(d) How do your answers change if the positions of you and your roommate are not vertically aligned?

Answer: (c) 2 m/s (d) no change

In Figure 3.4, the paths of the two bodies (indicated by the dotted lines) intersect. Does this mean that the bodies crash into one another at the point of intersection? Not necessarily. They will make contact only if they arrive at the intersection point *at the same time*. Mathematically, this requires that $\vec{r}_{21}(t) = 0$ for some value of t, or $(\vec{r}_{20} - \vec{r}_{10}) + (\vec{v}_2 - \vec{v}_1)t = 0$. This condition can be satisfied only if $(\vec{r}_{20} - \vec{r}_{10})$ is antiparallel to $\vec{v}_{21}$. Writing $\vec{r}_1(t) = x_1(t)\hat{i} + y_1(t)\hat{j} + z_1(t)\hat{k}$, and likewise for $\vec{r}_2(t)$, a crash will occur if there is a time t when $x_1(t) = x_2(t)$, and, *at the same time*, $y_1(t) = y_2(t)$ and $z_1(t) = z_2(t)$. All three equalities must be satisfied simultaneously. Otherwise the paths may cross but there will be no collision.

EXERCISE 3.10

The positions and velocities of two ships are shown on the diagram to the right.

(a) Calculate $(\vec{r}_{20} - \vec{r}_{10})$ and $\vec{v}_{21}$.
(b) Do the ships collide? If so, at what time?

Answer (a) $\vec{r}_{20} - \vec{r}_{10} = 7\hat{i} - 5\hat{j}$ km, $\vec{v}_{21} = -14\hat{i} + 10\hat{j}$ km/h
(b) yes, at $t = 0.5$ h

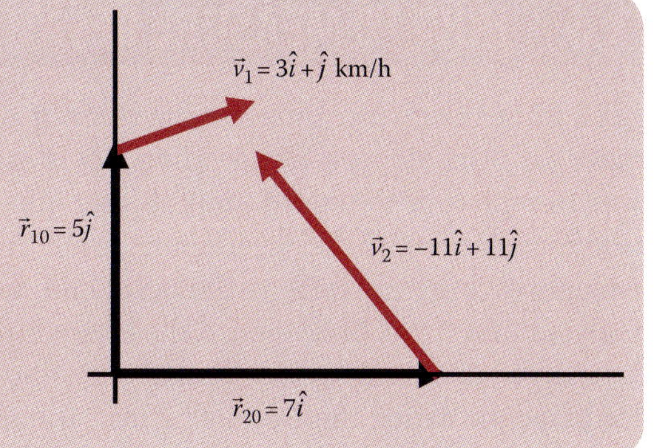

When we describe the motion of a car, we give its speed *relative* to the road. Similarly, an airplane's airspeed or groundspeed describes how fast it is moving *relative* to the surrounding air or the Earth's surface directly below. But Earth's surface and atmosphere are rotating about the planet's center with a speed of nearly 500 m/s (1000 mile/h) at the Equator, and the planet itself orbits the Sun with a speed of 30 km/s. Moreover, the Sun orbits the center of the Milky Way Galaxy with a speed of 200 km/s, and if that isn't enough, galaxies move at astonishing speeds relative to one another (as you will see shortly). Is there some cosmic coordinate system, or *reference frame*, that is truly stationary in which we can—at least in principle—measure the absolute velocity of our galaxy and from that determine the absolute velocity of the Sun, then the Earth, etc.? Einstein's theory of relativity provides an unequivocal answer: No! A body does not have an absolute velocity; neither can one say it is absolutely at rest. In this context, *absolute* has no physical meaning. *All motion is relative*, and the best we can do is to measure a body's velocity relative to other bodies.* This has profound implications for our understanding of the universe. In classical Greek astronomy, it was assumed that the Earth was motionless and located at the center of the universe. Epicycles were then required to generate the observed retrograde motion of the planets. In the sixteenth century, when it was recognized that Earth itself was in solar orbit, retrograde motion was easier to understand in terms of relative velocity. For example, Mars exhibits retrograde motion when it is in opposition to the Sun, when it is closest to Earth. At this time, the faster-moving Earth overtakes Mars, so that Mars appears to us to be moving backward, just like our example of the two airplanes discussed above. Earth is not at the center of the universe. So where *is* the center? As you will see, the lack of a cosmic reference frame, one which is at rest with respect to intergalactic *space*, provides surprising answer to this question.

* In fact, it was Galileo who first recognized this. In his *Dialogues*, he imagined being confined to a lower cabin of a ship. As long as the ship was moving with constant velocity, he wrote, there was no way within the cabin to determine if the ship were moving or standing still.

3.6 HOW TO MEASURE RELATIVE VELOCITY

Let's return—one last time—to the story of you and your classmate. While driving to the city, your classmate has been stopped for speeding. The state trooper used a *laser speed gun* to measure his speed, which was 10 mph over the speed limit. A laser gun works by emitting short pulses of infrared light in the direction of oncoming traffic (see Figure 3.5a). These pulses are spaced by Δt and travel at the speed of light. An approaching car reflects the light, and, because of the car's motion, the reflected pulses are bunched together and travel back toward the laser gun with a shorter spacing $\Delta t'$. The difference in spacing $\Delta t - \Delta t'$ is detected and used to calculate the car's line-of-sight speed. A *radar gun* (Figure 3.5b) operates on a similar principle called the *Doppler effect*. It emits microwaves (high frequency radio waves) of wavelength $\lambda \approx 3$ cm, which are reflected by the oncoming car. Due to the car's motion, the reflected waves are squeezed to a shorter wavelength λ' (just like the laser pulses) and the change in wavelength is used to calculate the car's speed. The Doppler effect is also used by astronomers to measure the relative motion of astronomical bodies. See Box 3.1 for an introduction to wave terminology and a derivation of the Doppler effect.

Most of our knowledge about distant physical objects—planets, stars, and galaxies—comes to us via the waves they emit, absorb, reflect, or scatter. A star emits light over a continuous range of wavelengths encompassing the ultraviolet, visible, infrared, and radio portions of the electromagnetic spectrum. When this radiation passes through the cooler atmosphere surrounding the star, energy is *absorbed* at certain wavelengths that depend on the chemical identity of the atmospheric gas. These absorption features are imprinted on the *transmitted* spectrum—the light we see—and they allow us to determine the radial (line-of-sight) velocity of the star. See Figure 3.6a and b.

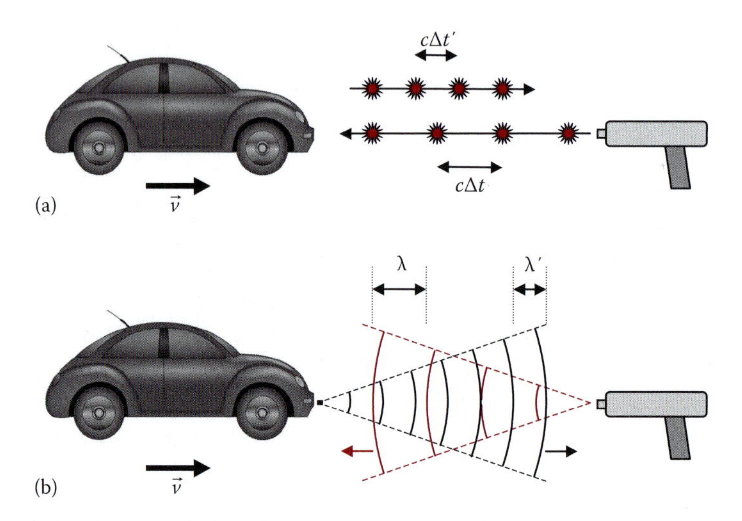

FIGURE 3.5 (a) A laser speed gun emits pulses of light at time intervals Δt. When reflected from an oncoming car, the spacing between pulses decreases to $\Delta t' < \Delta t$. (b) A radar speed gun emits microwaves of wavelength λ. Waves reflected from approaching traffic have a shorter wavelength λ' due to the Doppler effect.

BOX 3.1 WAVES AND THE DOPPLER EFFECT

The world is awash in electromagnetic waves. Besides visible light, we live in a vast sea of invisible waves emitted by radio and television broadcasters, cell phones, wireless routers, microwave ovens, GPS satellites, etc. Electromagnetic (EM) waves carry power to our homes via hundreds of miles of high voltage transmission lines. EM waves from the Sun streak across space to fuel all life processes on Earth. From the farthest reaches of the visible universe comes the *cosmic microwave background radiation*, faint radio signals that whisper the secrets of the birth of the cosmos. From sources near and far, EM waves carry vital information about the external world, data that is transmitted from our eyes, ears, and fingertips to our brains by electrochemical waves coursing along the pathways of our nervous systems.

The simplest type of EM wave is a *sine wave*, one which carries electric and magnetic fields that vary sinusoidally in time and space.* Monochromatic light is a good example.† The laser shown in Figure B3.1 emits a sine wave that propagates (travels) with speed $v_p = c = 3 \times 10^8$ m/s. Wave maxima indicate the locations where electric and magnetic fields have their peak amplitudes. The distance between two neighboring positive peaks is called the *wavelength* λ, and the number of positive peaks emitted by the laser in one second is called the *frequency* f. The units of frequency (peaks per second) are s^{-1} and are commonly expressed in physics as Hertz (Hz): $1\ s^{-1} = 1$ Hz.‡

If f peaks, spaced a distance λ apart, are emitted in 1 s, the speed of propagation is

$$v_p = \lambda f. \tag{B3.1}$$

This is a fundamental relation holding true for all sinusoidal waves. A simple analogy may be helpful. Imagine you are at a railroad crossing, watching a passing freight train. Each car has a length L meters, and if n cars pass by in 1 s, the train must be moving nL m/s relative to you: $v_{train} = nL$. Clearly, L, n, and v_{train} are the counterparts of λ, f, and v_p, so $v_p = \lambda f$.

Acoustic waves need a medium (such as air) to propagate, and v_p ($\approx$ 340 m/s in air) is measured relative to the medium. If the wind is blowing from west to east, sound travels faster in this direction, relative to the ground, than from east to west. Light waves are fundamentally different: they require no medium and travel easily through a vacuum. In Figure B3.1b, a light source and an observer are approaching one another with speed v. Relative to the source, what is the speed of propagation? Relative to the observer, what is the speed? In 1905, Albert Einstein proposed a revolutionary, counterintuitive answer: *all observers measure the same speed*, $v_p = c$, regardless of how an observer and source are moving. This is the starting point for the special theory of relativity; it leads directly to the phenomena of time dilation, length contraction, and ultimately, the earthshaking equivalence of mass and energy: $E = mc^2$.

If the laser in Figure B3.1a has been switched on for Δts, then $f\Delta t$ maxima have been emitted. The peaks are spaced λ apart, and the front of the wave train is now at a distance $v_p\Delta t$ from the laser.

---

* Mathematically, it can be shown (Fourier's Theorem) that all nonsinusoidal waves can be considered to be combinations of sinusoidal ones. Sine waves are therefore the basic ingredients of all waves.

† An *acoustic* sine wave produces a pure musical note (e.g., concert A).

‡ Named after Heinrich Hertz, who was the first (1886) to demonstrate the existence of electromagnetic waves.

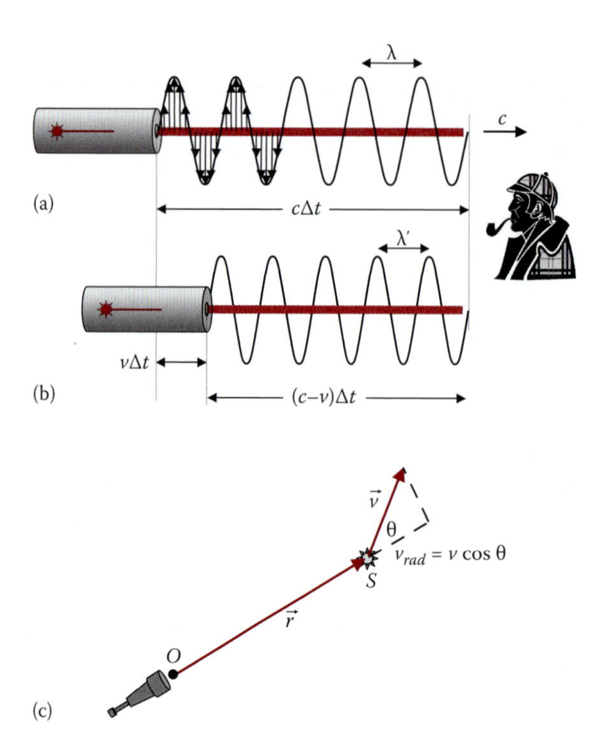

FIGURE B3.1 (a) An ideal laser emits a sinusoidal EM wave that propagates toward an observer at the speed of light c. The electric and magnetic fields carried by the wave vary sinusoidally in space, with successive maxima spaced by the wavelength λ. After a time Δt, the leading edge of the wave has traveled a distance $c\Delta t$ from the source. (b) The source is moving with speed v toward the observer, reducing the wavelength and increasing the frequency measured by the observer. (c) Let $\vec{r}$ be the position of the source S relative to the observer O. The radial velocity of the source is the component of its velocity $\vec{v}$ that is parallel to $\vec{r}$: $v_{rad} = v\cos\theta = \vec{v}\cdot\vec{r}/r$.

When the source is moving with speed v toward the observer (Figure B3.1b), the wave still travels with speed $v_p = c$ relative to the observer, and the same number of peaks are emitted per second.* The front of the wave is at a distance $v_p\Delta t$ from the *original* position of the source, but the rear of the wave is now located at the source's present position, which has moved a distance $v\Delta t$ toward the observer. The $f\Delta t$ peaks are thus squeezed into a distance $(v_p - v)\Delta t$, so the wavelength is reduced to a new value λ' given by

$$\frac{\lambda'}{\lambda} = \frac{(v_p - v)\Delta t}{v_p \Delta t} = 1 - \frac{v}{v_p}.$$

By the same reasoning, if the source is moving *away* from the observer, the wavelength is stretched to $\lambda' = \lambda(1 + v/v_p)$.

* We are ignoring the complication of time dilation. This is permissible for $v \ll v_p$.

Since v_p does not change, a shorter wavelength λ' means that more peaks will reach the observer per second,* so the frequency f' measured by the observer will be higher than f: $\lambda' f' = \lambda f = v_p$. Defining $\lambda' = \lambda + \Delta\lambda$ and $f' = f + \Delta f$, we can easily show (using the binomial theorem) that $\Delta\lambda/\lambda = -\Delta f/f$. For example, a 1% decrease in wavelength is accompanied by a 1% increase in frequency. Combining these results, we obtain

$$\frac{\Delta\lambda}{\lambda} = -\frac{\Delta f}{f} = \mp\frac{v}{v_p}, \tag{B3.2}$$

where the minus (plus) sign is used when the source is approaching (receding from) the observer. This change in wavelength and frequency, due to the relative motion v of the source and observer, is called the *Doppler effect*. As long as $v \ll v_p$, Equation B3.2 is valid for all waves (light, sound, water, etc.).

For simplicity, the analysis above has been carried out in one dimension. In the real three-dimensional world, the relative velocity $\vec{v}$ may point in any direction, but only the component of $\vec{v}$ that is parallel to the line of sight between the observer and source plays a role in the Doppler shift. This component is called the *radial* velocity. Let $\vec{r} = \vec{r}_S - \vec{r}_O$ be the displacement of the source from the observer and θ be the angle between $\vec{r}$ and $\vec{v}$ (see Figure B3.1c). The radial velocity is

$$v_{rad} = v\cos\theta = \frac{\vec{v}\cdot\vec{r}}{r},$$

and it is this radial component of the velocity that is to be used in Equation B3.2.

Exercise B3.1: A satellite in low earth orbit is passing overhead, moving at a speed of 7.9 km/s relative to a tracking station on Earth's surface. The satellite emits a radio signal at a frequency of 300 MHz.

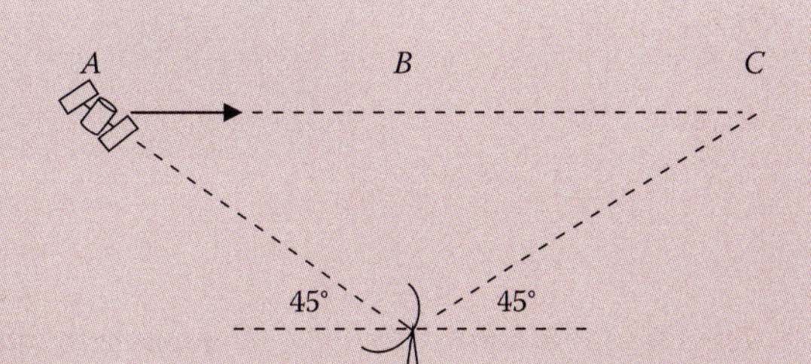

(a) What wavelength does this correspond to?
(b) What are the fractional changes in wavelength recorded by the tracking station when the satellite is at positions *A*, *B*, and *C*?
(c) What are the fractional changes in frequency when the satellite is at these positions?

* In our railroad analogy, the car length *L* is shorter while the train speed remains the same.

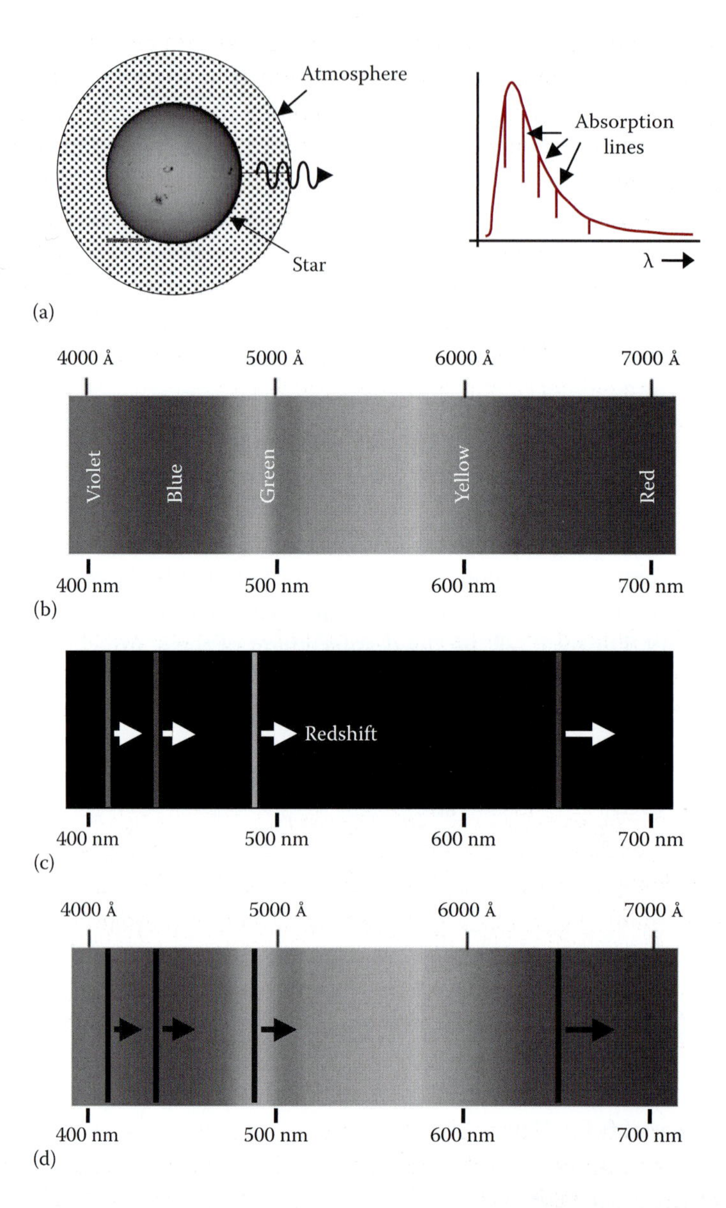

FIGURE 3.6 (a) A star emits light over a continuum of wavelengths that includes (b) the entire visible spectrum from about 400 nm (violet) to 700 nm (red). As light passes through the star's atmosphere, it is absorbed at wavelengths λ_i ($i = 1, 2,...$) that depend on the chemical identity of the atmospheric gases. (c) Hot hydrogen gas atoms emit visible light at discrete wavelengths collectively called the Balmer spectrum. When the star is moving radially away from us, the emitted light is *redshifted* to longer wavelengths. (d) Cooler hydrogen gas in the star's atmosphere *absorb* light at the same Balmer wavelengths, reducing the light reaching us at these wavelengths. These absorption lines are Doppler shifted in the same manner as the emission lines shown in (c).

In the laboratory, hot gas atoms* emit light at well-defined *characteristic* wavelengths λ_i ($i = 1, 2, \ldots$). For example, hydrogen emits visible light at $\lambda_1 = 656.3$ nm (the so-called Hα line, colored red), $\lambda_2 = 486.1$ nm (cyan), $\lambda_3 = 434.1$ nm (blue), and $\lambda_4 = 410.2$ nm (violet), collectively called the Balmer spectrum of hydrogen.† Hot *interstellar* hydrogen gas emits light at the same wavelengths, but if the gas cloud is moving away from us, the Balmer wavelengths are stretched (shifted toward the red end of the visible spectrum, or *red-shifted*). See Figure 3.6c. If the gas is instead moving toward us, the same wavelengths are squeezed (shifted toward the blue end of the visible spectrum, or *blue-shifted*). As noted above, the *cooler* gases that surround stars *absorb* some of their radiation (Figure 3.6d). Absorption—like emission—occurs at a well-defined set of wavelengths that is unique to the chemical identity of the gas. Absorption lines are redshifted or blueshifted in the same way as emission lines, and the full pattern of emission or absorption lines tells us how a galaxy or star is moving relative to us.

The change in wavelength $\Delta\lambda$ due to the relative velocity between the radiation source (star or galaxy) and the observer is called the *Doppler effect*. For speeds $v \ll c$,

$$\frac{\Delta\lambda}{\lambda} = \mp\frac{v}{c}, \tag{3.11}$$

where the minus (plus) sign is used when the source and observer are approaching (moving away from) each other. As explained in Box 3.1, the speed v in Equation 3.11 is the *radial* velocity, the component of the relative velocity vector that is parallel to the line of sight from the observer to the source.

EXERCISE 3.11

In 1913, the astronomer Vesto Slipher reported that the Andromeda galaxy is approaching us at a speed of 300 km/s. His conclusion was based on Doppler measurements of light emitted by the galaxy.

(a) By how much is the violet line of the hydrogen Balmer spectrum shifted?
(b) Is this a redshift or a blueshift?

Answer (a) $\Delta\lambda/\lambda = 0.001$ (b) blueshifted

The Doppler effect is a powerful astronomical tool. Using Equation 3.11, astronomers detect the relative motion and rotation of planets, stars, and galaxies. In 1970, measurements of galactic rotation led to the discovery of *dark matter*, the invisible and as yet unidentified substance that permeates interstellar space and is five times more abundant than ordinary matter. More recently, scientists

* By *hot* we mean atoms that are not in their lowest energy electronic states. These *excited* atoms lower their energy by emitting electromagnetic radiation.

† *Helium* (after *helios*, the Greek word for the Sun) was first detected during the solar eclipse of 1868. Astronomer Norman Lockyer detected a solar emission line of wavelength 587.6 nm (yellow) that could not be associated with the emission spectrum of any known element. Twenty-seven years later the chemist William Ramsey discovered helium here on Earth.

have used Doppler measurements of nearby stars to discover hundreds of *extrasolar planets*, and entire solar systems, in orbit about stars other than our Sun. This breathtaking work has revitalized the search for extraterrestrial life. These and other applications of the Doppler effect will be discussed in later chapters.

3.7 HUBBLE'S LAW

At the beginning of the twentieth century, it was widely accepted that the Milky Way Galaxy encompassed the entire universe. All of the visible objects in the sky, including the mysterious, indistinct patches of light called nebulae were assumed to be inhabitants of our home galaxy. In 1912, the astronomer Vesto Slipher measured light spectra from 40 nebulae and reported huge redshifts, indicating that most of the nebulae were racing away from us at incredible speeds (as much as 0.01*c*!), much faster than the measured speeds of ordinary Milky Way stars. Nebulae were exceptional objects, and scientists speculated (but hotly contested!) that they were extragalactic. The debate was settled by Edwin Hubble in 1924. Using the 100 in. (diameter) telescope—the largest of its day—at Mount Wilson Observatory in California, Hubble measured the period and apparent brightness of Cepheid variable stars in the Andromeda nebulae. Employing the Leavitt Law,* he then calculated the distance to Andromeda as 900,000 ly (0.275 Mpc), three times the estimated diameter of the Milky Way.† Without a doubt, Andromeda lay well beyond the borders of our galaxy. Slipher's other nebulae were later discovered to be even farther away. We now know they are "island universes", or galaxies, of their own. Hubble's groundbreaking work, which occupied him for nearly a decade, forever changed our conception of the cosmos.

Hubble was aware of two mathematical models of the universe, both based on simplified solutions to Einstein's equations of general relativity. The first, proposed by Einstein, described a static (unchanging) universe. To produce a steady state, Einstein introduced an unidentified repulsive force—a fudge factor, really—to counterbalance the gravitational attraction between galaxies. The second model, due to the Dutch astronomer Willem de Sitter, described an *empty* expanding universe in which gravitational interactions between galaxies were negligible. According to de Sitter's model, light reaching us from distant galaxies would be redshifted by an amount proportional to distance. Einstein's model, on the other hand, predicted no systematic redshift. From 1924 to 1929, Hubble and his assistant Milton Humason labored to measure the distance to many of the galaxies whose redshifts had been measured earlier by Slipher. Their results (Figure 3.7a)

* In 2009, the American Astronomical Society Council voted to refer to the Cepheid period-luminosity relation as the Leavitt Law, in honor of Henrietta Leavitt's pioneering work 100 years earlier.

† We now know that the distance to Andromeda is 2.5 Mly (0.770 Mpc), while the diameter of the Milky Way Galaxy is "only" about 100,000 ly (0.03 Mpc).

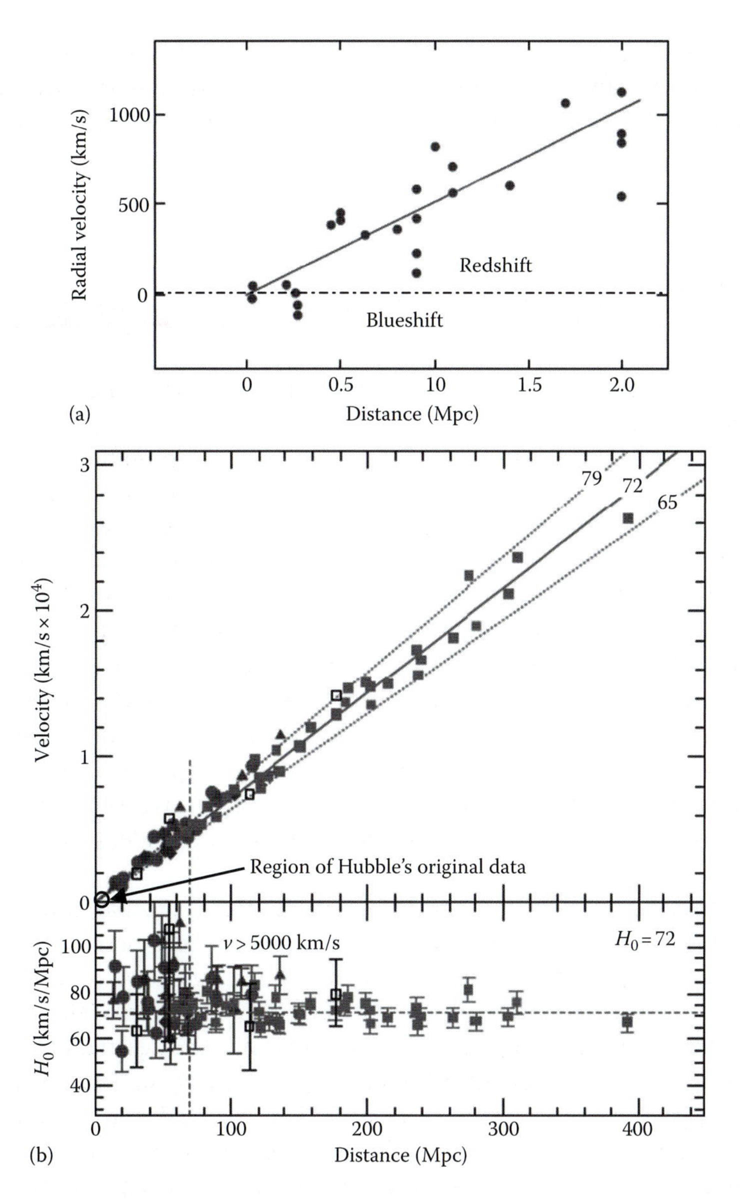

FIGURE 3.7 (a) Hubble's original 1929 data relating the recessional speed of galaxies to their distance from the Milky Way. Note that nearest galaxies are blueshifted, due to their gravitational attraction to our home galaxy. Hubble's estimates of distance were too small by a factor of about 7. (b) A recent (2010) data set using the Hubble space telescope. The upper graph is a plot of velocity vs. distance as in (a). The slope of the best-fitting straight line ("72") is our best estimate of the Hubble constant H_0. (Hubble's original data would be confined to a tiny region close to the origin.) In the lower graph, the value of H_0 is calculated from Equation 3.12 for each galaxy, and these values are plotted vs. distance. The agreement between these values of H_0 and the value of H_0 derived from the best-fitting straight line improves with distance. (From Freedman, W. and Maldore, B.F., *Annu. Rev. Astron, and Astrophys, 48,* 673–710 [2010]. With permission of Annual Reviews.)

unequivocally supported the picture of an expanding universe.* Attributing redshift to the classical Doppler effect (Equation 3.11), Hubble concluded that

$$v = H_0 r, \tag{3.12}$$

where r and v are the galaxy's distance and radial velocity relative to the Milky Way.

Equation 3.12, called *Hubble's Law*, is the cornerstone of modern cosmology. If the *Hubble constant* H_0 is accurately known, then, by measuring the redshift of light emitted by a galaxy, one can calculate v and use Equation 3.12 to determine the galactic distance. H_0 sets the size scale of the universe, and—as you will see—also provides an estimate for the age of the universe. But finding H_0 is no easy matter. Hubble's original data (Figure 3.7a) look none too promising. Cepheids—the most reliable distance indicators—were found in only the nearest few galaxies, and the alternate strategies used by Hubble to estimate galactic distance were unreliable and prone to large error. (In fact, the distances to most of the galaxies shown in Figure 3.7a are too small by about a factor of 7!) Moreover, light from the nearest galaxies is blueshifted rather than redshifted. This is because nearby galaxies (including Andromeda) are strongly attracted gravitationally to the Milky Way, so their behavior is distinctly different from galaxies farther away. In spite of these complications, Hubble confidently fit a straight line to his data and estimated its slope H_0 to be 500 km/(s Mpc),† a factor of 7 higher than present-day estimates.

Much like the quest to measure the Earth–Sun distance, the campaign to measure H_0 has occupied generations of astronomers. The best contemporary measurements of H_0 take advantage of telescopes in low Earth orbit. Just as in Hubble's day, the primary distance calibration relies on the Leavitt Law applied to Cepheid variables in nearby galaxies. To measure the distance to more distant (and therefore dimmer) galaxies, astronomers must abandon the Cepheids in favor of brighter objects with known luminosities. These brighter *standard candles* are provided by the spectacular events called *Type 1a supernovae*, referring to the violent self-destruction of a late-stage star (called a white dwarf) that has grown too massive to support itself against gravitational collapse. The star's collapse, and subsequent explosion, releases a huge amount of radiant energy with a peak luminosity $L \approx 10^{10} L_{Sun}$. Often, the dying star will outshine the rest of the galaxy for several weeks! All Type 1a supernovae are remarkably similar, allowing them to be used as standard candles to determine the distance to galaxies hundreds of times farther away than the ones studied by Hubble. One of the highest priority missions of the Hubble space telescope—named in honor of Edwin Hubble—was to measure H_0 to an accuracy of ±10%. This project was completed successfully in 2009 (see Figure 3.7b) with the result

$$H_0 = 72 \pm 7 \text{ km/(s Mpc)}. \tag{3.13}$$

Notice that the dimensions of H_0 are distance/(distance × time) = 1/time.

---

* After hearing of Hubble's results, Einstein called his repulsive force—or cosmological constant—the *biggest blunder* of his life. In recent years, however, it has reemerged as a possible model for the *dark energy* fueling the acceleration of the universe's expansion.

† Meaning that a galaxy 1 Mpc away is moving away from us at 500 km/s. 1 parsec (pc) = 3.26 light years (ly) and 1 Mpc = 10^6 pc.

EXERCISE 3.12

Calculate the distance to galaxy NGC 2276 using the Doppler shift of the Hα emission line shown in Figure 3.8 and the value of H_0 given above.

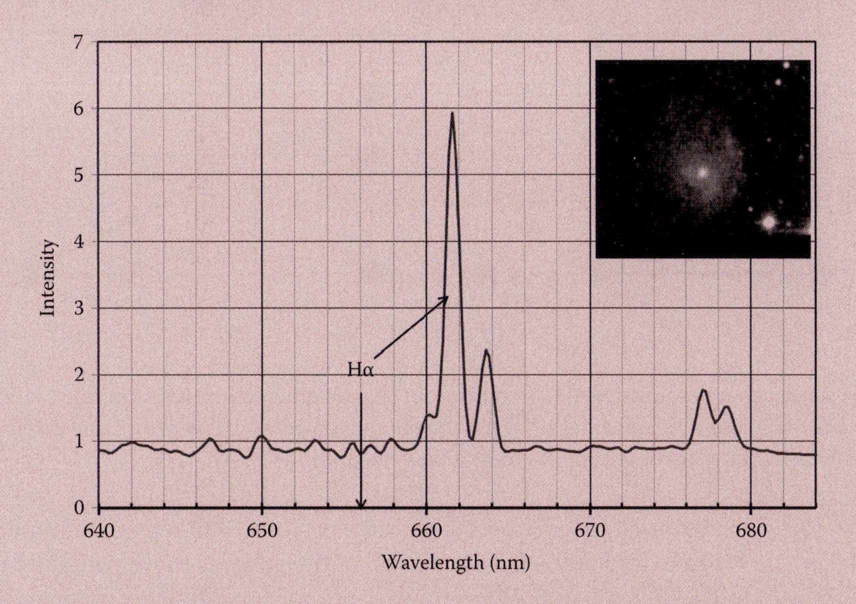

FIGURE 3.8 The hydrogen emission spectrum of the galaxy NGC 2276 for wavelengths near the hydrogen Hα line. The vertical axis is light intensity. The Hα line is redshifted by about 6 nm. (From R.C. Kennicutt, Jr., *ApJ Supp. Ser.* 79, 255 [1992]. Graph adapted from data posted on NASA/IPAC Extragalactic Database [NED].)

3.8 HUBBLE'S LAW AND THE EXPANSION OF THE UNIVERSE

Hubble's Law shattered the notion of a static, ageless universe, and launched the era of modern cosmology. Equation 3.12 is consistent with the *big bang* model of the early universe, wherein the cosmos erupted from a tiny, inconceivably hot and dense ball that then swelled to its present size. Yet the modern interpretation of Hubble's Law is based—not on the classical Doppler effect—but on the *expansion of intergalactic space* described by the equations of general relativity. As space expands, it drags the galaxies along with it. Imagine a two-dimensional universe represented by the surface of a balloon (space), populated by bugs (galaxies) resting at random locations on the balloon's surface. As the balloon is inflated, the distance between any two bugs increases, even though they are not creeping over the surface. Expansion of space, not the motion of galaxies relative to their local surroundings, is the source of the redshifts measured by Hubble. In Figure 3.9a, the solid grid represents an x–y coordinate system with its origin centered on the Milky Way Galaxy A at time t_1. (Imagine this grid painted on the surface of a gigantic balloon.) At a later time $t_2 = t_1 + \Delta t$, the grid has expanded (dotted lines) along with space, carrying galaxies B, C, and D along with it. For example, at t_1 galaxy

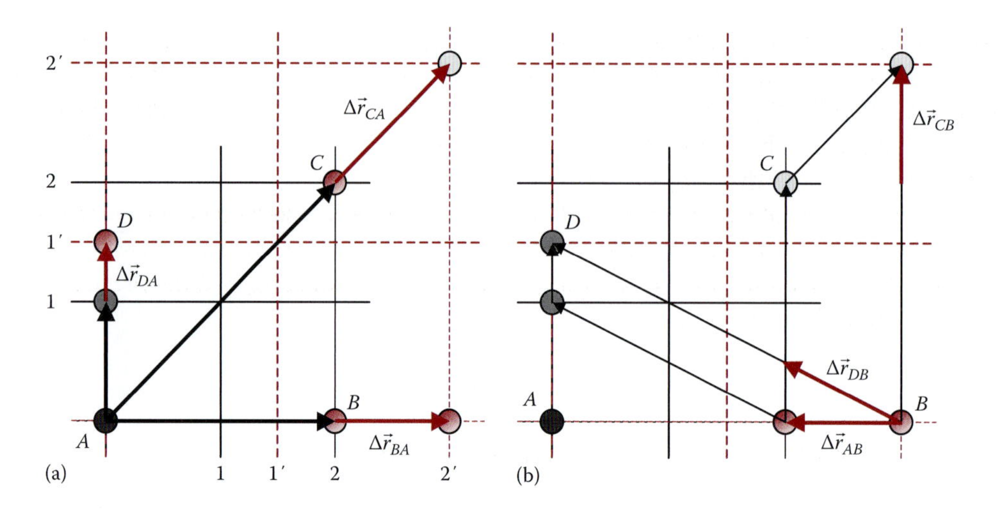

FIGURE 3.9 (a) As space expands, galaxies B, C, and D retain their spatial coordinates yet move *radially* away from A by amounts proportional to their distances to A. (b) An astronomer on galaxy B sees the same thing: A, C, and D move radially away from B by amounts proportional to their distances from B.

C is at coordinates (2, 2); in expanded space (time t_2), it is farther from A but remains at the same coordinates (2′, 2′). The same is true for galaxies B and D. Note that all of the galaxies move *radially away* from A and that their displacements in the time interval Δt are directly proportional to their original distances from A. (Check this by taking direct measurements off the figure.) This behavior is consistent with Hubble's Law.

Let's reexpress these ideas using vector algebra. Imagine a three-dimensional universe centered on the Milky Way, populated by N galaxies that are presently located at $\vec{r}_1, \vec{r}_2, \vec{r}_3, \ldots, \vec{r}_N$ relative to our galaxy. In the expanding universe described, each displacement vector $\vec{r}_i$ grows at the same rate, or

$$\vec{r}_i(t) = a(t)\vec{r}_i(t_0), \tag{3.14}$$

where t_0 indicates the present time and $a(t)$ is called the scale factor.

Obviously, $a(t_0) = 1$ and $a(t) > 1$ for $t > t_0$. The velocity of galaxy i relative to us is given by the time derivative of $\vec{r}_i$:

$$\vec{v}_i = \frac{d\vec{r}_i}{dt} = \left.\frac{da}{dt}\right|_0 \vec{r}_i(t_0). \tag{3.15}$$

Over a short time interval Δt, in which da/dt remains constant, the change in intergalactic distance is

$$\Delta\vec{r}_i = \vec{v}_i\Delta t = \left.\frac{da}{dt}\right|_0 \vec{r}_i\Delta t. \tag{3.16}$$

In these two equations, we have introduced the notation $|_0$ to indicate that the expansion rate da/dt is to be evaluated at the present time t_0. Note that $\vec{v}_i$ is directly proportional to $\vec{r}_i$ and that it is a purely radial velocity (i.e., parallel to $\vec{r}_i$). Equation 3.16 is identical to Hubble's Law if $H_0 = (da/dt)|_0$. As space expands, so does the wavelength of light propagating from the source to Earth: $\Delta\lambda = (da/dt)|_0\lambda\Delta t$. This is the source of the redshifts measured for distant galaxies by astronomers.

If the expansion rate has always been equal to its present value H_0, then $a(t) = a(t_0) + H_0(t - t_0)$. (Check: $a(t = t_0) = ?$ $da/dt = ?$) As we trace time backwards ($t < t_0$), the universe shrinks and its density grows. Let $t = 0$ be the time of the big bang, when $a(t) \ll a(t_0)$. Setting $a(0) = 0$ and $a(t_0) = 1$, the present age of the universe (the elapsed time since the big bang) is

$$t_0 = \frac{1}{H_0}. \tag{3.17}$$

When we measure H_0 we are also estimating the age of the universe.

EXERCISE 3.13

Use Equation 3.17 to estimate the age of the universe. The numbers you will need are 1 pc = 3.09×10^{13} km, 1 year = 3.16×10^{7} s. Express your answer in billions of years.

But isn't there something fishy about this picture? From our vantage point in the Milky Way, all galaxies appear to be rushing away from us (Equation 3.15) no matter what direction of the sky we look to. Doesn't this imply that we are at the center of the universe? Were Aristotle and Ptolemy right after all? Before answering this question, try the following exercise.

EXERCISE 3.14

Imagine a one-dimensional universe. You are in galaxy A and other galaxies are spaced evenly to your right and left, as far as you can see. Based on redshift data, the nearest galaxies (B and C) are moving away from you with speeds 100 km/s. The next nearest galaxies (D and E) are moving away with speeds 200 km/s, and so on. What would an observer in galaxy B report for the recessional speeds of its nearest, and next nearest, galaxies? Are you at the center of this universe? Is galaxy B at the center? What's going on?

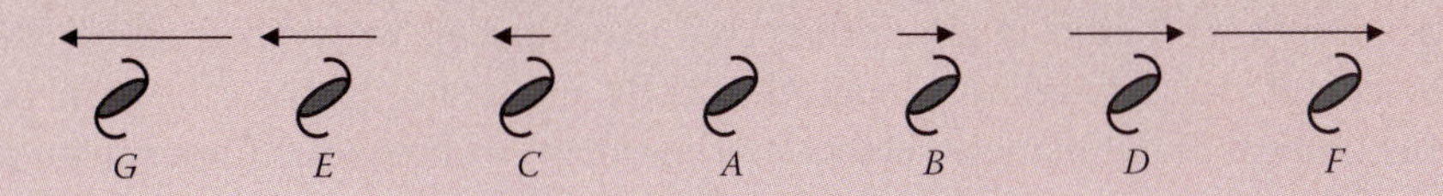

Let's now return to the three-dimensional world and our own Milky Way Galaxy. From where we stand, the universe is expanding in agreement with Hubble's Law and Equation 3.15: $\vec{v}_i = H_0\vec{r}_i$, where $\vec{v}_i$ and $\vec{r}_i$ denote the velocity and position of the ith galaxy relative to us. Choose any two galaxies, say $i = 1$ and $i = 2$. Then, $\vec{v}_1 = H_0\vec{r}_1$ and $\vec{v}_2 = H_0\vec{r}_2$. Subtracting these two equations, we get

$$\vec{v}_2 - \vec{v}_1 = H_0(\vec{r}_2 - \vec{r}_1).$$

But $\vec{r}_2 - \vec{r}_1$ is the position of galaxy 2 relative to galaxy 1, and, likewise, $\vec{v}_2 - \vec{v}_1$ is the velocity of galaxy 2 relative to galaxy 1. So, to an astronomer in galaxy 1, galaxy 2 is receding from it with a radial velocity proportional to its distance from galaxy 1. This holds for all galaxies (i = 2, 3, 4, ...) observable from galaxy 1. In other words, an observer in galaxy 1 sees *exactly* the same behavior as we do: Hubble's law is satisfied for this observer with the same constant of proportionality H_0 that we measure from our own galaxy (see Figure 3.9b). The same result is found by observers in galaxy 2, or galaxy 3, or any other galaxy in the universe. Astronomers everywhere in the universe see the same behavior, so none of them can rightfully claim the privileged spot at the center of the universe. It is not just that *we're* not at the center. *There is no center*! This conclusion is consistent with the *cosmological* (*or Copernican*) *principle*, a pillar of present-day cosmology, which asserts that there is nothing unique or special about our location in the universe. It was bad enough, in the sixteenth century, for Copernicus to dislodge us from the center of the universe. Now, in the twentieth and twenty-first centuries, astronomers have cast us adrift in a centerless cosmos.

EXERCISE 3.15

In the preamble to this chapter, Edwin Hubble said, "We are, by definition, in the very center of the observable region." What did he mean?

3.9 SUMMARY

In this chapter, we learned how to use vectors to describe the two simplest—most fundamental—types of motion: constant velocity motion and uniform circular motion. We began with precise mathematical definitions of the vector displacement $\Delta\vec{r}$ and velocity $\vec{v}$ and found that $\vec{v}$ was merely the first derivative of $\Delta\vec{r}$. The vector equations for both types of motion were derived, and we learned that $\vec{v}$ was always tangent to the trajectory. For uniform circular motion, we introduced the angular speed ω and the rotation period T and then derived the relationship between the two. As a first application, we studied the *relative* motion of two bodies and argued that *all motion is relative*; that is, whenever we measure a body's velocity, we are inescapably measuring its velocity relative to another body. For example, the Ptolemaic picture of the universe describes the motion of the planets relative to the Earth, whereas the Copernican picture describes planetary motion relative to the Sun. We learned how to use redshifts to measure the (relative) radial velocity of far off galaxies and found undeniable evidence for an expanding universe. Finally, we explored the implications of Hubble's law for the birth and age of the cosmos.

Just like the Greeks, however, we are still merely *describing* motion rather than attempting to understand it. It is now time to dig deeper, to look for underlying *laws* that govern the motion of all bodies, from gigantic galaxies to the smallest subatomic particles. We are now ready to begin our study of physics.

PROBLEMS

3.1 An air traffic controller uses a radar display to track an approaching airplane. The position of the aircraft vs. time is given in the table below ($x = 0$, $y = 0$ is the location of the radar tower):

Time (s)	x (km)	y (km)
0	0.6	0.2
1	0.4	0.5
2	0.2	0.8
3	0.0	1.1
4	−0.2	1.4
5	−0.4	1.7

(a) Find the average velocity during each second of elapsed time (i.e., from $t = 0$ to $t = 1$, $t = 1$ to $t = 2$, etc.). What do you conclude regarding the motion of the airplane?

(b) Is the airplane supersonic? (The speed of sound is 340 m/s.)

(c) Make graphs to scale of x vs. t and y vs. t.

(d) From each graph, extract the equation for $x(t)$ and $y(t)$ and give an expression for $\vec{r}(t)$.

(e) From your result in (d), obtain the instantaneous vector *velocity* at $t = 2.5$ s.

(f) If the airplane continues its motion, what will its position be at $t = 10$ s?

3.2 The position of an object is given by the vector $\vec{r} = (8t - t^2)\hat{i} + (12t - 3t^2)\hat{j}$ m.

(a) What is the vector displacement between $t = 0$ s and $t = 1$ s?

(b) What is the velocity vector at $t = 0$ s and $t = 1$ s? Sketch the two vectors.

(c) When is the motion entirely in the y-direction?

(d) What is the maximum value of the y-coordinate?

(e) What is the velocity vector when y is a maximum?

3.3 The positions of two bodies at $t = 0$ are given by the following vectors:

$$\vec{r}_{10} = 2\hat{i} + 2\hat{j}\text{ m} \quad \text{and} \quad \vec{r}_{20} = 6\hat{i} + \hat{j}\text{ m}.$$

Their velocities are given by $\vec{v}_1 = 4\hat{i} - 3\hat{j}$ m/s and $\vec{v}_2 = -4\hat{i} - \hat{j}$ m/s.

(a) Make a drawing showing the initial position vectors and the directions of the velocity vectors.

(b) What is the velocity of body 2 relative to body 1? Express your answer in $\hat{i}, \hat{j}$ notation and sketch the direction of the relative velocity.

(c) Using the dot product, calculate the angle between $\vec{v}_1$ and $\vec{v}_2$.

(d) Do the bodies collide? If so, at what time?

3.4 A body is moving in the x–y plane with a constant velocity $\vec{v} = -4\hat{i} + 3\hat{j}$ m/s. Its position at $t = 0$ is $\vec{r}_0 = 12\hat{i} + 5\hat{j}$ m. (See the figure below.)

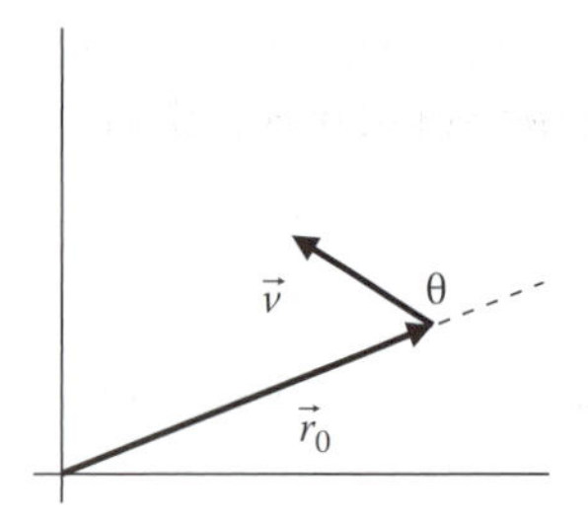

(a) Write the position vector $\vec{r}(t)$ in $\hat{i},\hat{j}$ notation.

(b) Using the dot product, find the angle between $\vec{r}_0$ and $\vec{v}$.

(c) Write a unit vector parallel to $\vec{r}_0$ in $\hat{i},\hat{j}$ notation.

(d) You are located at $(x,y) = (0,0)$ and $\hat{j}$ is in the vertical direction. At what time is the body directly overhead? What is its distance from you at this time?

3.5 The position of an object is given by $\vec{r} = 2t\hat{i} + (5-3t)\hat{j}$ m.

(a) Find the position of the object at t = 0 s, t = 1 s, and t = 2 s and draw its trajectory on an x–y graph.

(b) What is the vector velocity of the object? Indicate its direction on the above graph.

(c) What is the *speed* of the object?

3.6 The second largest near-Earth asteroid, 433 Eros, is peculiar because of its very irregular shape. A study of its geology may shed light on the origins of the solar system. The panel of pictures below was taken from the spacecraft NEAR-Shoemaker (Near Earth Asteroid Rendezvous) on February 12, 2000, from a distance of about 1800 km.

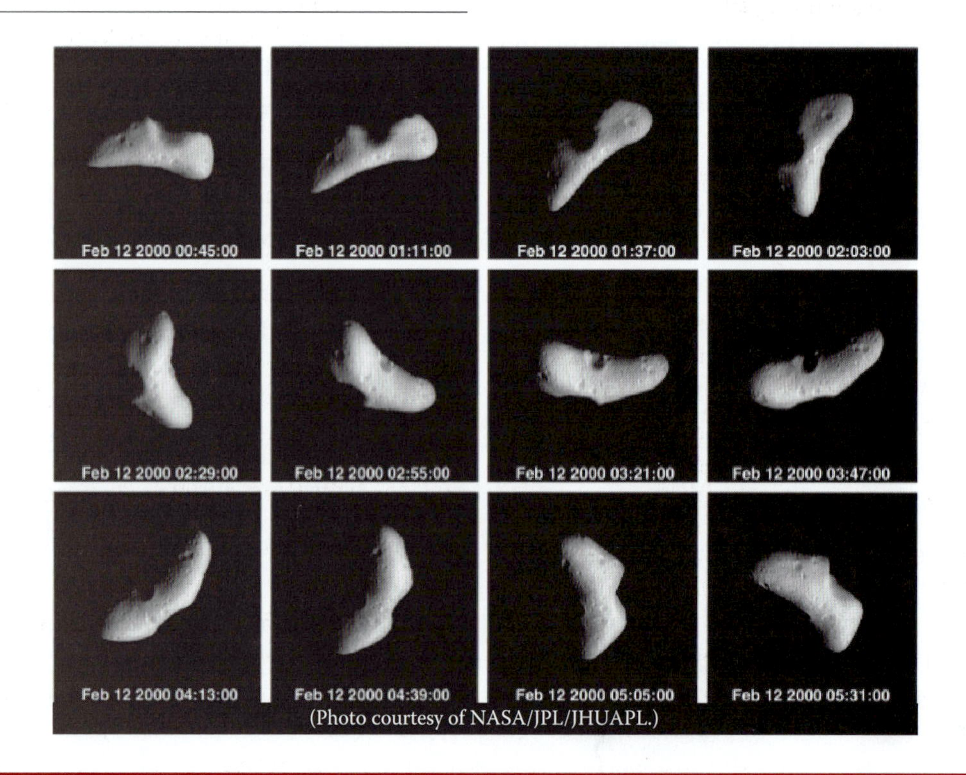

(Photo courtesy of NASA/JPL/JHUAPL.)

(a) Analyze the images carefully, and from the times and angles find the period and angular velocity ω of the asteroid's rotation.

(b) 433 Eros is 35 km long. If it revolves around its geometric center, what would be the rotational speed of one of the extreme points of the asteroid?

(c) On February 14, 2000, the spacecraft entered an orbit around the asteroid, and 1 year later it landed successfully on the surface. It transmitted signals back to Earth at a frequency of 2.0 GHz (1 GHz = 10^9 Hz). Suppose that on Earth, NEAR's beacon signal was detected with a maximum Doppler shift of 18.0 Hz due to the asteroid's rotation. Assuming that the asteroid is rotating in the plane of the solar system, how far from the axis of rotation was the spacecraft's landing site?

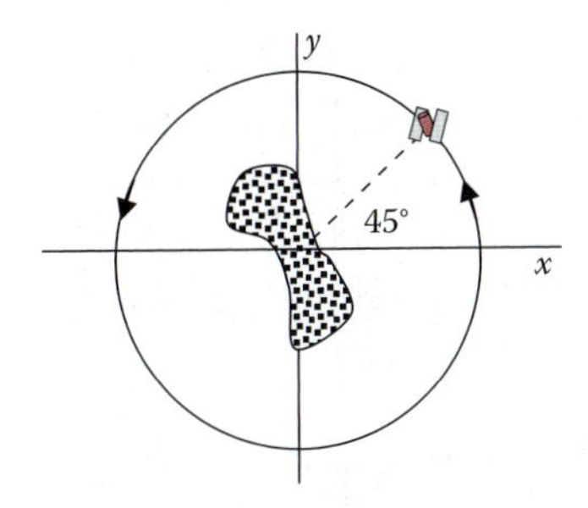

Before its historic landing on 433 Eros, NEAR circled the asteroid with an orbit radius of 36 km and a period of 0.7 days.

(d) Find the orbital speed and angular speed of NEAR.

(e) Find the spacecraft's instantaneous velocity at the position shown in the figure above.

3.7 A bicycle wheel of radius 0.3 m rolls without slipping on a horizontal surface. The wheel's hub moves with a translational velocity $\vec{v} = 9.0\hat{i}$ m/s.

(a) Find the period of rotation of the wheel, and the angular velocity ω.

(b) What is the rotational *speed* of a point on the wheel's rim relative to the hub?

(c) What is the instantaneous vector velocity of points A, B, and C relative to the hub?

(d) What is the *relative* velocity between point D and the *surface*?

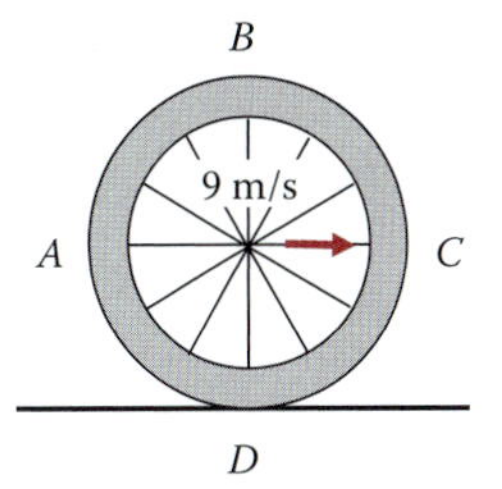

3.8 In the laboratory, hot calcium gas emits visible light at many wavelengths, one of which is its "K" line at λ = 393.3 nm. When viewed in the spectrum of a distant galaxy, this same line has an apparent wavelength λ' = 476.0 nm.

(a) Is the galaxy approaching us or moving away from us?

(b) What is the speed of approach or recession?

(c) A second line, whose apparent wavelength λ' = 508.0 nm, is also observed in the spectrum of the same galaxy. What would be the wavelength of this line if it were produced in the laboratory?

(d) The Hubble constant H_0 = 72 ± 7 km/(s Mpc). Find the distance to this galaxy.

3.9 The K spectral line (λ = 393.3 nm) from the left edge of a rotating galaxy is blueshifted by 0.58 nm. The same spectral line from the right edge is redshifted by 0.10 nm. Assume the galaxy has a circular disk shape.

(a) Is the galaxy approaching us or moving away from us?

(b) What is the translational speed of the galaxy relative to us?

(c) What is the rotational velocity of the edges of the galaxy?

3.10 The figure shows the Sun, Earth, and Mars at opposition, that is, when the Sun and Mars are in opposite directions when viewed from Earth. The orbital period of Earth is 365 days and that of Mars is 687 days.

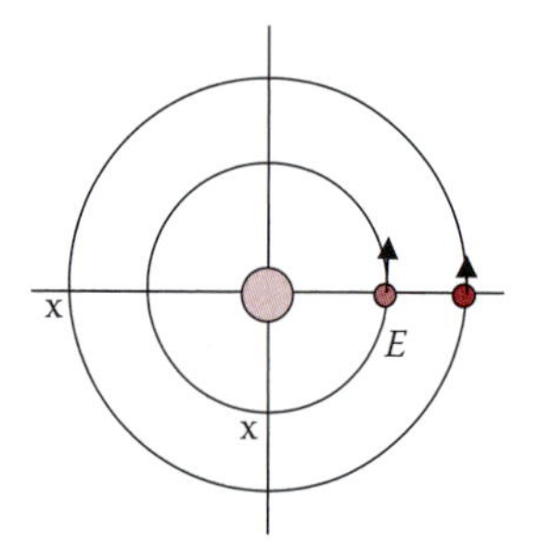

(a) How many days will elapse before Mars is once again in opposition to the Sun? (This time interval is called the *synodic period* of Mars.)

(b) Derive the general expression for the time found in (a) in terms of the periods of the two planets T_E and T_M.

(*Hint:* The angular velocity of Earth ω_E is greater than the angular velocity of Mars ω_M. Earth will circulate the Sun ahead of Mars, completing its orbit and then *catching up* to Mars at opposition.)

(c) Find the orbital speed of both planets in km/s. Mars' orbital radius is 1.52 AU. (A handy approximation is 1 year $\simeq \pi \times 10^7$ s.)

(d) Find the velocity of Mars relative to Earth when the two planets are at the positions marked with x's in the above figure.

Mars "Rover" *Opportunity*. (Photo courtesy of NASA.)

(e) In 2004, the "rovers" *Spirit* and *Opportunity* landed on the Martian surface to study the planet's geochemistry and atmosphere. Their operational life expectancy was 3 months, but *Spirit* functioned for 6 years, and (as of this writing) *Opportunity* is still collecting and transmitting data to scientists on Earth using a frequency of 10 GHz. When the planets are at the positions marked by x's in the above figure, are their signals detected on Earth with a positive, negative, or zero Doppler shift in frequency?

(f) While *Spirit* and *Opportunity* were getting all the press attention, the spacecraft *Odyssey* was faithfully orbiting the planet serving as a vital communications link from Mars back to Earth. *Odyssey* was in a nearly circular orbit with an altitude of 400 km and a period of 2 h. Find its speed and angular velocity ω (R_{Mars} = 3400 km).

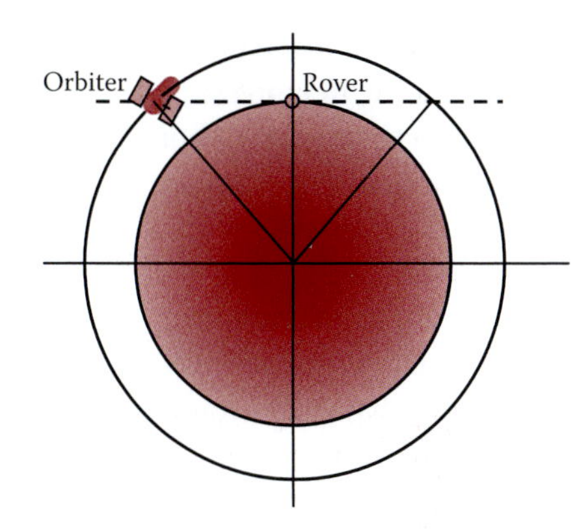

(g) Assuming the orbiter passes directly over a rover, for how long a time is it in view during each pass? See the figure above.

3.11 A satellite moves in a circular orbit around Earth, as shown in the figure. The satellite is used for television broadcasting so it must be in a geostationary orbit, that is, it orbits the planet once a day.

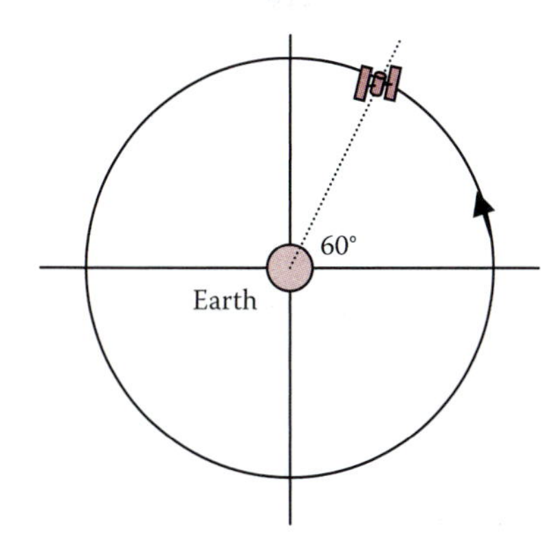

(a) Calculate the angular velocity ω in rad/s.

(b) At time $t = 0$, the satellite is in the position shown. The radius of its circular orbit is 42,000 km. Write an expression for the angle (rad) that the satellite forms with the x-axis as a function of time.

(c) Write the position vector in $\hat{i}, \hat{j}$ notation as a function of time t.

(d) Find the velocity vector in $\hat{i}, \hat{j}$ notation at $t = 2$ h.

3.12 The K spectral line (λ = 393.3 nm) from the left edge of a rotating galaxy is blueshifted by 0.393 nm. The same spectral line from the right edge is redshifted by the same amount.

(a) What is the speed of the galactic matter at the edges of the galaxy?

(b) If the radius of the galaxy is 11 kpc (1 pc = 3.09×10^{16} m), what is the period of rotation of the galaxy?

3.13 The figure shows part of the radio spectrum of galaxy NGC 918. It is the famous *21 cm line* emitted by neutral hydrogen. (A more precise value is λ = 21.106 cm.) Because of galaxy rotation and recession the emission is not a sharp line but is redshifted and broad.

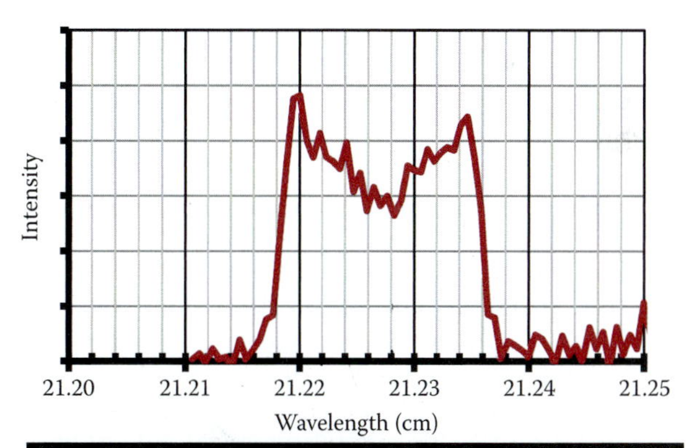

(From C.M. Springob et al., *Astroph. J. Supp. Ser.* 160(1), 149, 2005.)

(a) Find the radial velocity of the galaxy.

(b) Assume you are viewing the galaxy edge on. Calculate the rotational velocity at the edge of the galaxy.

(c) If the radius of the galaxy is 10 kpc (1 pc = 3.09×10^{16} m), find the time (in millions of years) for the galaxy to make one complete rotation.

3.14 Galaxies *A*, *B*, and *C* are visible from our planet within the Milky Way. Viewed from Earth, the potassium (K) spectral line (λ = 393 nm) from galaxy *A* is redshifted by 0.39 nm and the same line from galaxy *B* is redshifted by 0.78 nm.

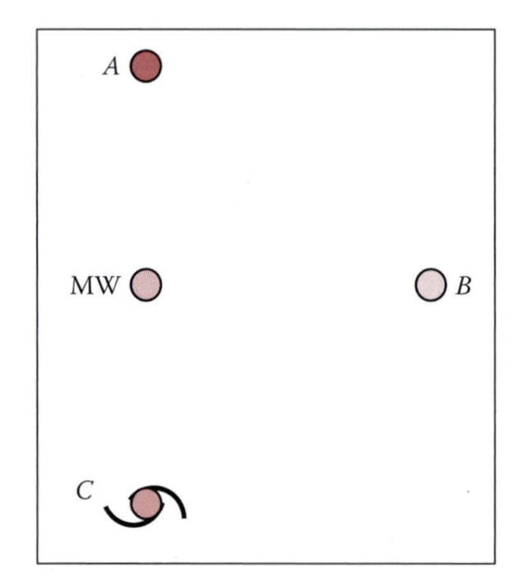

(a) Which galaxy, *A* or *B*, is farther from us? Explain your choice clearly.

(b) Find the radial speed of galaxy *A* relative to us.

(c) If you were on galaxy *A*, looking at galaxy *B*, what would you measure for the redshift $\Delta\lambda$ of the K line?

(d) Galaxy *C* is a rotating spiral galaxy with a radius of 9.46×10^{16} km (10^4 ly). Light emitted from the left edge of galaxy *C* is redshifted by 0.29 nm and light from the right edge is redshifted by 0.49 nm. How long does it take for this galaxy to make one full rotation?

3.15 *Great Scott*! You are captain of the starship *Raider*, cruising at constant velocity through the solar system somewhere between the orbits of Mars and Jupiter. Your navigator reports the approach of a large asteroid which appears to be on a collision course with the starship! Onboard computers set the positions and velocities of the centers of the spacecraft and asteroid as shown in the following diagram.

(a) Write vector expressions for the positions of the centers of the two bodies $\vec{r}_S(t)$ and $\vec{r}_A(t)$.

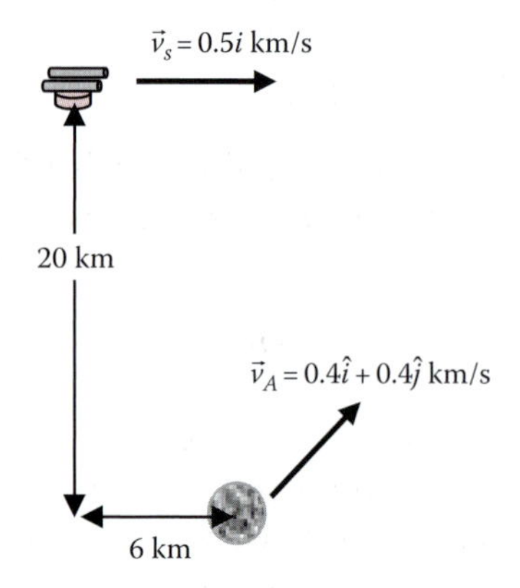

(b) Will the asteroid collide with the spacecraft? If so, how much time do you have to prepare for the collision? Assume the asteroid has a radius of 1.2 km.

(*Hint:* Write an expression for the center-to-center distance between the two bodies. Minimize that expression by differentiation.)

3.16 Particle 1 is initially at a position $\vec{r}_{10} = 1\hat{i} + 2\hat{j}$ m, while particle 2 is initially at a position $\vec{r}_{20} = 3\hat{i} + 1\hat{j}$ m (where m indicates meters). Particle 1 moves directly to the right with a speed of 5 m/s, whereas particle 2 moves radially away from the origin at a speed v_2.

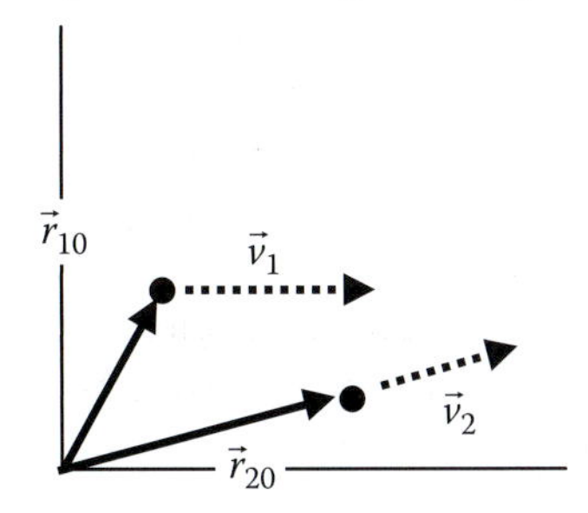

(a) Write the velocity vector $\vec{v}_1$ and the position vector $\vec{r}_1(t)$.

(b) The two particles collide at a position $\vec{r} = x\hat{i} + y\hat{j}$. Find the vector $\vec{r}$. At what time t do they collide?

(c) What is the speed v_2 needed for the particles to collide at this time t?

(d) Sketch the direction of the relative velocity vector $\vec{v}_{21} = \vec{v}_2 - \vec{v}_1$.

3.17 A spy satellite in a low Earth orbit, at an altitude of 300 km, passes directly overhead an observer on the Earth's surface. The satellite is transmitting a beacon signal at a frequency f = 3000.00 MHz (1 MHz = 10^6 Hz). When $\theta = 49.5°$, the measured frequency of the satellite's beacon is 3000.05 MHz (see the figure).

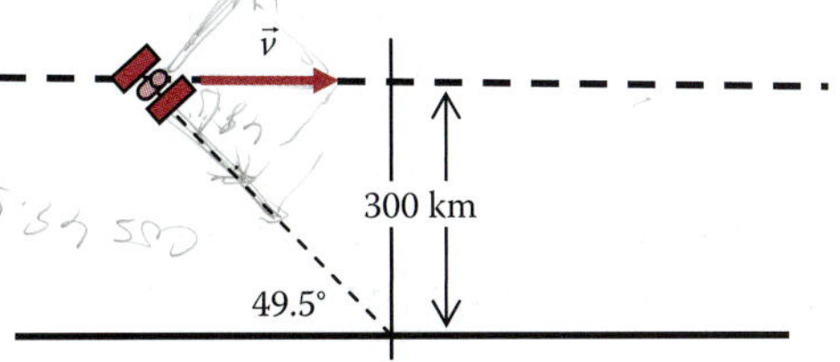

(a) Find the speed v of the satellite.

(b) Estimate the overall change in the measured frequency of the satellite (i.e., the highest minus the lowest frequency measured by the observer on Earth). You may neglect the curvature of the satellite's orbit.

(c) At time t = 0, the satellite is in the position shown in the figure below, 300 km above the Earth's surface (R_E = 6400 km), moving with the same speed v found in part (a). At what time will its velocity be given by $\vec{v} = -v\cos(\pi/4)\hat{i} + v\sin(\pi/4)\hat{j}$?

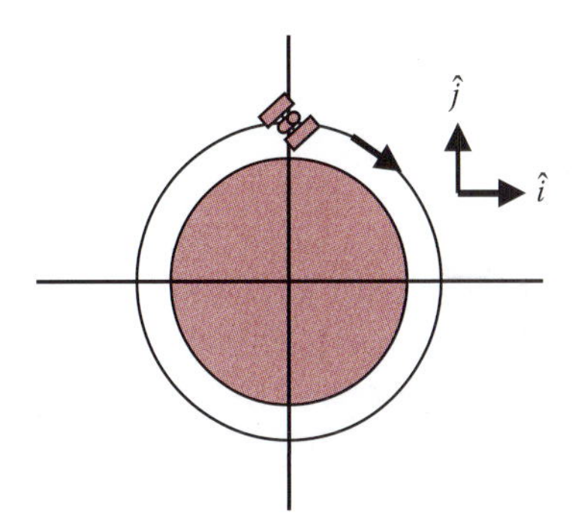

3.18 Mercury's rotation rate was not well known until 1965, when it was measured by radar. Astronomers at the Arecibo Observatory in Puerto Rico bounced a 430.00 MHz radar signal from the planet and measured the Doppler shift of the reflected echo (the same technique police use to measure your highway speed). Mercury is 0.39 AU from the Sun and has an orbital speed v_M = 48 km/s. Its radius is 2440 km.

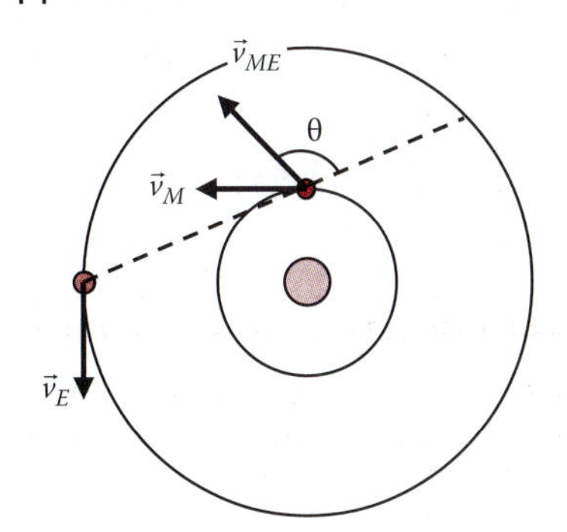

(a) Write vector equations for the velocity of Mercury relative to Earth ($\vec{v}_{ME} = \vec{v}_M - \vec{v}_E$) and the position of Mercury relative to Earth ($\vec{r}_{ME} = \vec{r}_M - \vec{r}_E$) at the moment shown in the figure above. Use appropriate units.

(b) What is the angle θ between $\vec{r}_{ME}$ and $\vec{v}_{ME}$?

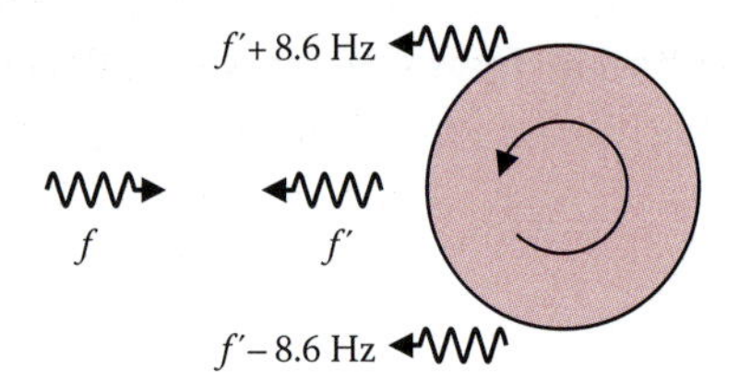

(c) What is the Doppler shift $\Delta f = f' - f$ of the return signal assuming that Mercury is not rotating? Is this frequency shift positive or negative? (*Note:* the Doppler shift of a *reflected* signal is *twice* as large as it would be if the signal originated on Mercury. Reflection can be understood as a two-step process: the radar signal from Earth is first detected—absorbed—by the moving surface of Mercury and then it is reradiated from the moving surface.)

(d) If the signal from Mercury's *east* limb is shifted from its *west* limb by 17.2 Hz, what is Mercury's rotational speed at its equator?

(e) Show that your answer to (d) corresponds to a rotation period of 59 days. Mercury makes exactly three rotations for every two orbits around the Sun, a consequence of gravitational *tidal* forces.

3.19 On February 4, 2011, asteroid 2011 CQ1, a 1 m diameter rock, came within 0.85 Earth radii of striking the planet. At that time, its speed (in the Sun's reference frame) was $v_a = 26.8$ km/s and its speed relative to Earth was 5.17 km/s. What was the angle between the velocity vectors $\vec{v}_A$ and $\vec{v}_E$?

3.20 Imagine that you are on a planet that has just exploded. Miraculously, you are unhurt and carried away on an outgoing planetary fragment. Being an aspiring physicist, you calmly proceed to analyze the situation and take some measurements. You reason that each fragment is receding from the source of the explosion at constant velocity, so at time t after the explosion the position of the nth fragment is related to its velocity as

$$\vec{r}_n = \vec{v}_n t,$$

where $\vec{r} = 0$ is the site of the explosion. In the figure, the numbered fragments are at distances r, $2r$, and $3r$ from the point of the explosion. Assume you are on fragment 1, a distance r from the explosion.

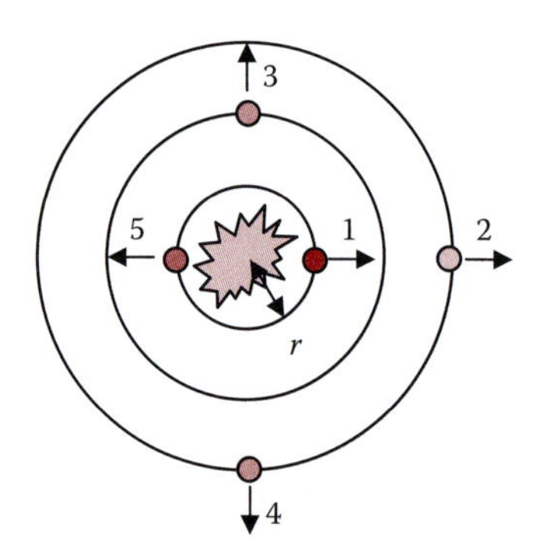

(a) What is the ratio of the magnitude of $\vec{v}_5$ to the magnitude of $\vec{v}_4$?

(b) Carefully sketch relative position vectors $\vec{r}_{21}, \ldots, \vec{r}_{51}$ and show that the relative velocities of the fragments 2, ..., 5 are directed radially away from you, with speeds proportional to the distance from you.

(c) You measure the distance to fragment 2 to be 1×10^5 km and its relative velocity to be 2×10^5 km/year. What is the speed of fragment 3 with respect to you? How long ago did the explosion occur?

(d) From your vantage point on fragment 1, you are surrounded by fragments which are all moving radially away from you. Can you conclude that you are at the point of the explosion? Explain your answer. What would an observer on fragment 4 report?

The universe exhibits a similar situation, but instead of exploding bodies it is space itself that is expanding. Astronomers find that other galaxies are receding from our own galaxy with relative velocities that obey Hubble's Law:

$$\vec{v}_{rel} = H_0 \vec{r}_{rel},$$

where H_0 is the Hubble constant. The precise value of H_0 is critical to knowing the size, age, and ultimate fate of the universe. Recent observations of the Virgo cluster of galaxies indicate that its distance from us is 18.0 ± 1.2 Mpc. Redshift measurements establish that its speed of recession is 1296 km/s.

(e) Using this data, estimate H_0 and its uncertainty in units of km/(s Mpc).

(f) Using this estimate, find the age of the universe (with uncertainty). Express your answer in years. (*Note:* 1 year $\approx \pi \times 10^7$ s, 1 pc $= 3.09 \times 10^{16}$ m.)

Section II

CONSERVATION OF MOMENTUM

4

THE FIRST CONSERVATION LAW
Mass, Momentum, and Rocketry

Robert Goddard at the historic launch of the first liquid-fueled rocket, on March 16, 1926. The rocket rose to a height of 41 ft, and landed 2.5 s later in a cabbage patch 184 ft from the launch site. (Photo courtesy of NASA, Washington, DC.)

Every vision is a joke until the first man accomplishes it. After that it becomes commonplace.

Robert Goddard (1920)

third law follows from it rather than the other way around. If you were to meet extraterrestrial physicists, they would more likely be familiar with the principle of momentum conservation than with the concept of *force*.

In our idealized two-cart experiments, we carefully leveled the track and eliminated friction to prevent anything other than the interaction between the carts from significantly influencing their motion. In this sense, the motion of the two-cart system was *isolated* from outside influences. Only for isolated systems is momentum conserved.

EXERCISE 4.3

A 5.0 g bullet moving at 300 m/s passes through a 500 g block of wood that was initially at rest, causing the block to slide over a horizontal, frictionless surface. The bullet emerges from the block at 100 m/s. What is the final speed v of the block?

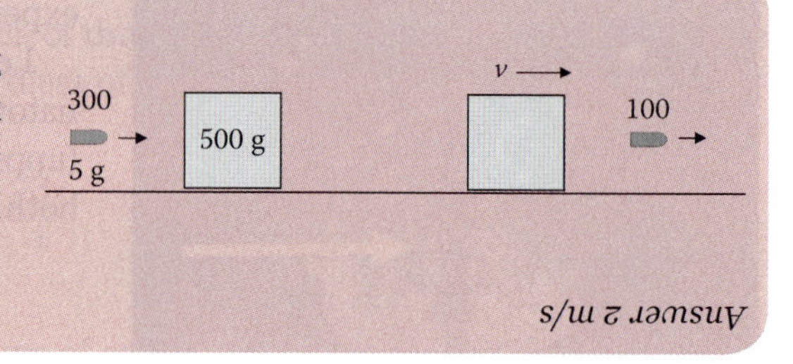

Answer 2 m/s

4.3 ELASTIC COLLISIONS AND ENERGY

There is another important result that holds only for elastic collisions. Rearranging Equation 4.3, we find that

$$m_1(v_1 - v_1') = m_2(v_2' - v_2).$$

But for an elastic collision, Equation 4.1 tells us that $v_2' - v_1' = -(v_2 - v_1)$, or $v_1 + v_1' = v_2' + v_2$, so

$$m_1(v_1 - v_1')(v_1 + v_1') = m_2(v_2' - v_2)(v_2' + v_2).$$

Clearing the parentheses and multiplying both sides by 1/2, we obtain

$$\frac{1}{2}m_1v_1^2 + \frac{1}{2}m_2v_2^2 = \frac{1}{2}m_1v_1'^2 + \frac{1}{2}m_2v_2'^2. \tag{4.5}$$

Defining a body's *kinetic energy* to be $K = (1/2)mv^2$, Equation 4.5 can be rewritten as

$$K_{tot} = K_1 + K_2 = K_1' + K_2' = K_{tot}'.$$

The total kinetic energy is the same before and after an elastic collision. But note well: while Equations 4.3 and 4.4 are *always* true, Equation 4.5 holds only for the special case of an elastic collision. You have surely heard of the law of conservation of energy. That law is much more subtle and far reaching than what we have derived here. Conservation of energy will be developed with care in Chapter 9.

EXERCISE 4.4

Two carts collide while rolling without friction on a horizontal track. Before the collision, cart 1 of mass $m_1 = 2$ kg is rolling with velocity $v_1 = 3$ m/s; cart 2 ($m_2 = 1$ kg) is initially stationary. After the collision, $v_2' = 4$ m/s.

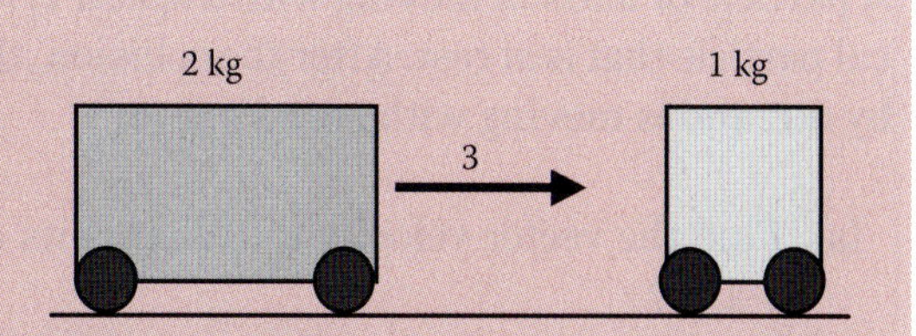

(a) Find v_1'.
(b) Is the collision elastic? Show this in two ways.

Answer (a) 1 m/s (b) Yes $v'_{rel} = -v_{rel}$ and $K'_{tot} = K_{tot}$

4.4 THREE-DIMENSIONAL COLLISIONS

When studying collisions in more than one dimension, we must use vectors to express velocities and momenta: $\vec{p} = m\vec{v}$. Conservation of momentum is then expressed by the vector equivalents of Equations 4.3 and 4.4:

$$m_1\vec{v}_1 + m_2\vec{v}_2 = m_1\vec{v}_1' + m_2\vec{v}_2', \tag{4.6}$$

$$\vec{p}_{tot} = \vec{p}_1 + \vec{p}_2 = \vec{p}_1' + \vec{p}_2' = \vec{p}_{tot}'. \tag{4.7}$$

When applying Equation 4.6 or 4.7, remember that the momenta are to be added vectorially. Each of the above vector equations is shorthand for three separate expressions of conservation, all of which must be satisfied separately. For example, Equation 4.6 is equivalent to

$$m_1 v_{1x} + m_2 v_{2x} = m_1 v_{1x}' + m_2 v_{2x}',$$
$$m_1 v_{1y} + m_2 v_{2y} = m_1 v_{1y}' + m_2 v_{2y}',$$
$$m_1 v_{1z} + m_2 v_{2z} = m_1 v_{1z}' + m_2 v_{2z}',$$

and likewise for Equation 4.7. See Exercise 4.5.

Let's now specialize to the case of elastic collisions. Just as in the one-dimensional case, an elastic collision in three dimensions is defined as one where the relative speed of approach between the two bodies is equal to the relative speed of separation after the collision. In three-dimensions, this is expressed as

$$|\vec{v}_r'| = |\vec{v}_r| \quad \text{or} \quad |\vec{v}_2' - \vec{v}_1'| = |\vec{v}_2 - \vec{v}_1|. \tag{4.8}$$

EXERCISE 4.7

Consider a *head-on* elastic collision between two identical spherical bodies, one of which is initially at rest. Symmetry requires that neither body will be deflected sideways by the collision, so $\vec{v}_2'$ must be parallel to $\vec{v}_1$. The preceding analysis requires $\vec{p}_1' \cdot \vec{p}_2' = 0$. How is this possible?

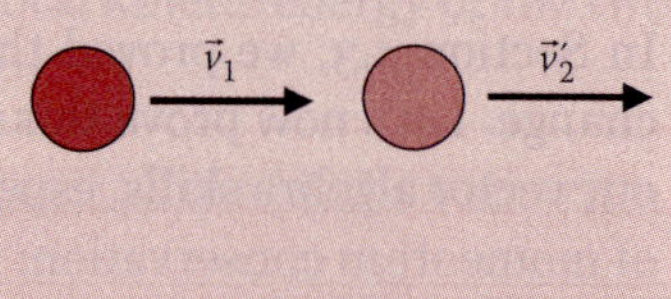

Answer: $\vec{v}_1' = 0$, $\vec{v}_2' = \vec{v}_1$

What if there are more than two interacting bodies in the system? For an N-body system, where the number of bodies N can be any number (N = 2, 3, 4, ..., 10^{23}, ...), the total momentum of the system is $\vec{p}_{tot} = \vec{p}_1 + \vec{p}_2 + \cdots + \vec{p}_N$. No matter how many bodies there are, or how many times they collide with one another, momentum is conserved in each collision, so the total momentum of the system remains constant, just as in the simpler cases described earlier. Remember, though, that momentum conservation holds only for *isolated* systems: bodies that are *not* included in the system we are studying must not influence the bodies within the system in any way.

4.5 MOMENTUM CONSERVATION AND THE PRINCIPLE OF RELATIVITY

Galileo was perhaps the first to understand the full significance of the relativity of motion. In his daring defense of Copernicus,* he envisioned a passenger confined to a windowless cabin within a large ship. There was no experiment, he argued, the passenger could perform that could sense whether the ship was at rest or *moving with constant velocity* relative to the sea. A tossed ball, a swinging pendulum, a collision between two bodies—all would behave *normally* according to the passenger, whether the ship were moving or not.† Now imagine two identical ships moving with constant velocity relative to one another. Each vessel carries an inquisitive passenger confined to a windowless cabin. Whatever experiments the passengers perform, the balls, pendula, or colliding bodies they study will behave in exactly the same way, regardless of the ships' motion. This observation has profound significance for physical theory. If motion is governed by physical law, then the *laws* of motion must be the same for observers moving with constant velocity relative to one another. This is called the *principle of Galilean relativity*. Here is a simple example.

The law of inertia (better known as Newton's first law) asserts that *an isolated body moving with velocity $\vec{v}$ will remain moving with constant velocity $\vec{v}$ indefinitely*. Suppose that two observers, Alice and Bob, are watching a ball roll along a horizontal track. Alice notes that the ball is moving with constant velocity $\vec{v}_A$, as predicted by the law of inertia. Bob, who is moving with velocity $\vec{V}$ relative to Alice, determines the ball's velocity to be $\vec{v}_B = \vec{v}_A - \vec{V}$. Although Alice and Bob measure different velocities, they agree that the ball is moving with constant velocity and conclude that the law of inertia obeys the principle of relativity. Any observer who finds that the law of inertia is obeyed is said to be in an *inertial reference frame*. Alice's and Bob's reference frames are both inertial and so is any other reference frame moving with constant

* Galileo Galilei, *Dialogues Concerning the Two Chief World Systems* (1632).

† Today, it might be easier to imagine traveling in a plane with all of the window shades pulled down.

velocity relative to them. In the language of reference frames, the principle of Galilean relativity is this: *the laws of physics are the same in all inertial reference frames.*

If conservation of momentum is a valid law of physics, it too must obey the principle of relativity. Imagine our same observers studying a two-body collision. Before the collision, Alice measures initial masses m_1 and m_2 and initial velocities $\vec{v}_1$ and $\vec{v}_2$. After the collision, she measures masses[*] m_1' and m_2' and velocities $\vec{v}_1'$ and $\vec{v}_2'$. Calculating initial and final momenta, she confirms that momentum is conserved:

$$m_1\vec{v}_1 + m_2\vec{v}_2 - m_1'\vec{v}_1' - m_2'\vec{v}_2' = 0.$$

Bob is moving at constant velocity $\vec{V}$ relative to Alice. He measures velocities $\vec{u}_i$ before the collision and $\vec{u}_i'$ after the collision. While these velocities (and momenta) are not the same as Alice's, he will conclude that momentum is conserved if $m_1\vec{u}_1 + m_2\vec{u}_2 - m_1'\vec{u}_1' - m_2'\vec{u}_2' = 0$. Assuming $\vec{u}_1 = \vec{v}_1 - \vec{V}$, etc., momentum is conserved in Bob's reference frame if

$$\left(m_1\vec{v}_1 + m_2\vec{v}_2 - m_1'\vec{v}_1' - m_2'\vec{v}'\right) - \left(m_1 + m_2 - m_1' - m_2'\right)\vec{V} = 0. \tag{4.10}$$

Based on Alice's measurements the left hand term equals zero. Bob will conclude that momentum is conserved if the right hand term equals zero, that is, if *total mass is conserved.* This makes perfect sense for ordinary collisions between lab-scale bodies, but quite commonly, mass is *not* conserved in subatomic processes. (A good example is shown in Figure 4.2b, where $\pi^- + p \rightarrow \pi^- + p + \pi^0$.) A second, equally disturbing concern arises when velocities become comparable with the speed of light, that is, when they are *relativistic.* At high speeds, the simple relation $\vec{u}_i = \vec{v}_i - \vec{V}$ between Alice's and Bob's measured velocities breaks down, further threatening the law of momentum conservation. Does this mean that the law—as expressed earlier—is not universally true?

The trouble stems from the laws of electricity and magnetism, called Maxwell's Equations.[†] Galileo and Newton had no knowledge of these laws, but Einstein certainly did, and he insisted that they must also comply with the principle of relativity. Maxwell's Equations predict that all electromagnetic waves travel through a vacuum with the same speed c. Einstein added that the principle of relativity requires that *all observers measure the same speed of light,* regardless of their motion relative to the light source or their motion relative to each other. For example, if Alice and Bob were to measure the speed of a passing light pulse (rather than a rolling ball), each would measure the same speed c independent of their relative velocity $\vec{V}$. The constancy of c is the starting point for Einstein's theory of special relativity, and it has been well-verified by over 100 years of painstaking experiment.[‡]

---

* For generality, we'll allow the masses to change. Perhaps a chunk of m_1 breaks off and becomes attached to m_2.

† First stated by James Clerk Maxwell in 1862.

‡ The opposing theory held that c was to be measured relative to an undetectable medium—*ether*—that permeated all space. This view was widely accepted in the latter part of the nineteenth century. The American physicists Albert Michelson and Edward Morley offered strong experimental evidence refuting the ether theory in 1887.

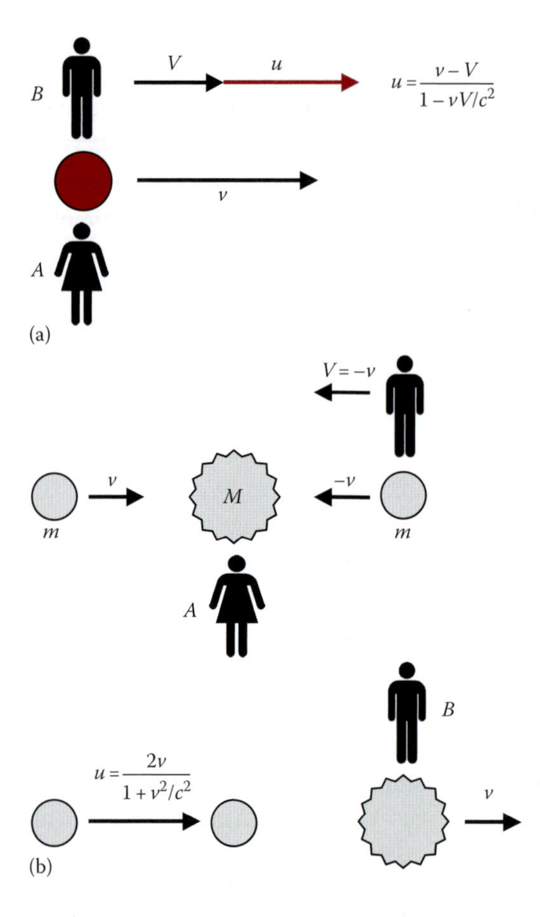

FIGURE 4.5 (a) The transformation of parallel velocities according to special relativity. The lengths of the vectors are approximately correct for $v = c/2$. (b) A totally inelastic collision between two identical bodies. Observer A sees a symmetric collision with $p_{tot} = 0$. In B's reference frame, the initial momentum is smaller than the final momentum.

Figure 4.5a shows a body moving with velocity $\vec{v}$ relative to Alice. In this one-dimensional example, $\vec{v}$ is *parallel* to $\vec{V}$, the velocity of Bob relative to Alice. According to special relativity, the velocity $\vec{u}$ measured by Bob is also parallel to $\vec{V}$ with magnitude (see Box 4.2 for a straightforward derivation)

$$u = \frac{v - V}{1 - (vV/c^2)}, \tag{4.11}$$

where we have suppressed vector notation. (Note that, for ordinary velocities, where $v, V \ll c$, we recover the familiar result $u = v - V$.) At relativistic speeds, even if total mass is conserved, Equation 4.11 spells trouble for the law of momentum conservation.

For example, consider the totally inelastic collision shown in Figure 4.5b. Two identical bodies ($m_1 = m_2 = m$) collide and—assuming conservation of mass—combine to form a new body of mass $M = 2m$.

BOX 4.2 RELATIVISTIC ADDITION OF VELOCITIES

The constancy of the speed of light forces us to re-derive how velocities transform between different reference frames. Imagine an experiment* that takes place on a train traveling with velocity V (see Figure B4.1a). A subatomic particle and a pulse of light are emitted simultaneously from the rear of a passenger car toward the front of the car. An observer, Alice (A), is standing alongside the railroad track. She measures the velocity of the particle v (relative to her) and observes that the light pulse travels with velocity c ($v < c$). After a time interval Δt_1, the pulse reaches the front of the car where it is instantly reflected by a mirror. The reflected pulse intercepts the oncoming particle a short time Δt_2 later. Let L be the length of the car and xL be the distance from the front of the car to the spot where the particle and reflected pulse meet. According to Alice, the light pulse travels a distance $c\Delta t_1$ in the

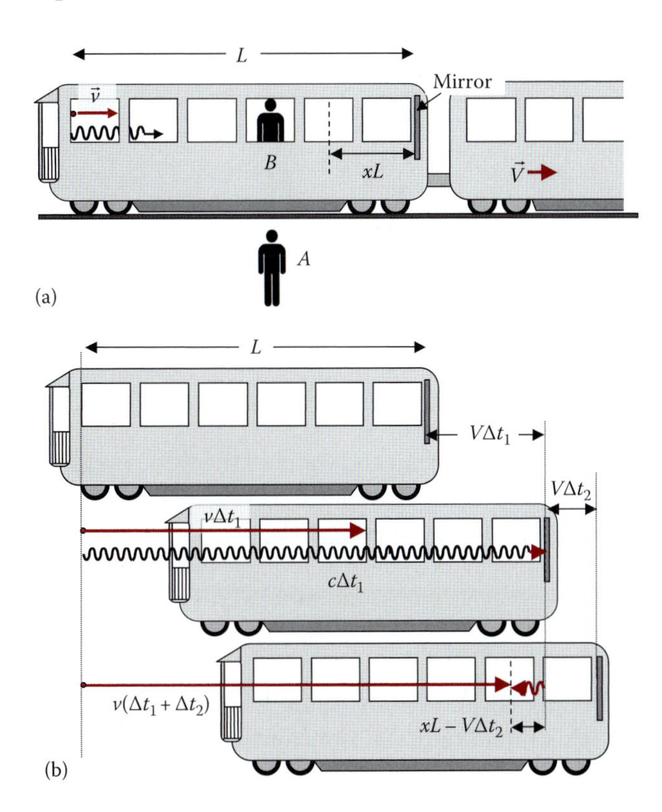

FIGURE B4.1 (a) A passenger train carrying observer B passes a stationary observer A. A light pulse and a high-speed particle are emitted simultaneously from the rear of the car. (b) The light pulse is reflected by the mirror at time Δt_1 and meets the particle Δt_2 later at a point xL from the front of the car. During the intervals Δt_1 and Δt_2, the train advances to the right by $V\Delta t_1$ and $V\Delta t_2$.

* The source of this derivation is N. David Mermin, Relativistic addition of velocities directly from the constancy of the speed of light, *American Journal of Physics* **51**(12), 1130–1131 (1983).

forward direction and $c\Delta t_2$ backwards. The spot where the pulse and particle meet is thus $c(\Delta t_1 - \Delta t_2)$ from the emission point (see Figure B4.1b). Because the particle travels only in the forward direction, it is $v(\Delta t_1 + \Delta t_2)$ from the starting point when it meets the reflected pulse. Equating these two distances, $c(\Delta t_1 - \Delta t_2) = v(\Delta t_1 + \Delta t_2)$, or

$$\frac{\Delta t_2}{\Delta t_1} = \frac{c-v}{c+v}.$$

During the time interval Δt_1, the train moves forward by $V\Delta t_1$, so the light pulse must travel $L + V\Delta t_1$ before striking the mirror: $c\Delta t_1 = L + V\Delta t_1$, or $\Delta t_1 = L/(c - V)$. Similarly, during the interval Δt_2, the train travels $V\Delta t_2$ farther while the pulse travels *backwards*, so the distance between the reflection point and the interception point is $c\Delta t_2 = xL - V\Delta t_2$, or $\Delta t_2 = xL/(c + V)$. Taking the ratio of time intervals,

$$\frac{\Delta t_2}{\Delta t_1} = x\frac{c-V}{c+V}.$$

Equating the two expressions for $\Delta t_2/\Delta t_1$ and solving for x,

$$x = \frac{(c+V)(c-v)}{(c-V)(c+v)}. \tag{B4.3}$$

A second observer, Bob (B), is riding on the train and watches the same events. In his reference frame the particle has a velocity u, while once again the light pulse travels with speed c. Using the same reasoning as mentioned earlier, Bob determines the spot x' where the pulse and particle meet:

$$x' = \frac{c-u}{c+u}. \tag{B4.4}$$

This follows because Equation B4.3 is valid for any value $V < v$. The train is at rest relative to Bob; setting $V = 0$ in Equation B4.3 yields Equation B4.4. Now for the important point: Suppose pulse-particle pairs are emitted at regular intervals, while a detector is moved along the aisle until it finds the exact spot where particles and pulses arrive simultaneously. Let's say that this happens when the apparatus is adjacent to the second row of seats. Alice and Bob will agree about the detector's location within the car, so $x' = x$. Eliminating x and x' in Equations B4.3 and B4.4,

$$\frac{c-u}{c+u} = \frac{(c+V)(c-v)}{(c-V)(c+v)}.$$

Solving for u (after some admittedly messy algebra), we obtain Equation 4.11:

$$u = \frac{v - V}{1 - (vV/c^2)}.$$

This is the way *parallel* velocities transform in special relativity. Perpendicular velocities (i.e., perpendicular to the relative velocity between observers A and B) transform in a distinctly different way.

Exercise B4.1: A body is moving relative to observer A with velocity $v_A = (2/3)c$. Observer B is moving in the opposite direction with velocity $V = -(2/3)c$ relative to A.

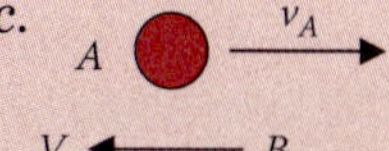

(a) What does B measure for the ball's velocity?
(b) Now replace the body by a light pulse moving to the right with speed c. What does B measure for the speed of the light pulse?

Answer: (a) $v_B = 12/13c$ (b) c

According to Alice, the two bodies initially have opposite velocities $v_2 = -v_1$, so after the collision (by symmetry) M must be stationary. Bob, who is moving with velocity $V = -v$ relative to Alice, measures initial velocities $u_2 = 0$ and, using Equation 4.11, $u_1 = 2v/(1 + v^2/c^2)$. Since M is at rest relative to Alice, it is moving with speed $V = v$ relative to Bob. Alice claims that momentum is conserved, since $p_{tot} = p'_{tot} = 0$. Bob measures the total initial momentum to be $2mv/(1 + v^2/c^2)$, while the total final momentum is $2mv$. For $v \ll c$, he will agree with Alice that momentum is conserved. But as $v \to c$, Alice will maintain that momentum is conserved while Bob will insist that it is not. Since the two observers are in inertial reference frames, the conservation law—as we have presented it above—does not satisfy the principle of relativity, and so it cannot be a valid law of nature.

Einstein rescued the law of momentum conservation by adding a speed-dependent factor to the definition of momentum, $\vec{p} = \gamma(v)m\vec{v}$, where $\gamma(v) = 1/\sqrt{1 - v^2/c^2}$ is called the *Lorentz factor*. (For $v \ll c$, $\gamma \to 1$ and the familiar expression for momentum is recovered.) With this modification, the law of momentum conservation is obeyed for all velocities in all inertial reference frames. We will have much more to say about inertial reference frames and relativistic momentum and energy in later chapters. But for now, we want to stress that Einstein's theory of Relativity is not so much a *replacement*, or a *rejection*, of Newtonian physics; rather, it is a bold extension of Newton's laws necessitated by the laws of electromagnetism and principle of relativity first stated by Galileo in the early seventeenth century.

4.6 APPLICATION: ROCKET PROPULSION

Rocket propulsion is a fascinating but often misunderstood phenomenon. The underlying physics is easily explained by invoking momentum conservation. Imagine a girl (mass m) wearing roller skates who is standing still on a horizontal sidewalk. In each hand she is holding a heavy ball (each of mass $\Delta m/2$). Since

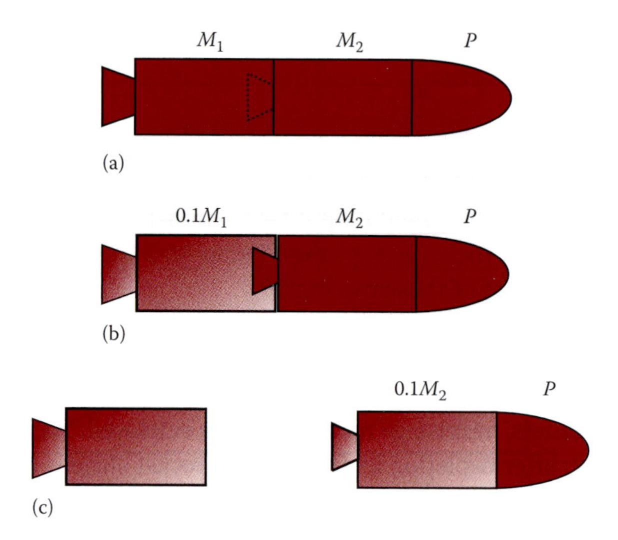

FIGURE 4.9 A two-stage rocket. (a) Prior to ignition, the total mass is $M_1 + M_2 + P$. (b) When the first stage's fuel is depleted, the rocket's mass is $0.1M_1 + M_2 + P$. The stages now separate. (c) At ignition of the second stage, the rocket's mass is $M_2 + P$. When the second stage fuel is depleted, the rocket's mass is $0.1M_2 + P$.

To appreciate the advantage of this arrangement, consider the two-stage rocket shown in Figure 4.9. For simplicity, assume that the two stages have identical mass $M_1 = M_2$ and exhaust velocity. Let's also say that when stage 1 (stage 2) has depleted its fuel, its mass is $0.1M_1(0.1M_2)$, about the best we can hope for. At liftoff, the total mass of the rocket, including the payload P, is $M_0 = M_1 + M_2 + P$. Let the payload mass be $P = 0.04M_0$ (so $M_1 = M_2 = 0.48M_0$). Just before stage 1 separates, the rocket's mass is $0.1M_1 + M_2 + P = 0.568M_0$; just after separation it is $M_2 + P = 0.520M_0$. Later, when the second stage has depleted its fuel, the remaining mass of the rocket is $0.1M_2 + P = 0.088M_0$. Using Equation 4.15 to calculate the velocity contributed by each stage we find

$$v_f = \Delta v_1 + \Delta v_2 = v_{ex}\ln\frac{M_0}{0.1M_1 + M_2 + P} + v_{ex}\ln\frac{M_2 + P}{0.1M_2 + P} = v_{ex}\left(\ln\frac{1}{0.568} + \ln\frac{0.52}{0.088}\right)$$
$$= (0.57 + 1.77)v_{ex} = 2.34v_{ex}.$$

Compare this to a single stage rocket with the same parameters, that is, 90% of M_1 is allocated to fuel. In this case, $M_1 = 0.96M_0$ and $M_f = 0.1M_1 + P = 0.136M_0$, so $v_f = v_{ex}\ln(1/0.136) = 2.0v_{ex}$. The two stage rocket delivers a 17% increase in the final velocity.

But we can do even better. Figure 4.10 is a plot of the final velocity as a function of how the initial rocket mass is partitioned between the two stages. If $M_1 = 0$, then $\Delta v_1 = 0$ and the rocket is simply a single-stage vehicle with the final velocity given above: $v_f = \Delta v_2 = 2.0v_{ex}$. As M_1 is increased (at the expense of M_2) $\Delta v_1/\Delta v_2$ increases and so does the final velocity, reaching a peak value $v_f = 2.55v_{ex}$ when $\Delta v_1/\Delta v_2 = 1$. This a 27% improvement over the single-stage rocket.

Remarkably, this last result has a general validity. Given a rocket with N stages, all having the same fuel-to-total-mass ratio and the same exhaust velocity, the largest final velocity will be attained when each stage contributes an equal velocity boost, that is, $\Delta v_1 = \Delta v_2 = \cdots = \Delta v_N$. In practice, this is difficult to achieve

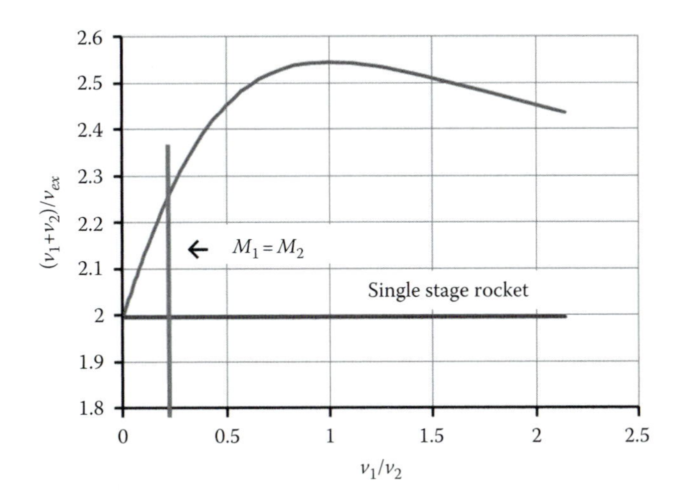

FIGURE 4.10 The curve shows the final speed of a two-stage rocket, as a function of the ratio of velocity boosts due to each stage, also shown are the final velocity of a single-stage rocket with the same initial mass and amount of fuel as the two-stage rocket (horizontal line); and the final rocket speed when the two stages initially have equal mass (vertical line).

because the lower stages must be more robust to support the weight of the upper stages. Nevertheless, real multistage rockets are proportioned with the equal velocity rule in mind.

EXERCISE 4.13

A two-stage rocket is used to lift a satellite of mass P = 1000 kg into low Earth orbit. Each stage is propelled by reacting a mixture of liquified O_2 and H_2 (LOX/LH2) to produce high temperature water vapor with an exhaust velocity of 4.0 km/s. To achieve orbit, the final speed of the rocket must be 8 km/s. If the fuel-to-total-mass ratio of each stage is 7/8, what are the minimum masses of each stage? (*Hint:* For minimum mass, choose $\Delta v_1 = \Delta v_2$ = 4 km/s. Start by calculating the second stage mass.)

Answer $M_2 = 2.6P$, $M_1 = 2.6(M_2 + P) = 9.4P$, M_1 is significantly larger than M_2

4.8 SUMMARY

In this chapter, we have seen how simple collision experiments led seventeenth century scientists to discover the law of conservation of momentum. Since then, physicists have amassed sufficient evidence to be convinced that the law is valid for all isolated systems irrespective of their physical size, the number of bodies within them, or the type of interactions that take place between the bodies (i.e., electric, magnetic, gravitational, nuclear, or otherwise.)* To conform to the principle of relativity, the definition of momentum $\vec{p} = m\vec{v}$ must be modified to include the Lorentz factor: $\vec{p} = \gamma m\vec{v}$. The distinction becomes important when $v \to c$. When an apparent violation of the law is encountered, we *know* that momentum is hiding from us and a

* *Simple* contact forces are actually electric forces.

new particle or other carrier* of momentum is waiting to be discovered. The law of momentum conservation explains other phenomena, like rocket propulsion, which at first glance may seem completely unrelated collisions. We have unshakable faith that the conservation law is valid, not just on Earth, but everywhere in the universe for all time. As you will see, faith in physical law allows us to execute interplanetary travel; discover the properties of moons, planets, stars and galaxies; and investigate the workings of the cosmos.

> "On this day [October 19, 1899] I climbed a tall cherry tree at the back of the barn... as I looked toward the fields at the east, I imagined how wonderful it would be to make some device which had even the possibility of ascending to Mars...
>
> I was a different boy when I descended the tree... Existence at last seemed very purposive [purposeful]."†
>
> **Robert Goddard**

* For example, light (electromagnetic radiation) carries momentum.

† As quoted in M. Lehman, *Robert H. Goddard: A Pioneer of Space Research*, Da Capo Press, New York, 1988.

PROBLEMS

4.1 Complete the following table, where m is in kg and v is in m/s. Each row summarizes the conditions before (unprimed) and after (primed) a collision between masses m_1 and m_2. Each collision is characterized as elastic, totally inelastic, or neither.

m_1	m_2	v_1	v_2	v_1'	v_2'	
3	2	+4	−2			Inelastic
2	4	+4	0		+1	Neither
3	2			−1	+4	Elastic

4.2 Two bodies m_1 and m_2, move without friction along a long straight track and collide with each other. Their velocities before and after the collision are given below in m/s.

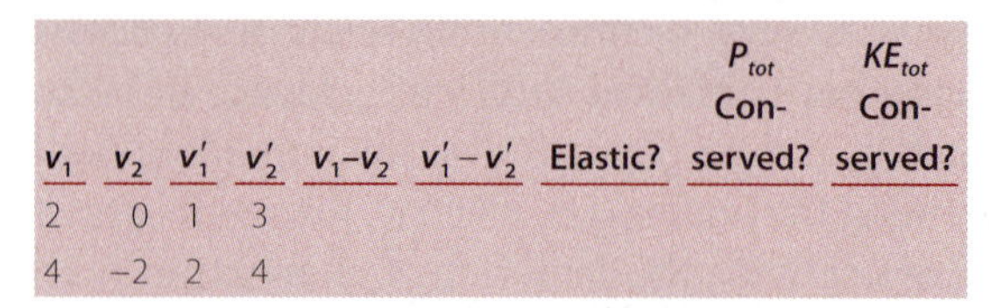

v_1	v_2	v_1'	v_2'	v_1-v_2	$v_1'-v_2'$	Elastic?	P_{tot} Conserved?	KE_{tot} Conserved?
2	0	1	3					
4	−2	2	4					

(a) Complete the table above.

(b) Find the ratio of the masses: $m_1/m_2 = ?$

4.3 The blocks shown in the following figure slide without friction.

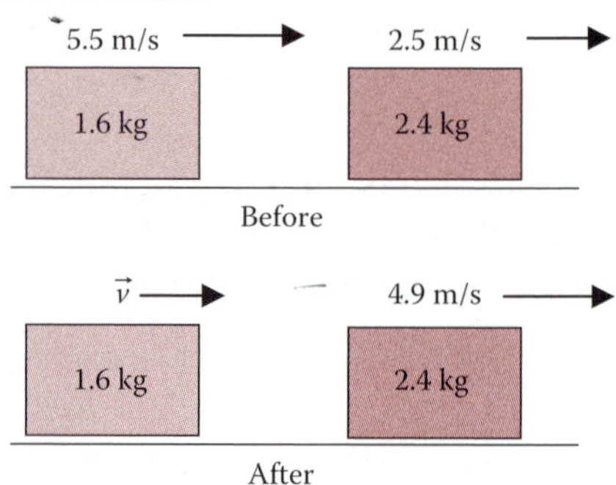

(a) What is the velocity of the 1.6 kg block after the collision?

(b) Is the collision elastic?

4.4 Masses m_1 and m_2 move without friction along a horizontal straight track. They collide, and their velocities before and after the collision are recorded in the table.

Before Collision		After Collision	
v_1 (m/s)	v_2 (m/s)	v_1' (m/s)	v_2' (m/s)
5	−4	−1	6

(a) If, at $t = 0$, the two bodies are 10 m apart, at what time do they collide?

(b) Is the collision elastic?

(c) What is the ratio of the masses m_2/m_1? Support your answer with a calculation.

(d) Which of the following is(are) true for the above collision?

(i) Momentum is conserved.

(ii) Kinetic energy is conserved.

(iii) The magnitude of the relative velocity is the same before and after the collision.

(e) Imagine a different collision between the same two bodies, with $v_1 = 5$ m/s but $v_2 = 0$. Using the mass ratio you calculated above, is it possible to have a collision such that $v_1' = 0$? Support your answer mathematically.

4.5 A ball of mass $m_2 = 0.4$ kg is at rest on a horizontal table. A second ball of mass m, initially moving with a speed 1.0 m/s, strikes m_2 head on. After the collision, the balls stick together and slide at a speed of 0.6 m/s. Neglect friction.

(a) What is the mass m of the second ball?

In a second collision (shown in the following figure), a ball of mass $m_1 = m_2$, initially moving at a speed of 1.0 m/s, strikes m_2 (initially at rest). After the collision, m_1 moves with a speed of 0.5 m/s at an angle of 60° relative to its initial direction.

(b) Find the final velocity of m_2.

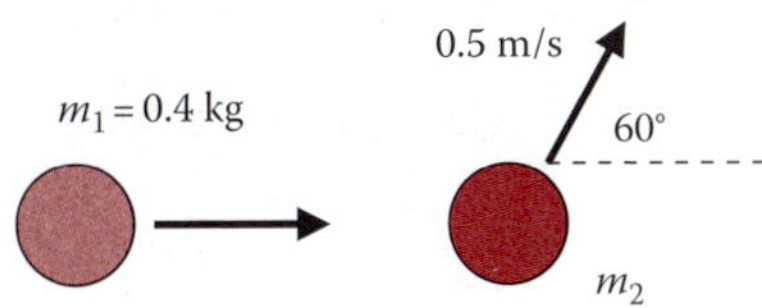

(c) Is this second collision elastic? Prove your answer.

(d) For the second collision, show that $\vec{v}_2' \cdot \vec{v}_1' = 0$.

4.6 A nucleus of mass $M = 200u$ is moving at a speed $v = 1.00 \times 10^7$ m/s in the direction shown when it fissions (splits) into two smaller nuclei of mass $m_1 = 83u$ and $m_2 = 117u$. The two fission fragments travel at speeds v_1 and v_2 in the directions shown. Assume $v \ll c$ so you can ignore relativistic effects. (*Note:* A proton has a mass of about $1.007u$.)

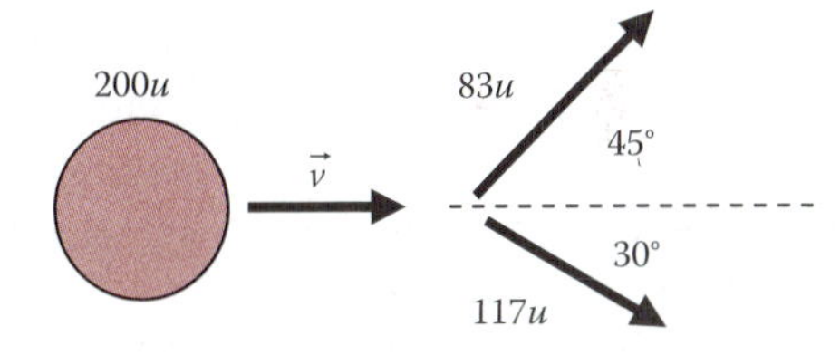

(a) Which of the following are true statements?

(i) Momentum is conserved.

(ii) Kinetic energy is the same before and after the fission.

(iii) The relative velocity of m_1 and m_2 must be equal in magnitude to v.

(b) If the speed of m_1 is $v_1 = 1.25 \times 10^7$ m/s, find the speed v_2 of m_2.

4.7 An alpha particle ($m = 4u$) moving at 2.0×10^7 m/s collides with the nucleus of a gold atom ($M = 197u$) at rest. Find the recoil speed of the gold nucleus if

(a) The α rebounds from the collision moving in the opposite direction with (essentially) its initial speed.

(b) The α particle rebounds from the collision moving in a direction 30° from its original direction with its initial speed.

4.8 A car of mass 2000 kg traveling south collides with a truck of mass 6000 kg traveling west. The speed limit on both roads is 40 mph and the car failed to stop at a traffic light. After the collision the vehicles lock together and skid along a line in the southwest direction. Using Doppler radar, a police officer traveling behind the truck confirms that the truck was traveling with a speed of about 30 mph. Aside from giving the car driver a ticket for running a red light, should the police officer issue a speeding ticket?

4.9 Consider a collision between two identical particles ($m_1 = m_2 = m$), one of which (m_2) is initially at rest. These *particles* might represent marbles, billiard balls, atoms, nuclei, or even stars and galaxies. After the collision, particle 1 is deflected by 45° and is moving with speed $v_1' = (1/2)v_1$.

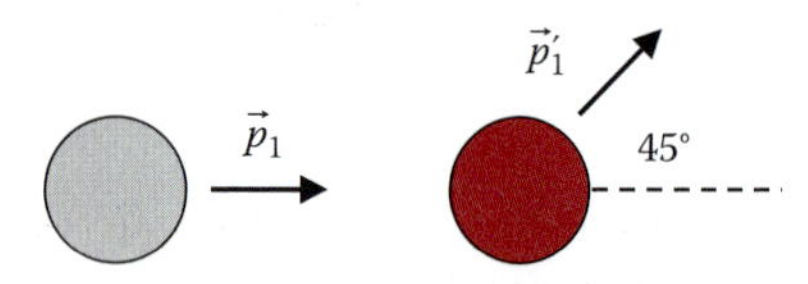

(a) Find the speed v_2' and direction of particle 2 after the collision. (Express the speed v_2' as some fraction of the initial speed v_1.)

(b) Show that this collision is inelastic.

(c) If instead the balls collide *elastically*, and m_1 is again deflected by 45°, what must be the final speed v_1'?

4.10 A certain nucleus, at rest, spontaneously decays into three particles. Two of them are detected and their masses and velocities are shown in the following figure. What is the velocity of the third particle, which is known to have a mass of 7500 MeV?

(*Note:* In this problem, mass is expressed in terms of its equivalent amount of energy using Einstein's famous equation, $E = Mc^2$. The mass of a proton is about 938 MeV. In the atomic and subatomic worlds, a convenient unit of energy is the electron volt [eV]. 1 eV = 1.6×10^{-19} J, 1 MeV is therefore 1.6×10^{-13} J.)

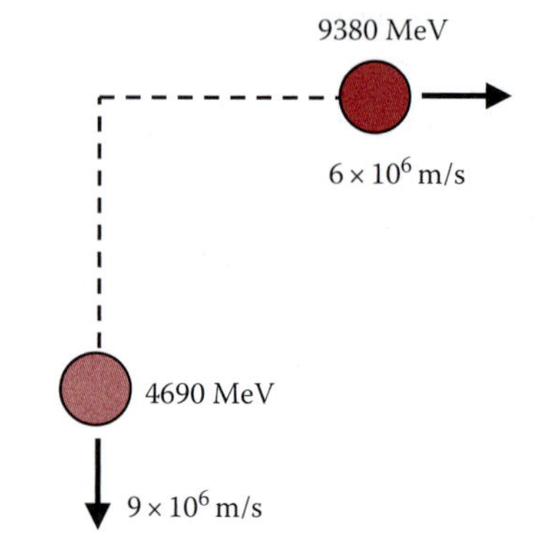

4.11 A projectile of mass m moving with a speed of 10.0 m/s strikes a stationary target of the same mass. After the collision, the projectile is deflected by 60° and is moving with a speed of 5.00 m/s.

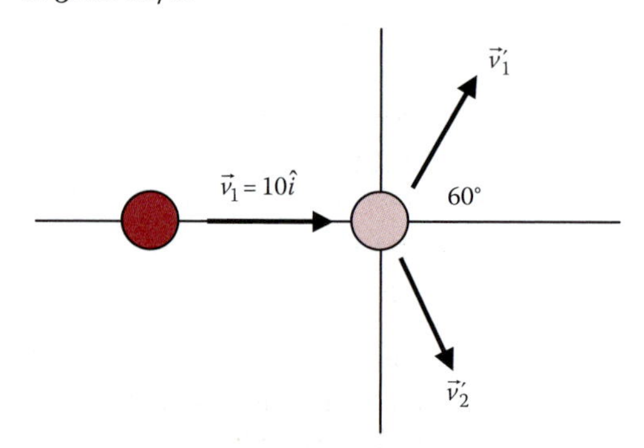

(a) Write the vector velocity $\vec{v}_1'$ of the projectile after the collision.

(b) Find the vector velocity $\vec{v}_2'$.

(c) What is the direction of $\vec{v}_2'$?

(d) Find the relative velocity between the particles before ($\vec{v}_{21}$) and after ($\vec{v}_{21}'$) the collision. Is the collision elastic?

4.12 In orbit, the space shuttle has a mass $M = 6.9 \times 10^4$ kg. To prepare for reentry into Earth's atmosphere, the shuttle *deorbits* by first slowing itself from its orbital speed of 7780 to 7700 m/s. If the rocket exhaust speed is 4400 m/s, what is the mass of the fuel that must be *burned* in this maneuver? (What is the simplest way to do this problem?)

4.13 A 6000 kg spacecraft heading toward Jupiter at 100 m/s relative to the Sun speeds up by firing its rocket engine, ejecting 100 kg of exhaust at a speed of 2500 m/s relative to the spaceprobe. What is the final speed of the craft? (Treat this as a single short burn and then as a continuous burn—your two answers should not be significantly different.)

4.14 In the movie *Armageddon*, a giant asteroid *the size of Texas* is on course for a head on collision with Earth. The mass of the asteroid is 6×10^{20} kg, and it is moving toward Earth with a speed of 10 km/s relative to the Sun. Earth's mass is 6×10^{24} kg, and its orbital speed around the Sun is 30 km/s. You are a member of the NASA team assigned to assess the consequences of the collision.

(a) If the collision with Earth is fully inelastic, what will be the fractional change in the speed of Earth?

(b) For a fully inelastic collision, how much kinetic energy will be converted into heat? Compare this to the energy released by a large nuclear weapon (3.5 Mton $\approx 10^{16}$ J).

4.15 A spacecraft is in deep space, far from the gravitational influence of other bodies. It is initially at rest relative to you. The spacecraft carries a payload of 1000 kg and a fuel pack of mass 100 kg. On your command, the craft's rocket engine is ignited and all of the fuel is exhausted. The payload acquires a velocity of 200 m/s.

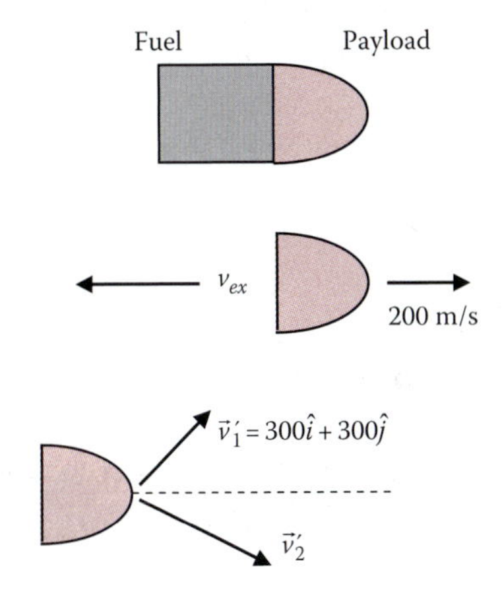

(a) Is momentum conserved when the rocket engine is ignited? Explain carefully.

(b) What is the exhaust speed (relative to the spacecraft)?

(c) A malfunction occurs and an explosion splits the payload into two fragments of *equal mass*. The final velocity of one fragment is $300\hat{i} + 100\hat{j}$ m/s. Find the velocity of the second fragment.

4.16 An 8000 kg spacecraft is equipped with two rocket engines mounted at 90° with respect to each other. The exhaust speed from either engine is 4.0 km/s. Initially the spacecraft is at rest in deep space. Both engines are ignited simultaneously; engine X burns 10 kg of fuel and engine Y burns 20 kg before shutting off.

(a) Three rockets of initial mass M are fired in unison.

(b) First two rockets are fired in unison; then they are jettisoned and the third is fired.

(c) Three rockets are fired in sequence with each spent rocket dropping off before the next one is fired.

(d) One larger rocket of initial mass $3M$ is fired for three times the time of the smaller rockets used.

Assume that the residual mass of each rocket is 10% of its initial mass and the exhaust speeds of all rockets are the same. Rank the four schemes in terms of the final velocity of the payload.

4.24 *Oh no!* You are an astronaut performing an EVA (extravehicular activity, aka spacewalk) outside the space shuttle. You weren't paying attention and have drifted away from the shuttle. When you notice this, the shuttle is at a position $\vec{r}_{so} = 300\hat{i} - 100\hat{j}$ m relative to you and is moving with a velocity $\vec{v}_s = 4\hat{i} + 2\hat{j}$ m/s, as shown in the figure. You have a *jet pack* on your spacesuit, which contains 2 kg of compressed CO_2 gas, which, when released through a nozzle, has a speed of 400 m/s relative to the nozzle. *Can you save yourself?*

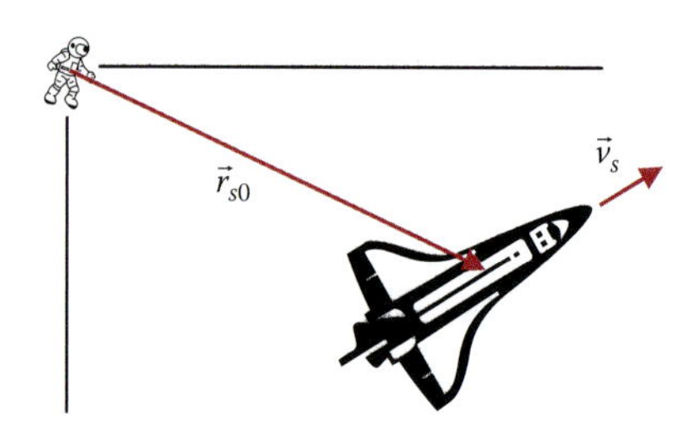

(a) Using the dot product, find the angle θ between $\vec{r}_{so}$ and $\vec{v}_s$.

(b) Write a vector equation for $\vec{r}_s(t)$, the position of the shuttle as a function of time. At what time does the shuttle cross the x-axis?

(c) You activate your jet pack, releasing CO_2 in a very short burst which can be considered instantaneous. What vector velocity must you attain in order to intercept the space shuttle when it crosses the x-axis?

(d) Your total mass with space suit is 90 kg. With the mass of CO_2 you have in your jet pack, can you reach the shuttle as it crosses the axis, or are you stranded in space?

4.25 Another mishap! (Some people never learn!) While on a spacewalk, you've become separated from the space shuttle. It is 30 m away from you and you are moving away from it at a speed of 0.1 m/s. Worse, you forgot your jet pack, but you are carrying a 2 kg hammer (used to repair the shuttle). Your mass with space suit is 80 kg.

(a) Can you save yourself by returning to the shuttle safely? Explain how or why not. Use reasonable numbers. (Example: a professional baseball pitcher can throw a baseball at 44 m/s.)

(b) How long will it take to return to the shuttle, if this is possible?

4.26 Having depleted its dilithium crystals, the starship *Enterprise* is drifting through space with a constant velocity. You are on the rescue ship *Raider*, and are attempting to *rendezvous* with *Enterprise* to resupply the starship with dilithium and essential food. At time t = 0, *Raider* is 1000 km from *Enterprise*, which is moving away from you with velocity $\vec{v}_{rel} = 2.0\hat{i}$ km/s (see the following figure).

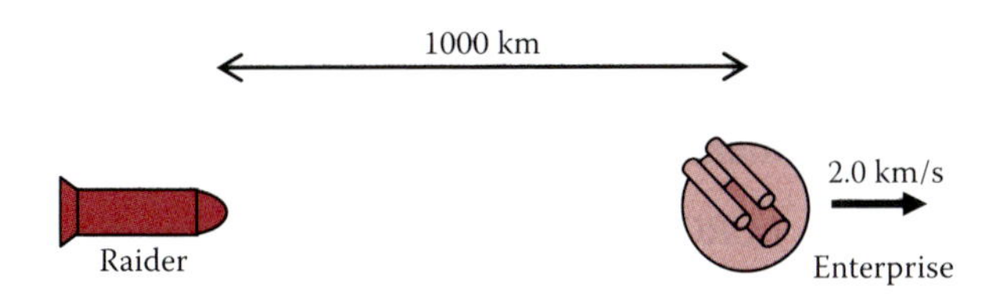

(a) *Raider* has a mass of 1.00×10^4 kg, 60% of which is aluminum + ammonium perchlorate fuel, which has an exhaust speed of 2.70 km/s. What is the minimum time needed for *Raider* to reach the starship?

(b) What is the minimum amount of fuel (kg) needed to eventually reach *Enterprise*?

(c) To avoid crashing into *Enterprise*, the two spacecraft must be traveling at nearly the same speed when they come together. (Note that 0.1 km/s = 100 m/s = 225 mile/h!) This means that you must use your retro rockets and some of your fuel to slow down *Raider* as you approach the star ship. After firing your engines, you are moving at a velocity $2.3\hat{i}$ km/s. Will you have enough fuel left to dock safely?

4.27 You are attempting to dock with the stricken starship *Enterprise*, to resupply its weary crew with desperately needed oxygen, water, food, and DVDs. At the present moment, *Enterprise* is 11,000 km from you and is coasting with a velocity 2.20 km/s in the x-direction (as shown in the following figure). Your rescue vessel *Falcon* has a mass M_i = 10,000 kg, 80% of which is fuel. The exhaust speed of the fuel is 2.50 km/s. To save the crew, you must reach *Enterprise* within 2 h.

(a) You ignite your rocket engine when *Enterprise* is in the position shown ($x = 0$, $y = 11{,}000$ km) and *Falcon* is at ($x = 0$, $y = 0$). You fire the engine for a very short time, and then coast at constant velocity until you reach the starship in 2 h. Write vector equations for the positions of the starship ($\vec{r}_E = ?$) and your rescue vessel ($\vec{r}_F = ?$).

(b) What mass of fuel must be burned to intercept the starship at $t = 2.00$ h?

(c) When you reach *Enterprise*, you must slow down your spacecraft so that your velocity relative to *Enterprise* is ≈0. (*Note:* 0.01 km/s = 22.5 mph = crash landing!). Suppose you reach the starship in 1.5 h burning 70% of your mass (7000 kg) as fuel and you arrive with a relative velocity of 2.00 km/s. Do you have enough remaining fuel to dock safely with *Enterprise*?

(d) If you do not care how long it takes to intercept with *Enterprise*, what is the *minimum* amount of fuel needed to perform this maneuver? Assume the same starting configuration as shown in the figure.

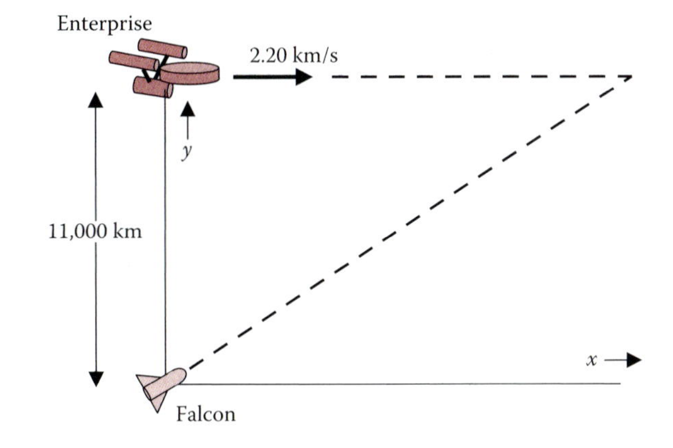

4.28 The 2009 mission of NASA's Lunar Crater Observation and Sensing Satellite (LCROSS) was to search for the presence of water on the Moon. To do this, a two part spacecraft was

The second equation follows because the magnitude of the relative velocity is preserved in an elastic collision. Eliminating mass (using $m_1 = m_2$), and solving each equation for v_1, we find $v_2' - v_1' = v_1' + v_2'$, or $v_1' = 0$. Body 1 comes to rest after the collision ($v_1' = 0$), transferring all of its momentum and kinetic energy to body 2, so $v_2' = v_1$. After the collision, then,

$$V_{CM}' = \frac{(m_1 0 + m_2 v_2')}{(m_1 + m_2)} = \frac{v_2'}{2} = \frac{v_1}{2} = V_{CM}.$$

Consequently, the CM velocity is unchanged, even though the velocity of each body has changed radically.

In Figure 5.2b, the same two bodies experience a totally *in*elastic collision: they stick together after they collide, so $v_1' = v_2'$. To conserve momentum, $m_1 v_1 = (m_1 + m_2) v_1'$, or $v_1' (= v_2') = v_1/2$. Once again, $V_{CM}' = v_1/2 = V_{CM}$; that is, the CM velocity is unchanged from its value before the collision. The constancy of the CM velocity is a consequence of momentum conservation, and it provides a simple strategy for analyzing collisions in one, two, or three dimensions.

EXERCISE 5.3

Two bodies of equal mass undergo a collision, as shown in the following figure. Treat the bodies as point particles, with negligible radii. Before the collision, $\vec{v}_1 = 10\hat{i}$ m/s and $\vec{v}_2 = -10\hat{j}$ m/s.

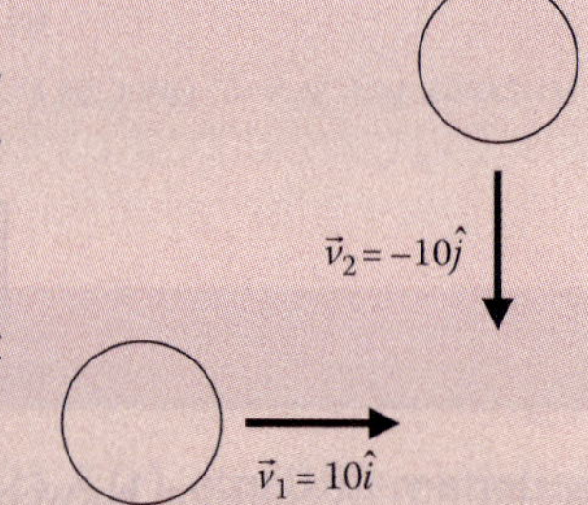

(a) Where is the CM at the time shown in the following figure? In what direction is it moving?
(b) What is $\vec{V}_{CM}$?
(c) Where is the CM 2 s after the collision?

Answer (a) *The CM is located halfway between the two bodies* (b) $5\hat{i} - 5\hat{j}$ *m/s* (c) $10\hat{i} - 10\hat{j}$ *from the collision point*

5.4 USING THE CM TO ANALYZE COLLISIONS

We now know where the center of mass is located and how it moves. Imagine that you are observing a two-body collision while stationed on the CM; that is, imagine that you are witnessing the collision from the *center of mass reference frame*. What would you see? Figure 5.2c and d illustrates the same one-dimensional collisions discussed above, but now shown as they would appear to you while traveling on the moving CM. Figure 5.3 shows a collision in two-dimensions for the case $m_2 = 2m_1$. From Section 5.2, we know that for this two-dimensional collision, (a) the positions of m_1, m_2, and the CM are collinear, i.e., the CM lies along the line joining the two masses. We also know that throughout the collision, (b) the CM is located twice as far from m_1 as from m_2, and (c) the CM moves with constant velocity.

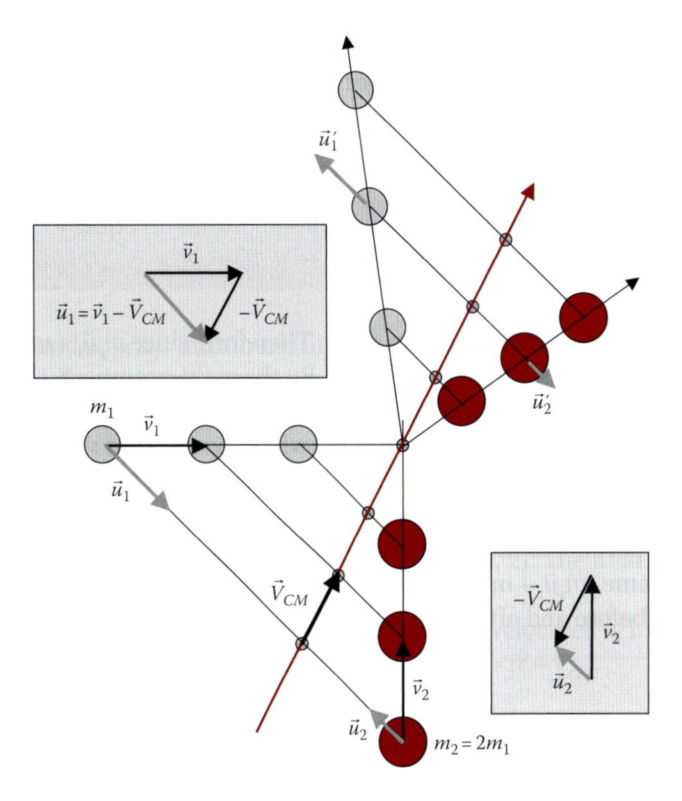

FIGURE 5.3 A two-dimensional collision between two bodies with $m_2 = 2m_1$. Throughout the collision, the CM moves with constant velocity and remains collinear with the two bodies, two-thirds of the distance from m_1 to m_2. Insets: In the CM reference frame, the velocities $\vec{u}_1$ and $\vec{u}_2$ are antiparallel and point directly toward the CM; likewise, $\vec{u}_1'$ and $\vec{u}_2'$ are antiparallel and point directly away from the CM.

EXERCISE 5.4

Verify that these three conditions are rendered correctly in Figure 5.3.

Now study the figure closely. From your observation point on the (moving) CM, the two bodies are traveling *directly toward you*—in opposite directions—before the collision. After the collision, they are moving *directly away from you*, once again in opposite directions. In this sense, collisions viewed in the CM reference frame are remarkably similar to one-dimensional collisions: the two bodies approach each other along a single line of motion, and separate along another (generally different) line of motion. Let's prove this mathematically. Let $\vec{v}_1(\vec{v}_2)$ be the velocity of $m_1(m_2)$ in the lab frame before the collision, and let $\vec{V}_{CM}$ be the velocity of the CM. The initial velocities of the two bodies relative to the CM are $\vec{u}_1 \equiv \vec{v}_1 - \vec{V}_{CM}$, and $\vec{u}_2 \equiv \vec{v}_2 - \vec{V}_{CM}$. The total momentum measured relative to the CM is

$$m_1\vec{u}_1 + m_2\vec{u}_2 = m_1\vec{v}_1 + m_2\vec{v}_2 - (m_1 + m_2)\vec{V}_{CM}.$$

For this idealized 180° maneuver, the planetary fly-by adds twice the planet's speed to the initial speed of the spacecraft. Note that this is exactly what happened during the collision of the superball and basketball, where both balls were moving in opposite directions with speed v at the time of collision.

EXERCISE 5.8

A space probe is executing a close fly-by of Jupiter, approaching the planet in the opposite direction from its orbital velocity about the Sun. Jupiter's orbital speed is $V_J = 13$ km/s, and the speed of the probe is $v_S = 10$ km/s when it is still far enough away from the planet so that gravitational effects are insignificant.

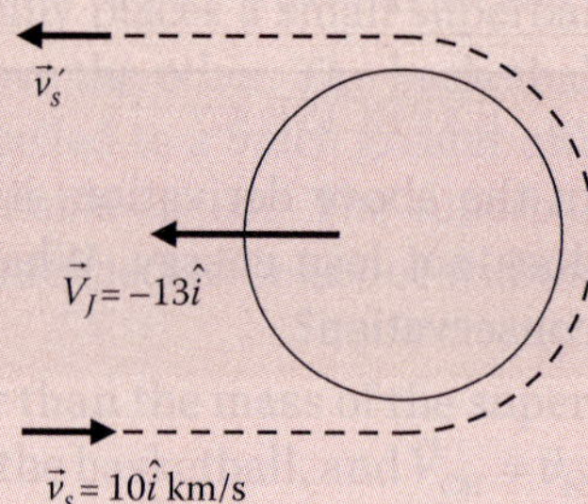

(a) What is the initial velocity $\vec{u}_S$ relative to the CM?
(b) What is the final relative velocity $\vec{u}_S'$?
(c) What is the final speed of the probe relative to the Sun?

Answer: (a) $23\hat{i}$ km/s (b) $-23\hat{i}$ km/s (c) 36 km/s

In Exercise 5.8, the spacecraft's speed (relative to the Sun) increases by 26 km/s, with no expenditure of rocket fuel.* This is a very substantial increase. Recall the discussion of multistage rockets in Chapter 4. Using chemical fuel with an exhaust velocity of 4 km/s, a three stage rocket would be needed to raise the spacecraft's speed by this amount. If the probe's mass is 2,400 kg (the approximate mass of the *Galileo* spacecraft), 2,400,000 kg (2,400 tons) of fuel would be required to reach this speed. Furthermore, this fuel must first be lofted from Earth using additional rocket stages of even greater mass, making the mission prohibitively expensive. But with careful planning (i.e., finding the optimum "launch window"), plus precise tracking and control of space vehicles, flight engineers can take advantage of planetary fly-bys to boost spacecraft to speeds well beyond what can be achieved using chemical fuels alone. Planetary fly-bys, or gravity assists, allow spacecraft to explore the most distant planets and moons of the solar system, or even—like *Pioneer* and *Voyager* launched nearly 40 years ago—to escape the solar system entirely and plunge into the depths of interstellar space.

* Actually, some fuel would be required to coax the spacecraft into its close encounter with Jupiter. But by careful timing of the launch from Earth, this amount of fuel can be kept small.

5.7 PLANETARY FLY-BYS IN TWO DIMENSIONS

The one-dimensional example discussed above is unrealistic. It is unlikely that the spacecraft's approach velocity $\vec{v}_S$ is antiparallel to the planet's velocity $\vec{V}_P$, and rarely could it be steered close enough to the planet's surface (without crashing) to execute a complete U-turn. Let's now treat the more realistic case where

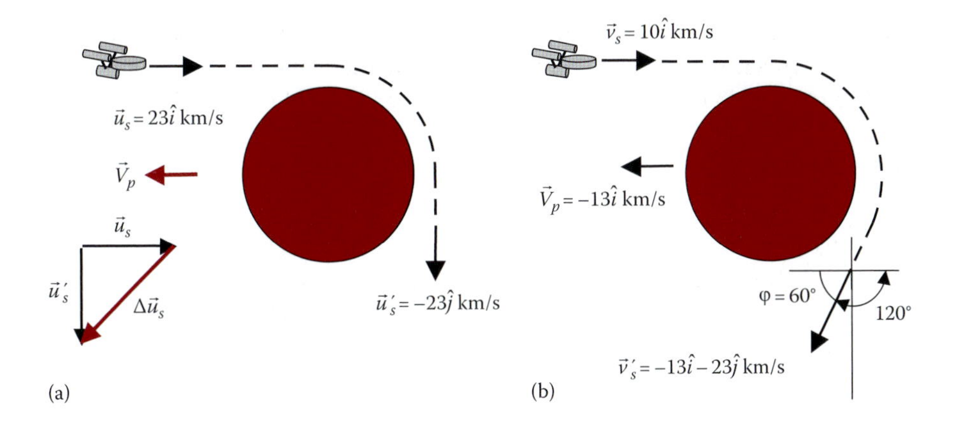

FIGURE 5.7 A more general gravity-assist trajectory. The initial velocities of the spacecraft and planet are $\vec{v}_S = 10\hat{i}$ km/s and $\vec{V}_P = -13\hat{i}$ km/s, respectively. (a) In the planet's (CM) reference frame, the spacecraft is deflected by 90°, but its initial and final speeds are the same: $u'_S = u_S$. (b) In the Sun's reference frame, the spacecraft is deflected by 120° and acquires a velocity $\vec{v}'_S = \vec{u}'_S + \vec{V}_P$.

the velocities $\vec{v}_S$, $\vec{v}'_S$, and $\vec{V}_P$ are not all parallel. As you will see, this makes the analysis only slightly more complicated than the one-dimensional example discussed above.

Regardless of initial and final directions of motion, the spacecraft and planet undergo an *elastic* collision, and the CM method of analysis remains applicable. In particular, the initial and final speeds of the spacecraft in the CM (planet's) reference frame are identical: $u_S = u'_S$. Figure 5.7a shows a hypothetical planetary fly-by in the *planet's* reference frame. The angle of deflection (i.e., the angle between $\vec{u}_S$ and $\vec{u}'_S$) depends sensitively on the planet's mass, the spacecraft's speed, and its closest approach to the planet. In Chapter 12, we will calculate this angle, but for now, let's simply choose a value (such as 90° in Figure 5.7a) and calculate $\vec{v}'_S$, the spacecraft's final velocity relative to the Sun.

It's easiest to use a numerical example. Let $\vec{V}_P = -13\hat{i}$ km/s and $\vec{v}_S = 10\hat{i}$ km/s, as in Exercise 5.8. Then, $\vec{u}_S = \vec{v}_S - \vec{V}_P = 23\hat{i}$ km/s, and assuming a 90° deflection of the spacecraft (in the CM reference frame), $\vec{u}'_S = -23\hat{j}$ km/s. To return to the Sun's reference frame, simply add $\vec{V}_P$ to this result: $\vec{v}'_S = \vec{u}'_S + \vec{V}_P = -13\hat{i} - 23\hat{j}$ km/s. By the Pythagorean theorem, $v'_S = \sqrt{13^2 + 23^2} = 26.4$ km/s, and the direction of the final velocity is given by the angle $\varphi = \tan^{-1}(23/13) \approx 60°$ (Figure 5.7b). In the Sun's reference frame, the spacecraft is deflected by $180° - 60° = 120°$, and its speed has increased by 16.4 km/s. This is a smaller speed boost than we found in the previous example, showing that the change of speed depends strongly on the directions of $\vec{v}_S$, $\vec{V}_P$, and the angle of deflection. A more detailed treatment shows that the spacecraft's speed will be increased by the fly-by maneuver if the direction of $\Delta\vec{v}_S$, the spacecraft's change of velocity, is such that $\Delta\vec{v}_S \cdot \vec{V}_P > 0$. See Box 5.1.

BOX 5.1 AN ALTERNATE APPROACH TO GRAVITY-ASSISTED SPACEFLIGHT

Given the deflection angle measured in the CM reference frame, we can derive a general equation for the final speed v'_S of a spacecraft undergoing a "slingshot" maneuver. Since $u_S = u'_S$, then

$$u'^2_S = \vec{u}'_S \cdot \vec{u}'_S = \left(\vec{v}'_S - \vec{V}_P\right)^2 = v'^2_S + V_P^2 - 2\vec{v}'_S \cdot \vec{V}_P,$$

and

$$u^2_S = \vec{u}_S \cdot \vec{u}_S = \left(\vec{v}_S - \vec{V}_P\right)^2 = v^2_S + V_P^2 - 2\vec{v}_S \cdot \vec{V}_P.$$

Equating these two expressions and simplifying, $v'^2_S = v^2_S + 2(\vec{v}'_S - \vec{v}_S)\cdot\vec{V}_P = v^2_S + 2\Delta\vec{u}_S \cdot \vec{V}_P$, where $\Delta\vec{u}_S = \vec{u}'_S - \vec{u}_S = \vec{v}'_S - \vec{v}_S = \Delta\vec{v}_S$. (Why?) The final speed of the spacecraft can therefore be expressed as

$$v'_S = \sqrt{v^2_S + 2\Delta\vec{u}_S \cdot \vec{V}_P}. \tag{B5.1}$$

For an increase in speed, the dot product in Equation B5.1 must be greater than zero. This means that the angle between $\Delta\vec{u}_S$ (or $\Delta\vec{v}_S$) and the planet velocity $\vec{V}_P$ must be less than 90°. In Figure 5.7a, for example, $\Delta\vec{u}_S$ makes an angle of 45° with $\vec{V}_P$, so the speed of the spacecraft is increased. Using Equation B5.1 and $\Delta\vec{u}_S = -23\hat{i} - 23\hat{j}$, $|\Delta\vec{u}_S| = \sqrt{2}u_S = 32.5$ km/s,

$$v'_S = \sqrt{10^2 + 2(32.5)(13)\cos 45°} = 26.4 \text{ km/s},$$

exactly as we found previously.

Exercise B5.1: The trajectories of a spacecraft and planet are shown in the reference frame of the planet.

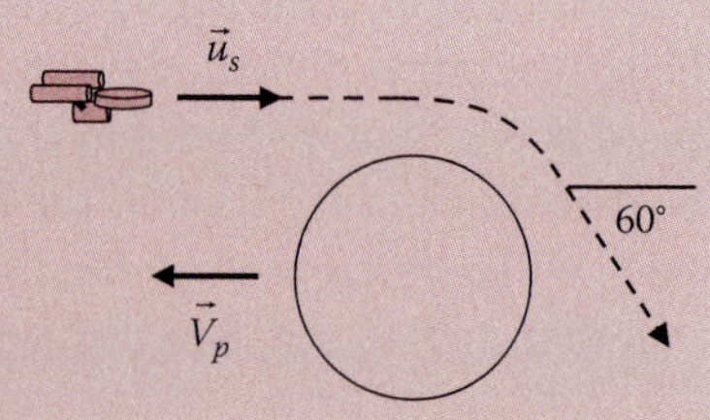

a. What is the direction of $\Delta\vec{u}_S$?
b. Is the spacecraft's speed increased by this encounter? Answer this without calculation.
c. Sketch a trajectory where the spacecraft's speed is reduced.

EXERCISE 5.9

What if the angle of deflection (in the CM reference frame) were only 45°, as shown in the figure to the right?

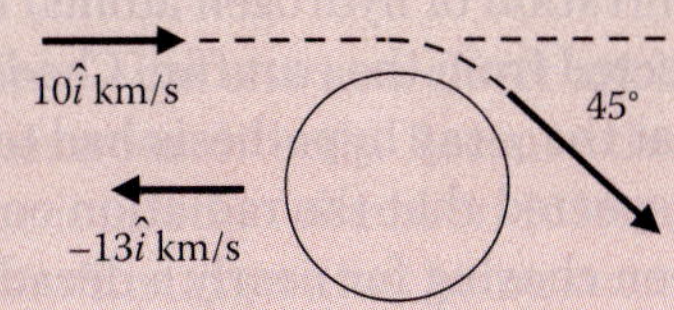

(a) What is the initial velocity of the spacecraft in the CM reference frame?

(b) What is the final velocity $\vec{u}_S'$ in the CM frame?

(c) What is the final velocity of the spacecraft in the Sun's reference frame?

Answer (a) $\vec{u}_s = 23\hat{i}$ km/s (b) $\vec{u}_s' = 16.3\hat{i} - 16.3\hat{j}$ km/s (c) $\vec{v}_s' = 3.3\hat{i} - 16.3\hat{j}$ km/s

5.8 CHADWICK'S DISCOVERY OF THE NEUTRON

In 1911, Ernest Rutherford discovered that nearly all of an atom's mass is concentrated in a point-like *nucleus* whose radius is 10^5 times smaller than the size of the electronic cloud surrounding it. Soon afterwards, his student Henry Moseley showed that the chemical identity of an atom (e.g., H, He, Li, etc.) is conferred by the total electric charge contained in its nucleus. Only two subatomic particles were known at that time: the electron and the proton. But protons could account for only about half the mass of the nucleus; for example, the mass of a helium nucleus is $4u$,* while its nuclear charge is only $+2e$.† Where was the missing mass? Furthermore, what were *isotopes*—those atoms with identical chemical properties but distinctly different masses? Rutherford proposed that the nucleus contained electrons as well as protons. The nuclear electrons cancelled some of the protons' charge, reducing the total nuclear charge with negligible change of mass.‡ According to this hypothesis, a helium nucleus (or α-particle) contained 4 protons and 2 electrons, giving its observed mass $4u$ and charge $(4 - 2)e = +2e$. Rutherford also suggested that within the nucleus, an electron may bind to a proton so tightly that the combination of particles, which he dubbed a *neutron*, forms "an atom of mass 1 with zero ... charge." Because it lacked electric charge, Rutherford continued, the neutron "should be able to pass freely through matter." In the ensuing decade, he and his collaborator James Chadwick searched for neutrons without success. Then, in 1932, a curious discovery made by the French physicists Irène Curie (daughter of Marie Curie) and Frédéric Joliot caught Chadwick's eye.

It was well known that when a light element such as aluminum or beryllium was bombarded with α-particles, it emitted a highly penetrating radiation—so penetrating that it could pass easily through a lead barrier many centimeters thick. Neither electrons nor protons can possibly do this, because—as charged particles—they interact strongly with the electrons and protons of the lead atoms. The mysterious radiation was assumed to be gamma rays (γ-rays), higher energy cousins of x-rays that might

* The unified atomic mass unit, or u, is approximately equal to the mass of a proton, $1u \approx 1.66 \times 10^{-27}$ kg.

† $e = 1.60 \times 10^{-19}$ C, the charge of a proton; $-e$ is the charge of an electron.

‡ This was a very reasonable hypothesis. It provided an intuitively simple picture of β-decay, wherein a radioactive nucleus changes its chemical identity by ejecting an electron.

Assuming an *elastic* collision, we know that $u'_M = u_M$, no matter what direction M is scattered into. The final velocity of the recoil nucleus in the lab frame is found in the usual way: $\vec{v}'_M = \vec{u}'_M + \vec{V}_{CM}$. Clearly, the maximum final speed of M (see Figure 5.9b) is obtained when M is scattered in the forward direction, parallel to $\vec{V}_{CM}$:

$$v'_M(\text{max}) = \frac{m_n}{m_n + M} v_n + V_{CM} = \frac{2m_n}{m_n + M} v_n. \tag{5.5}$$

For a paraffin target, the ejected nuclei were protons of mass $M_H = 1u$, and Chadwick measured their maximum speed to be $v'_H(\text{max}) = 3.7 \times 10^7$ m/s. When the paraffin target was replaced by paracyanogen (C_2N_2, a solid substance containing a copious supply of nitrogen), the ejected particles were nitrogen nuclei of mass $M_N = 14u$. Together with Feather's measurements using a cloud chamber* instead of the Geiger counter, Chadwick concluded that $v'_N(\text{max}) = 4.7 \times 10^6$ m/s.† Summarizing these results,

$$v'_H(\text{max}) = \frac{2m_n}{m_n + 1} v_n = 3.7 \times 10^7 \text{ m/s},$$

and

$$v'_N(\text{max}) = \frac{2m_n}{m_n + 14} v_n = 4.7 \times 10^6 \text{ m/s}.$$

Eliminating v_n and solving for m_n, he found $m_n = 0.9u \pm 10\%$, consistent with the expectation that the neutron and proton have approximately the same mass: $m_n = 1u$.

EXERCISE 5.10

(a) Verify the preceding statement using Chadwick's data. (b) If $m_n = 1u$, what was the neutron speed v_n? Note that both m_n and v_n can be obtained without ever detecting the neutrons directly.

Answer $v_n = 3.7 \times 10^7$ m/s

* The cloud chamber is an early version of the bubble chamber described in Chapter 4.

† These numbers are quoted from Chadwick's 1935 Nobel Lecture, *The Neutron and Its Properties*, December 12, 1935.

5.9 RELATIVISTIC COLLISIONS

In Chapter 4, we asked if the law of momentum conservation is consistent with the principle of relativity; that is, is the law true in all inertial reference frames? We found (Equation 4.10) that for collisions at nonrelativistic velocities ($v \ll c$), total momentum $\vec{p}_{tot} = \sum_i m_i \vec{v}_i$ is conserved in all inertial frames *provided* that the total mass of the interacting bodies does not change. For everyday collisions between macroscopic bodies, there is no need to question this assumption. But in the realm of nuclear and subatomic processes, total mass nearly always changes!* In Chapter 4, we also learned that relativistic velocities do not transform between inertial frames in the simple way we have assumed above. Let's rejoin our observers Alice (A) and Bob (B), who are moving relative to one another with velocity $\vec{V}$ (Figure 4.5a). Let $\vec{v}$ be the velocity of a body measured by Alice, and $\vec{u}$ be the velocity measured by Bob. If $\vec{v}$ is parallel to $\vec{V}$, then $\vec{u}$ is also parallel to $\vec{V}$, and (suppressing vector notation)

$$u = \frac{v - V}{1 - \left(vV/c^2\right)}. \tag{4.11}$$

In Section 4.5, we used this result to examine an inelastic one-dimensional collision, and found that for relativistic velocities, Alice and Bob disagree about whether $\vec{p}_{tot} = \sum_i m_i \vec{v}_i$ is conserved. When total mass changes, or when relativistic speeds are involved, the law of momentum conservation—as expressed above—does not satisfy the principle of relativity; it is not true in all inertial frames.

In 1905, Albert Einstein showed that the momentum conservation law could be retained by adding the speed-dependent *Lorentz* factor $\gamma(v) = 1/\sqrt{1 - v^2/c^2}$ to the definition of momentum: $\vec{p} = \gamma(v)m\vec{v}$. Total momentum is then conserved in all inertial reference frames, for all velocities,

$$\vec{p} = \sum_i \gamma(v_i) m_i \vec{v}_i = \sum_i \gamma(v_i') m_i' \vec{v}_i' = \vec{p}', \tag{5.6}$$

provided that a second quantity is also conserved:

$$\sum_i \gamma(v_i) m_i = \sum_i \gamma(v_i') m_i'. \tag{5.7}$$

As usual, the unprimed and primed quantities indicate values before and after the collision. Equation 5.6 is the relativistic expression for momentum conservation, and Equation 5.7 replaces the requirement of mass conservation introduced in Equation 4.10. See Box 5.2 for further discussion of these two results. Once again, note that for $v_i, V \ll c$, $\gamma(v) \simeq 1$ and the familiar conservation expressions are recovered.

Equation 5.7 is a new conservation law that goes hand-in-hand with momentum conservation. To see what it means, multiply $\gamma(v_i)m_i$ by c^2 and use the binomial approximation to evaluate the product for $v_i \ll c$. For a body of mass m moving with speed v,

$$\gamma(v)mc^2 = \frac{mc^2}{\sqrt{1 - v^2/c^2}} \simeq mc^2\left(1 + \frac{1}{2}\frac{v^2}{c^2}\right) = mc^2 + \frac{1}{2}mv^2.$$

* The Sun shines by nuclear fusion, converting hydrogen into helium. It emits energy at a rate of 3.9×10^{26} W, losing 4.3×10^{6} tons of mass each *second*!

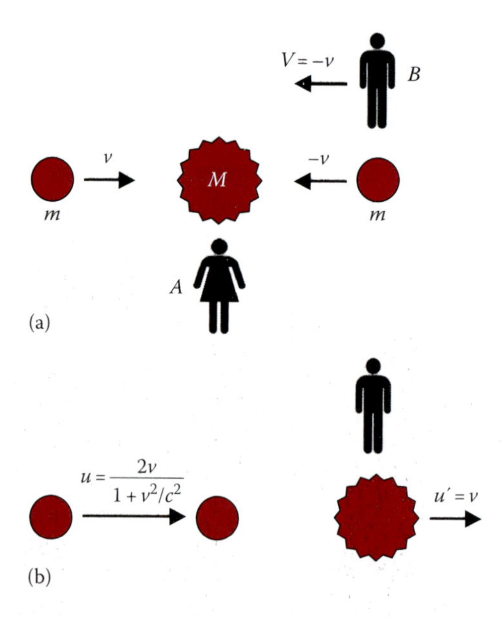

FIGURE 5.10 A totally inelastic collision viewed in (a) Alice's and (b) in Bob's frame of reference. The initial velocity of m_1 in Bob's frame is found from Equation 4.11.

> **EXERCISE 5.11**
>
> Finish the above calculation for the case $v = (1/2)c$. (a) Show that $u = (4/5)c$. (b) Find M in terms of m. (*Hint:* $M > 2m$.) (c) Does Alice agree that energy is conserved? (Write the expression for energy conservation as it appears to Alice. Use the value for M found in part [b].)
>
> *Answer* (b) $M = 4m/\sqrt{3} = 2.31m$ (c) $E_A = 4mc^2/\sqrt{3}$.

The relativistic corrections introduced above are important only when speeds approach c, or when the total mass of the interacting bodies does not remain constant. For ordinary collisions between lab-sized bodies, $\gamma(v) \simeq 1$ and the non-relativistic expression $\vec{p} = m\vec{v}$ is perfectly adequate. For example, James Chadwick's neutrons were traveling at 3.7×10^7 m/s = 0.12c. Even at this blistering speed, their Lorentz factor was only $(1 - 0.12^2)^{-1/2} = 1.01$, and the 10% experimental uncertainty of Chadwick's measurements made relativistic corrections unnecessary. In today's world of high energy physics, $\gamma \gg 1$ typically, and the relativistic expressions for mass and energy are vitally important. Let's now look at a very important recent example of this.

5.10 APPLICATION: PARTICLE ACCELERATORS AND COLLIDERS

The Large Hadron Collider (LHC) is the world's biggest particle accelerator. Straddling the French-Swiss border near the city of Geneva, and boasting a circumference of 27 km (17 miles), it is likely to be the largest scientific instrument ever to be built on Earth. The LHC accelerates protons to speeds exceedingly

FIGURE 5.11 A view of the LHC tunnel. The curvature of the tunnel, with radius 4.3 km, is just visible. The long cylinders contain superconducting electromagnets that bend the ultrarelativistic protons into circular orbits. Each magnet carries a current of 11,850 A, and is cooled by liquid helium to a temperature of 1.9 K (above absolute zero). (Courtesy of CERN, copyright CERN, Geneva, Switzerland.)

close to the speed of light, and confines them to circular paths by means of magnetic fields 170,000 times greater than the Earth's magnetic field (see Figure 5.11). These ultra-relativistic protons orbit the accelerator ring—half of them traveling clockwise and the other half counter-clockwise—to collide nearly head-on at one of four intersection points (Figure 5.12). A head-on inelastic collision between two counter-rotating protons can convert their total energy (4 TeV per particle,* or 8 TeV in total) into a teeming cauldron of exotic particles, among them the *Higgs boson*, the long sought-after particle that imparts mass to all other fundamental particles. The LHC was designed to hunt for the Higgs boson, which was until recently the only missing piece of the "Standard Model" of particle physics. Its discovery may eventually help physicists understand dark energy, dark matter, and the conditions that existed at the birth of the universe.

A proton *collider* is much more complex (and expensive) than a fixed target accelerator, wherein high speed particles travel in only one direction around the ring and smash into a stationary target (such as the paraffin wax used by Chadwick.) What is the advantage of using two counter-rotating beams rather than a single beam? The results of the preceding section allow us to compare the two approaches.

Let's start with the collider. Once more, imagine a head-on, totally inelastic collision between two identical particles ($m_1 = m_2 \equiv m$) in which a short-lived composite body of mass M is formed. According to an observer who is at rest relative to the collider, the two colliding particles have opposite velocities ($v_1 = -v_2 \equiv v$) and total momentum $p_A = \sum_i \gamma(v_i) m v_i = 0$. This observer is in the CM (zero momentum) frame of the particles. After the collision, $p'_A = p_A = 0$, so $v' = 0$ and M is at rest. Conservation of energy requires that $\sum_i \gamma(v_i) mc^2 = \gamma(0) Mc^2$, or $Mc^2 = 2\gamma(v) mc^2$.

* 1 electron volt, or 1 eV = 1.6×10^{-19} J. 1 tera-electron volt, or 1 TeV = 10^{12} eV = 1.6×10^{-7} J. The design energy of the LHC is 7 TeV per particle, but it is presently running at the lower energy quoted above.

shows velocities *relative to the CM*.) What is the superball's final velocity in the lab frame?

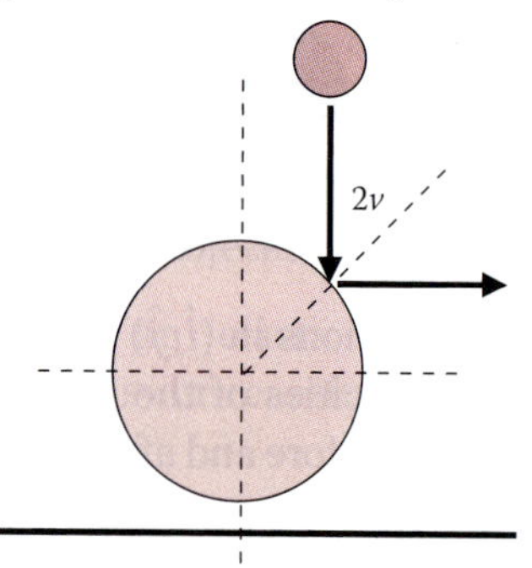

5.20 A planetary flyby is an elastic "collision" just like the examples considered above. Rather than bouncing off the surface of the planet, the satellite travels around the planet, being deflected by the planet's gravitational force. Let the spacecraft's initial velocity be $\vec{v}_s = 10\hat{i}$ km/s, and the planet's velocity be $\vec{V}_p = -14.1\hat{i}$ km/s. Three flyby's are shown in the following figure. *All are depicted in the CM (planet) reference frame.*

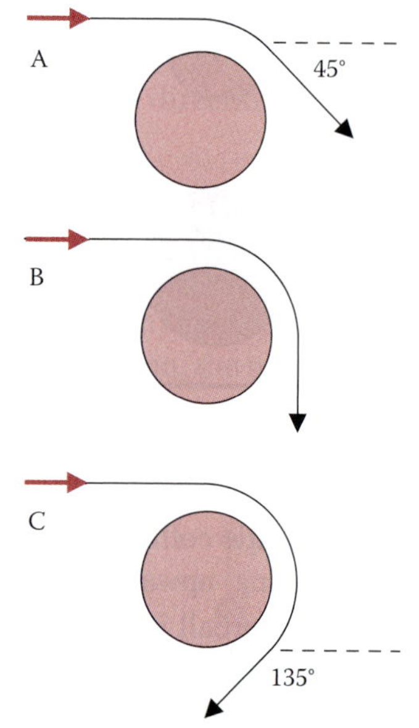

(a) In each case, what are the magnitudes of u_s and u_s'?

(b) For each path, sketch the vector $\Delta\vec{u} = \vec{u}_s' - \vec{u}_s$.

(c) Which of the three fly-by's gives the highest boost in speed? Calculate the speed for this case.

(d) Fly-by trajectories can look very different in the reference frames of the Sun and the planet. Imagine a spacecraft moving along trajectory *B*. Let the spacecraft velocity $\vec{v}_s = 35\hat{i}$ km/s, and the planet velocity $\vec{v}_s = 30\hat{i}$ km/s, as shown below. By what angle φ is the spacecraft deflected in the Sun's frame of reference?

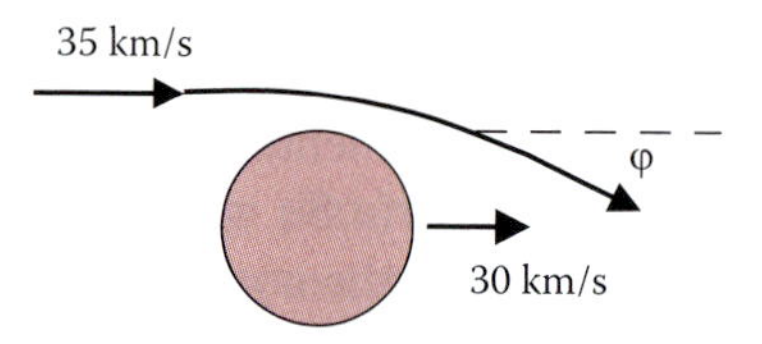

5.21 After being deployed by the Space Shuttle, the spaceprobe *Galileo* was intended to reach Jupiter by means of on-board cryogenic fuels (liquid hydrogen and oxygen). However, the use of cryogenic fuel was cancelled for safety reasons, so it was equipped instead with conventional chemical fuel that was by itself insufficient to reach Jupiter. Flight planners instead the probe on a flyby of Venus to get a gravity boost allowing it to reach Jupiter. (There were also two flyby's of Earth, comprising a so-called Venus–Earth–Earth-Gravity-Assist, or VEEGA, trajectory.)

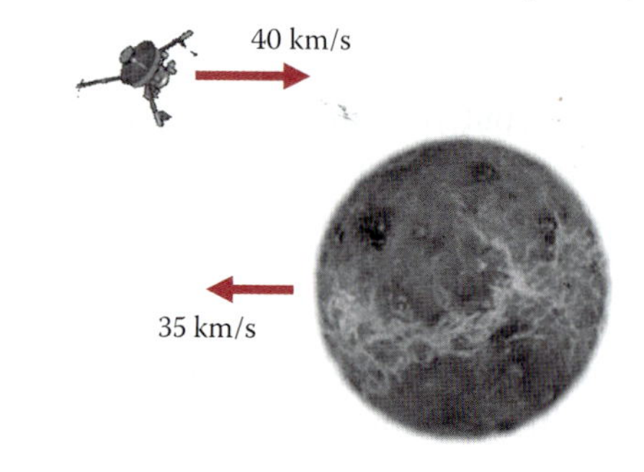

Assume that Galileo's velocity is 40 km/s, Venus' orbital speed is 35 km/s, and that the two bodies are approaching each other in opposite directions, as shown above.

(a) What is the vector velocity of Galileo relative to Venus?

(b) If Venus deflects Galileo by an angle of 150° (in the planet's reference frame), what is the final velocity of Galileo (relative to the Sun)?

(c) Suppose the spacecraft has a velocity of 40 km/s (relative to the Sun) in the *same* direction as the planet's velocity (35 km/s). If the deflection of the spacecraft relative to the planet is 90° (measured relative to the planet), what is the final speed of the spacecraft in the Sun's reference frame?

5.22 Spacecraft *Cassini* successfully reached Saturn in 2004. Like Galileo, Cassini needed a gravity assist from Venus. The following figure shows the encounter between the spacecraft and the planet in the reference frame of the Sun. As Cassini approached Venus, it was moving with a speed of 37.7 km/s at an angle of 15° to the direction of the planet's motion. After the flyby, Cassini and Venus were moving in the same direction.

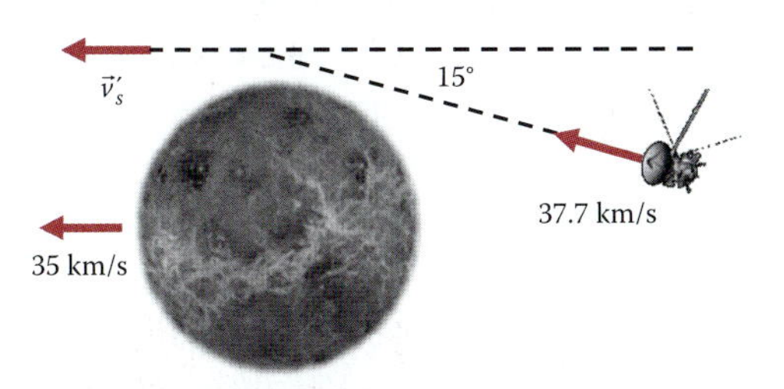

(a) Calculate the initial vector velocity of the spacecraft relative to Venus. What is the angle of approach as seen from the planet? Draw a figure in the planet's frame of reference.

(b) What was the approach *speed*?

(c) What was the final velocity of the spacecraft relative to Venus?

(d) What was the final spacecraft velocity relative to the Sun?

5.23 A satellite S passes near a planet P as shown in the following figure. The velocities shown in the figure are the initial velocities of the two bodies *in the Sun's reference frame,* before the fly-by. The final speed of the satellite, relative to the Sun, is 50.8 km/s.

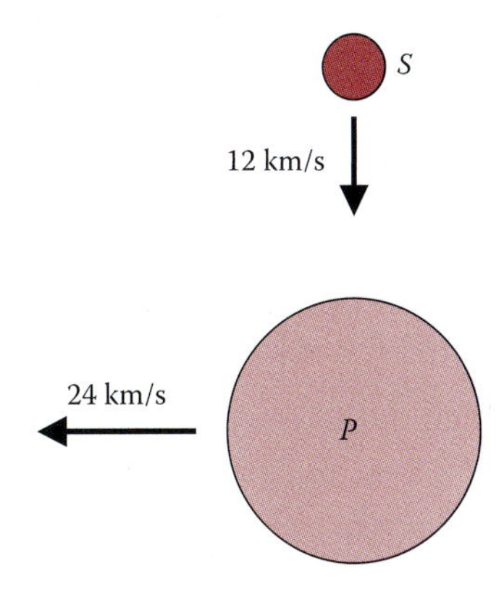

(a) Find the vector velocity of the satellite relative to the center of mass of the satellite-planet system at the time shown. Sketch the relative velocity vector.

(b) After encountering the planet, what is the speed of the satellite relative to the planet?

(c) What is the direction of the satellite's final velocity (in the Sun's reference frame)?

5.24 In Chadwick's experiment, neutrons having an initial velocity $\vec{v}_n$ collided elastically with protons that were initially at rest. After such a collision, the proton was moving with velocity $\vec{v}_p'$ (in the lab frame) at an angle θ with respect to the neutron's initial direction of motion. Assuming that $m_n = m_p$, prove that $v_p' = v_n \cos\theta$. (*Hint:* Use the CM reference frame and the law of cosines.)

5.25 The Relativistic Heavy Ion Collider (RHIC) is located at the Brookhaven National Laboratory on Long Island in NY. It is a circular accelerator with a circumference of 3.8 km. Gold nuclei ($M = 197u$) are accelerated to 99.9995% of the speed of light and circle the ring in both directions, colliding head-on at one of six intersection points. The kinetic energy absorbed by the colliding particles is sufficient to "evaporate" the protons and neutrons within each nucleus, briefly separating their constituent quarks and gluons (gluons bind the quarks together to form nucleons) to re-establish the conditions that existed during the first few microseconds of the universe.

(a) What is the value of the Lorentz factor γ for a ^{197}Au nucleus in the RHIC?

(b) The colliding nuclei briefly form a composite body before shattering into thousands of elementary particles. What is the rest mass of this short-lived composite body? Express your answer in TeV and also joules. (For a rest mass of $1u$, $E = m_0 c^2 = 932$ MeV, where 1 eV = 1.6×10^{-19} J. 1 TeV = 10^3 GeV = 10^6 MeV.)

(c) Compare this energy with the kinetic energy in a head-on inelastic collision of two houseflies ($m_f \approx 20$ mg) moving in opposite directions at 1 m/s.

5.26 The relativistic relationship between energy and momentum is more complicated than the familiar low velocity expression $K = (1/2)mv^2 = p^2/2m$. To derive it, start from $E = \gamma(v)mc^2 = mc^2 + K$, and $p = \gamma(v)mv$.

(a) Begin by multiplying the momentum by c. The units of pc are the same as energy.

(b) Square the momentum equation, and square the energy equation, and subtract one from the other, to get $E^2 - p^2c^2 = \cdots$

(c) Use the definition of γ to eliminate it from the right hand side of the equation derived in (b). You should find that $E = \sqrt{p^2c^2 + m^2c^4}$.

(d) In the low speed limit, $pc \ll mc^2$. Apply the binomial approximation to the above result to find K in this limit.

(e) Show that $E = pc$ for ultra-relativistic particles ($\gamma \gg 1$) or massless particles ($m = 0$). How fast would a massless particle be moving if its energy $E > 0$? (What would be its Lorentz factor?) An example of a massless "particle" is a photon, the quantum of light.

5.27 Two identical sub-atomic particles $m_1 = m_2 = m$ experience a totally inelastic, head-on collision, forming a new particle of mass M. According to observer A, particle 1 had an initial velocity $v_1 = (4/5)c$, while particle 2 was stationary.

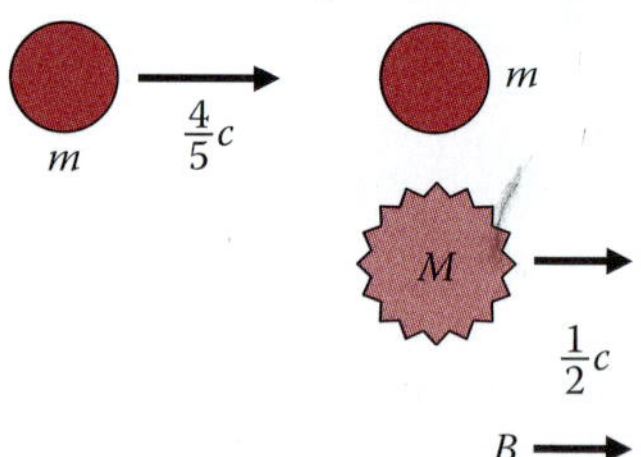

(a) Calculate the Lorentz factor $\gamma(v)$ for both particles.

(b) What is the total relativistic momentum of the two particles before the collision?

(c) By conserving momentum and energy show that after the collision, the new particle M has a velocity $v' = (1/2)c$.

(d) Calculate M'/m.

(e) Observer B is moving in the same direction as particle 1 with velocity $V = v' = (1/2)c$. Calculate the initial velocities u_1 and u_2, and the final velocity u', as measured by B. Show that momentum is conserved in B's reference frame.

5.28 Figure 4.2a shows an elastic collision between a pion and a stationary proton. The proton's mass ($m_p c^2 = 938$ MeV) is 6.7 times greater than the pion's mass ($m_\pi c^2 = 140$ MeV). After the collision, the two particles move in perpendicular directions. Show that the pion was initially moving with speed $0.99c$.

6

ACCELERATION, FORCE, AND NEWTON'S LAWS

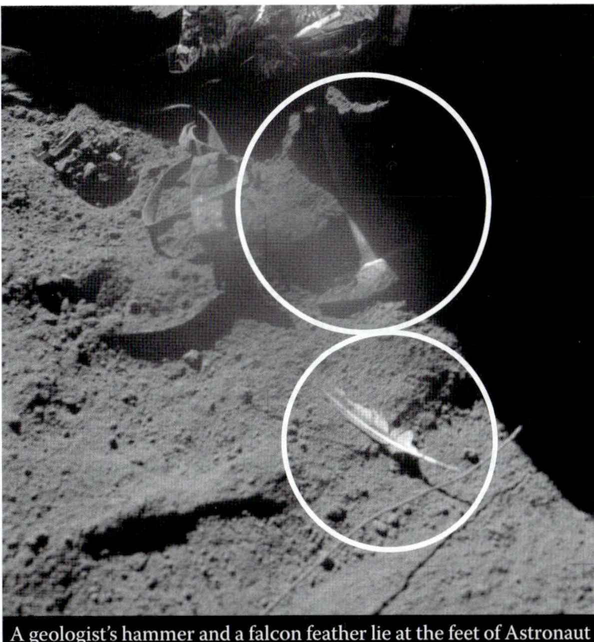

A geologist's hammer and a falcon feather lie at the feet of Astronaut David Scott during the Apollo 15 moon landing in 1971. The mission's lunar module was named *Falcon*. (Image scanned by Kipp Teague; Courtesy of NASA, Washington, DC.)

Well, in my left hand, I have a feather; in my right hand, a hammer. And I guess one of the reasons we got here today was because of a gentleman named Galileo, a long time ago, who made a rather significant discovery about falling objects in gravity fields. And we thought: where would be a better place to confirm his findings than on the Moon? ... And I'll drop the two of them here and, hopefully, they'll hit the ground at the same time ... (He drops them.) ... How about that! Which proves that Mr. Galileo was correct in his findings.

Astronaut David R. Scott, Apollo 15 Mission (1971)

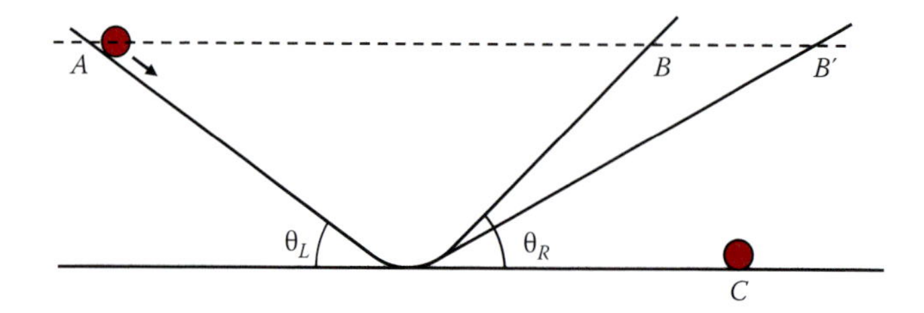

FIGURE 6.2 A ball rolls along a V-shaped track defined by angles θ_L and θ_R. Regardless of their values, the ball always rises to the same height (points B, B', etc.) as A, the point where it was released. As $\theta_R \to 0$, the speed of the ball along the right hand slope approaches a constant value. But if the ball is released from rest at C, it remains at rest.

farther along the track before stopping at point B', returning again to the same height as A. As $\theta_R \to 0$, the ball travels farther and farther, taking longer and longer to return to rest, until at $\theta_R = 0$ it travels indefinitely at *constant velocity* along the now-horizontal section of the track. If the ball is instead placed *at rest* on a horizontal track, it will remain at rest, indicating that along the horizontal track there is nothing pushing it in either direction. Hence, in the *absence* of external influences (a.k.a., "forces"), a stationary body will remain at rest and a moving body will remain moving with constant velocity. This is called the *law of inertia*. (Later it would be called Newton's first law of motion.) Galileo proved that the true measure of force is not velocity (as postulated by Aristotle), but *change* in velocity, or *acceleration*.

6.2 ACCELERATION

Acceleration is the time rate of change, or derivative, of the velocity $\vec{v}(t)$:

$$\vec{a}(t) = \lim_{\Delta t \to 0} \frac{\vec{v}(t+\Delta t) - \vec{v}(t)}{\Delta t} = \frac{d\vec{v}}{dt}. \tag{6.1}$$

Since velocity is a vector, acceleration is also a vector. A nonzero acceleration implies a change in speed and/or a change in the direction of travel. Moreover, since $\vec{v}(t)$ is the first derivative of the position vector $\vec{r}(t)$, the acceleration $\vec{a}(t)$ may also be written as the *second* derivative of position:

$$\vec{a}(t) = \frac{d\vec{v}}{dt} = \frac{d}{dt}\frac{d\vec{r}}{dt} = \frac{d^2\vec{r}}{dt^2}. \tag{6.2}$$

Recall that a vector equation is shorthand for three component equations. Equation 6.1 implies $a_x = dv_x/dt$, $a_y = dv_y/dt$, and $a_z = dv_z/dt$. In this chapter, we will restrict our attention almost exclusively to the special case of *constant* acceleration: $\vec{a}(t) = \vec{a} = \text{const}$. In this case, each of these three equations can be integrated easily to find the velocity in the x-, y-, or z-directions. For example,

$$v_x(t) = \int \frac{dv_x}{dt} dt = \int a_x \, dt = a_x \int dt = a_x t + c_x,$$

where c_x is an integration constant. At $t = 0$, $v_x(0) = c_x \equiv v_{x0}$, the *initial velocity* in the x-direction. Hence, $v_x(t) = v_{x0} + a_x t$. Expressions for $v_y(t)$ and $v_z(t)$ are found in the same manner. Combining these three results into a single vector relation, we obtain

$$\begin{aligned} \vec{v}(t) &= v_x(t)\hat{i} + v_y(t)\hat{j} + v_z(t)\hat{k} \\ &= (v_{x0}\hat{i} + v_{y0}\hat{j} + v_{z0}\hat{k}) + (a_x t\hat{i} + a_y t\hat{j} + a_z t\hat{k}) \\ &= \vec{v}_0 + \vec{a}t, \end{aligned} \tag{6.3}$$

where $\vec{v}_0$ is the body's initial velocity. But it is unnecessary to first split Equation 6.1 into its x-, y-, and z-components to derive Equation 6.3. We can integrate the vectors in Equation 6.1 directly to obtain Equation 6.3. Let's use this strategy to find the expression for the position vector $\vec{r}(t)$. Since $\vec{v}(t) = d\vec{r}/dt$,

$$\begin{aligned} \vec{r}(t) &= \int \vec{v}(t)dt = \int (\vec{v}_0 + \vec{a}t)dt \\ &= \vec{r}_0 + \vec{v}_0 t + \frac{1}{2}\vec{a}t^2, \end{aligned} \tag{6.4}$$

where the integration constant $\vec{r}_0$ is the body's *initial position*, that is, its position at time $t = 0$.

EXERCISE 6.2

(a) Check the validity of Equation 6.4 by working backward, differentiating once to recover the velocity $\vec{v}(t)$ given in Equation 6.3 and once again to recover the acceleration $\vec{a}$. (b) Show directly that $\vec{v}_0$ and $\vec{r}_0$ are the initial velocity $\vec{v}(0)$ and position $\vec{r}(0)$.

Suppose that at $t = 0$, a body is at rest at the origin of our coordinate system, so $\vec{r}_0 = \vec{v}_0 = 0$. If the acceleration is constant, Equation 6.4 tells us that $\vec{r}(t) = ½\vec{a}t^2$; that is, the body moves in the direction of the acceleration with a displacement that grows *quadratically* with time. Conversely, if the body's displacement is proportional to t^2, we know its acceleration is constant. Galileo used this argument to demonstrate that a body near the Earth's surface falls with constant acceleration. But since freely falling bodies dropped too fast for him to measure accurately, he instead studied the slower movement of a ball rolling down an inclined track (see Figure 6.3a). To justify this approach, he first showed that the speed of the ball at the bottom of the track by Guillaume Amontons, and further elaborated in the eighteenth century by Charles Coulomb.* Consider a block resting on, or sliding along, a flat surface. Experimentally, it is found that F_f is proportional to the normal force N, and is independent of the apparent area of contact between the block and the surface. If the block is *sliding*, then $F_f = \mu_k N$, where μ_k is called the coefficient of *kinetic* friction. The direction of $\vec{F}_f$ is opposite to the block's velocity and its magnitude is independent of the speed. If the block is *at rest* relative to the surface, it must overcome a static friction force equal to $\mu_s N$ before it begins to move. Mathematically, this is expressed as $F_f \leq \mu_s N$, where μ_s, the coefficient of *static* friction, is larger than μ_k. Since F_f and N are both forces, the two friction coefficients are dimensionless. For unlubricated surfaces, μ_k and μ_s are typically in the range 0.1–1.0. Figure 6.10b illustrates the behavior of the friction force. A horizontal force $\vec{F}_a(t)$ is applied to a block of mass m that is at rest on a horizontal surface. $\vec{F}_a(t)$ increases with time, starting from 0, but the block does not move until $F_a(t) \geq \mu_s N$. Since $\vec{a} = 0$, the static friction force is equal but opposite to $F_a(t)$, and it grows until it reaches its maximum value $F_f = \mu_s N$. Then the block begins to move and F_f drops to a lower value $F_f = \mu_k N$.

The friction laws are deceptively simple looking. But even today, their underlying atomic-scale origins are complicated and inadequately understood. Given their importance to the world's economy, and spurred by the recent development of new research instruments† and techniques, tribology—the study of friction—has reemerged as a vital and challenging field of physics research.

EXERCISE 6.11

A 2 kg block rests on a horizontal surface. The coefficients of static and kinetic friction between the block and surface are $\mu_s = 0.5$ and $\mu_k = 0.4$. A horizontal force $\vec{F}(t)$ is applied to the block. $\vec{F}(t)$ increases from 0 until the block begins to move, after which it remains constant.

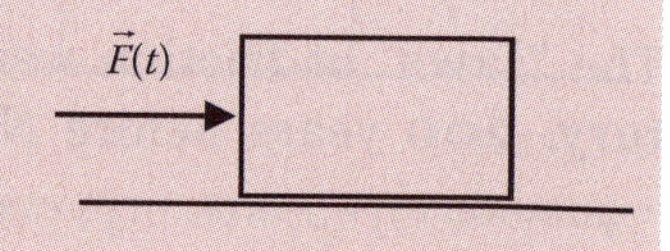

(a) What is the value of $F(t)$ when the block begins to move?
(b) What is the acceleration of the block after that time?

Answer: (a) 9.8 N (b) 0.98 m/s²

* Better known as the discoverer of "Coulomb's Law" describing the force between two electrically charged bodies.
† One such instrument is the atomic force microscope (AFM), wherein a tiny, extremely sharp needle is dragged across a flat surface while F_f and N are measured. The needle's tip radius can be as small as 10 nm, so that atomic-scale friction processes can be studied.

6.10 APPLICATION: DRAG FORCES

Projectiles follow parabolic trajectories if gravity is the sole significant force acting on them. This is not the case for objects moving through gaseous or liquid media ("fluids") at speeds high enough for *drag forces* to be appreciable. For example, as a body moves through the Earth's atmosphere, it collides with the air molecules in its path. By imparting momentum to these molecules, it experiences a reaction force that opposes its own motion. This force, called *air resistance* or *air drag*, is commonly written as

$$F_d = \frac{1}{2} C_D \rho A v^2 \tag{6.11}$$

where

ρ is the density of the gas (kg/m^3) through which the body is moving
A is the cross-sectional area of the body's leading surface
v is its speed relative to the gas*

The dimensionless coefficient C_D is called the *drag coefficient*, and its value depends strongly on the shape of the body as well as its speed

Typical values of C_D vary from about 1.3 for a flat plate (oriented with its flat face perpendicular to $\vec{v}$) to 0.04 for a streamlined "teardrop." The direction of $\vec{F}_d$ is opposite to $\vec{v}$.

EXERCISE 6.12

A circular cylinder having a radius R and a length h is moving through a fluid with a velocity $\vec{v}$. If its axis is parallel to $\vec{v}$, the cross-sectional area A in Equation 6.11 is πR^2. What is the cross-sectional area A if the cylinder's axis is perpendicular to $\vec{v}$?

Answer $2Rh$

Air resistance can be appreciable even at everyday speeds. A typical automobile has a drag coefficient $C_D \approx 0.3$ and an effective frontal area $A \approx 2.5\ m^2$. Using Equation 6.11, and noting that the atmospheric density ρ on Earth is $1.23\ kg/m^3$, it is easy to show that a car traveling at 30 m/s (67 mph) experiences a drag force of about 400 N. As a comparison, this is the force needed to pull an 80 kg (175 lb) block up a frictionless 30° slope or over a horizontal surface with $\mu_k = 0.5$. For another example, consider the trajectory of a cannonball with radius 0.1 m traveling with a horizontal speed of 200 m/s. The drag coefficient C_D of a sphere at this speed is approximately 0.5, and once again $F_d \approx 400$ N. If the ball is iron, its mass is about 32 kg, so it experiences a *de*celeration in the x-direction $a_x = -400/32 \approx -12.5\ m/s^2$ (=1.3 g). The ball's horizontal velocity will decrease continually over its time of flight, and, after reaching its maximum height, the ball will plunge toward the Earth at an steeper angle than its launch angle (see Figure 6.11). This was recognized—if not understood—by military engineers well before the time of Newton (see Figure 6.1).

* The drag force is proportional to v^2 when the speed is high enough to cause *turbulence* in the surrounding fluid. At lower speeds, the drag is proportional to v.

Now imagine a skydiver falling from a helicopter hovering at a high altitude (see Figure 6.12). While falling, she is subject to two forces: a downward gravitational force $\vec{F}_{grav} = -mg\hat{k}$ plus an upward drag force $\vec{F}_{drag} = ½ C_D \rho A v^2 \hat{k}$. According to Newton's Second Law, her equation of motion is

$$ma_z = m\frac{dv_z}{dt} = -mg + \frac{1}{2}C_D\rho A v^2. \tag{6.12}$$

Immediately after stepping from the helicopter, the skydiver's velocity is zero, and so her acceleration is $-g$. As she gains speed, the drag force increases, reducing the magnitude of her acceleration. As long as $F_d < mg$, her speed continues to increase. When $F_d = mg$, the total force acting on the skydiver is zero, so her acceleration is also zero. From that time on, she will fall with a constant *terminal speed* v_t, found by setting $a_z = 0$ in Equation 6.12:

$$v_t = \sqrt{\frac{2mg}{C_D\rho A}}. \tag{6.13}$$

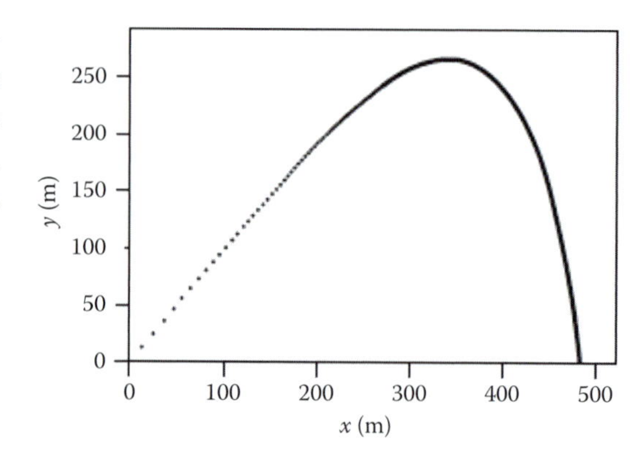

FIGURE 6.11 The trajectory of a high speed cannonball through the Earth's atmosphere, calculated using a drag force proportional to v^2. The individual points are plotted at equal time intervals. Note that the angle of descent is significantly steeper than the initial slope of the trajectory. Compare this to Figure 6.1. (From Rocca, P.L. and Riggi, F., Projectile motion with a drag force: Were the Medievals right after all? *Phys. Educ.*, 44(4), 398, 2009.)

EXERCISE 6.13

Treat the skydiver as a cylinder of water ($\rho_{H_2O} = 10^3$ kg/m^3) with radius r = 0.3 m and height h = 1.8 m. The drag coefficient varies with the cylinder's orientation. If its axis of symmetry is parallel to $\vec{v}$, $C_D \approx 0.8$. If the axis is perpendicular to $\vec{v}$, $C_D \approx 1.2$.

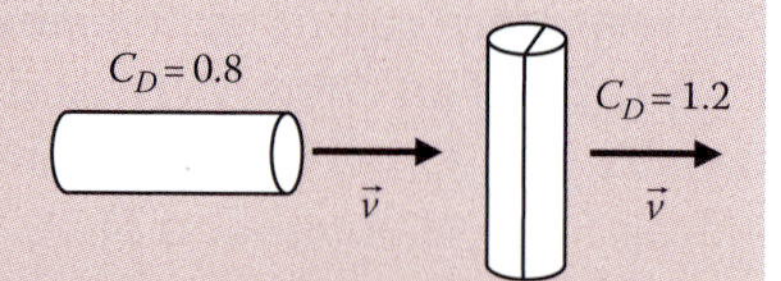

(a) Derive equations for v_t for the two orientations.
(b) Find v_t for both orientations.

Answer (a) $v_{t,\parallel} = \sqrt{2h\rho_{H_2O}g / C_D\rho_{air}}$, $v_{t,\perp} = \sqrt{\pi r\rho_{H_2O}g / C_D\rho_{air}}$ (b) $v_{t,\parallel} = 190$ m/s, $v_{t,\perp} = 80$ m/s

Of course, a human being could not survive a landing at either of the two terminal speeds you just calculated (80 m/s ≈ 180 mph!). To reduce speeds to a safe level, large-area parachutes are needed. The familiar dome-shaped parachute has a drag coefficient $C_D \approx 1.5$ and an effective area $A \approx 50$ m^2. Using Equation 6.13, the terminal speed of an 80 kg parachutist is about $v_t \approx 4$ m/s.

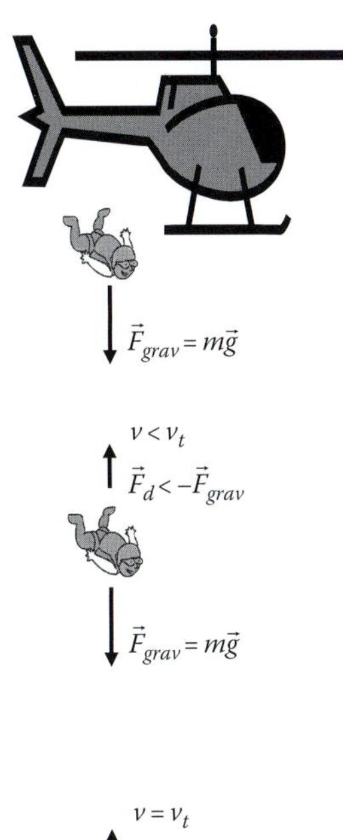

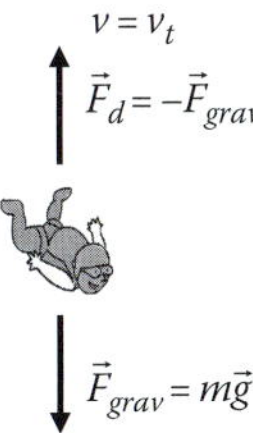

FIGURE 6.12 A skydiver experiences a gravitational force $\vec{F}_{grav} = m\vec{g}$ as well as an opposing drag force proportional to v^2. For $v < v_t$, the net force is downward, and the skydiver falls with increasing speed. When $v = v_t$, $\vec{F}_{drag} = -\vec{F}_{grav}$, and the skydiver's acceleration is zero.

EXERCISE 6.14

A parachutist of mass m_1 is falling vertically at a steady speed. The mass of the parachute is m_2.

(a) In terms of m_1, m_2, and g, what is the upward force exerted on the parachute by the air?
(b) What is the downward force exerted on the parachute by the parachutist?
(c) What was the acceleration earlier in the fall, when $v = v_t/2$?

Answer (a) $F_\uparrow = (m_1 + m_2)g$ (b) $F_\downarrow = m_1 g$ (c) $a = 3g/4$

The first known design for a parachute appeared in the fifteenth century notebooks of Leonardo da Vinci, but the idea was probably conceived well before this time (see Figure 6.13a). In the modern era, parachutes have played essential roles in science as well as warfare. The 1944 "D-Day" invasion of Europe began with the midnight drop of nearly 20,000 American, British, Canadian, and French paratroopers behind the

(a)

(b)

FIGURE 6.13 (a) *Homo Volans* (Flying Man), a sketch by Fausto Veranzio (1551–1617), adapted from drawings by Leonardo da Vinci. (From the public domain.) (b) Splashdown of a solid rocket booster after a Space Shuttle launch from Cape Canaveral. (Courtesy of NASA, Washington, DC.)

beaches in Normandy, France. At the height of the USA–USSR "cold war," before the era of digital cameras, American spy satellites periodically ejected canisters containing exposed rolls of film. These canisters were slowed by parachutes and snatched in mid-air by waiting aircraft. Manned spacecraft such as *Mercury*, *Gemini*, and *Apollo* returned safely to Earth by parachute. More recently, parachutes have provided soft landings on the surface of Mars for the "rovers" *Sojourner*, *Spirit*, *Opportunity*, and *Curiosity*. The largest parachutes ever deployed slowed the fall of the two solid rocket boosters used to launch the space shuttle, allowing the boosters to be recycled for future launches (see Figure 6.13b).

6.11 OBLIQUE MOTION

Consider a block of mass m sliding along a frictionless surface that is *tilted* by an angle θ from the horizontal plane (Figure 6.14a). For $0 < \theta < 90°$, the body moves vertically as well as horizontally. Our task is to solve for the block's acceleration. Because the friction force $\vec{F}_f = 0$, we know from Section 6.8 that the force exerted on the body by the surface is normal to the surface: $\vec{F}_{sur} = \vec{N}$. An FBD of the system (block) is shown in Figure 6.14b, and Newton's Second Law reads, $m\vec{a} = \vec{F}_{grav} + \vec{N}$. Aligning our x- and y-axes in the usual horizontal and vertical directions, we find $\vec{N} = N\sin\theta\hat{i} + N\cos\theta\hat{j}$, and

$$ma_x = N\sin\theta$$
$$ma_y = -mg + N\cos\theta.$$

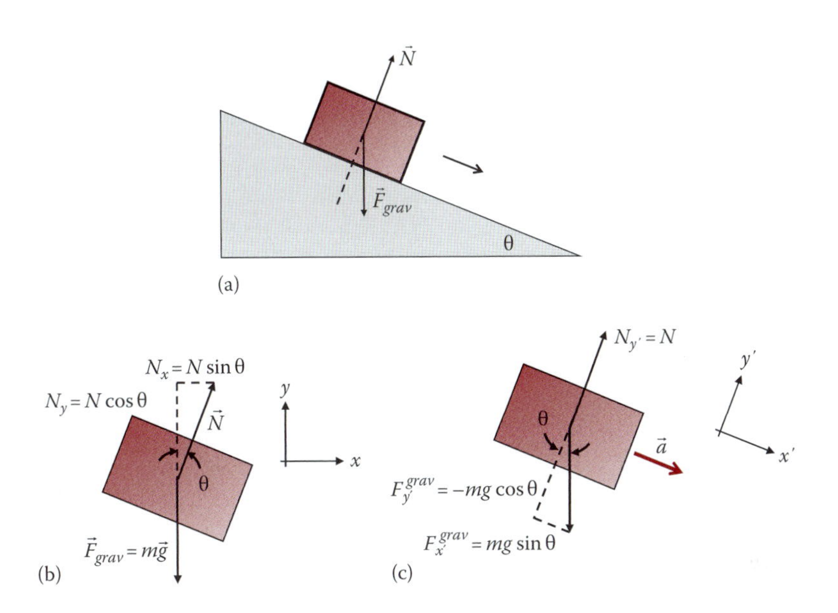

FIGURE 6.14 (a) A block slides down a frictionless surface inclined by an angle θ. (b) FBD for the block, with forces decomposed along the x- and y-axes. (c) The same FBD with forces decomposed along the x'- and y'-axes.

There are three unknowns (a_x,a_y,N) in the above two equations, so they cannot be solved without additional information. The block's acceleration is parallel to the surface, so $\vec{a} = a\cos\theta\hat{i} - a\sin\theta\hat{j}$ or $a_y/a_x = -\tan\theta$. This is the third equation that we need to solve the problem. But what a mess! While we could persist with this approach (and eventually solve the problem), it is much more convenient to define new x'- and y'-axes parallel and perpendicular *to the motion*, as shown in Figure 6.14c. Newton's Second Law still reads $m\vec{a} = \vec{F}_{grav} + \vec{N}$, but now $\vec{F}_{grav}$ has nonzero components in both x'- and y'-directions: $\vec{F}_{grav} = mg\sin\theta\hat{i}' - mg\cos\theta\hat{j}'$. The normal force is still perpendicular to the surface, so it is parallel to the y'-axis, or $\vec{N} = N\hat{j}'$. Our equations of motion for the two directions are

$$ma_{x'} = mg\sin\theta$$
$$ma_{y'} = -mg\cos\theta + N.$$

But we know that $a_{y'} = 0$, so there are only two unknowns ($a_{x'}$ and N) and two equations. The first equation yields the acceleration directly, $a_{x'} = g\sin\theta$. The second equation can be solved for the normal force, $N = mg\cos\theta$. And we're done! By aligning our axes parallel and perpendicular to the acceleration vector $\vec{a}$, we have sidestepped much of the work and reduced the potential for error.

Whenever possible, it is a good idea to check your results by asking if they make sense in cases where the answer is already known. For example, when $\theta = 0$, we know the block does not accelerate ($a = 0$). Conversely, when $\theta = 90°$, the block falls freely with downward acceleration g. Both of these "obvious" answers agree with our solution $a(\theta) = g\sin\theta$ derived.

EXERCISE 6.15

What about our solution for the normal force: $N = mg\cos\theta$? Does it also make sense in the limiting cases $\theta = 0$ and $\theta = 90°$?

EXERCISE 6.16

A 1.0 kg block slides down a plane inclined at an angle $\theta = 30°$. The coefficient of kinetic friction between the block and surface is $\mu_k = 0.2$.

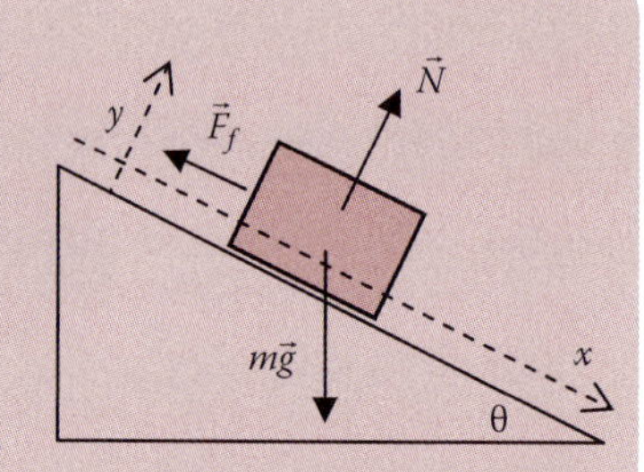

(a) Calculate the normal force N.

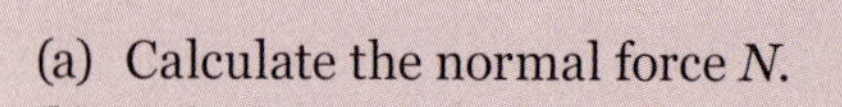

(b) Calculate the friction force F_f.
(c) Find the acceleration a_x along the surface.
(d) If the coefficient of static friction $\mu_s = 0.4$, what is the smallest angle θ for which a block at rest would begin to move?

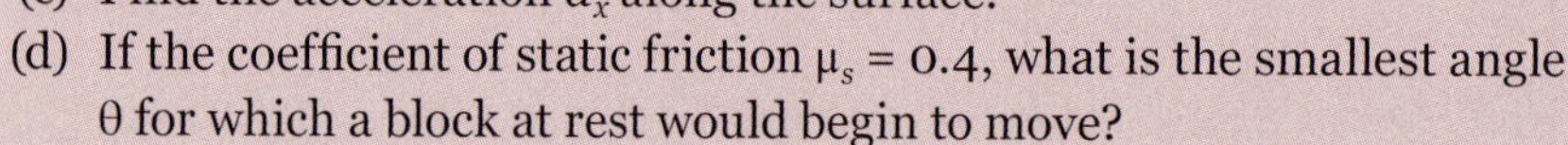

Answer (a) $N = 8.5$ N (b) $F_f = 1.70$ N (c) $a_x = 3.2$ m/s² (d) 21.8°

6.12 MULTIPLE BODIES: CHOOSING THE "BEST" SYSTEM

Sometimes it is not clear how to choose the system. The "best" choice depends on the goal: it is often the most convenient system, the one that leads to a solution with the least amount of fuss; or it may be the system that illuminates the physics most clearly; or it may be one chosen for some other reason. We might even define several systems within the same problem! Regardless of our procedure, Newton's Laws must always lead to the same solution. Consider the body m shown in Figure 6.15. It hangs from a lightweight string which is itself attached to two other strings. Our task is to determine the tension T_i in each of the strings. If we first choose m as our system (Figure 6.15a), the Second Law gives us $\vec{F}_3' + \vec{F}_{grav} = m\vec{a} = 0$ or $F_3'\hat{k} - mg\hat{k} = 0$, where $\vec{F}_3' = F_3'\hat{k}$ is the force exerted on m by the attached string. In this case, $F_3' = T_3 = mg$. To find the tensions in the other two strings, choose the *junction* of the three strings (circled) as the system (Figure 6.15b). Since $\vec{a} = 0$, then $\vec{F}_1 + \vec{F}_2 + \vec{F}_3 = 0$, where $\vec{F}_i$ is the force exerted by each string on the junction. Ignoring string mass, $\vec{F}_3 = -\vec{F}_3'$, and the Second Law for the x- and y-directions is

$$F_2\cos 45° - F_1 = 0$$
$$F_2\sin 45° - mg = 0.$$

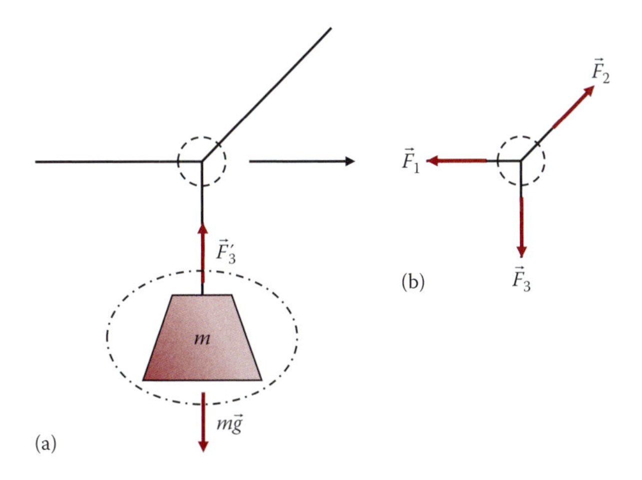

FIGURE 6.15 A body of mass m hangs from a string which is supported by two other strings. (a) FBD of the body as the system. (b) FBD of the junction of the three strings as the system.

EXERCISE 6.17

The junction of the three strings is indicated by the circle.

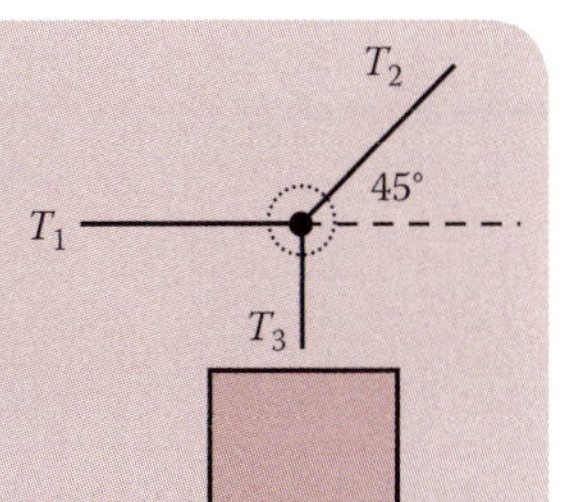

(a) Indicate by arrows the directions of the forces on the junction of the strings.
(b) Find the tension in each string, in terms of mg. If m is slowly increased, which string will break first?

Answer (b) $T_1 = T_3 = mg$, $T_2 = \sqrt{2}\, mg$

A more complicated example is shown in Figure 6.16a. Two "bodies," M_1 and M_2, are tied together by a lightweight string and slide along a horizontal frictionless surface. The leading body is pulled by a "worker" exerting a horizontal force F. Our task is to find the acceleration of the bodies and the tension T in the string between them. The forces in the vertical y-direction cancel because $a_y = 0$. Since the two bodies are tied together, they share the same acceleration $a_1 = a_2 \equiv a_x$. If we choose M_1 as our system, then $M_1\vec{a} = \vec{F}_{12}$, or

$$M_1 a_x = F_{12},$$

where $\vec{F}_{12}$ is the force on M_1 exerted by the string (Figure 6.16b). If we choose M_2 as our system, then $M_2\vec{a} = \vec{F} + \vec{F}_{21}$, or $M_2 a_x = F + F_{21}$.

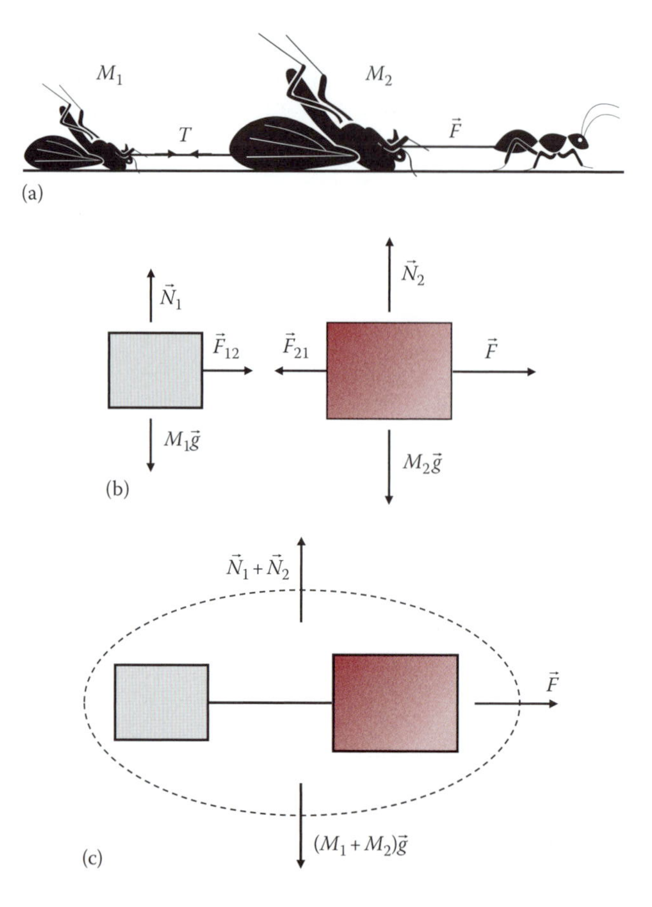

FIGURE 6.16 (a) Two "bodies" attached to one another are pulled to the right by a force $\vec{F}$. (b) Free body diagrams for each body. (c) Since they are attached, they share the same acceleration, and can be treated as a single system.

Adding these equations together, and noting that $\vec{F}_{21} = -\vec{F}_{12}$, we obtain

$$(M_1 + M_2)a_x = F,$$

or $a_x = F/(M_1 + M_2)$. Notice that the string forces have disappeared from the calculation, and also that the solution for a_x is the one we would have found by choosing the *pair* of bodies as our system (Figure 6.16c). In this case, the total mass of the system is $M_1 + M_2$, and the net force on the system is the *external* force $\vec{F}$. From Newton's Third Law, the *internal* forces *between* the two members of the system cancel and do not affect the acceleration. Whenever two or more bodies are constrained to move together with the same acceleration, they may be treated as a single body, and Newton's Third Law assures us that the internal forces between the bodies cancel.* Finally, the string tension, which is the *magnitude* of F_{12} or F_{21}, is found by substituting the solution for a_x into either of the two equations of motion above. Let's use the first equation:

$$T = F_{12} = M_1 a_x = F\frac{M_1}{M_1 + M_2}.$$

It's always a good idea to check your answers: let's substitute this expression for T into the second equation, using $F_{21} = -T$, to make sure it predicts the same acceleration for M_2:

$$a_2 = \frac{F - T}{M_2} = \frac{F}{M_2}\left(1 - \frac{M_1}{M_1 + M_2}\right) = \frac{F}{M_2}\frac{M_2}{M_1 + M_2} = \frac{F}{M_1 + M_2} = a_x,$$

just as we hoped.

EXERCISE 6.18

Four blocks, tied to each other by lightweight strings, are pulled along a horizontal frictionless surface by a force $F = 15$ N.

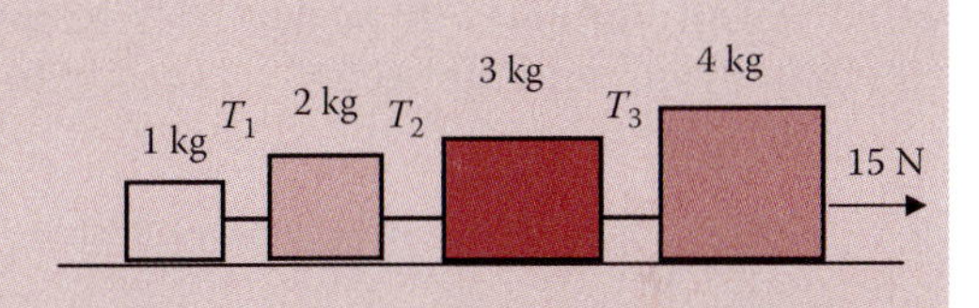

(a) Find the acceleration of each block.
(b) Find the tension in each string.

Answer (a) $a_x = 1.5$ m/s^2; (b) $T_1 = 1.5$ N, $T_2 = 4.5$ N, $T_3 = 9$ N

Finally, consider the apparatus† shown in Figure 6.17. Two masses, m_1 and m_2, are tied to the ends of a massless string and hung over a lightweight, frictionless pulley. What is the acceleration of each mass? Although the two bodies are tied together and move in concert, their accelerations are not the same,

* This is fortunate. Otherwise, when we calculate the acceleration of a ball, say, we would need to include the 10^{23} internal forces which bind the molecules of the ball to one other.

† This apparatus is commonly called Atwood's Machine, after its inventor Rev. George Atwood, who used it in the late eighteenth century to demonstrate uniformly accelerated motion.

so we cannot lump them together into a single system. Instead, define two systems, one for each body, then draw a FBD for each system. The two equations of motion for the two systems are:

$$m_1\vec{a}_1 = \vec{F}_1 + m_1\vec{g}$$
$$m_2\vec{a}_2 = \vec{F}_2 + m_2\vec{g},$$

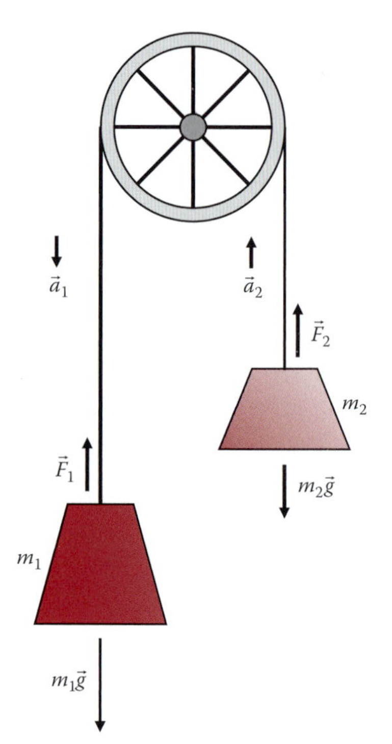

FIGURE 6.17 Atwood's Machine for studying uniform acceleration. Two masses are attached to a string that loops over a pulley. If the mass and friction associated with the pulley are negligible, then $\vec{F}_1 = \vec{F}_2$. The two bodies move in concert, with $\vec{a}_1 = -\vec{a}_2$.

where $\vec{F}_1(\vec{F}_2)$ is the force exerted by the string on $m_1(m_2)$. As the pulley's mass approaches zero, $\vec{F}_1$ and $\vec{F}_2$ must become equal. Otherwise, the pulley would "spin up" at an infinite rate. (This argument will be made rigorously when we discuss torque in Chapter 13.) Moreover, because the string has a constant length, when m_1 moves down (up) by Δz, m_2 must move up (down) by the same Δz, so $\vec{a}_1 = -\vec{a}_2$. Subtracting the second equation from the first, we can eliminate $\vec{F}_1$ and $\vec{F}_2$, to obtain

$$(m_1 + m_2)\vec{a}_1 = (m_1 - m_2)\vec{g}$$
$$\vec{a}_1 = -\vec{a}_2 = \frac{m_1 - m_2}{m_1 + m_2}\vec{g}. \qquad (6.14)$$

To calculate the string tension, eliminate the acceleration from the two equations of motion. The easiest way to do this is to multiply the first equation by m_2, and the second by m_1, obtaining

$$m_2 m_1\vec{a}_1 = m_2\vec{F}_1 + m_2 m_1\vec{g}$$
$$m_1 m_2\vec{a}_2 = m_1\vec{F}_2 + m_1 m_2\vec{g},$$

Adding these equations, with $\vec{F}_1 = \vec{F}_2$ and $\vec{a}_1 = -\vec{a}_2$, we obtain $0 = (m_1 + m_2)\vec{F}_1 + 2m_1 m_2\vec{g}$, or

$$T = |\vec{F}_1| = \frac{2m_1 m_2}{m_1 + m_2}g.$$

As always, it's prudent to ask if these answers make sense. If $m_1 = m_2 \equiv m$, then the apparatus is balanced and the acceleration of either mass should be zero. In addition, the forces on either body should cancel, so $T = mg$. This agrees with our result. If $m_1 \neq m_2$, Equation 6.14 correctly predicts that the heavier body accelerates downward. (Check this.) In the limiting case $m_1 \gg m_2$, the lighter mass is negligible, and the equation predicts $\vec{a}_1 \simeq \vec{g}$, so $T \simeq 2m_2 g \ll m_1 g$, which again makes perfect sense. We can thus be more confident that our solution is correct.

EXERCISE 6.19

Two masses, m_1 and $m_2 = 2m_1$, are attached to a string which hangs over a lightweight, frictionless pulley. The pulley itself is suspended by a vertical cord, as shown in the drawing.

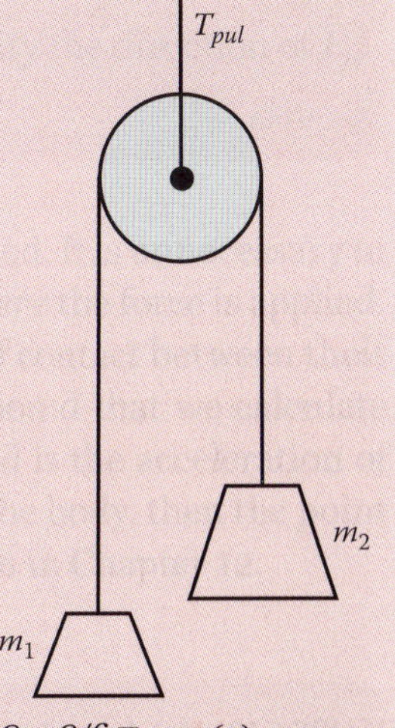

(a) What are the accelerations of the two masses?
(b) Find the tension T in the string attached to m_1 and m_2.
(c) Show that $T_{pul} < (m_1 + m_2)g$.

Answer: (a) $a = \mp g/3 = 3.3\ m/s^2$ (b) $T = \frac{2}{3}m_2 g = \frac{4}{3}m_1 g$ (c) $T_{pul} = 2T = \frac{4}{3}m_2 g = m_2 g + \frac{2}{3}m_1 g < (m_1 + m_2)g$

6.13 TWO COMMON ERRORS

There are many pitfalls that must be avoided when solving problems. Often they are due to unwarranted "common sense" assumptions that promise to make the solution easier. Other errors arise from misinterpretation of Newton's Laws. Two examples are presented here. Fortunately, they are easy to avoid. The first is the failure to distinguish between $\vec{F}$ and $m\vec{a}$. Although we say that $\vec{F}$ "equals" $m\vec{a}$, always remember that $m\vec{a}$ is not a force and so it should not be included in a free body diagram. Consider the following analogy. Let P be the cost of a large pepperoni pizza, and suppose that you order N of them for a dorm party. The total cost C of the food is $C = NP$. But this equation does not imply that C dollars are as tasty and edible as N pizzas! It merely says that with C dollars, you can buy N pizzas. Likewise, Newton's Second Law does not say that $m\vec{a}$ is a force; instead, it says that a force $\vec{F}$ will induce a mass m to accelerate at a rate $\vec{a}$. $\vec{F}$ is the cause, and $\vec{a}$ is the effect. But you may protest that we write $\vec{F}_{grav}$ as $m\vec{g}$, and indeed we do! But except for the special case of ballistic motion, the actual acceleration $\vec{a} \neq \vec{g}$. The constant $\vec{g}$ is the local gravitational *field*, the ratio between $\vec{F}_{grav}$ and m, just as the electric field is the ratio between electric force and charge.

The second example involves a misinterpretation of Newton's Third Law. Forces between bodies always appear in action–reaction pairs, but only one force of each pair acts *on* each body, and only forces that act on a body are to be included in its FBD. Figure 6.18a shows a tractor pulling a heavy wagon across a level surface. It is often argued (incorrectly) that while the tractor exerts a force $\vec{F}_{WT}$ on the wagon, the wagon exerts an equal and opposite force $\vec{F}_{TW}$ on the tractor, so no acceleration is possible ($\vec{F}_{WT} + \vec{F}_{TW} = 0$). But $\vec{F}_{TW}$

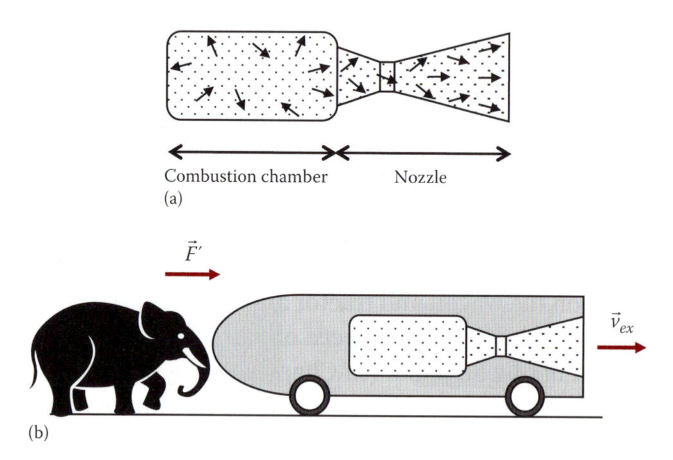

FIGURE 6.19 (a) Schematic representation of a rocket engine. Inside the combustion chamber, gas molecules move randomly, with zero net momentum. As the gas moves through the nozzle, the random molecular motion is converted into a directed stream of exhaust with high momentum. (b) A horizontally mounted rocket is held in place by a counterforce $\vec{F}' = -\vec{F}_{th}$.

A detailed analysis of this process requires the techniques of *fluid mechanics*, and is beyond the scope of this text. Fortunately, there is an easier strategy that not only yields the correct answer, but also illustrates the amazing versatility of Newton's Laws.

Imagine a rocket that is equipped with frictionless wheels and is mounted on a horizontal track. Because the track is level, gravity does not influence the rocket's motion. When the engine ignites, a stream of exhaust issues from the nozzle, and the force exerted by the gas on the nozzle wall propels the rocket forward. This force is called the rocket *thrust* $\vec{F}_{th}$. To prevent the rocket from moving, let's apply a counter force $\vec{F}' = -\vec{F}_{th}$ (see Figure 6.19b). If we define our system as the rocket plus its exhaust stream, then the *net external* force acting on the system is $\vec{F}'$. ($\vec{F}_{th}$ is an internal force between the rocket body and the exhaust.) Newton's Second Law equates the total force on the system with the rate of change of its momentum, or $\vec{F}' = -\vec{F}_{th} = d\vec{p}/dt$. Since the rocket is stationary, $d\vec{p}/dt$ must be the rate of change of the momentum due to the exhaust. During each interval of time Δt, an amount of gas Δm emerges from the nozzle with velocity $\vec{v}_{ex}$, to be replaced by an equal amount of gas entering the nozzle from the combustion chamber with negligible average velocity. The rocket engine generates momentum at a rate $\Delta\vec{p}/\Delta t = \vec{v}_{ex}\Delta m/\Delta t$, or, in the $\lim \Delta t \rightarrow 0$, $d\vec{p}/dt = \vec{v}_{ex} dm/dt$. Therefore, the magnitude of the rocket thrust is given by

$$F_{th} = v_{ex}\frac{dm}{dt}. \tag{6.15}$$

This is a rather amazing result. The counter force $\vec{F}'$ has no causal connection to the combustion and streaming of the rocket exhaust, yet Newton's Second Law still prevails. Without studying the inner workings of the rocket engine and nozzle, a judicious choice of system (the rocket body plus its gaseous contents) allows us to derive an important relation for rocketry.

EXERCISE 6.22

The Space Shuttle was launched with the aid of two solid rocket boosters (SRBs). Each SRB carried 5×10^5 kg of fuel which was exhausted with velocity 2.63 km/s, producing a maximum thrust of 13.8 MN.

(a) How much fuel mass was burned per second?
(b) How long could the SRB deliver maximum thrust?

Answer (a) 5250 kg/s (b) 95 s

EXERCISE 6.23

An automobile is propelled by explosive gases that ignite and expand rapidly within the cylinders of its engine. These gases push on pistons, which turn a crankshaft and cause the wheels to rotate. From the point of view of Newton's Laws, what *external* force causes the car to accelerate?

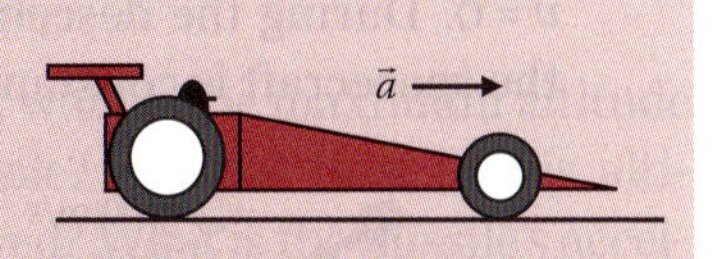

6.15 SUMMARY

Aristotle's domination of physics ended with Galileo, who gave us the law of falling bodies ($\vec{a} = \vec{g}$), the concept of superposition, the law of inertia (subsequently called Newton's First Law), the principle of relativity, and the recognition that force causes *acceleration*—not velocity. Newton seized on Galileo's insights, and clarified the key concepts of force and inertial mass. By defining these ideas precisely in his Second and Third Laws, he forged the theoretical framework of classical mechanics, which he exploited brilliantly to explain and predict a broad range of phenomena on Earth and in the heavens: parabolic paths of projectiles on Earth; ocean tides and shape of the planet; orbits of the Moon, of other planets of the Solar System, of the satellites of Jupiter and Saturn; comets. No new principles of mechanics were added for over 200 years, when Einstein's theories of Special and General Relativity appeared in the early part of the twentieth century. Even now, in our modern *quantum* world, Newton's Laws remain valid for the vast range of applications in which the corrections decreed by relativity and quantum theory are unimportant.

Yet Aristotle's physics remains with us. It is the instinctive refuge of those who have not mastered Newton's Laws. It often appeals to our intuition whereas Newton's Laws often do not. Were we not enthralled when astronaut David Scott released the hammer and feather on the Moon? Don't be discouraged if you are sometimes confused or frustrated by Newton's Laws. They are not "common sense." Louis Wolpert, an accomplished biologist and writer, offers the following observation:

> The physics of motion provides one of the clearest examples of the counter-intuitive and unexpected nature of science... The enormous conceptual change that the thinking of Galileo [and Newton] required shows that science is not just about accounting for the 'unfamiliar' in terms of the familiar. Quite the contrary: science often explains the familiar in terms of the unfamiliar.*

* Louis Wolpert, *The Unnatural Nature of Science*, Faber and Faber Ltd., London, U.K. (1992).

6.14 Consider the blocks shown in the figure below. A force F is applied to the 5 kg block and the whole contraption rolls horizontally with the same acceleration $\vec{a}$. There is no friction on any surface.

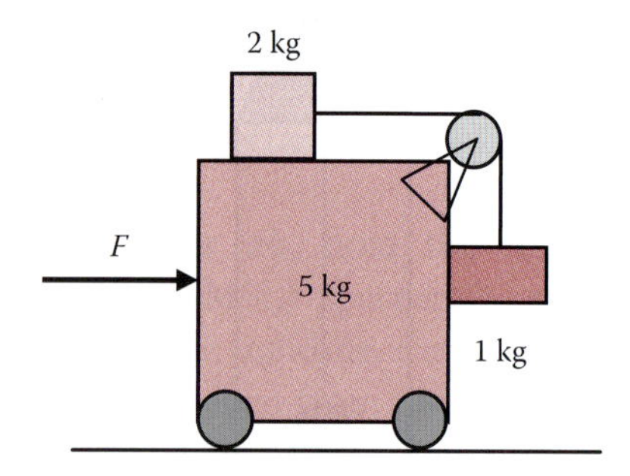

(a) Draw free body diagrams for the 1 and 2 kg masses.

(b) The three masses do not move relative to each other. Calculate the tension in the cable.

(c) What is the acceleration of the system?

(d) What is the value of F?

(e) Explain in words why the three masses do not move relative to each other. If F were increased, how would they move?

6.15 A drop tower is a facility for inexpensively inducing a "weightless" state for a brief amount of time. NASA maintains a drop tower at its Glenn Research Center near Cleveland, OH. Basically, the facility is a long vertical shaft in which a large container is dropped into an air bag. Inside the container there is an experimental apparatus which drops independently of the surrounding container. While the container feels the effects of air drag, the apparatus inside is almost perfectly shielded from air resistance. During a drop, the container falls 24.1 m in a time t, and the experimental apparatus falls 20 cm relative to the container, i.e., it falls a total distance of 24.3 m in the same time t.

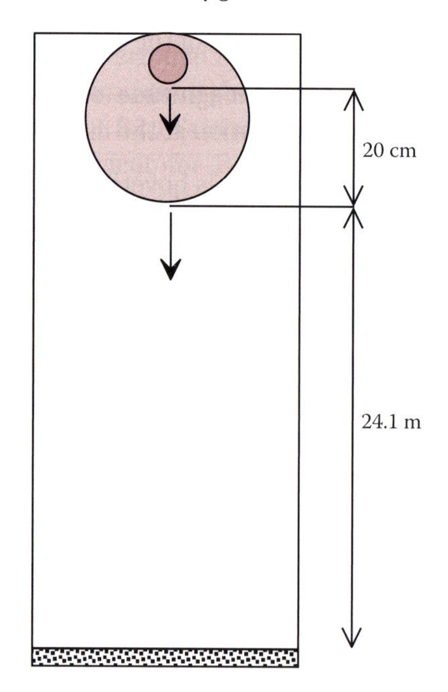

(a) What is the time t?

(b) It can be shown that for velocities much smaller than its terminal velocity v_t, a body falls a distance

$$s(t) = \frac{1}{2}gt^2 - \frac{1}{12}\frac{g^3t^4}{v_t^2}.$$

Find the terminal velocity of the container, assuming its speed $v(t) \ll v_t$ throughout the drop.

(c) If the air shield has a mass of 330 kg, what is the maximum value of the air drag force on the container? (*Hint:* Take the second derivative of $s(t)$ above.)

6.16 An "Atwood's Machine" consists of a 180 g ball attached by a lightweight string to a counterweight of mass m_2 = 200 g. The string passes over two frictionless pulleys, as shown in the figure. The ball is lowered into a tall cylinder containing a viscous fluid. When the ball moves through the fluid, it encounters a viscous drag force $F_v = 20.0v^2$ N in the direction opposite to the motion, where v is the ball's speed. (For the purposes of this question, ignore other forces on the ball, e.g., the buoyant force.)

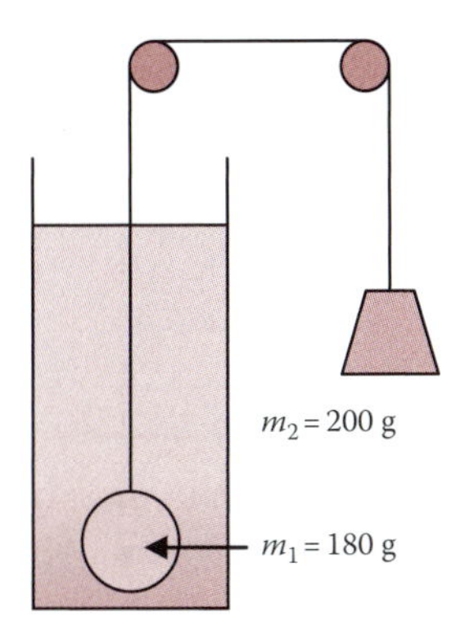

(a) Draw a free body diagram for the ball as it moves *upward* through the fluid. Also draw a free body diagram for the counterweight m_2. Identify all forces and indicate their proper directions.

(b) When m_2 is released, what is the *initial* acceleration of the ball?

(c) Find the tension in the string after the ball reaches its terminal speed. Does this tension increase or decrease while the ball is accelerating from 0 to v_t?

(d) What is the ball's terminal speed?

6.17 Here is a problem that emphasizes the difference between mass and apparent weight. An Atwood's machine consists of a mass m_1 = 25 g connected to a large helium-filled balloon by a string passing over pulleys. The balloon is sufficiently inflated so that it "floats," that is, the upward "buoyant" force exerted by the surrounding air on the balloon just balances the gravitational force. When m_1 is released, its initial acceleration is 1.63 m/s².

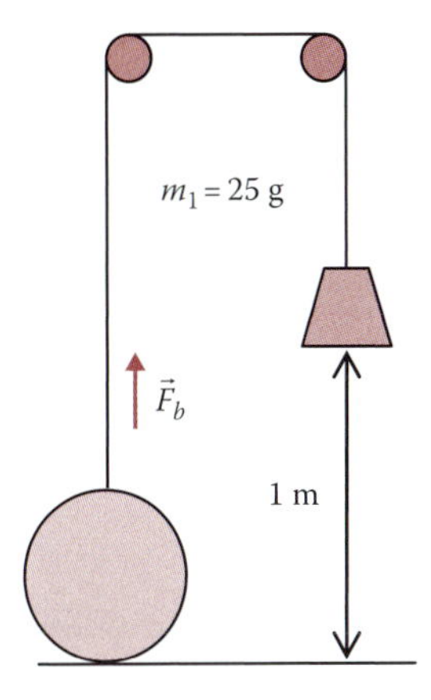

(a) Use this information to determine the mass of the helium balloon.

(b) What is the initial tension in the string?

(c) If m_1 is initially 1 m above the floor, what would be its highest speed if air drag on the balloon is neglected?

(d) Now consider air drag. Calculate v_t assuming that the drag coefficient C_D = 0.5, the balloon radius is 0.3 m, and the density of air is 1.23 kg/m³. Is air drag a significant factor in this problem?

6.18 A block of mass M = 2 kg rests on a frictionless inclined plane at a 30° angle with respect to the horizontal. The block is attached to two other blocks by a string passing over a pulley, as shown in the figure. Assume that the pulley is massless and frictionless. The blocks do not move when released. But after the string between m_1 and m_2 is cut, M moves with an acceleration $a = g/3$.

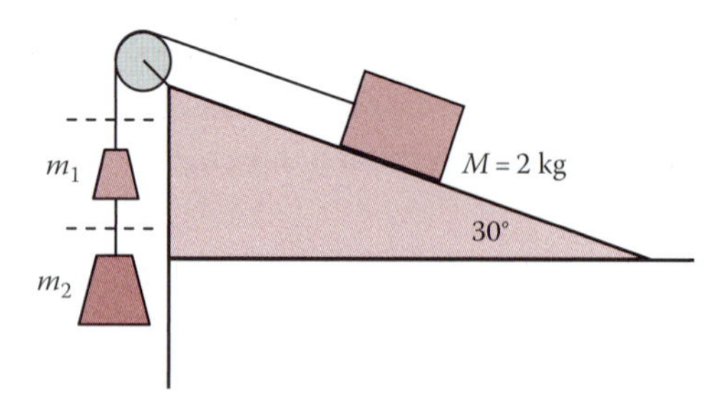

(a) Draw a free body diagram showing all forces on M. What is the tension in the string attached to M after m_2 is cut free?

(b) Find the masses m_1 and m_2.

6.19 A block of mass m is placed on the sloping surface of a triangular wedge. The wedge has a mass M and slope θ, and slides over a horizontal surface when a force F is impressed on it. Friction is negligible on all surfaces of the wedge. Derive an expression for F such that the block and wedge move together, with zero velocity relative to one another. Check to make sure your answer makes sense in the limiting cases of θ = 0 and θ = 90°.

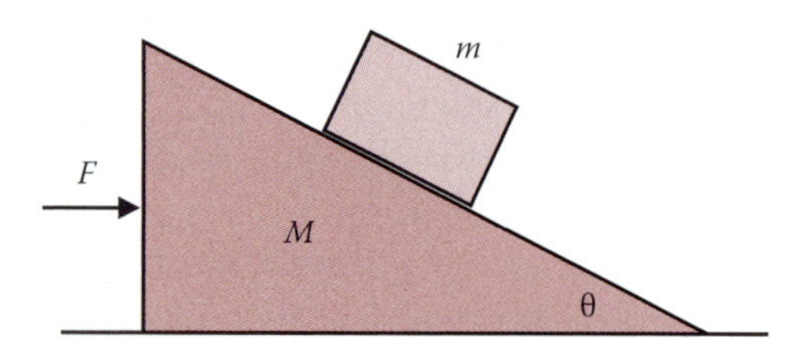

6.20 A block of mass m is resting on a surface that is inclined to the horizontal by an angle θ. The coefficients of friction between the block and the surface are $\mu_s = 0.5$ and $\mu_k = 0.4$.

(a) Derive an algebraic expression for the value of θ at which the block will begin to slide.

(b) Derive an expression for the block's acceleration a down the plane at this value of θ.

(c) Calculate the numerical values of θ and a.

6.21 An object of mass $m = 10$ kg is suspended by two cords, as shown in the figure.

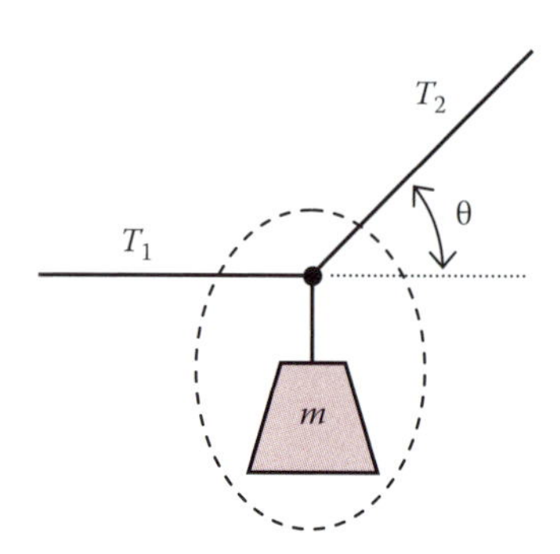

(a) Draw a FBD for the "free body" contained within the dotted oval.

(b) Derive expressions for the tension in each cord in terms of m, g, and the angle θ of cord 2. Calculate T_1 and T_2 for θ = 45°.

(c) If the breaking tension of either cord is 200 N, over what range of θ will the cords be able to support the mass? Which cord will break first?

6.22 A skydiver of mass $m = 70$ kg leaps from a plane. He gains speed until he reaches a speed of 60 m/s.

(a) Make rough sketches of the skydiver's speed v and acceleration a vs. time. Be sure that the two graphs agree with each other (i.e., $a = dv/dt$).

(b) If the air drag force is written as $F_d = Bv^2$, find the terminal speed v_t in terms of m,

g, and B, and then write an expression for the total force on the skydiver in terms of g, v, v_t and m. Find the numerical value of the acceleration when the speed is $v = 0.5v_t$ and $v = 0.9v_t$.

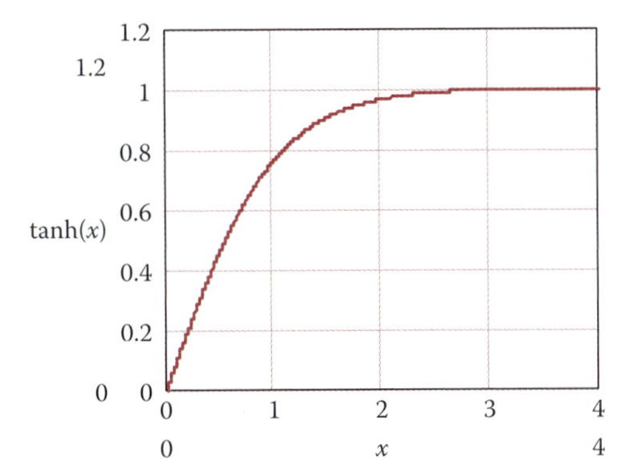

Taking the drag force into account, the expression for the speed of an object falling from rest can be shown to be $v(t) = v_t \tanh(gt/v_t)$, where the hyperbolic tangent function is given by

$$\tanh(x) = \frac{e^x - e^{-x}}{e^x + e^{-x}}$$

and is shown in the figure above.

(c) From the graph of tanh(x), estimate the time in seconds for the skydiver to reach a speed $v = 0.9v_t$, starting from rest.

(d) After the skydiver reaches terminal speed, he opens his parachute, and is quickly slowed to a new terminal speed $v_t' = 20\text{ m/s}$. What is the acceleration of the skydiver immediately after the parachute is fully deployed?

6.23 Two identical blocks $m_1 = m_2 = m$ are connected by a string that passes over a frictionless massless pulley. One of the blocks slides over a tabletop with negligible friction, while the other block hangs vertically. The tabletop is hinged, so that it can be inclined by an angle θ relative to the horizontal, where $0 \le \theta \le 90°$.

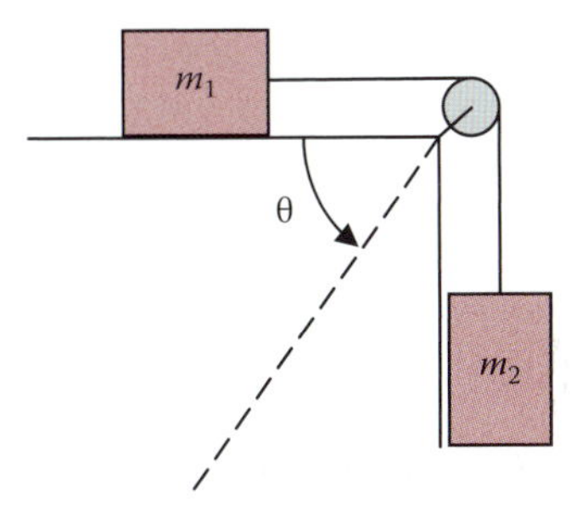

(a) What is the acceleration of the two blocks when the table surface is level ($\theta = 0$)?

(b) Derive an expression for the acceleration as a function of θ, g, and the block mass m.

(c) Check your answers by solving for the acceleration in "obvious" cases. For example, in (a), what would you get if $m_1 = 0$? If $m_2 = 0$? In (b), check by finding a for $\theta = 0$, $\theta = 90°$.

6.24 At liftoff, the Space Shuttle used two solid rocket boosters (SRBs) along with its three main engines to reach an altitude of 46 km. The reusable SRBs were then jettisoned, reaching a peak altitude of 67 km before parachutes were deployed to cushion their splashdown in the Atlantic Ocean. An SRB is 3.71 m in diameter, 45 m long, and after its fuel is exhausted, its mass is 91,000 kg.

(a) Without parachutes, what would have been the velocity of an SRB just before impact? Assume that the booster fell with its axis aligned vertically, and include the effects of air resistance.

(b) Each SRB used three large domed parachutes to slow its rate of descent to 23 m/s at splashdown. If the parachute'sdrag coefficient was 0.7, what was its diameter? You can check your answer by taking measurements directly from Figure 6.13b.

(c) Before the main chutes were opened, a smaller "drogue" parachute was used to align the booster's cylindrical axis vertically, so that it would splash down tail-first. If the main chutes were deployed when $v = 46$ m/s, what was the acceleration of the booster immediately after deployment?

6.25 An alternate universe contains an Earthlike planet inhabited by intelligent beings. In this universe, gravitational and inertial masses are not equal. Alien physicists erect an Atwood's machine (Figure 6.17) from a (massless, frictionless) pulley and two hanging bodies. In terms of the local value of g, find the acceleration of the bodies for the following two cases:

(a) $m_1^{grav} = m_1^{iner} = m_2^{iner}$ but $m_2^{grav} = 0$

(b) $m_1^{grav} = m_1^{iner} = m_2^{grav}$ but $m_2^{iner} = 0$

7

CIRCULAR MOTION, SIMPLE HARMONIC MOTION, AND TIME

John Harrison's first chronometer, called *H*1, stands 67 cm tall. Inset: *K*1, a copy of Harrison's prize-winning timekeeper *H*4, is only 10 cm in diameter. It was carried by Captain James Cook on his historic voyage to the South Pacific in 1773. (Copyright National Maritime Museum, Greenwich, London.)

In October 1707, a fleet of British warships was returning home after a military campaign in the Mediterranean Sea. Plagued by stormy weather, the fleet was off course by nearly 100 km, or 1.2° of longitude, as it approached the English Channel. Four of the 21 ships ran aground, sinking with the *loss of more than 1400 men. This disaster brought fresh urgency to the centuries-old problem of how to determine longitude while at sea. In the early seventeenth century, both Spain and Holland offered lucrative rewards for a practical solution, inspiring Galileo and other notables to study the problem. (His strategy, which relied on sighting eclipses of Jupiter's moons, was deemed impractical for use at sea.) In 1714, the British Parliament raised the stakes, offering the Longitude Prize of £20,000 (roughly $12 million today) for a practical solution having a precision better than 0.5°. The issue was time: to determine longitude, one needs to invent a clock that keeps accurate time during long voyages amid the harsh conditions aboard a sailing ship. An error of 1 min translates to (1/4)° of longitude, or 30 km at the equator. (This will be explained shortly.) Newton declared the task to be hopeless, but two of his most distinguished contemporaries, Christiaan Huygens and Robert Hooke, made pioneering contributions to clock technology. Yet it was a self-educated tradesman and clockmaker, John Harrison, who claimed the Longitude Prize in 1762 after more than 30 years of ingenious design and painstaking construction. A copy of Harrison's H4 chronometer was carried by Captain James Cook on his second voyage to the South Pacific in 1773. Upon his return to England, Cook exclaimed that the watch had "exceeded all expectations" and had been his "never failing guide" throughout his journey. In the eighteenth century, the ability to keep accurate time made it possible to navigate the world's oceans. In the twenty-first century, the same ability allows us to navigate the vast expanse of outer space.*

7.1 INTRODUCTION

Organized societies must keep track of time, and the more sophisticated the society, the more accurate its clocks must be. All methods of timekeeping are based on repetitive, or *periodic*, events: the first (yearly) sighting of Sirius on the dawn horizon, the daily passage of the Sun across the sky, the slow drip of water from a leaky container, the swing of a pendulum, the vibration of an atom. In the sixteenth century, each city or town kept its own unique time set by a clock mounted prominently in its central square. When cities were linked by railroads, it became necessary to synchronize their clocks. As long distance travel and communication connected ever more distant communities, the need for precision timekeeping and synchronization became more acute. John Harrison's chronometers solved the "longitude problem," enabling mariners to know their location on Earth to within about 50 miles. Today, the orbiting atomic clocks aboard the Global Positioning System (GPS) satellites allow us to navigate anywhere on Earth with an accuracy better than 10 m.

In Chapter 6, Newton's Laws were used to study motion with *constant acceleration*. Now we will study two new classes of motion: *uniform circular* and *simple harmonic* motion. The two are alike in that they are both periodic. Uniform circular and simple harmonic motion (SHM) are pervasive throughout the physical world. Together with constant-accelerated motion, these three types of motion underpin our understanding of all motion. Surprisingly, the mathematics describing uniform circular motion is nearly identical to the mathematics of SHM, and this affords us the opportunity to study them together. The simple systems we will examine in this chapter are prototypes for the wide variety of more complicated systems—from atoms and molecules to planets, stars, and galaxies—that exhibit the same basic behavior. Likewise, the mathematics we will develop to study these simpler systems is equally applicable to the more complex systems.

Periodic systems are important in their own right, but they are also essential for defining and maintaining the standard *second* and for measuring exceedingly fleet intervals of time. Both of these roles are vital in modern research and technology. Our study of oscillations and circular motion provides an opportunity to examine the concept of time. As you will see, even the simplest questions are difficult to answer.

7.2 UNIFORM CIRCULAR MOTION AND CENTRIPETAL ACCELERATION

A body in *uniform circular motion* is moving with constant speed, but since its direction of motion is continually changing, its acceleration is nonzero. Let's find the acceleration, first by means of geometry, and then by using vectors. Figure 7.1a shows the trajectory of a body moving with speed v along a circular path of radius r. At time t_1, the body's position and velocity are $\vec{r}_1 = \vec{r}(t_1)$ and $\vec{v}_1 = \vec{v}(t_1)$; at time $t_2 = t_1 + \Delta t$,

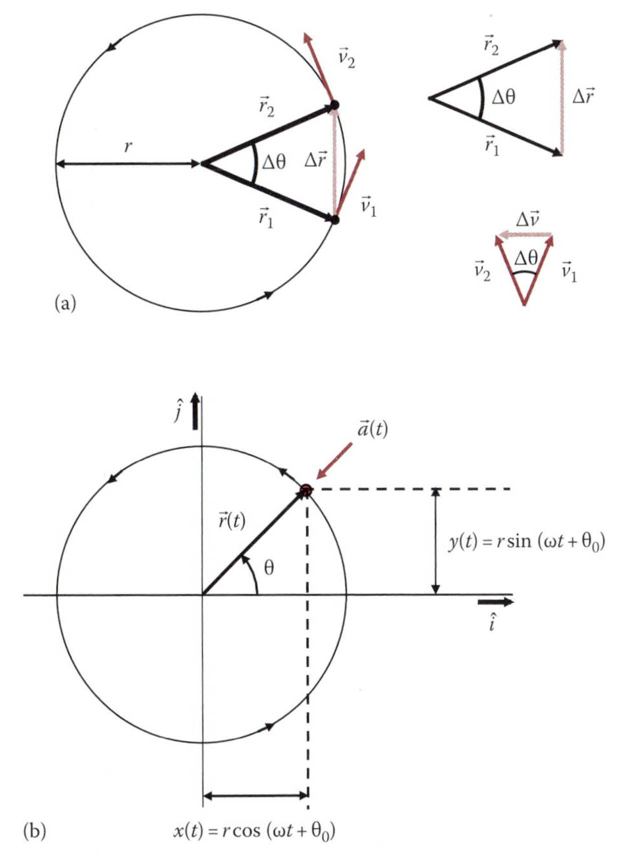

FIGURE 7.1 Position $\vec{r}$ and velocity $\vec{v}$ of a body in uniform circular motion, shown at times t_1 and $t_2 = t_1 + \Delta t$. Since $\vec{v}_1 \perp \vec{r}_1$ and $\vec{v}_2 \perp \vec{r}_2$, the vector triangle constructed from $\vec{r}_1$, $\vec{r}_2$ and $\Delta\vec{r}$ is similar to the triangle made from $\vec{v}_1$, $\vec{v}_2$ and $\Delta\vec{v}$. (b) x- and y-coordinates of $\vec{r}(t)$, with $\theta(t) = \omega t + \theta_0$.

its position and velocity are $\vec{r}_2 = \vec{r}_1 + \Delta\vec{r}$ and $\vec{v}_2 = \vec{v}_1 + \Delta\vec{v}$. Let $\Delta\theta$ be the angle between $\vec{r}_1$ and $\vec{r}_2$. Because the instantaneous velocity $\vec{v}(t)$ is always tangential to the trajectory, the angle between $\vec{v}_1$ and $\vec{v}_2$ is also $\Delta\theta$, so the isosceles triangle formed by the vectors $\vec{r}_1$, $\vec{r}_2$ and $\Delta\vec{r}$ is similar to the triangle formed by $\vec{v}_1$, $\vec{v}_2$, and $\Delta\vec{v}$. Hence, $\Delta r/r = \Delta v/v$. During the time interval Δt, the body moves along a circular arc of length $v\Delta t$. As Δt and $\Delta\theta \to 0$, this arc becomes indistinguishable from the chord Δr, so $v\Delta t \to \Delta r$ and

$$\frac{\Delta r}{r} = \frac{v\Delta t}{r} = \frac{\Delta v}{v}.$$

The magnitude of the acceleration a_c is then given by

$$a_c = \lim_{\Delta t \to 0} \frac{\Delta v}{\Delta t} = \frac{v^2}{r}. \tag{7.1}$$

As shown in Figure 7.1a, the direction of this acceleration is radially inward, or *centripetal*, pointing towards the center of the circle. In uniform circular motion, the magnitude of the acceleration remains equal to v^2/r, and the direction of the acceleration always points toward the center of the circle.

Now let's confirm this by using vectors. The vector expression for the position of a body in uniform circular motion is given by Equation 3.5, with $\theta = \omega t + \theta_0$:

$$\vec{r}(t) = r\cos(\omega t + \theta_0)\hat{i} + r\sin(\omega t + \theta_0)\hat{j}, \tag{7.2}$$

where

θ_0 is the angle between $\vec{r}(t)$ and the $\hat{i}$ direction at $t = 0$

ω is the (constant) angular velocity: $\omega = d\theta/dt$ (see Figure 7.1b)

To find the orbital velocity $\vec{v}(t)$, take the derivative of Equation 7.2 with respect to time:

$$\vec{v}(t) = \frac{d\vec{r}}{dt} = -\omega r\sin(\omega t + \theta_0)\hat{i} + \omega r\cos(\omega t + \theta_0)\hat{j}.$$

EXERCISE 7.1

As a quick review of vectors and the dot product, prove by evaluating $\vec{v}\cdot\vec{v}$ that the speed of the body is given by $v = \omega r$.

Differentiate once again to obtain the acceleration:

$$\vec{a}(t) = \frac{d\vec{v}}{dt} = \frac{d^2\vec{r}}{dt^2} = -\omega^2 r\cos(\omega t + \theta_0)\hat{i} - \omega^2 r\sin(\omega t + \theta_0)\hat{j}$$
$$= -\omega^2\left[r\cos(\omega t + \theta_0)\hat{i} + r\sin(\omega t + \theta_0)\hat{j}\right],$$

or

$$\vec{a}(t) = -\omega^2 \vec{r}(t). \tag{7.3}$$

Equation 7.3 confirms that $\vec{a}(t)$ is antiparallel to $\vec{r}(t)$, so it is indeed centripetal, pointing *radially inward* toward the center of the circle. The magnitude of the acceleration is $a_c = \omega^2 r$, or using $v = \omega r$, $a_c = v^2/r$, just as we found in our geometric solution. Finally, the body executes a full revolution in time $T = 2\pi r/v$, where T is the *period* of revolution, so its speed is given by $v = 2\pi r/T$. The centripetal acceleration can now be expressed in three equivalent ways:

$$a_c = \frac{v^2}{r} = \omega^2 r = \frac{4\pi^2 r}{T^2}. \tag{7.4}$$

Each of these expressions is important, and you should remember them all.

EXERCISE 7.2

Two racecars A and B are abreast of each other as they enter a curved section of the race track. They remain side by side as they navigate the curve.

(a) Which car has the higher speed?
(b) Which has the higher acceleration?
(c) Which car is more likely to skid out of control?

Answer B

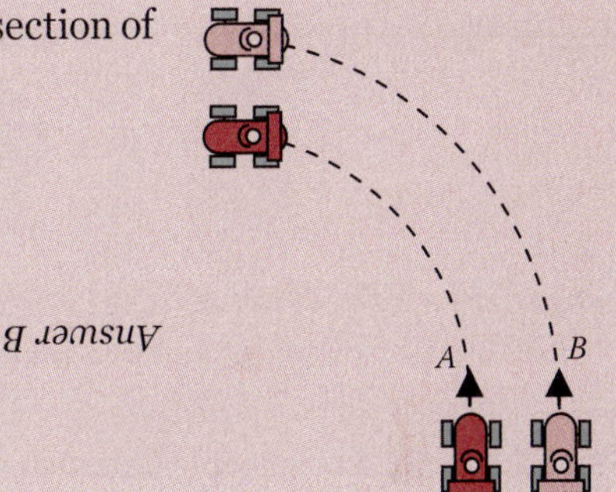

7.3 APPLYING NEWTON'S LAWS TO CIRCULAR MOTION

Uniform circular motion is similar to motion with constant acceleration because the *magnitude* of the acceleration does not change. It differs because the *direction* of the acceleration is always changing. According to Newton's Second Law, a centripetal acceleration requires a *centripetal force* $\vec{F}_c$. For the simple case of a ball twirled by a string, the centripetal force is supplied by the string's tension. For a car rounding a circular curve on a horizontal highway, the force is supplied by the sideways friction exerted on the car's tires by the road. For the protons racing around the Large Hadron Collider, the centripetal force is produced by 1200 bending magnets interspersed along the circumference of the giant ring.

Figure 7.2a shows a ball of mass m hanging by a lightweight string. The ball is drawn aside by an angle θ and given a push. If done just right, the ball will execute a circular orbit at constant speed v, while the

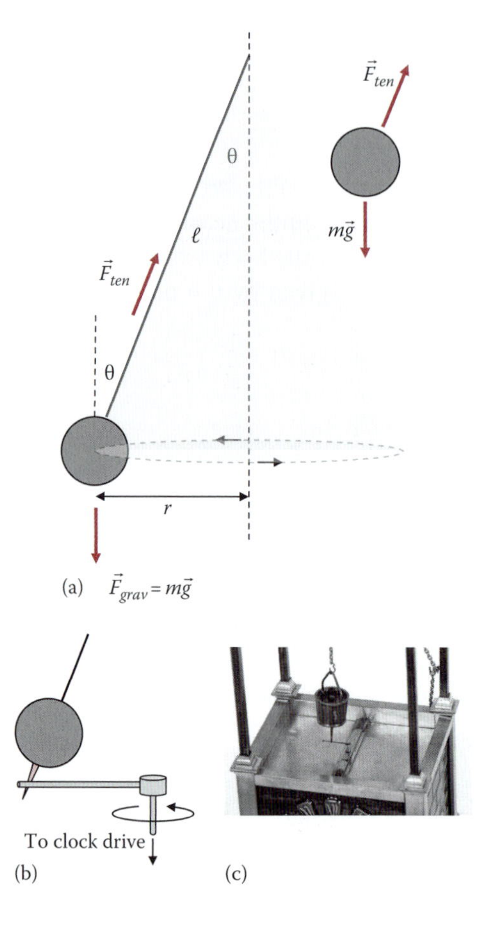

FIGURE 7.2 (a) A conical pendulum made from a string of length ℓ and a ball (bob) of mass m. The radius of the ball's circular path is $r = \ell \sin\theta$. (b) A pin attached to the bob is used to drive a clock mechanism. (c) The drive of an antique conical pendulum clock made in the late 1800s. (Guilmet "wishing well" clock. Image courtesy of liveauctioneers.com and Fontaine's Auction Gallery.)

string will trace out an inverted cone. This is called a *conal pendulum*. It was first studied by Robert Hooke in 1666, who proposed it as a scale model of planetary motion.* As shown in the figure, the ball is subject to two forces: the gravitational force $\vec{F}_{grav} = m\vec{g}$ and the string tension $\vec{F}_{ten}$.† (We will denote the tension by F_{ten} instead of T to avoid confusion with the orbital period.) Circling in the horizontal plane, the ball has zero vertical acceleration, so

$$F_{grav} = mg = F_{ten} \cos\theta. \tag{7.5}$$

* Hooke was mistaken. Mathematically, the conical pendulum is distinctly different from planetary motion.

† Note that $m\vec{a}_c$ is not a force, and so it is not included in the FBD! Centripetal acceleration is the *result* of the applied forces $\vec{F}_{ten}$ and $\vec{F}_{grav}$. This was discussed earlier in Section 6.13.

The ball's centripetal acceleration is due to the horizontal component of the string tension $F_{ten}\sin\theta$, so

$$ma_c = m\frac{v^2}{r} = F_{ten}\sin\theta, \tag{7.6}$$

where r is the radius of the orbit, is related to the string length ℓ by $r = \ell\sin\theta$. Dividing Equation 7.6 by Equation 7.5, we find

$$\tan\theta = \frac{a_c}{g} = \frac{v^2}{rg}. \tag{7.7}$$

EXERCISE 7.3

(a) Show that for small angles ($\theta \ll 1$ rad), θ is proportional to v. (*Hint:* Use the small angle approximations for $\sin\theta$ and $\cos\theta$, and do not forget that r depends on θ.) (b) At what speed would the string be horizontal, i.e., $\theta = 90°$?

Answer (b) $v = \infty$

Let's solve for the period T of the orbit. Using $v = 2\pi r/T$ and $r = \ell\sin\theta$, Equation 7.7 becomes

$$\tan\theta = \frac{4\pi^2\ell\sin\theta}{gT^2},$$

or

$$T = 2\pi\sqrt{\frac{\ell\cos\theta}{g}}.$$

For $\theta \ll 1$, $\cos\theta \simeq 1 - \frac{1}{2}\theta^2 \approx 1$, and the dependence of T on θ is quite weak. Thus, the conical pendulum is nearly *isochronous*, i.e., its period is nearly independent of how it is set into motion. This makes it a good candidate for regulating a clock, as proposed by Huygens in 1673 (see Figures 7.2b and 7.2c). Devices based on the conical pendulum have been used since the seventeenth century to govern the speed of windmills, steam engines, and internal combustion engines. More recently, the conical pendulum has been used to provide a smooth, continuous drive for astronomical telescopes.

The above analysis is common to a wide variety of physical situations. Figure 7.3a shows a race car negotiating a circular turn on a *banked* race track. The road surface is tilted by an angle θ to allow the car to make the turn at high speed without braking. The centripetal force acting on the car is the sum of the horizontal component of the friction $\vec{F}_f$ between the car's tires and the road surface, and the horizontal

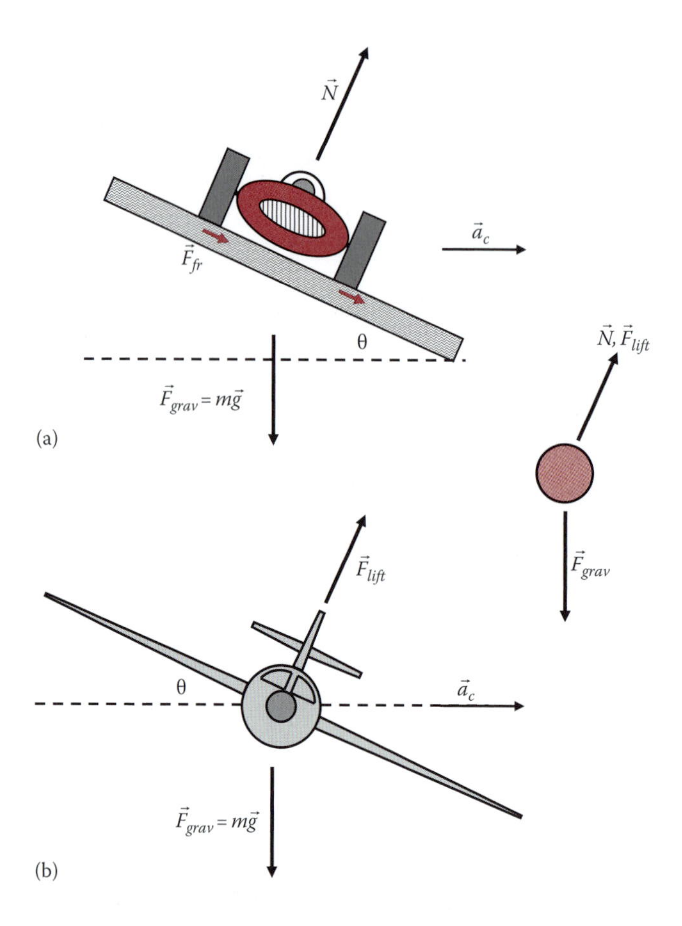

FIGURE 7.3 (a) A race car on a banked track making a circular turn of radius r at speed v. (b) An airplane making a turn of radius r at speed v. Note that the lift force is perpendicular to the wing surface. The FBD's of the two bodies are essentially the same.

component of the normal force $\vec{N}$. There is a speed at which the car can navigate the turn successfully even if the road surface is icy or slick with oil, i.e., even if $\vec{F}_f = 0$. Let's find this speed.

But do you see that we have already solved this problem? Compare Figures 7.2a and 7.3a. If $\vec{F}_{fr} = 0$, and $\vec{F}_{ten}$ is replaced by $\vec{N}$, the free body diagram for the race car looks exactly like the FBD for the ball. Replacing $\vec{F}_{ten}$ with $\vec{N}$ in Equations 7.5 and 7.6, we once again obtain Equation 7.7, where r is now the radius of the turn. The speed at which the car can make the turn without the aid of road friction is found from Equation 7.7 to be

$$v_{no\ friction} = \sqrt{rg\tan\theta}. \tag{7.8}$$

Of course, if $v \neq v_{no\ friction}$, friction forces are necessary to negotiate the curve safely.

EXERCISE 7.4

A circular turn at the Daytona International Speedway in Florida has a radius of 1000 ft (305 m) and is banked at an angle of 31°. (a) Calculate $v_{nofriction}$. (b) If a race car enters the turn with speed 120 mph (53.3 m/s), in what direction is the friction force acting on the tires? (c) What if the car's speed is 80 mph?

Answer (a) 42.4 m/s (b) inward (c) outward

An airplane is held aloft by the lift generated from the flow of air above and below its wings. This lift force $\vec{F}_{lift}$ is perpendicular to the wing surface. As you know from experience, the airplane must bank, or tilt its wings, in order to navigate a turn (see Figure 7.3b). The FBD for the airplane looks just like the one for the race car, with $\vec{F}_{fr} = 0$ and $\vec{F}_{lift}$ replacing $\vec{N}$. This leads to the same relation, Equation 7.8, between the bank angle θ and the speed and radius of the turn. (One slight complication is that, when banking, the vertical component of the lift force is reduced from its full value, so the nose of the plane must be raised slightly in order to maintain altitude.) Although the conical pendulum, the race car, and the airplane are distinctly different physical systems, *inspection* of their FBDs (Figures 7.2, 7.3a and b) shows that they are mathematically equivalent, and so their solutions have the same mathematical form.

EXERCISE 7.5

The stall speed of an Boeing 737 is roughly 70 m/s. What is the minimum turn radius of the aircraft if its bank angle must remain less than 30°?

Answer 870 m

7.4 NEWTON'S LAWS AT THE STATE FAIR

In the above examples, the motion was confined to the horizontal plane, perpendicular to the gravitational force. Now let's look at a case where the motion is in the vertical plane. Imagine that you are visiting the State Fair, where you are about to take a ride on the Ferris wheel. Intent on learning some new physics while enjoying the ride, you have brought along a bathroom scale that you place under you on your seat. The ride begins, and soon you are rotating at a constant speed v. Watching the scale (this may be awkward), you discover that its reading changes with time (see Figure 7.4). When you are at A, the highest point of the ride, the scale reads 450 N (101 lb). At the lowest point (C), it reads 750 N (168 lb). What's going on?

The forces acting on you are the force of gravity ($\vec{F}_{grav} = m\vec{g}$) and the force $\vec{F}_{scale}$ exerted on you by the scale: $m\vec{a} = \vec{F}_{grav} + \vec{F}_{scale}$. In the usual way, we can split the latter into its components parallel (friction) and perpendicular (normal) to the scale's surface: $\vec{F}_{scale} = \vec{F}_f + \vec{N}$. The reading on the scale indicates the magnitude of the normal force $\vec{N}$. At A, your acceleration is directed vertically downward, $\vec{a} = -a_c\hat{j}$, where $a_c = v^2/r$ is the centripetal acceleration, so $\vec{F}_f = 0$ and

$$-ma_c\hat{j} = N_A\hat{j} - mg\hat{j}.$$

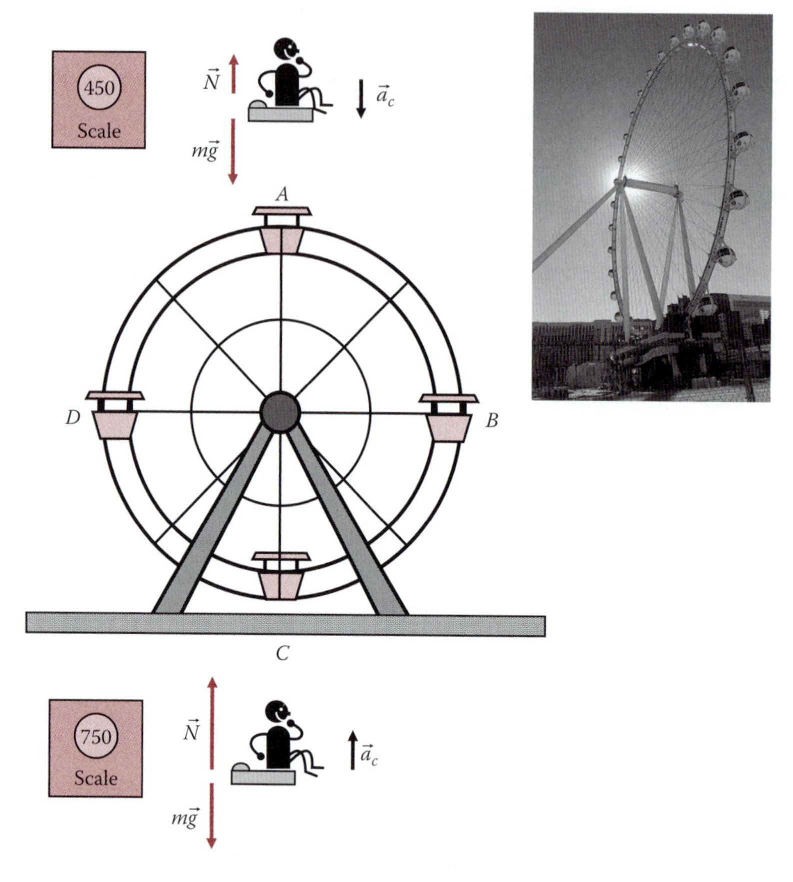

FIGURE 7.4 A student sits on a scale while riding a Ferris wheel. FBD's of the student at the highest (A) and lowest (C) points of the ride are shown. The normal force $\vec{N}$ is the force exerted by the scale on the student. Inset: the world's tallest Ferris wheel is the *High Roller* in Las Vegas, NV. Its diameter is 160 m. (From the public domain.)

At C, the centripetal acceleration is vertically upward ($\vec{a} = +a_c\hat{j}$), so

$$+ma_c\hat{j} = N_C\hat{j} - mg\hat{j}.$$

Eliminating the unit vectors and rearranging,

$$N_A = m(g - a_c),$$

and (7.9)

$$N_C = m(g + a_c).$$

EXERCISE 7.6

Finish the above analysis using $N_A = 450$ N and $N_C = 750$ N.

(a) What is your weight mg in Newtons?
(b) What is the magnitude of the centripetal acceleration? Express a_c as a fraction of g.
(c) If the speed of the Ferris wheel were doubled, what would be your apparent weight at the lowest point of the circle (point C)?

Answer (a) 600 N (b) g/4 (c) 1200 N

In this example, the centripetal acceleration is $a_c = g/4$. When the speed v is doubled, then $a_c = g$ and $N_A \to 0$. Your apparent weight at the top of the ride will be zero! For a brief time, the scale will not exert an upward force on you, so your acceleration will equal $\vec{g}$ and you will be in free fall—just like a bungee jumper! This is what we mean by being "weightless." If you are holding your physics textbook, it will feel weightless to you at point A. The *sensation* of weight does not arise directly from the force of gravity, but rather from the *counter* forces exerted on you (or by you) that prevent you (or your book) from falling with acceleration $\vec{g}$! This simple idea may seem like nothing more than an amusing curiosity, but to Albert Einstein, it was the key to understanding the space–time fabric of the universe!

EXERCISE 7.7

In a favorite lecture demonstration, your instructor pours water into a bucket until it is about half full. Then, holding the bucket by a short rope, he swings it in a vertical circle of radius 1 m. (a) What is the minimum speed of the bucket so that no water is spilled at the top of its swing? (b) Assuming uniform circular motion, what is the maximum period T of the motion? (c) What happens if the rope becomes slack near the top of the swing?

Answer (a) 3.1 m/s (b) 2 s (c) the water will spill

7.5 EINSTEIN'S HAPPIEST THOUGHT: WEIGHTLESSNESS

You have surely seen pictures of astronauts drifting within the International Space Station (ISS), apparently "weightless" even though they are only 400 km above the Earth's surface. You may also have seen images of scientists, astronauts in training, and sometimes students (often undergrads!) floating within NASA's zero-g aircraft* at an altitude of only about 10 km (see Figure 7.5). In either case, the force of gravity is nearly as strong as it is on the Earth's surface, so why do we describe what we see as "weightlessness?" The ISS is in a near circular orbit with centripetal acceleration $\vec{a} \simeq \vec{g}$, and NASA's plane achieves "zero-g" conditions by executing a parabolic arc with vertical acceleration $\vec{g}$. The astronauts in the ISS, and the passengers in the aircraft, are falling—along with their spacecraft or airplane—with acceleration $\vec{g}$. There is no force to

* Affectionately referred to as the "vomit comet."

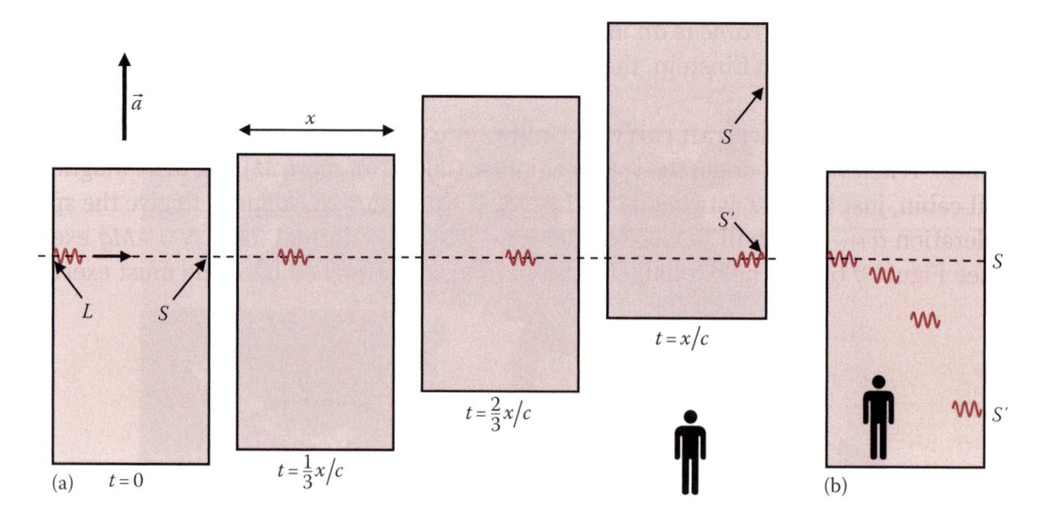

FIGURE 7.7 (a) A space capsule of width x accelerates upward, starting from rest. To an outside observer, a light pulse emitted from a laser L on the left wall propagates in a straight line relative to the observer. (b) To an observer within the spacecraft, the same light pulse follows a parabolic arc with acceleration $a_z = -a$.

force $\vec{F} = -m\vec{g}$ on the ball so that it will accelerate upward along with him. If he releases the ball, the force on it will vanish along with its acceleration. Thereafter, it will approach the cabin floor with relative acceleration $-\vec{a} = \vec{g}$. The ball's motion and the forces exerted on and by Galileo will be exactly as they would be on the Earth's surface (Figure 7.6b). If the cabin has no windows, Galileo will be unable to distinguish between an upward acceleration of the spacecraft (his reference frame) and a downward gravitational field $\vec{g}$. This is the essence of Einstein's *Principle of Equivalence.*

One consequence of the Equivalence Principle is that *gravity bends light.* Imagine a spacecraft drifting in deep space carrying an astronaut equipped with a laser pointer (Figure 7.7). If the craft is moving with constant velocity, the laser beam crosses the space capsule in a straight line, striking the wall at point S. But if the spacecraft has an upward acceleration $\vec{a} = -\vec{g}$, the beam strikes the wall at S', a distance $\Delta z = \frac{1}{2}at^2$ below S, where $t = x/c$ is the time taken by the light to cross the cabin. This is exactly what we would expect to happen. But here's the catch: if the spacecraft is at rest (on its launch pad, perhaps) or moving with constant velocity *in a gravitational field* $\vec{g}$, and if there is no way for the astronaut to distinguish between acceleration and gravity, then the astronaut will observe the same thing: the light beam will be deflected downward by $\Delta z = \frac{1}{2}gt^2$. Contemplating these things, Einstein began to doubt that gravity is an attraction between bodies in the same sense as electrically charged bodies are attracted to one another. What physical property (analogous to charge) would ensure that $m^{iner} = m^{grav}$ *exactly,* so that the free-fall acceleration of all bodies is *exactly* equal to $\vec{g}$? Perhaps, thought Einstein, $\vec{g}$ is better understood as a property of three-dimensional space and one-dimensional time, or four-dimensional *spacetime.* To paraphrase Princeton physicist John A. Wheeler, "Matter tells spacetime how to bend, and spacetime tells matter how to move." What could he mean by bending space and time? For a brief introduction, see Box 7.1.

BOX 7.1 RELATIVITY AND TIME

In the *Principia*, Isaac Newton declared his steadfast belief in *absolute space and time*; that is, that all observers everywhere will agree on the length of a body, or distance between points, and on the elapsed time between events. But the principles of relativity and equivalence are incompatible with this view. In the early twentieth century, Albert Einstein proved that an observer's perception of space and time is influenced by the observer's speed and location in a gravitational field. Since this chapter is devoted to periodic motion and time, let's concentrate on the concept of time in Einstein's theories of relativity.

Let's first ignore gravity and examine how speed affects the passage of time. Imagine an observer A, riding on a moving railroad car, equipped with a laser gun and a mirror mounted a distance ℓ vertically above the laser (see Figure B7.1a). The laser emits short pulses of light that are reflected by the mirror and return to the source a few nanoseconds after emission. According to A, the pulses travel a total distance 2ℓ at the speed of light c, so the total time needed for the round trip is $\Delta\tau = 2\ell/c$. This is the elapsed

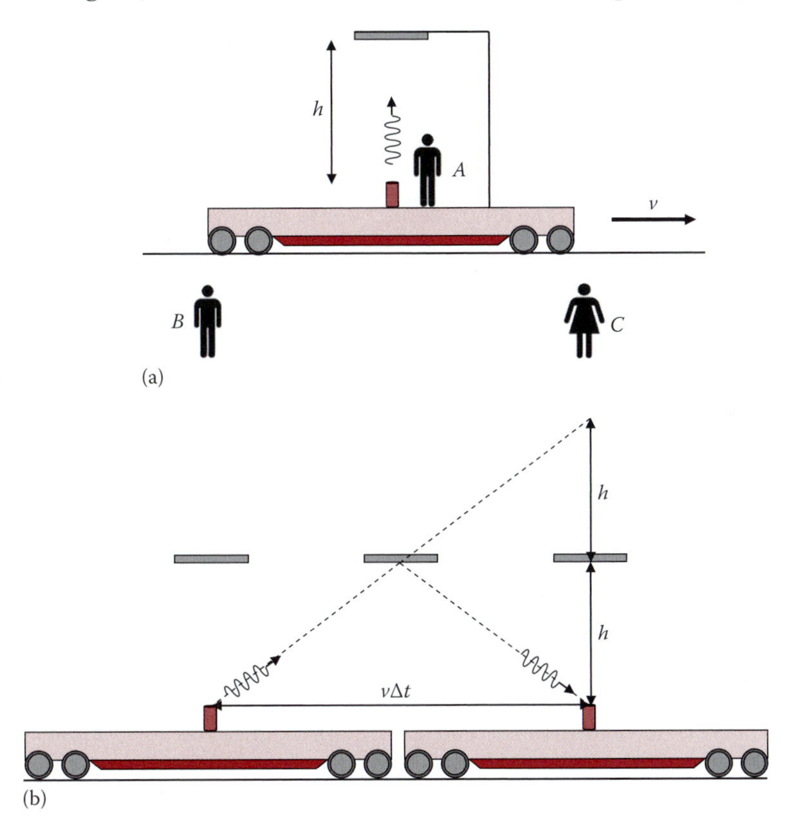

FIGURE B7.1 An experiment with light on a moving railroad car. (a) To observer A, moving with the light source and detector, the light pulses travel vertically, and after reflection from the mirror, return to the source after traveling a total distance $2h = c\Delta\tau$, where τ is the wristwatch time. (b) To ground-based observers B and C, the light travels a longer triangular path of length $\sqrt{(2h)^2 + v^2\Delta t^2} = c\Delta t$, where Δt is the time B and C calculated using their two clocks.

time recorded by A using a single clock, and is referred to as *proper* time or, in the more picturesque language of John Wheeler, *wristwatch* time. Now the train is moving with speed v relative to the tracks, so to ground-based observers B and C, the light pulses travel diagonally (Figure B7.1b) rather than vertically, arriving back at the laser after traveling a total distance $\sqrt{(2\ell)^2 + (v\Delta t)^2}$, where Δt is the elapsed time determined by B and C after comparing the readings on their separate clocks. The principle of relativity asserts that the speed of light is the same for all observers, so for our ground-based observers B and C, $c\Delta t = \sqrt{(2\ell)^2 + (v\Delta t)^2}$, or $\Delta t^2 = 4\ell^2/(c^2 - v^2)$. But $\Delta\tau^2 = 4\ell^2/c^2$, so the two measurements of time are not the same:

$$\Delta t = \frac{\Delta\tau}{\sqrt{1 - v^2/c^2}} = \gamma\Delta\tau, \qquad \text{(B7.1)}$$

where γ is the Lorentz factor introduced in Chapter 4. Since $\gamma \geq 1$, $\Delta t \geq \Delta\tau$, an effect known as relativistic time dilation, and described succinctly as "Moving clocks run slow."

At low speeds, time dilation is not important. But even at zero velocity, the equivalence principle implies that gravity influences the passage of time. Imagine that you (observer A) are in a space capsule in deep space, floating freely in the absence of gravity (Figure B7.2). Your fellow crew members B and C are stationed at the bottom and top of the ship's cabin, where they are equipped with a light source (at B) and a light detector (at C). The height of the cabin is h. The ship's rocket engines are ignited, and the spacecraft and your two companions are propelled upward with acceleration $\vec{a} = -\vec{g}$. At time t_B, B transmits a light wave to C, which arrives at time t_C, where $t_C - t_B = h/c$. At t_B, when the light wave was emitted, the source was moving toward you with speed $v_B = gt_B$, and at t_C, when the light wave was detected, the detector was moving away from you with speed $v_C = gt_C = g(t_B + h/c) > v_B$. Therefore, the light is detected by C with lower frequency, Doppler shifted by $\Delta f/f = -(v_C - v_B)/c$, or

$$\frac{\Delta f}{f} = \frac{f_C - f_B}{f_B} = -\frac{gh}{c^2}, \qquad \text{(B7.2)}$$

where $f_B(f_C)$ is the light frequency measured at the source B (detector C). So far, this is just the standard Doppler shift equation for the case where source and detector are accelerated. But here's the twist. Suppose the spacecraft were still on its launch pad on Earth, and you were falling freely from the top to the bottom of the cabin. According to the equivalence principle, you cannot distinguish between the upward acceleration $\vec{a} = -\vec{g}$ of the spacecraft and your own free fall with acceleration $\vec{g}$. Therefore, you must measure the same Doppler shift given by Equation B7.2, even though *you* are the one moving, not the spacecraft! This is called the *gravitational red shift.*

The frequency of a wave is the number of wavecrests N emitted (or detected) per second. In a time interval Δt_B, the number of wavecrests counted by B is $N = f_B\Delta t_B$. These must all propagate to C, where they are detected with lower frequency f_C over a time interval Δt_C. Since none of the wavecrests are lost

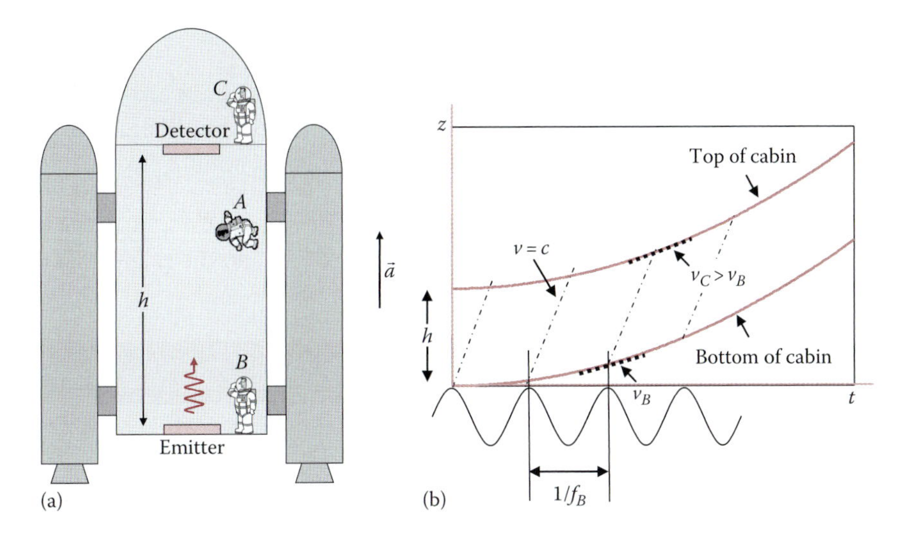

FIGURE B7.2 (a) A spacecraft located in deep space is equipped with a light (or radio) emitter and detector, separated by a vertical distance h. The free-floating astronaut concludes that the light (or radio wave) is redshifted as it propagates from the emitter to detector. She must reach the same conclusion in a gravitational field $\vec{g}$ with zero acceleration. (b) The velocity v_B of the source at the time of emission is less than the velocity v_C of the detector at the time of detection. According to the free-floating astronaut, this is the reason for the red shift.

along the way, $N = f_B \Delta t_B = f_C \Delta t_C$, or $\Delta t_C / \Delta t_B = f_B / f_C$. In other words, $\Delta t_C > \Delta t_B$, which implies that the higher altitude clock must be "ticking" faster than the lower one! Using $f_C = f_B + \Delta f$ and $gh/c^2 \ll 1$,

$$\frac{\Delta t_C}{\Delta t_B} = 1 + \frac{gh}{c^2}. \tag{B7.3}$$

These results do not depend on the construction of the instruments used to measure time or red shift. The rate of all physical processes—including biological processes—is governed by Equations B7.1 and B7.3. In other words, these equations are not about clock rates, but about the advance of time itself. Newton's assumption of absolute time, advancing at the same rate for all observers, is simply untrue. As you will see, this has deep significance for how we interpret gravity, and is also critically important for space-age navigation.

Exercise B7.1: In September 2005, software engineer Tom Van Baak loaded his minivan with his three kids, plus three cesium atomic clocks, batteries and controlling electronics, and headed for a weekend on Mount Rainier in Washington. Overall, he spent roughly 42 h at a lodge high on the mountain. Upon his return home, he compared his traveling clocks to his stay-at-home clocks, and found that the traveling clocks were ahead by about 22 ns. What was the altitude of the lodge relative to his home?

Answer 1340 m

EXERCISE 7.8

Why don't we notice the gravitational bending of light? Imagine a laser beam crossing a room 10 m wide. (a) How long does it take the beam to cross the room? (b) By how much is the beam deflected? A hydrogen atom has a diameter of about 0.1 nm, and the diameter of its nucleus is about 1 fm = 1×10^{-15} m. Compare your answer to (b) with these numbers.

Answer (a) 33 ns (b) 5.4 fm

7.6 SIMPLE HARMONIC MOTION

Today's world requires highly sophisticated clocks. Sporting events (like the 100 m dash) are timed in hundredths of a second; high speed photography can capture microsecond events; and computer instructions are spooled at nanosecond intervals. At the heart of any clock lies an oscillator, and the more precisely repeatable its behavior, the more accurate the clock. Oscillators are found in all fields of physics and in all branches of science and engineering. They abound on all size scales, from subatomic to galactic, with *periods* that range from 10^{-22} s for nuclei, to 10^{-13} s for molecules, to millions of years for galaxies. Electromagnetic waves are generated by oscillating electric charges, and sound waves by vibrating reeds, strings, vocal chords, etc. In astronomy, the Sun's surface pulsates with periods of several minutes, and the Cepheid variable stars—which form a vital rung on the cosmological distance ladder—swell and shrink with periods of a few days to weeks. Indeed, the existence of life on Earth may be synchronized to the rhythm of a cosmic-scale oscillation: the motion of our Solar System perpendicular to the plane of the Milky Way galaxy. This oscillation is suspected to be linked to the mass extinctions of life that occur every 62 million years.

The oscillating systems we will study in this chapter are simpler ones: lab-scale mechanical devices with periods of the order of 1 s, such as a mass attached to a spring, or a simple pendulum. Even though these systems differ radically from nuclear, molecular and stellar oscillators, they are mathematically equivalent to their more exotic cousins. Within limits, the mass–spring system—or the pendulum—serves as a model for all other oscillating systems. As an added bonus, the mathematics of SHM is closely related to the mathematics of uniform circular motion.

7.7 PROTOTYPE OSCILLATOR: THE MASS–SPRING SYSTEM

An *ideal* spring is one that changes its length by an amount $\Delta\ell$ that is proportional to an applied force F. Whether the spring is stretched ($\Delta\ell > 0$) or compressed ($\Delta\ell < 0$), it obeys *Hooke's Law* $F = k\Delta\ell$, where k is called the *spring constant*, or stiffness, and has the dimensions of N/m* (see Figure 7.8a). This idealized

* Discovered by Newton's archrival Robert Hooke in 1660, and called *Hooke's Law.*

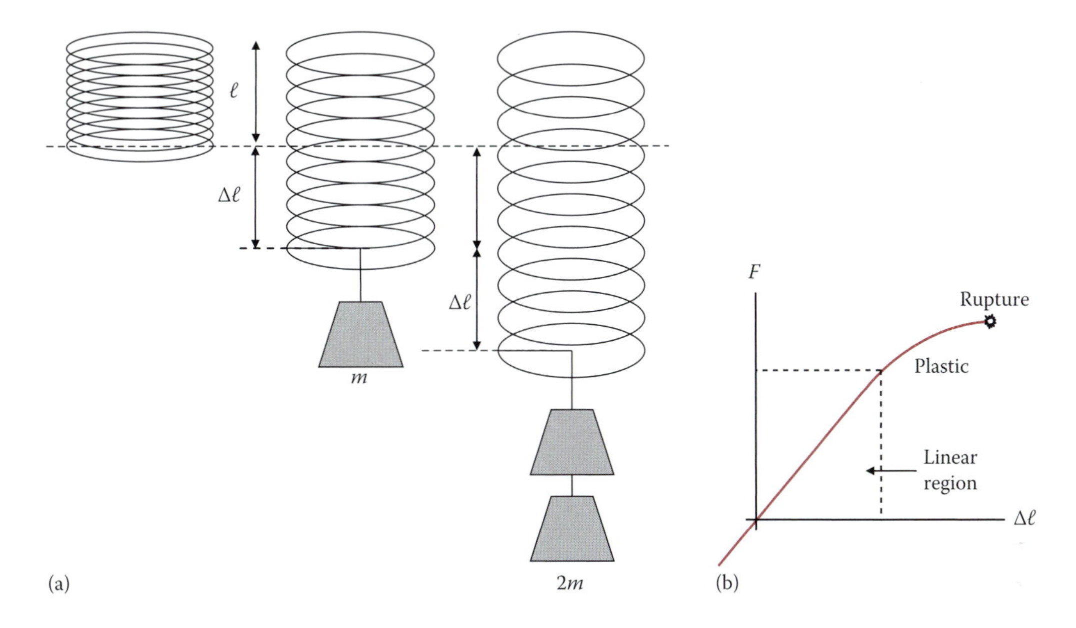

FIGURE 7.8 (a) An ideal spring stretches by an amount proportional to the force applied to it. A hanging mass m stretches the spring by an amount $\Delta\ell$. If the mass is doubled, the stretch of the spring is doubled. (b) For a real spring, $F_{sp} = -k\Delta\ell$ over a limited range of $\Delta\ell$. Beyond that, the spring's response is *nonlinear.*

linear relationship breaks down when $\Delta\ell$ becomes large, but faithfully describes the behavior of most real springs in the limit of small $\Delta\ell$. Throughout this chapter, we will always assume $\Delta\ell$ is small enough that Hooke's Law is obeyed (Figure 7.8b).

Imagine a block of mass m placed on a horizontal frictionless tabletop. The block is attached to one end of an ideal spring, and the other end is secured to a rigid support (see Figure 7.9). When $\Delta\ell = 0$, the spring is neither stretched nor squeezed, so it exerts no force on m. The position where the total force on m is zero is called its *equilibrium position*, and we will locate our origin of coordinates ($x = 0$) there. If we apply a force $F = kx$ to the block, pulling it to the right by a distance $x > 0$, the spring will stretch by the same amount $\Delta\ell = x$ and will exert a counterforce in the negative x-direction: $F_{sp} = -F = -kx$. If we instead compress the spring by pushing the block to the left ($x < 0$), the spring will exert a force to the right ($F_{sp} > 0$) obeying the same equation: $F_{sp} = -F = -kx$. The spring force $\vec{F}_{sp}$ always attempts to restore the spring to its relaxed length, and because it is directly proportional to the displacement x from equilibrium, it is called a "*linear restoring force.*" In terms of vectors, the force exerted by the spring is

$$\vec{F}_{sp} = -kx\hat{\imath}. \tag{7.10}$$

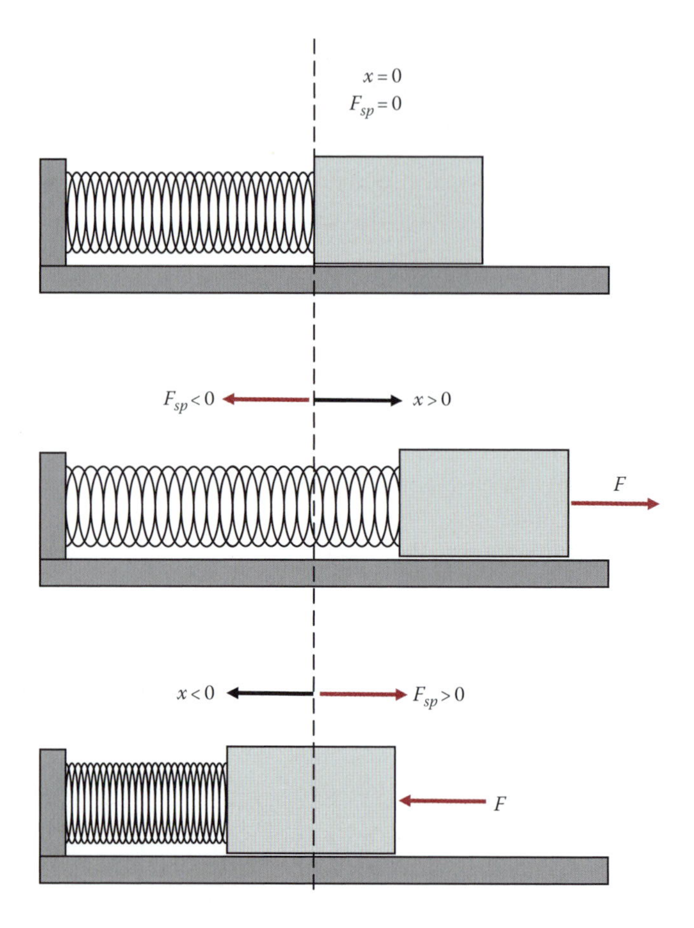

FIGURE 7.9 The spring force is a restoring force, always pointing toward the equilibrium position $x = 0$, where $\vec{F}_{tot} = 0$.

The negative sign in Equation 7.10 indicates that $\vec{F}_{sp}$ is a restoring force. This negative sign, plus the linear dependence of $\vec{F}_{sp}$ on the displacement from equilibrium, are critical to the discussion below.

EXERCISE 7.9

When a block of mass m is hung vertically from a certain coil spring, it stretches the spring by 8 cm. The same block is now placed on a frictionless surface inclined from the horizontal by 30°. When the block is at its equilibrium position (where the *total* force on it is zero), by how much is the spring stretched?

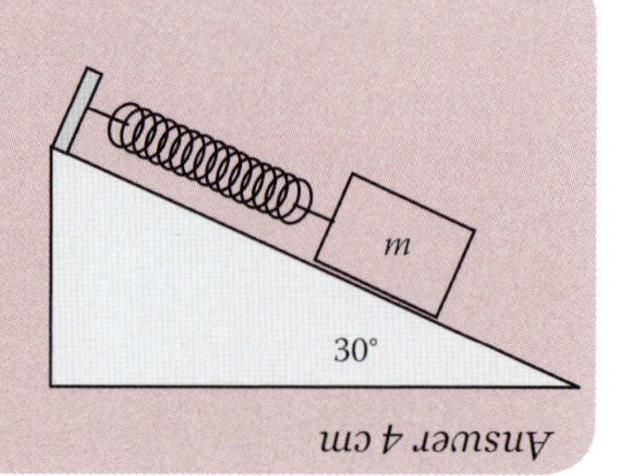

Answer 4 cm

What happens when we set the mass in motion, with no horizontal force other than $\vec{F}_{sp}$ acting on it? As usual, we will use Newton's Laws to predict the motion. But now the task is more complicated: even though the motion is only one dimensional, the applied force and thus the acceleration change with position and time. From Equation 7.10, $F_{sp} = ma_x = -kx$, or

$$a_x(t) = \frac{d^2x}{dt^2} = -\frac{k}{m}x(t). \tag{7.11}$$

This is a new kind of mathematical expression. Not only does it contain the variable $x(t)$, it also contains the second derivative of x. Equations of this sort, which contain variables and their derivatives, are called *differential equations*, and they are ubiquitous in the physical sciences. Equation 7.11 encapsulates all of the physics of the mass–spring oscillator, and—by extension—governs the behavior of a huge variety of oscillating systems on all size and time scales. It is therefore an important equation, and we will derive its solution in appropriate detail.

How do you solve a differential equation? The answer is easy: you *guess*! Just like solving a crossword puzzle, you search for solutions that fit the clue, and choose the one that fits in with the rest of the puzzle. The clue is the equation itself: Equation 7.11 asks us to find a function of time $x(t)$ whose second derivative is equal to the function itself multiplied by a negative constant $(-k/m)$. Since the motion is periodic, the function we are looking for must also be periodic. There are two familiar functions that fit these clues: the sine and the cosine: $(d^2/d\theta^2)\cos\theta = (-1)\cos\theta$, and likewise for the sine. Both functions are periodic in 2π (rad), e.g., $\cos(\theta + 2\pi) = \cos\theta$, and in either case the negative constant multiplying the function is equal to (-1). Let's make an "educated" guess, and ask if the function $x(t) = A\cos(\omega t + \theta_0)$ satisfies Equation 7.11,* where A, ω and θ_0 are unknown constants. Let $u(t) = \omega t + \theta_0$, and use the chain rule to find the first and second derivatives of this trial solution:

$$\frac{d}{dt}A\cos u(t) = \frac{d}{du}(A\cos u)\frac{du}{dt} = -A\sin u\frac{du}{dt} = -\omega A\sin(\omega t + \theta_0),$$

and

$$\frac{d^2}{dt^2}A\cos u(t) = \frac{d}{du}(-\omega A\sin u)\frac{du}{dt} = -\omega^2 A\cos(\omega t + \theta_0),$$

* Notice that the argument of the cosine, $\theta(t) = \omega t + \theta_0$, is familiar from our discussion of circular motion.

where we have used $du/dt = \omega$. Since $x(t) = A\cos(\omega t + \theta_0)$, the last equation may be rewritten as

$$\frac{d^2x}{dt^2} = -\omega^2 x.$$

If our trial solution is to satisfy Equation 7.11, then $\omega^2 = k/m$. The solution to Equation 7.11 is therefore given by:

$$x(t) = A\cos(\omega t + \theta_0),$$

where (7.12)

$$\omega = \sqrt{\frac{k}{m}},$$

and the two constants A and θ_0 are yet to be determined.

EXERCISE 7.10

Show that $\sqrt{k/m}$ has the same dimensions as ω.

If $\theta_0 = 0$, $x(t)$ is the simple cosine curve shown in Figure 7.10a. The maximum value of $x(t)$, i.e., the largest displacement of m from its equilibrium position, is called the *amplitude* of the oscillation and is equal to A. $x(t) = A$ whenever $\cos(\omega t) = 1$, that is, when $\omega t = 0, 2\pi, 4\pi, \ldots, n\pi$. The *period* T of the oscillation is the time required for the argument of the cosine to change by 2π rad, $\omega(t + T) = \omega t + 2\pi$, or

$$T = \frac{2\pi}{\omega} \equiv \frac{1}{f}, \tag{7.13}$$

where f is the *frequency* of the oscillation,* the number of repetitions per second, measured in hertz (Hz) or s^{-1}.

For $\theta_0 = 0$, the initial displacement is $x_0 = x(0) = A\cos(0) = A$, and the initial velocity (the slope of $x(t)$ at $t = 0$) is $v_0 = v(0) = -\omega A\sin(0) = 0$. Together, x_0 and v_0 are called the *initial conditions* of the problem. They reflect how the motion was initiated at $t = 0$. For $\theta_0 = 0$, the body was pulled aside by a distance $x_0 = A$ and then released from rest ($v_0 = 0$).

If $\theta_0 \neq 0$, then the cosine curve is shifted to the right (for $\theta_0 < 0$) or left (for $\theta_0 > 0$) along the time axis (Figure 7.10b). The initial conditions are now $x_0 = A\cos\theta_0$ and $v_0 = -\omega A\sin\theta_0$, but ω and the period T

* Both f and ω are commonly called the frequency of oscillation, although ω is more properly called the *angular* frequency or angular velocity.

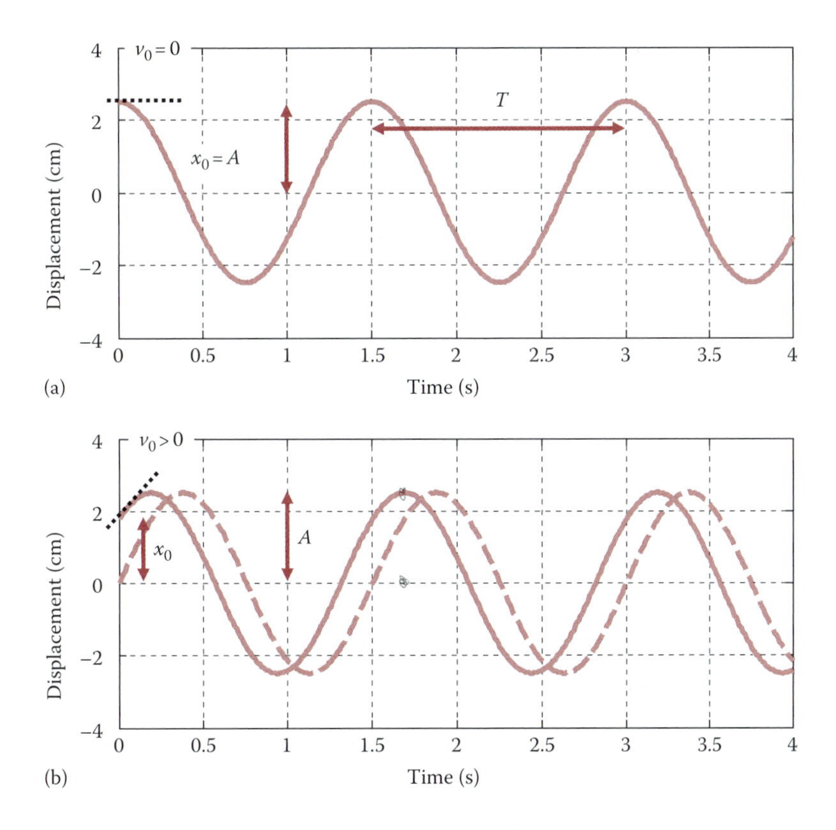

FIGURE 7.10 (a) SHM with $x(t) = A\cos\omega t$, for $\theta_0 = 0$, $A = 2.5$ cm, and $\omega = 2\pi/T = 2\pi/1.5 = 4\pi/3\ \text{s}^{-1}$. The initial velocity (slope at $t = 0$) $v_0 = 0$. (b) SHO with (solid line) $x(t) = A\cos(\omega t - \pi/4)$ and (dashed line) $x(t) = A\cos(\omega t - \pi/2) = A\sin(\omega t)$, with $A = 2.5$ cm and $\omega = 4\pi/3$. In the latter case (dashed line), $x_0 = x(0) = 0$.

remain the same. Knowing the initial conditions, we can derive the values of A and θ_0. To see how, try the following exercise.

EXERCISE 7.11

A system consists of an ideal spring with stiffness $k = 8$ N/m attached to a 0.5 kg mass that slides without friction over a horizontal surface, with $x(t) = A\cos(\omega t + \theta_0)$. At $t = 0$, the displacement and velocity are equal to 0 and 5 cm/s.

(a) What are the values of ω and the period T?
(b) Find the value of θ_0.
(c) What is the amplitude A of the oscillation?

Answer (a) $\omega = 4\ \text{s}^{-1}$, $T = \pi/2$ s (b) $\theta_0 = -\pi/2$ (c) $A = 1.25$ cm

Notice that if $\theta_0 = -\pi/2$, as in the above example, the original cosine curve shown in Figure 7.10a shifts to the right by one quarter of a cycle, and transforms into a sine curve: $x(t) = A\cos(\omega t - \pi/2) = A\sin\omega t$ (Figure 7.10b). This suggests that we could have chosen $x(t) = A\sin(\omega t + \theta_0')$ for our trial function, where $\theta_0' = \theta_0 + \pi/2$, and we would have arrived at an equivalent solution to the problem.

EXERCISE 7.12

The motion of an oscillating body is given by $x(t) = A\cos(\omega t - \pi/3)$. An equivalent expression for the displacement is $x(t) = A\sin(\omega t + \pi/6)$. Show that x_0 and v_0 are the same for both expressions.

It is instructive to compare the expressions for $x(t)$, $v(t)$ and $a(t)$. For any system undergoing SHM, $x(t) = A\cos(\omega t + \theta_0)$, and

$$v(t) = \frac{dx}{dt} = -\omega A\sin(\omega t + \theta_0),$$

and (7.14)

$$a(t) = \frac{dv}{dt} = -\omega^2 A\cos(\omega t + \theta_0) = -\omega^2 x(t).$$

Since $\sin\theta = \pm 1$ when $\cos\theta = 0$, the speed $|v(t)|$ is a maximum when $x(t) = 0$, and the acceleration $a(t) \propto -x(t)$ is maximally positive (negative) when $x(t)$ is maximally negative (positive) (see Figure 7.11). These statements are true for all SHM. No matter what system you are studying, you can be sure that it undergoes sinusoidal oscillation described by $x(t) = A\cos(\omega t + \theta_0)$ if it is subject to a linear restoring force (e.g., $\vec{F}_{sp} = -kx\hat{i}$), so that its differential equation of motion has the form of Equation 7.11.

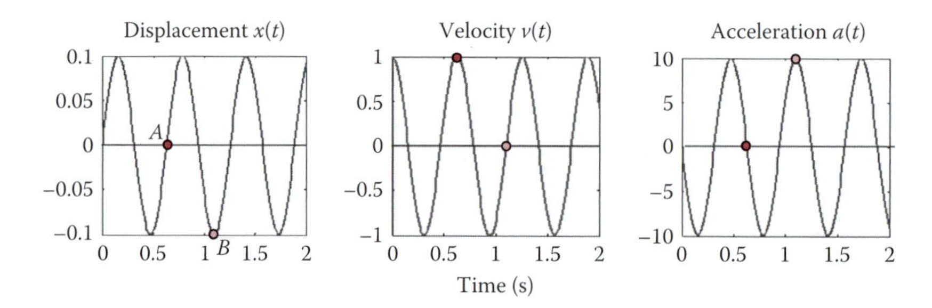

FIGURE 7.11 The displacement, velocity, and acceleration of a body during SHM. At t_A, when $x(t) = 0$, the acceleration is also zero but the speed is a maximum. At t_B, when $|x(t)| = x_{max}$, the velocity is zero but the acceleration is a maximum.

EXERCISE 7.13

The motion of an oscillating body is given by $x(t) = 10\cos(30t - \pi/6)$ cm, where x is its displacement from equilibrium.

(a) Estimate the initial position x_0 of the mass.
(b) Estimate the initial velocity v_0.
(c) What is the *position* when the acceleration has its maximum positive value?
(d) What is (are) the position(s) when the speed of the body is maximum?
(e) What is (are) the position(s) when the velocity equals zero?

Answer (a) ≈8.7 cm (b) 150 cm/s (c) −10 cm (d) 0 cm (e) ±10 cm

7.8 EXAMPLE: COLLISION WITH AN OSCILLATOR

Let's tackle a more interesting problem, one that combines SHM with the physics of collisions. Imagine a 100 g block (m_1) that is attached to a spring of stiffness k = 10 N/m and rests on a horizontal frictionless surface. A 50 g block (m_2), sliding along the same surface with speed 1.5 m/s, collides elastically with the first block (see Figure 7.12a). Let's find the ensuing motion of both bodies.

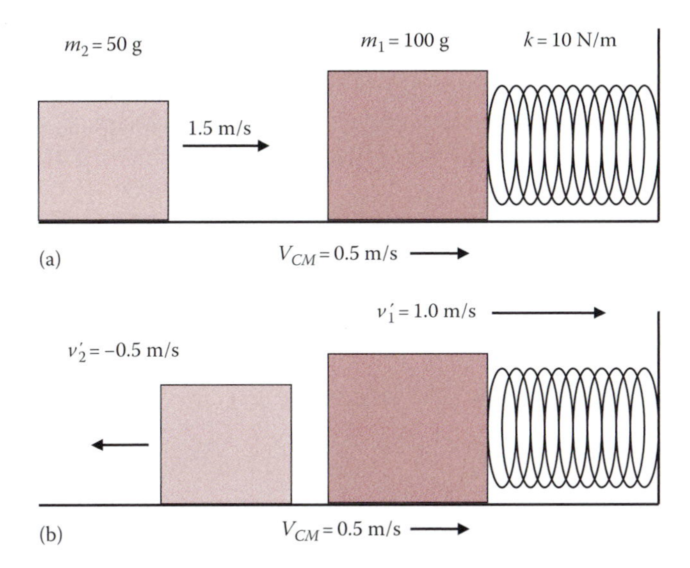

FIGURE 7.12 (a) A sliding block of mass m_2 = 50 g collides with a stationary block of mass m_1 = 100 g, initiating SHM in the latter block. The CM velocity before the collision, and immediately after the collision, is V_{CM} = 0.5 m/s. (b) Velocities of the two blocks immediately after the collision.

Begin by considering just the collision. Since it is nearly instantaneous, m_1 hardly moves while the blocks are in contact, so the spring remains at its relaxed length and exerts no force on m_1 during the collision. Therefore the spring has no influence on the velocity of either block *immediately* after the impact, so the CM velocity just after the collision is the same as it was initially:

$$V_{CM} = \frac{(50\times 1.5 + 100\times 0)}{(50+100)} = 0.5\ \text{m/s}.$$

Before the collision, $u_1 = v_1 - V_{CM} = -0.5$ m/s and $u_2 = +1.0$ m/s. Afterwards, the velocities relative to the CM reverse themselves, so $u_1' = +0.5$ m/s and $u_2' = -1.0$ m/s. Transforming back to the lab frame, $v_1' = V_{CM} + u_1' = 1.0$ m/s and $v_2' = -0.5$ m/s.

Once the two bodies separate, m_1 undergoes SHM. Its position is given by $x(t) = A\cos(\omega t + \theta_0)$, where $\omega = \sqrt{k/m_1} = \sqrt{10/0.1} = 10\ \text{s}^{-1}$. Let $t = 0$ be the time of impact, so the initial conditions for the motion of m_1 are: $x_0 = 0$ and $v_0 = v_1' = 1.0$ m/s. Now we can solve for A and θ_0,

$$x_0 = x(0) = 0 = A\cos\theta_0,$$

and

$$v_0 = v(0) = 1.0 = -\omega A\sin\theta_0.$$

From the first equation, $\theta_0 = \pm\pi/2$, but since $v_0 > 0$, the second equation compels us to select $\theta_0 = -\pi/2$ (as in Exercise 7.11). The amplitude can now be found from the initial velocity: $v_0 = 1.0\ \text{m/s} = -\omega A\sin(-\pi/2) = \omega A$, so $A = 0.1\ \text{m} = 10\ \text{cm}$. The motion of the 100 g block after the collision is therefore given by

$$x(t) = 0.1\cos\left(10t - \frac{\pi}{2}\right)\ \text{m},$$

or equivalently,

$$x(t) = 0.1\sin(10t).$$

EXERCISE 7.14

Suppose the collision is totally inelastic. (a) Find the values of x_0 and v_0, and from them find (b) θ_0 and A. (c) What is the frequency ω?

Answer (a) $x_0 = 0$, $v_0 = 0.5$ m/s (b) $\theta_0 = -\pi/2$, $A = 6.1$ cm (c) $\omega = 8.2\ \text{s}^{-1}$

7.9 "SIMPLE" PENDULUM

The simple pendulum is defined as a small object of negligible spatial extent (called a "bob") hanging from a massless string. In contrast to the mass–spring system, it has no spring or other elastic element, and the restoring force acting on the bob is due to the string tension $\vec{F}_{ten}$ which arises in opposition to gravity. In the horizontal mass–spring oscillator, gravity was not a factor. Yet despite their stark physical differences, the two systems behave the same: they each undergo SHM. Figure 7.13 shows the FBD of a bob m suspended from a string of length ℓ. At time t, the string makes an angle $\varphi(t)$ with respect to the vertical, displacing the bob horizontally by $x(t) = \ell \sin\varphi(t)$ from its position when the string is vertical. The forces acting on m are gravity and the string tension, so by Newton's Second Law, $m\vec{a} = \vec{F}_{ten} + m\vec{g}$, with x- and y-components

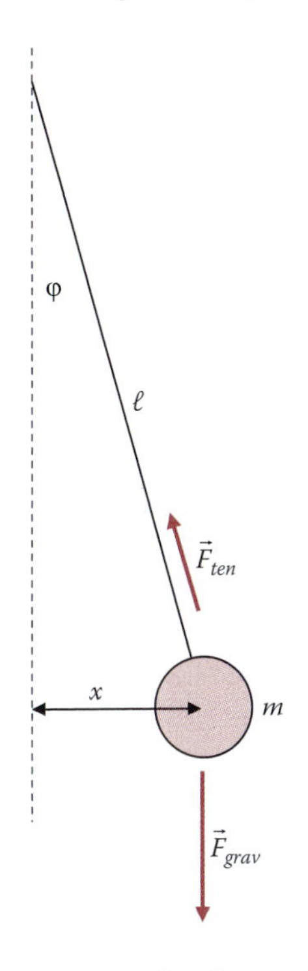

FIGURE 7.13 A simple pendulum consisting of a string of length ℓ and a ball of mass m. Unlike the conical pendulum, the ball swings in a vertical plane. The restoring force is supplied by the string tension $\vec{F}_{ten}$. Do not confuse the geometric angle φ with the phase angle θ_0 that appears in the equation $x(t) = A\cos(\omega t + \theta_0)$.

$$ma_x = -F_{ten}\sin\varphi = -F_{ten}\frac{x}{\ell},$$

and

$$ma_y = F_{ten}\cos\varphi - mg.$$

In general, $\vec{F}_{ten}$ depends on φ, and the two equations are difficult to solve exactly. But if $\varphi_{max} \ll 1$, the motion of the pendulum bob is almost entirely horizontal, so $a_y \simeq 0$ and $F_{ten}\cos\varphi \simeq mg$. In the same small angle limit, $\cos\varphi \simeq 1 - \frac{1}{2}\varphi^2 \simeq 1$ (see Box 1.1), so $F_{ten} \simeq mg = \text{const}$. This approximation greatly simplifies the analysis. Substituting for F_{ten} in the first equation,

$$a_x = \frac{d^2x}{dt^2} = -\frac{F_{ten}x}{m\ell} = -\frac{g}{\ell}x. \quad (7.15)$$

A quick inspection shows that Equation 7.15 has the same mathematical form as Equation 7.11. The only difference is the substitution of g/ℓ for k/m. Therefore, the solution to Equation 7.15 is the same as the solution to Equation 7.11, $x(t) = A\cos(\omega t + \theta_0)$, with

$$\omega = \sqrt{\frac{g}{l}} \quad \text{and} \quad T = 2\pi\sqrt{\frac{l}{g}}. \quad (7.16)$$

EXERCISE 7.15

A "grandfather" clock uses a pendulum with a period of 2 s to regulate its movement. What is the length of the pendulum?

Answer: 99.3 cm

Any system whose differential equation of motion can be written in the form $a_x = -C^2x$, where C^2 is a positive constant (for example, $C^2 = k/m$ or $C^2 = g/\ell$), is mathematically equivalent to the mass–spring system. Its motion will be simple harmonic, or *sinusoidal*, with frequency $\omega = C$. (We will encounter another example of this in the following section.) Remember, though, that we used the small angle approximation for $\cos\varphi$ in the derivation of Equation 7.15; the pendulum's motion is sinusoidal only in the limit of small deflections $\varphi_{max} \ll 1$ or $x_{max} = A \ll \ell$.*

EXERCISE 7.16

A pendulum of mass m and length ℓ is attached to a horizontal spring with spring constant k. When the bob is displaced from the vertical by a small distance x, where $x \ll \ell$, the force on the bob is $F = ma_x = -kx - (mg/\ell)x$. By inspection, find an expression for the frequency of oscillation ω.

Answer: $\omega = \sqrt{\frac{k}{m} + \frac{g}{\ell}}$

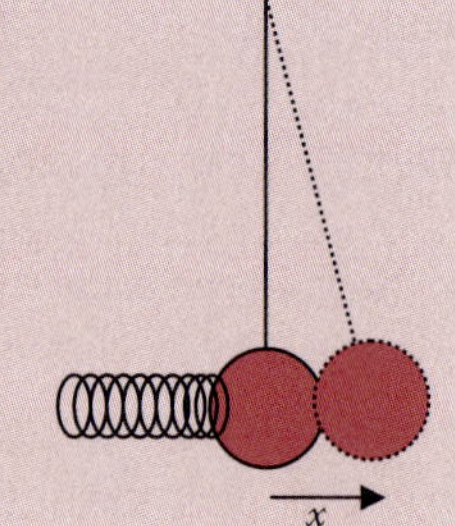

The pendulum plays an important role in the history in physics and astronomy. When Galileo was an undergraduate at the University of Pisa, he was intrigued by the swaying of a chandelier in the church he attended. Timing its oscillations by means of his pulse, he concluded that the motion was *isochronous*, i.e., the period did not depend on the amplitude of the motion. He later used pendula extensively to time his accelerated-motion experiments, and correctly deduced that the period of oscillation was proportional to the square root of the length. The first pendulum *clock* was constructed by Christiaan Huygens in 1656, likely motivated by the problem of determining longitude at sea. Over the next two decades, Huygens dramatically improved the accuracy of timekeepers from about 15 min/day to about 15 s/day. In 1671, Jean Richer carried a pendulum clock with him to Cayenne (French Guiana—near the Equator) during his expedition to measure the parallax of Mars (to determine the Earth–Sun distance—see Chapter 1). In Paris, his clock kept time accurately, but in Cayenne, Richer discovered that it lost $2\frac{1}{2}$ min/day. This implied that the effective acceleration of gravity g at the equator is less than its value in Paris (latitude 49°). Isaac Newton used this

* Do not confuse the geometric angle φ with the initial phase angle θ_0 appearing in $x(t) = A\cos(\omega t + \theta_0)$. Just as for the mass-spring system, θ_0 is a constant that depends on the initial displacement and velocity of the system, and shifts the cosine curve right or left along the time axis to match the initial conditions.

discovery to deduce that the Earth is not a perfect sphere, but a "prolate ellipsoid," slightly flattened at the poles and bulging at the Equator due to the planet's rotation. In 1851, Léon Foucault swung a heavy pendulum bob from the ceiling of the Panthéon in Paris.* Crowds of Parisians watched intently as the pendulum's path slowly rotated—or, as emphasized by Foucault—as the Earth slowly turned beneath the pendulum. Never before had there been a direct, nonastronomical demonstration of Earth's rotation.

EXERCISE 7.17

Richer found that he needed to shorten his pendulum ($\ell \simeq 1$ m) by 2.6 mm in order for it to keep time accurately. By how much does the effective value of g at Cayenne differ from its value in Paris?

Answer 0.026 m/s^2

The accuracy of pendulum clocks improved steadily during the eighteenth and nineteenth centuries, and they remained the reigning champions of timekeeping well into the twentieth century. In 1929, the U.S. National Bureau of Standards (presently called the National Institute of Standards and Technology) kept "official" time by a sophisticated pendulum clock having an accuracy of 1 ms/day. This was replaced by electronic clocks (quartz crystal oscillators) and later by atomic clocks (atomic oscillators). The current time standard is kept by a cesium atomic clock housed in Boulder, CO. It will not gain or lose one second in 100 million years!

* The Pantheon is the ceremonial burial place of famous French authors and scientists, including the physicists Pierre and Marie Curie. The top of the building's dome is 67 m above the floor.

7.10 COUPLED OSCILLATORS

The mathematical approach presented above can be adapted to more complex systems, even those in which more than one body is oscillating. The two-body system shown in Figure 7.14 is the analogue of a diatomic molecule. The two blocks (or atoms) are coupled together by a spring† and slide with negligible friction over a horizontal surface. Since there are no external forces acting on the blocks in the horizontal plane, their total momentum is conserved, and their center of mass (CM) moves with constant velocity. Let's examine the motion in the CM reference frame. Denote the position of m_1 (relative to the CM) by x_1, and that of m_2 by x_2. The separation of the blocks (equal to the instantaneous length of the spring) is $x_2 - x_1$. If the relaxed length of the spring is ℓ, then the amount by which the spring is stretched is $\Delta\ell = x_2 - x_1 - \ell$. Applying Newton's second law to each block, and ignoring the mass of the spring,

$$F_2 = m_2 \frac{d^2 x_2}{dt^2} = -k(x_2 - x_1 - \ell),$$

and (7.17)

$$F_1 = m_1 \frac{d^2 x_1}{dt^2} = +k(x_2 - x_1 - \ell).$$

† In a molecule, electromagnetic forces replace the spring force.

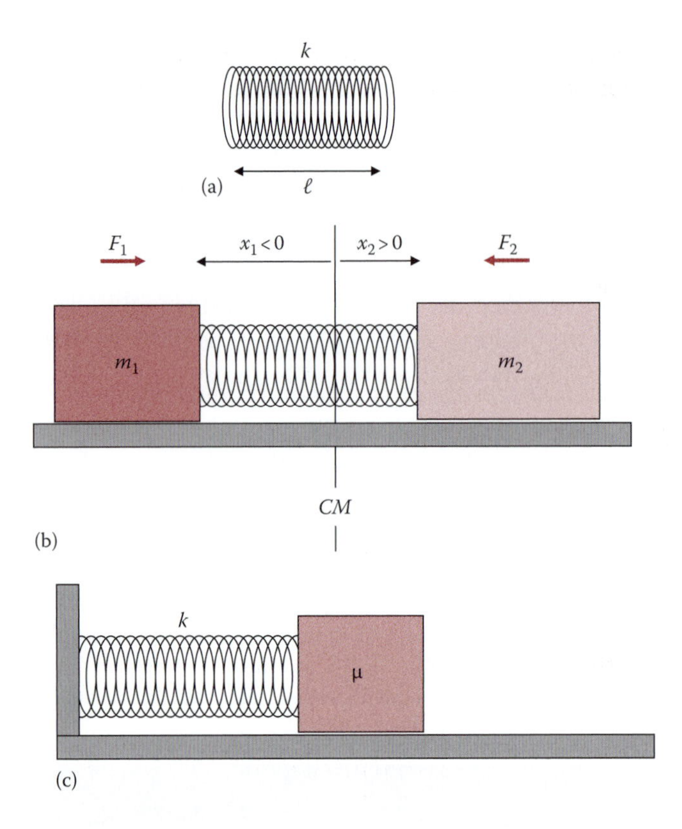

FIGURE 7.14 (a, b) A spring of relaxed length ℓ is attached to blocks m_1 and m_2. The positions of the blocks relative to the system's CM are denoted by x_1 and x_2. In the CM reference frame, $m_1x_1 + m_2x_2 = 0$. (c) The equivalent one-body problem of a block having a "reduced mass" μ attached to the same spring.

If the spring is stretched ($\Delta\ell > 0$), it pulls m_2 back to the left (so $F_2 < 0$) and pulls m_1 to the right ($F_1 > 0$). Note that $F_2 = -F_1$, as required by Newton's Third Law. To solve Equation 7.17, multiply both sides of the first equation by m_1, and both sides of the second equation by m_2:

$$m_1 m_2 \frac{d^2x_2}{dt^2} = -km_1(x_2 - x_1 - \ell),$$

$$m_2 m_1 \frac{d^2x_1}{dt^2} = +km_2(x_2 - x_1 - \ell).$$

Subtracting the second equation from the first,

$$m_1 m_2 \left(\frac{d^2x_2}{dt^2} - \frac{d^2x_1}{dt^2} \right) = -k(m_1 + m_2)(x_2 - x_1 - \ell). \tag{7.18}$$

At first glance, this equation looks hopelessly difficult to solve. But it can be simplified greatly by expressing it in terms of the change in spring length, $\Delta\ell = x_2 - x_1 - \ell$. Since ℓ is a constant,

$$\frac{d^2x_2}{dt^2} - \frac{d^2x_1}{dt^2} = \frac{d^2}{dt^2}(x_2 - x_1 - \ell) = \frac{d^2}{dt^2}\Delta\ell,$$

so we can rewrite Equation 7.18 as

$$m_1 m_2 \frac{d^2}{dt^2}\Delta\ell = -k(m_1 + m_2)\Delta\ell,$$

or

$$\frac{d^2}{dt^2}\Delta\ell = -k\frac{m_1 + m_2}{m_1 m_2}\Delta\ell \equiv -\frac{k}{\mu}\Delta\ell, \tag{7.19}$$

where we have introduced μ, the reduced mass of the system, defined as

$$\mu \equiv \frac{m_1 m_2}{m_1 + m_2}.$$

Equation 7.19 looks *exactly* like Equation 7.11 or 7.15 for the mass–spring system or the pendulum, with the substitution k/μ for k/m or g/ℓ (see Figure 7.14c). Consequently, the solution to Equation 7.19 must be:

$$\Delta\ell(t) = A\cos(\omega t + \theta_0),$$

with (7.20)

$$\omega = \sqrt{\frac{k}{\mu}}.$$

EXERCISE 7.18

Compare the reduced mass μ to the individual masses m_1 and m_2. Which of the following is correct? (a) $\mu < m_1, m_2$, (b) $m_2 < \mu < m_1$, and (c) $m_2, m_1 < \mu$.

Answer (a)

Equation 7.20 tells us that the two-body system undergoes sinusoidal oscillation with frequency $\sqrt{k/\mu}$. But μ is a fictitious mass, so the solution above does not tell us how either m_1 or m_2 moves. Since we are in the CM reference frame, where $x_{CM} = 0$, then $m_1x_1 + m_2x_2 = (m_1 + m_2)x_{CM} = 0$. Differentiating with respect to time, we obtain $m_1u_1 = -m_2u_2$, which expresses conservation of momentum in the CM—or zero momentum—frame. So whenever m_1 is moving to the right, m_2 is moving to the left, and vice versa. The only way for this to be true at all times is if both blocks are oscillating sinusoidally with the same frequency ω, moving in opposite directions with amplitudes of oscillation that are inversely proportional to their masses: $A_2/A_1 = m_1/m_2$.

EXERCISE 7.19

Two masses, m_a and $m_b = 3m_a$, are connected by an ideal spring. When m_a is immobilized, the oscillation frequency of m_b is ω. (a) What is the oscillation frequency when both masses are free to move? (b) If m_a oscillates with amplitude 0.6 cm, what is the amplitude of m_b's motion?

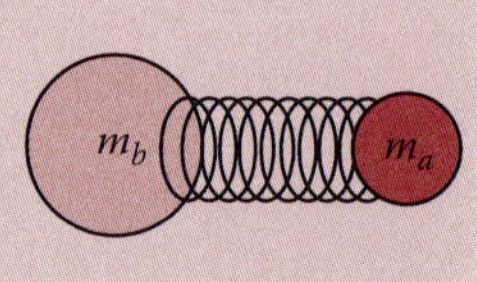

Answer (a) 2ω (b) 0.2 cm

To recap, we solved for the motion of a two-body system by combining two differential equations into a single equation for the motion of the reduced mass μ. This new equation had the same mathematical form as our earlier expressions for the pendulum and the single mass–spring system: $a_x = d^2x/dt^2 = -C^2x$, so we knew immediately that the motion of the system was simple harmonic. No matter how complicated a system may be, whenever a differential equation of this form governs the system's motion, we can be sure that the system undergoes SHM, or sinusoidal oscillation.

7.11 RELATION BETWEEN SINUSOIDAL OSCILLATION AND CIRCULAR MOTION

The position of a body in uniform circular motion is given by Equation 7.2,

$$\vec{r}(t) = r\cos(\omega t + \theta_0)\hat{i} + r\sin(\omega t + \theta_0)\hat{j}.$$

This looks remarkably similar to the expression for the position $x(t) = A\cos(\omega t + \theta_0)$ in SHM, Equation 7.12. To understand the similarity, differentiate $\vec{r}(t)$ twice to obtain the acceleration vector of uniform circular motion:

$$\vec{a}(t) = \frac{d^2\vec{r}}{dt^2} = -\omega^2\vec{r}(t). \tag{7.3}$$

Rewriting $\vec{r}$ and $\vec{a}$ in terms of their Cartesian coordinates, $\vec{r} = x\hat{i} + y\hat{j}$ and $\vec{a} = a_x\hat{i} + a_y\hat{j}$,

$$a_x = \frac{d^2x}{dt^2} = -\omega^2 x,$$

and

$$a_y = \frac{d^2y}{dt^2} = -\omega^2 y.$$

Each of these equations is identical to the equation defining SHM, Equation 7.11 or 7.15, so it should not be surprising that the expression for $x(t)$ or $y(t)$ in uniform circular motion looks exactly like the solution for $x(t)$ in SHM. Figure 7.15 compares the two types of motion graphically. We might describe SHM as one-dimensional circular motion, or conversely, circular motion as two-dimensional simple harmonic oscillation.

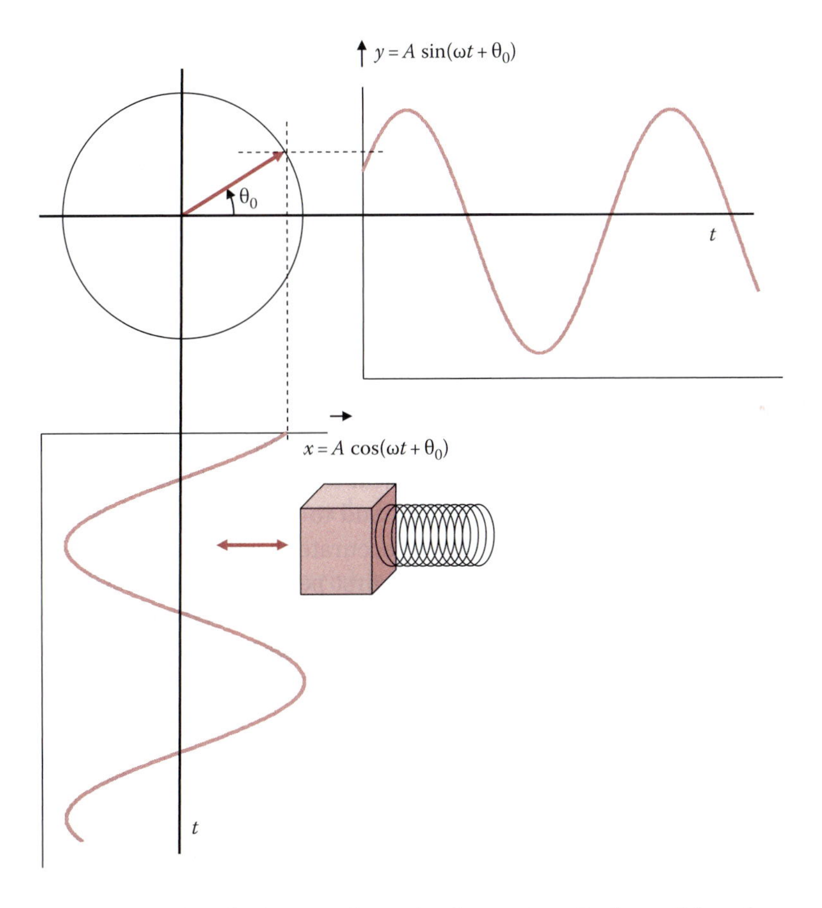

FIGURE 7.15 Graphs of $x(t)$ and $y(t)$ for a body in uniform circular motion. Each graph has the same sinusoidal dependence on time as a body undergoing SHM.

solar day (MSD) is 3.94 min (3 m 56 s) longer than a sidereal day. In 1895, the Canadian–American astronomer Simon Newcomb used nearly 150 years of astronomical data, dating from 1750 onwards, to establish the unit of time in terms of astronomical events. His "ephemeris second" was effectively defined as 1/86,400 of an MSD in the year 1820. In 1968, 1 s of atomic time was defined to be equal to Newcomb's ephemeris second.

But that's *still* not the whole story. Because of Earth's gradually decreasing rotation rate, the MSD is presently 2 ms longer than the one calculated by Newcomb. For each interval of 500 days, our ^{133}Cs atomic clock will be an additional 1 s ahead of the time based on astronomical events. In less than 5000 years, "midnight" will occur at 1:00 AM if UTC (atomic time) is left "uncorrected." Beginning in 1972, this discrepancy has been removed by introducing "leap seconds," extra seconds that are added periodically to synchronize atomic time with astronomical time: sunrise, sunset, and the seasons. Twenty-five leap seconds have been added since 1972. But the sporadic insertion of leap seconds is a nuisance for worldwide systems—such as computer networks—that must maintain precise synchronization. (GPS satellites do not use leap second corrections, and as a result, they are presently running 15 s ahead of corrected-UTC.) In January 2012, delegates from around the world convened in Geneva, Switzerland to debate a proposal to abandon the practice of leap seconds. If adopted, the proposal would sever official "civic" time from astronomical time—*forever*. No decision was reached, and the world is presently left with two standards of time: UTC, based on the radio waves emitted by ^{133}Cs, and ephemeris time, linked to the rotation of the planet.*

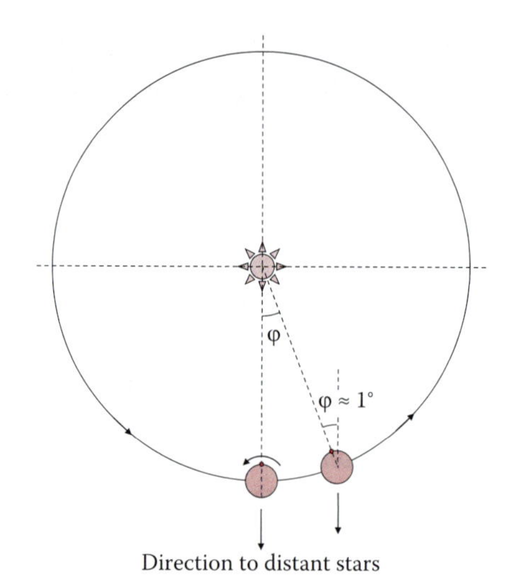

FIGURE 7.16 To an observer on Earth (indicated by the dot), high noon is when the Sun is at its highest point above the horizon. A solar day is the time from noon 1 day to noon the next day. A sidereal day is the time needed for a 360° rotation of the Earth, relative to the stars. Because of the Earth's revolution about the Sun, the planet must rotate about 361° (=360° + 360°/365.25°) in a solar day. The angle φ (≈1°) in the figure is greatly exaggerated.

EXERCISE 7.22

Verify that from 1972 to 2012, the difference between atomic time (UTC) and ephemeris time amounted to about 25 s.

* See *A Second Here, a Second There, May Just Be a Waste of Time*, by Kenneth Chang, *The New York Times*, January 18, 2012.

7.14 SUMMARY

The theme of this chapter is *periodic motion*: uniform circular and SHM. Together with constant-accelerated motion, these are the three most basic—and most important—types of motion found in the physical world. The subtheme of this chapter is *time*: how it is defined and how we measure it. A clock is a device that "ticks" at a well-defined rate regulated by a periodic system, or *oscillator*. Throughout history, advancements in science and technology necessitated more accurate clocks, ones with rates that were more precisely invariant with age or with changes in their surroundings. Water clocks were replaced by crude mechanical clocks, then by pendulum clocks, quartz crystal clocks, and presently atomic clocks.

A body in uniform circular motion moves at constant speed but experiences a centripetal acceleration of constant magnitude $a_c = v^2/r$ whose direction changes continually with time. A body undergoing simple SHM, or *sinusoidal oscillation*, experiences a linear restoring force proportional to its displacement x from its stable equilibrium position: $F_x(t) = -kx(t)$. In either case, the motion of the body is periodic and can be described by trigonometric functions. For SHM, $x(t) = A\cos(\omega t + \theta_0)$ or equivalently $x(t) = A\sin(\omega t + \theta_0')$, where A is the amplitude of the oscillation and ω is the angular frequency. The initial phase angle $\theta_0 (= \theta_0' + \pi/2)$ is determined by the position x_0 and velocity v_0 of the body at $t = 0$. Periodic systems that are not inherently sinusoidal (such as the pendulum) behave like simple harmonic oscillators in the limit of small displacements from equilibrium.

The mathematics of the mass–spring system or the pendulum is applicable to all sinusoidal oscillators, on all time and length scales, throughout the universe. Whenever a system's differential equation of motion takes the form $a_x = d^2x/dt^2 = -C^2x$, the system undergoes SHM, $x(t) = A\cos(\omega t + \theta_0)$ with angular frequency $\omega = C$. For the one-body mass–spring system, $\omega = \sqrt{k/m}$, while for the pendulum $\omega = \sqrt{g/l}$. For the two-body mass–spring oscillator, the equations of motion can be combined to form a new differential equation that predicts SHM for the *reduced mass* $\mu = m_1m_2/(m_1 + m_2)$ with frequency $\omega = \sqrt{k/\mu}$. This procedure can be extended to systems with much greater complexity.

The sensation of weightlessness led Einstein to propose his Equivalence Principle, and from it, the General Theory of Relativity. Using the principle, we found that a gravitational field deflects light, and also that it causes clocks (time) at different altitudes to advance at different rates, an effect clearly seen in the orbiting atomic clocks of the Global Positional System. In the *Principia*, Isaac Newton took great pains to define his scientific terms precisely—with several notable exceptions. Here is what he says at the end of Book 1:

> Hitherto I have laid down the definitions of such words as are less [well] known, and explained the sense in which I would have them to be understood... *I do not define time, space, place and motion*, as being known to all.

The very things that Newton did not question were the ones that perplexed Albert Einstein. Ironically, the falling-body experiments of Galileo, and the pendulum experiments of Newton himself (proving that

$m_{iner} = m_{grav}$), ultimately laid the foundation for Einstein's theories of relativity, in which the Newtonian concepts of absolute space and time were rejected. If Newton were to encounter relativity, what would be his reaction? We suspect he would be delighted and enthralled.

PROBLEMS

7.1 You are sledding with your little brother on the hill shown in the figure. The snow is smooth and frictionless. Your brother, whom you are holding, has a mass of 10 kg. The slope of the hill at A is $\theta = 30°$. The radii of curvature of the hill at B and C are 10 and 15 m, respectively. The hill is horizontal at B and C.

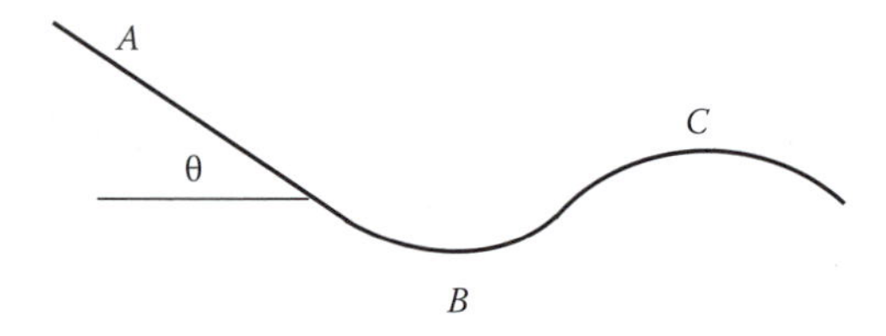

(a) Find your acceleration down the hill at point A.

(b) If you are moving at 10 m/s at point B, how heavy does your brother feel to you?

(c) If the sled is moving at 6 m/s at point C, how heavy does your brother feel?

(d) How fast would the sled need to be traveling to become airborne at C?

7.2 The radius of curvature at the bottom and top of a roller coaster is 10 m. The contact force between the car and the rail must not exceed six times the weight of the car.

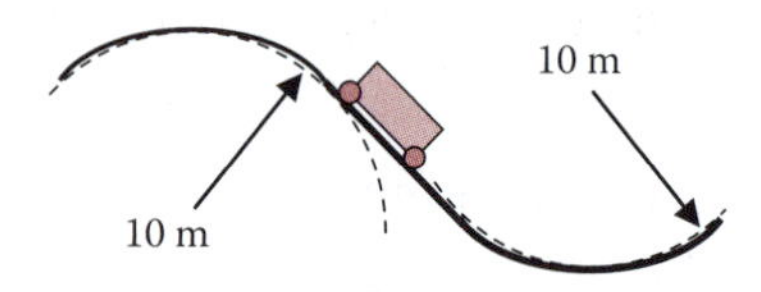

(a) What is the maximum speed of the car at the bottom?

(b) What is the highest speed at the top in order that the car remain in contact with the rail?

7.3 A mass m_1 undergoes circular motion of radius R on a horizontal frictionless table, and is connected by a massless string through a hole in the table to a hanging mass m_2. If m_2 is stationary, find expressions for:

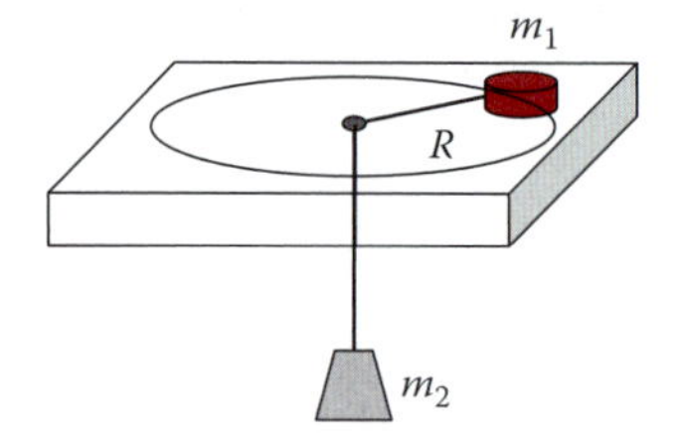

(a) The tension in the string.

(b) The period of the circular motion.

7.4 Two conical pendula of different length are hung from the same support. They rotate with their bobs moving in the same horizontal plane. Show that their periods are equal.

7.5 A space colony is in the shape of a hollow annulus with outer diameter 500 m. To simulate Earth's gravity, the colony is rotated so that the force exerted on a resident astronaut at the outer diameter of the annulus is the same as the force of gravity on Earth. Find the number of revolutions per minute.

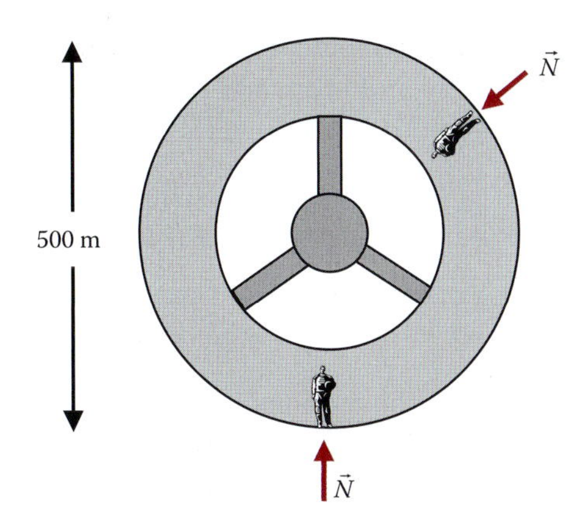

7.6 A ball of mass m = 0.1 kg rolls back and forth along a V-shaped track, as shown in the figure. Each side of the track makes an angle $\theta = 4.1°$ with respect to the horizontal. The ball is released from rest 1 m to the left of the center of the track. It rolls without slipping, and the magnitude of its acceleration is given by $a = \frac{5}{7} g \sin\theta$. The subsequent motion is periodic but not sinusoidal.

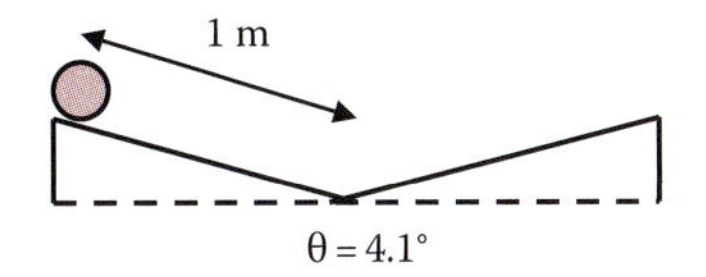

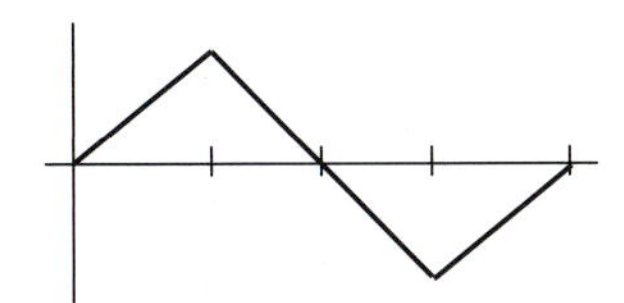

(a) Calculate the period of the motion.

(b) For such a small slope, the vertical motion can be ignored. The initial position is $x(0) = -1$ m, and $x = 0$ at the center of the track. A graph of velocity vs. time is shown. *Label the time axis at the times shown*. Plot the *acceleration* and *position* of the ball vs. time for one full period of motion. (*Hint:* Where is the ball when its velocity is a maximum? Where is it when $v = 0$?)

(c) If the V-shaped track were replaced by a *parabolic* track, the oscillation of the ball would be sinusoidal, $x(t) = A\cos\omega t$ for small amplitudes. In the figure, the track makes an angle $\phi = 5.1°$ with the horizontal at $x = 1$ m. The x-acceleration at this position is -0.62 m/s² and $x_{max} = 1$ m. Find the maximum speed of the ball.

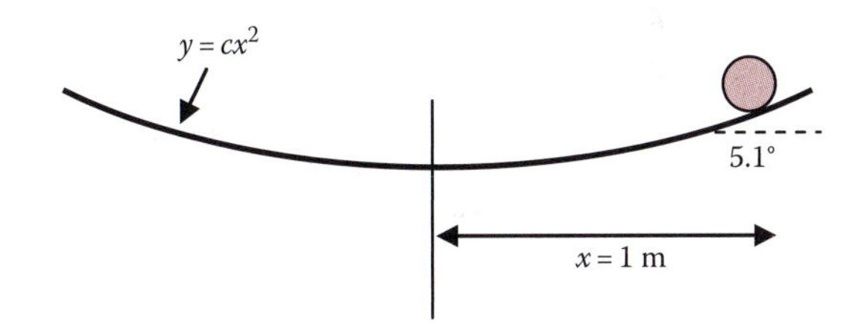

7.7 A 2.0 kg mass attached to a spring slides horizontally over a frictionless surface. The equation of motion of the mass is

$$x(t) = 0.1\cos\left(10t + \frac{\pi}{2}\right) \text{ m}$$

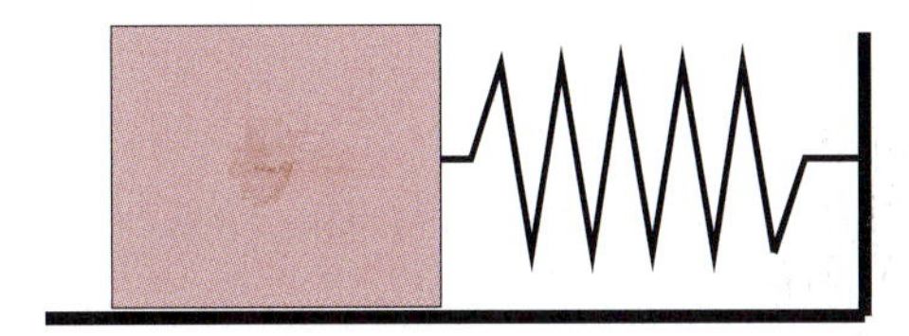

(a) What are the angular velocity ω, the frequency f, and the period T?

(b) Differentiate the above equation to derive equations for $v(t)$ and $a(t)$.

(c) What is the maximum velocity of the block? At what time is v maximized? What is the position x of the block when v reaches its maximum value?

(d) What is the maximum acceleration of the block? What is the position x of the block when this maximum acceleration is reached?

(e) The figure below shows $x(t)$. Sketch $v(t)$ and $a(t)$. Your curves must be consistent with your answers to (c) and (d).

(f) What is the maximum force exerted on the block by the spring?

(g) What is the spring constant k?

(h) What are the initial conditions $x(0)$ and $v(0)$?

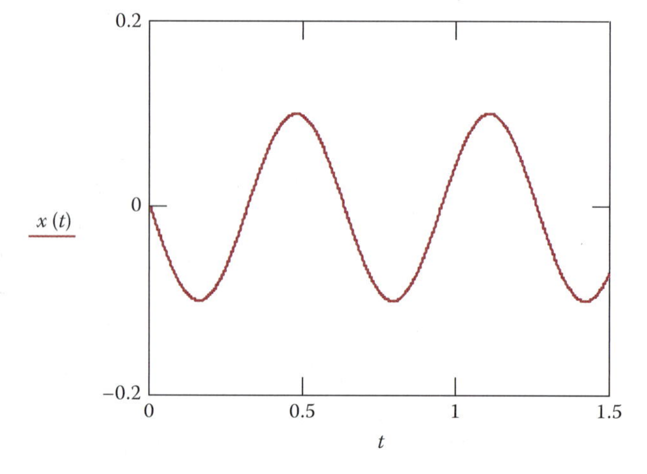

7.8 The same oscillator is now hung vertically, stretched and released.

(a) When the mass is at its equilibrium point (where $F_{tot} = 0$), by how much is the spring stretched?

(b) Is the motion still simple harmonic? If so, what is the angular velocity?

7.9 A simple harmonic oscillator has a period $T = 0.4\pi$ s, and is started from an initial displacement $x(0) = 16$ cm with an initial velocity $v(0) = 60$ cm/s.

(a) What is the angular velocity?

(b) Let $x(t) = A\cos(\omega t + \theta_0)$. Find values of A and θ_0.

(c) Sketch the motion.

7.10 The figure shows a mass m attached to two springs, of spring constants k_1 and k_2. In terms of these three variables, find the angular velocity ω.

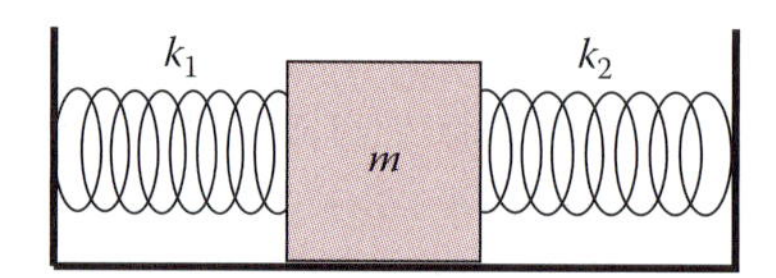

7.11 You are in the Space Shuttle, watching the motion of a pendulum of mass 0.6 kg and length 1.0 m. During launch, you measure its period to be 1.00 s, from liftoff to the time when the rocket engines cut off. (Assume constant acceleration during this time period.)

(a) Draw a free body diagram for the pendulum bob. Find an expression for the string tension F_{ten} in terms of the vertical acceleration of the Shuttle a_{sh}, the mass m, and the gravitational constant g. Assume small oscillations ($\theta \ll 1$ rad).

(b) Using $\sin\theta = x/L$ (L = length), derive an expression for the horizontal acceleration a_x of the bob as a function of x.

(c) From (b), find expressions for ω and the period T.

(d) What is the vertical acceleration a_y of the Shuttle? (How many g's?)

(e) After the rocket engines shut off, what is the period of the pendulum? (Careful!)

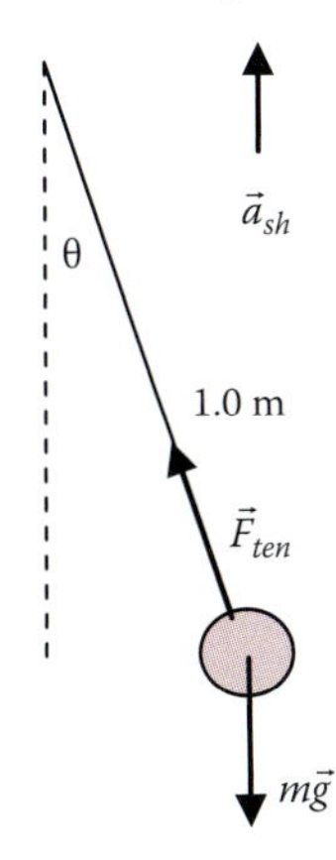

7.12 A pendulum swings in a vertical plane, as shown.

(a) At what point does the acceleration a_x have its largest negative value?

(b) At what point is a_x positive and the velocity v_x negative?

(c) What is the direction of the acceleration at point C?

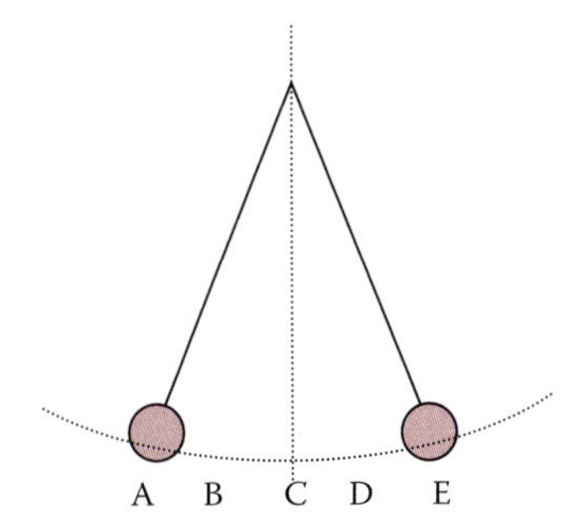

7.13 A conical pendulum is shown in the figure. Mass m is suspended from a lightweight string of length ℓ, and moves in a horizontal circle of radius $r = \ell \sin\theta$, where θ is the angle of the string relative to the vertical.

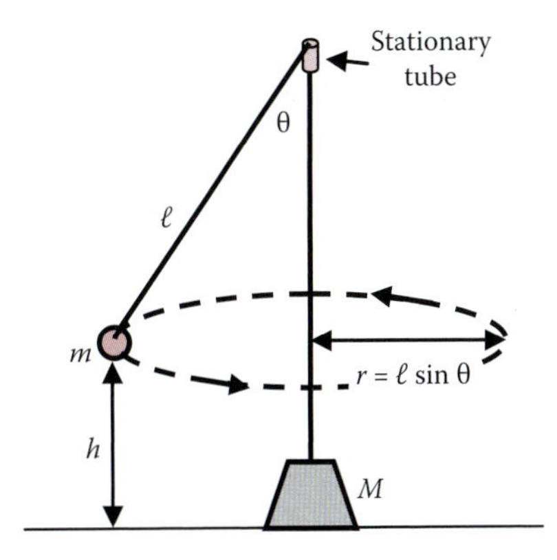

(a) Draw a free body diagram for the rotating mass m. On the diagram, clearly indicate the direction of all forces acting on m, and also the direction of the acceleration $\vec{a}$ at the moment shown in the figure.

(b) As shown in the figure, the string passes through a small stationary tube and is attached to mass M which rests on the floor. If $M = 2m$, what is the maximum angle θ such that M remains in contact with the floor?

(c) Let $M \gg m$. Derive an expression for the period of the pendulum (how long it takes m to execute a complete circle), in terms of ℓ, θ, m, and g. Explain how your answer is reasonable in the limits $\theta \to 0°$ and $\theta \to 90°$.

(d) Assume that M does not move, and let $\ell = 1$ m. At $\theta = 45°$, the string suddenly breaks. If m is initially at a height $h = 0.5$ m above the floor (when the string breaks), how far does it travel in the horizontal direction before hitting the floor? (*Hint:* What is the ball's initial speed?)

7.14 A small bead of mass 200 g is threaded onto a circular wire hoop of radius 20 cm. The hoop is rotated about a vertical axis at a rate of 1.5 rev/s.

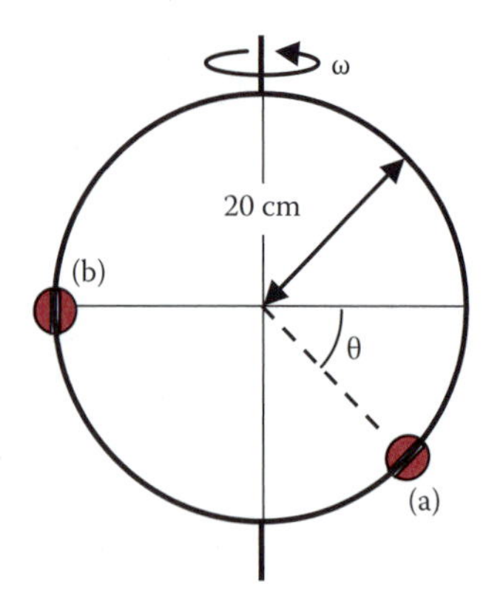

(a) If there is no friction between the wire and the bead, find the angle θ at which the bead will not move relative to the hoop.

(b) A second identical bead experiences friction with the wire. When placed at the same height as the center of the hoop (θ = 0), it remains in place until the rotation rate is reduced to 1.4 rev/s. Find the relevant coefficient of friction. Is this the static or kinetic friction coefficient?

7.15 Over periods of a few months, astronauts experiencing weightlessness can suffer muscle atrophy and other health problems. One solution is to rotate the space capsule carrying the astronauts. Imagine a crew in deep space, occupying a cylindrical "space colony" of radius *R*. The angular velocity of the rotating cylinder is ω.

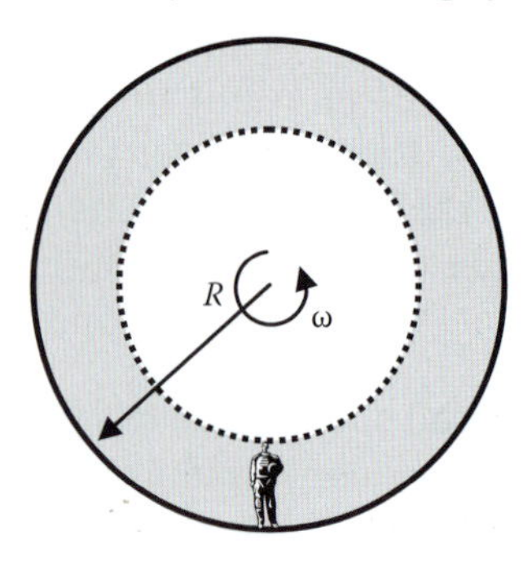

(a) Suppose you are an astronaut on the rotating space station, and that you are holding a heavy radio transmitter. Would it be easier to hold the transmitter at shoulder height or at waist height?

(b) Find the smallest value of *R* so that the value of "*g*" does not change by more than 10% from the head to foot of an average-sized astronaut.

(c) What value of ω would you use to simulate $g = 10.0\ \mathrm{m/s^2}$ at foot level?

(d) On the rotating capsule, you drop your cell phone from a height of 2 m. Describe its path from the point of view of an inertial (nonrotating) observer. Will the phone land at your feet, behind your feet, or in front of your feet? Show your calculations.

7.16 Two identical 0.5 kg masses are joined by a lightweight string and swung by a second string in a vertical circle, as shown in the figure. Both masses execute uniform circular motion with radii $r_1 = 0.5$ m, $r_2 = 1.0$ m.

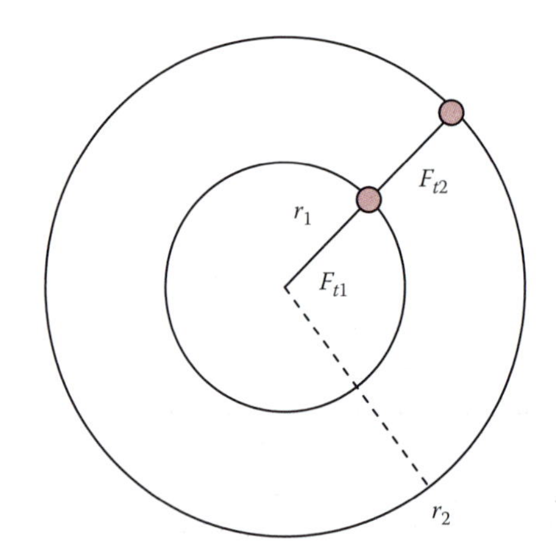

(a) Ignoring gravity, find expressions for the string tensions F_{t_1} and F_{t_2}, in terms of m, ω, r_1 and r_2. (Draw a free body diagram for each mass.)

(b) As the masses are spun faster and faster, which string breaks first? If the string can withstand 20 N tension before breaking, what is the minimum rotation period?

(c) Now include gravity. When the string breaks, what will be the positions of the two masses? Assume they lie in line with the center of the circle.

7.17 George Clooney and Sandra Bullock are not the only ones concerned with space junk. At this time, NASA tracks over 20,000 orbiting bodies—mostly defunct satellites or upper-stage rocket bodies—that are hazardous to future missions. Left to itself, a body in a circular orbit of altitude 800 km takes about 100 years to "deorbit," i.e., to lose altitude and burn up during reentry into the planet's atmosphere. One proposal to hasten that process is to equip the body with a long conducting wire called a *space tether* that hangs vertically below or above the body. As the wire sweeps through the Earth's magnetic field, a current is induced in the wire that robs the body of kinetic energy.

In the figure, a large defunct satellite m_1 is in a circular orbit of altitude of 800 km, where the local value of the gravitational field is $g_1 = 7.720\ \text{m/s}^2$. A tether of length $\ell = 20\ \text{km}$ is deployed and kept taut and vertical by a small end mass $m_2 = 100$ kg. The gravitational field at the lower altitude of m_2 is $g_2 = 7.763\ \text{m/s}^2$. Because $m_2 \ll m_1$, it does not affect the motion of the satellite.

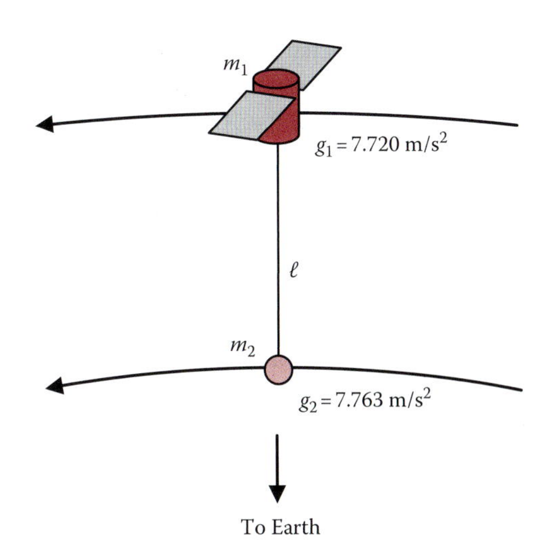

(a) Using just the information given above, find the orbital speed of the satellite.

(b) Find the tension of the tether wire. (Ignore the tether's mass.)

(c) If m_2 is displaced from its equilibrium vertically below the satellite, it will undergo SHM. Find the period of this motion.

7.18 A block of mass $M = 0.4$ kg is attached to a massless spring of stiffness k. The block slides without friction over a smooth horizontal surface. The maximum velocity of the block is 1.0 m/s and its maximum displacement from equilibrium is 0.2 m.

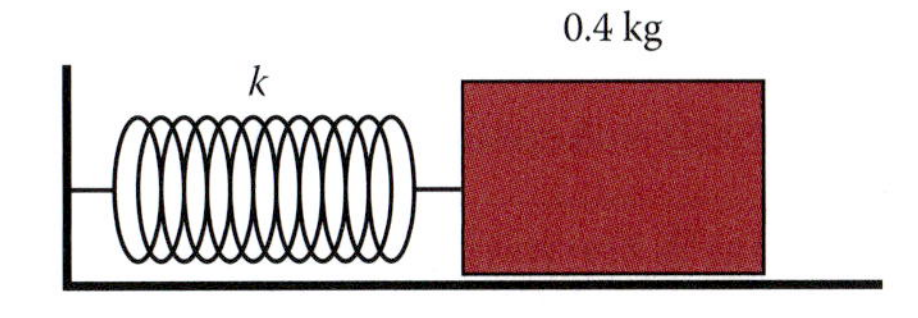

(a) Find the spring constant k, the angular frequency ω and the period T of the oscillation.

(a) What is the acceleration of the spacecraft during the time its engine is firing?

(b) Inside the craft, a 0.5 kg mass is attached to a spring with a spring constant k = 20 N/m. During the rocket burst, the spring is stretched by a constant amount. What is this amount?

(c) After the burst, describe the motion of the mass (relative to the spacecraft) by writing an expression for its displacement vs. time: $x(t) = \cdots$ (Define the time when the engines are shut off as $t = 0$.)

7.24 Two identical masses $m_1 = m_2 = m$ are connected by a lightweight thread; one mass is on a frictionless plane and is attached to a spring. The plane makes an angle θ with the horizontal.

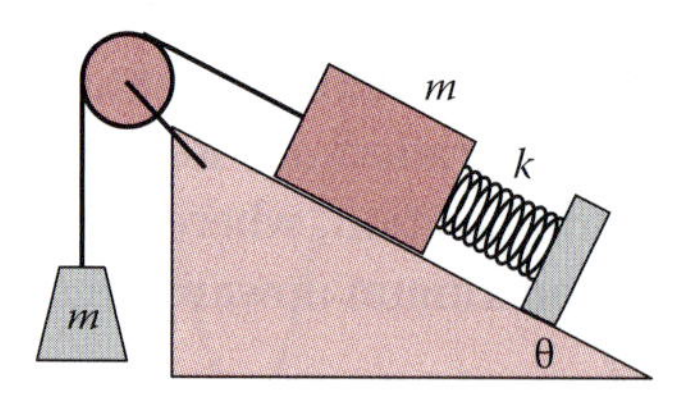

(a) Draw a free body diagram for each mass, and write the equations of motion ($F = ma$) for each. Choose appropriate axes!

(b) Find an expression for the stretch of the spring when the bodies are at their equilibrium positions.

(c) Show that the system undergoes SHM. Do this by writing an equation of motion in SHM form: $a = -C^2x$, where x is the displacement from the equilibrium position.

7.25 A 4.0 kg wood block hangs from a spring of stiffness k = 500 N/m. The block is struck by a 50 g bullet moving vertically upward with a speed of 150 m/s. After the collision, the block and bullet stick together.

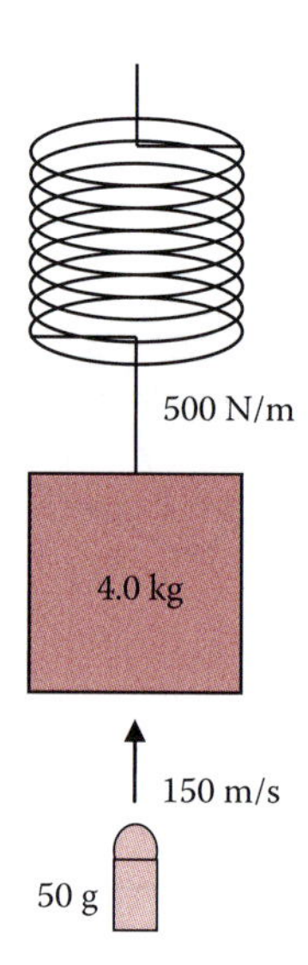

(a) What is the initial velocity of the mass *immediately* after the collision, i.e., before the mass has had a chance to move a significant distance from its equilibrium position?

(b) Write an equation for $y(t)$, the vertical displacement of the block from its equilibrium position. Evaluate the maximum displacement of the block, its angular velocity, and the phase constant of the motion.

(c) The block experiences gravitational and spring forces. What is the maximum *total* force exerted on the block?

(d) What fraction of the initial kinetic energy was lost in the collision?

7.26 A body m hung from a vertically suspended spring stretches the spring by 5 cm.

(a) What is the frequency of oscillation of this system?

(b) If a second, identical spring is attached end to end (see figure) to the first string, how much would each spring be stretched by the same body m?

(c) What would be the frequency of oscillation for this second case?

7.27 Two blocks are stacked on top of each other and then mounted on a lightweight platform suspended by a spring. The top and bottom masses are $m_1 = 1.0$ kg and $m_2 = 4.0$ kg, respectively. The spring constant is 500 N/m.

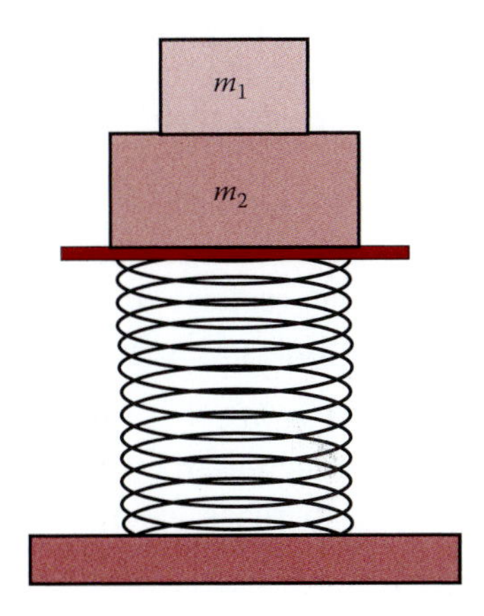

(a) When this system is set into oscillation, what is its frequency ω?

(b) At a certain amplitude of oscillation, the top block loses contact with the bottom block. What is the minimum amplitude for this to happen?

(c) For the amplitude found in part (b), at what point of the oscillation do the blocks first lose contact with each other?

(d) If the heavier block were on top, would the two blocks ever lose contact with each other? Would they ever lose contact with the platform? Explain carefully.

7.28 The 5600 kg *Cassini* spacecraft was launched October 15, 1997 aboard a Titan IV rocket. Inside the rocket *Cassini* was supported by a spring-mounted baseplate designed to absorb the shocks occurring during launch. When the rocket engines were ignited, the rocket slowly increased its acceleration from 0 to 6.25g.

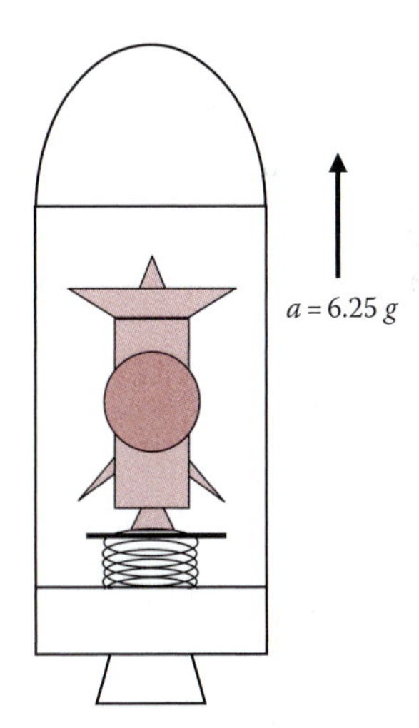

(a) Draw a free-body diagram of *Cassini* during launch.

(b) Find the force that the baseplate exerts on the spacecraft at maximum acceleration.

(c) The baseplate's suspension system can be modeled by a very stiff ideal spring. On the launch pad this spring was compressed by 5.00 mm. By how much was it compressed at maximum acceleration?

(d) After reaching full acceleration, the rocket engines are suddenly cut off, setting *Cassini* into oscillation relative to the rocket body. Find the amplitude and period of this oscillation, assuming the spacecraft is not far above the Earth's surface.

7.29 Consider the axial oscillations of a carbon monoxide molecule. The masses of O and C are $16u$ and $12u$, respectively. When the oxygen atom is bonded to a surface, the frequency of oscillation is 9.0×10^{13} Hz.

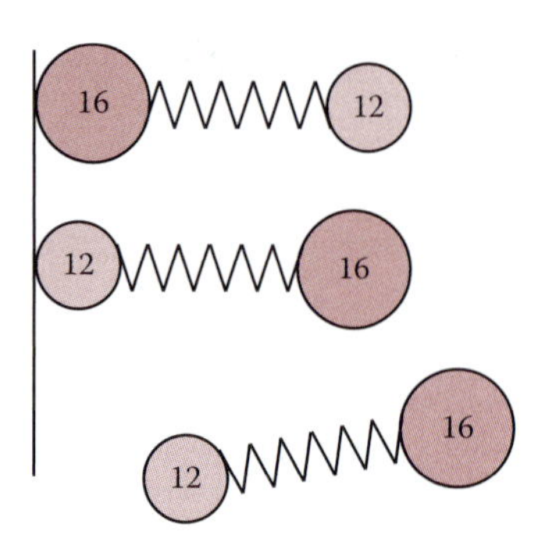

(a) What is the frequency of oscillation when the carbon atom is bonded to the surface?

(b) What is the frequency of oscillation when the molecule is not in contact with any surface?

7.30 A pendulum of length ℓ is mounted on a rocket. What is its period if the rocket is (a) at rest on its launch pad; (b) accelerating *upward* with $a = ½g$; (c) accelerating *downward* with $a = ½g$; and (d) in free fall?

7.31 You are in a space capsule approaching the surface of Mars, where the gravitational acceleration $g = 3.75$ m/s^2. Hanging inside the space capsule is a simple pendulum consisting of a string of length 0.75 m and mass 0.5 kg.

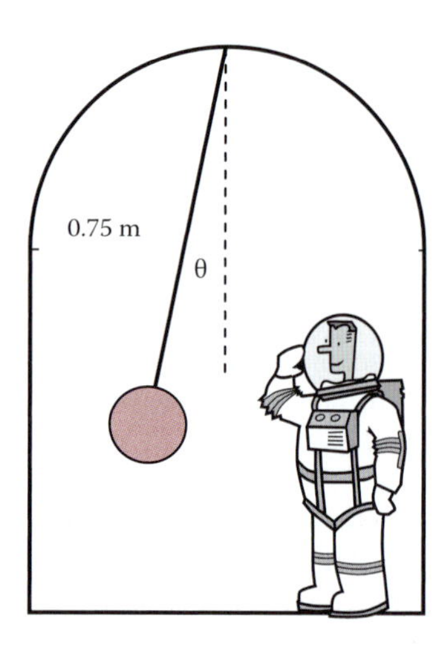

(a) As you descend with acceleration $\vec{a} = a_x\hat{i} + a_y\hat{j}$, you note that the period of the pendulum is 2.0 s. Show that your vertical acceleration is $+g$.

(b) If your initial height was 750 m and your initial vertical velocity was −75 m/s, will you land safely?

(c) As you descend, the pendulum's equilibrium position is at an angle $\theta = 5.73°$ relative to the vertical direction. Calculate the magnitude of the horizontal acceleration a_x.

(d) If a_y were increased by 4 m/s^2, would the period T increase, decrease, or stay the same?

(e) Similarly, if a_y were increased by 4 m/s^2, would the angle θ in part (c) increase, decrease, or stay the same?

7.32 A pendulum bob swings back and forth along the x-axis, with $x(t) = A\cos(\omega t + \theta_0)$. Its maximum velocity is 40 cm/s and its maximum acceleration is 160 cm/s² (1.6 m/s²).

(a) What are the period and amplitude of the motion?

(b) Find the position x when the acceleration $a = -80$ cm/s.

(c) If the initial velocity is 20 cm/s, and the initial displacement is positive, find the value of θ_0. Be sure to indicate whether your answer is in radians or degrees.

(d) What is the maximum angle that the string makes with the vertical?

7.33 Tarzan wishes to rescue Jane by swinging from a long massless vine across a deep ravine rife with alligators. The length of the vine is 30 m, the width of the ravine is 10 m, and the King of the Jungle has a mass of 80 kg. The vine will break if its tension exceeds 800 N.

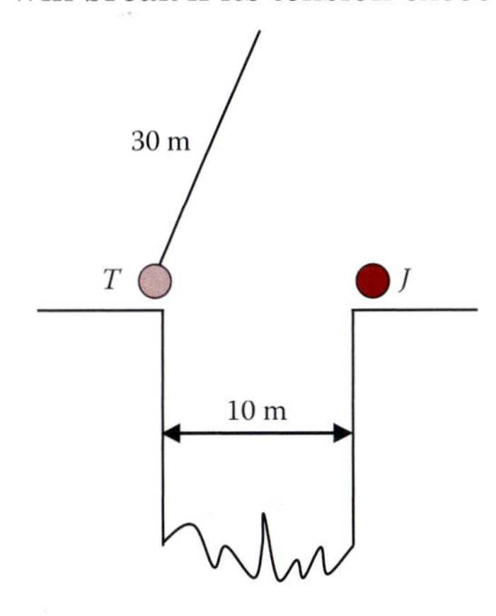

(a) How long does it take Tarzan to reach Jane, assuming the vine does not break?

(b) What is his maximum speed as he crosses the ravine?

(c) Will he rescue Jane or will the vine snap?

7.34 The figure shows an ingenious device used by Galileo to measure free fall times accurately. Ball m_1 and pendulum bob m_2 are released at the same time (by cutting the string shown). The length ℓ of the pendulum is adjusted so that it hits the wall at exactly the same time as the ball strikes the floor. (You can hear the coincidence quite easily.) Assume that the initial angle of the pendulum is $\theta = 10.0°$, and that its length $\ell = 0.5$ m.

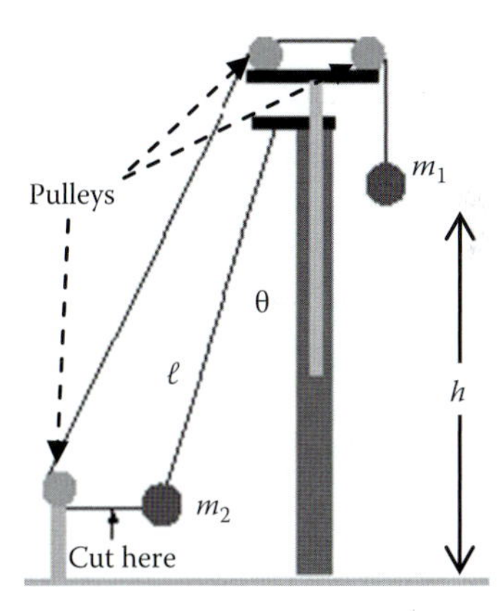

(a) If it is released at $t = 0$, at what time t does the pendulum strike the wall?

(b) What is the horizontal acceleration of the pendulum bob m_2 at the moment it is released?

(c) What is the height h? Galileo knew that the period of the pendulum was proportional to $\sqrt{\ell}$.

(d) What is the speed of the bob just before it strikes the wall?

7.35 The Sun and Solar System oscillate perpendicularly to the plane of the Milky Way Galaxy with a period of about 62 Myr. This motion has long been suspected to be linked to periodic extinctions of life-forms here on Earth that are revealed by fossil records. The mechanism for these oscillations will be examined in a later chapter, but for now it is sufficient to know that the Sun's equilibrium position is in the midplane of the galaxy, and there is a gravitational restoring force on the Sun that is roughly proportional to its distance from the midplane. Let $z(t)$ be the Sun's distance from equilibrium at time t.

(a) At present, the Sun is about 25 pc from the midplane and moving away from it with a relative speed of 7 km/s. Use this information to estimate z_{max}. (1 pc = 3.09×10^{16} m.)

(b) Roughly when were we last at the galactic midplane?

7.36 As a Cepheid variable star brightens and dims, it also swells and shrinks with the same period. Quite recently, a technique called *long baseline interferometry* has allowed astronomers to measure directly the angle $\theta(t)$ subtended by a Cepheid as it pulsates. Simultaneously, Doppler shift measurements of the light emitted by the star indicate the speed $v(t)$ with which the star's surface is moving relative to its center. By integrating the velocity curve, the change in the star's radius Δr can be found: given $\Delta r(t)$ and $\theta(t)$, the distance d to the Cepheid can be determined. (See Figure 1.4 and also the figure to the right.)

The Cepheid FF Aql is in the constellation Aquila. Its period of oscillation is 4.47 days, and the maximum radial velocity of its surface is 11.4 km/s. Assume that the surface motion is sinusoidal.

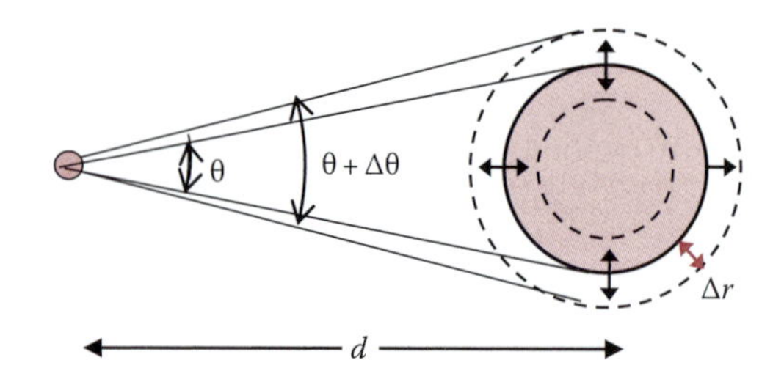

(a) What is the amplitude of the change in radius Δr_{max}? Express your answer in terms of the Sun's radius $R_{Sun} = 7 \times 10^8$ m.

(b) The average angle subtended by FF Aql is 0.878 milliarcsecond (1 milliarcsecond = 4.848×10^{-9} rad), and the amplitude of the change in angle $\Delta\theta_{max}$ is 0.026 milliarcsecond. Find the distance to the star. Express this distance in pc. (1 pc = 3.09×10^{16} m.)

(c) What is the average radius of the FF Aql? Express your answer in terms of R_{Sun}.

7.37 The current generation of cesium-133 atomic clocks boast an accuracy of 1 part in 10^{16}. In 2012, Campbell and coworkers in the United States and Australia proposed a *nuclear* clock that could theoretically achieve an accuracy of $1{:}10^{19}$. Since nuclei are nestled deep inside the electronic clouds of atoms, they are far less sensitive to the stray electromagnetic fields that limit the accuracy of atomic clocks.

(a) Show that the accuracy of a ^{133}Cs clock is "limited" to about a minute over the entire history of the universe. How many seconds might a nuclear clock gain or lose in the same time period?

(b) Imagine that "official" time is kept by a nuclear clock situated at sea level. A portable nuclear clock is synchronized with the official clock, and then carried to a location at a different altitude. What is the maximum change in altitude that can be tolerated if we want the two clocks to agree to within 1 ps (10^{-12} s) for at least 1 day?

7.38 The figure illustrates the mechanism of a centripetal (flyball) governor. The two balls, of mass m, are attached to the top of a rotating shaft by arms of length ℓ. They are also attached to a lower mass M by arms of the same length. Each of the four arms is hinged at both ends. The position of the lower mass changes with the angular velocity ω, and can be used to regulate the power delivered by the engine driving the shaft, to keep ω constant. For simplicity, treat the two balls as point masses. Then when $\omega = 0$, M is 2ℓ below the top of the shaft.

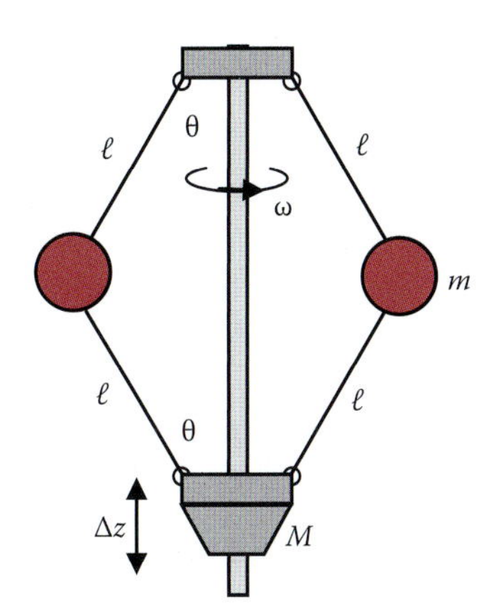

(a) Write Newton's second law for m (horizontal and vertical directions) and M (vertical only). From these three expressions, derive an expression for angle θ as a function of ω, g, ℓ, m, and M.

(b) Derive an expression for Δz, the change in height of M as a function of ω.

(c) What is the minimum value of ω needed to lift M from its vertical position when $\Delta z = l\omega = 0$?

7.39 The 24 satellites of the GPS orbit Earth twice per day at an altitude $h = 20{,}200$ km. Because of their speed and altitude, their onboard clocks do not tick at the same rate as ground-based clocks, and daily corrections are required to keep the system operational.

(a) Calculate the speed of a satellite. (Don't forget to add the Earth's radius to the altitude to find the orbit radius.)

(b) Show that due to time dilation ("moving clocks run slow"), the orbiting clocks lose 7.2 μs/day. (*Hint:* $1/\gamma = \sqrt{1 - v^2/c^2} \simeq -v^2/2c^2$ by the binomial approximation.)

(c) Show that due to gravitational effects, the orbiting clocks gain about 45 μs/day. (Since the gravitational field diminishes with distance from the Earth's center, we must use an average value of g to compute the difference in time. Assume that this average value is $\bar{g} = 2.36\ \text{m/s}^2$.)

7.40 Tatooine, Luke Skywalker's home, is a circumbinary planet (having two suns). The Kepler Space Telescope recently discovered the real Tatooine, a circumbinary planet (dubbed Kepler 16-B) with a mass $m = \frac{1}{3} m_J$ and radius $r \simeq \frac{3}{4} r_J$, where J stands for Jupiter. Its surface gravity is $g = 14.5 \pm 0.7\ \text{m/s}^2$. Standing on the planet's surface, Luke swings his lightsaber from a string held shoulder high. The weapon swings with a period of 2.00 s, and nearly grazes the ground at the lowest point of its arc. Estimate Luke's total height.

7.41 A linear triatomic molecule is composed of an atom M sandwiched between two identical atoms $m_1 = m_2 = m$ that interact with M via identical electromagnetic "springs" of effective stiffness k. At equilibrium, the force on each atom is zero and the springs are relaxed. Let x_1 (x_2) be the displacement of m_1 (m_2) from its equilibrium position, and similarly, let x_0 be the displacement of M from equilibrium. In this problem, we will only be concerned with motion along the axis of the molecule.

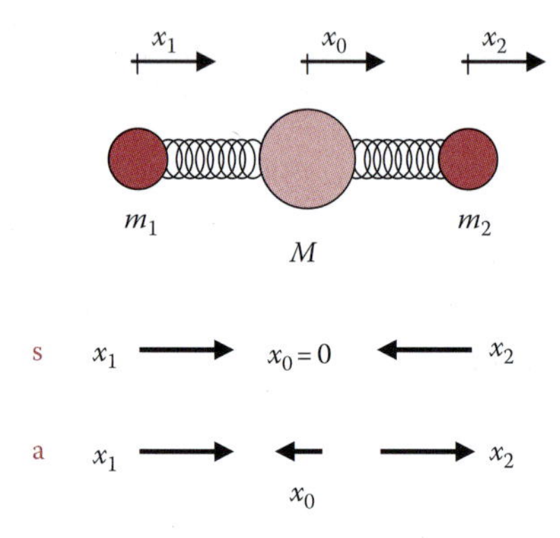

(a) Show that $m_1x_1 + m_2x_2 + Mx_0 = 0$ is the CM is to remain stationary.

The left–right symmetry of the molecule, and the condition given in (a), indicate that the molecule has two modes of oscillation. In the first (symmetric stretch mode), $x_1(t) = -x_2(t)$ and $x_0(t) = 0$, and in the second (asymmetric stretch mode), $x_1(t) = x_2(t)$ while $x_0(t) = -(2m/M)x_1(t)$. These modes of vibration are shown in the figure.

(b) Derive an expression for the frequency ω_s of the first (symmetric) vibrational mode.

(c) Write the differential equation of motion for m_2, $m_2a_2(t) = \cdots$, in terms of $x_i(t)$, k, and m. Then, using the information given above, find an expression for the frequency ω_a of the second (asymmetric) vibrational mode.

(d) The vibrational frequencies of the carbon dioxide (CO_2) are found experimentally to be $\omega_a = 7.05 \times 10^{13}$ s^{-1}, and $\omega_s = 3.86 \times 10^{13}$ s^{-1}. Show that the ratio ω_a/ω_s agrees *roughly* with your answers to parts (b) and (c).

(e) In parts (b) and (c), we ignored the direct interaction between the two oxygen atoms. Which of the two calculations is this likely to influence, and how would it be influenced? (Could this explain the difference between your calculated value of ω_a/ω_s and the experimental data?)

8

KEPLER'S LAWS AND NEWTON'S DISCOVERY OF UNIVERSAL GRAVITATION

A statue of Giordano Bruno stands in the Campo de' Fiori (field of flowers), Rome, on the site of his execution in 1600. (From the public domain.)

There are countless suns and countless earths all rotating around their suns in exactly the same way as the seven planets of our system. We see only the suns [stars] because they are the largest bodies and are luminous, but their planets remain invisible to us because they are smaller and non-luminous.

Giordano Bruno (1584)

The sixteenth century was a time of intense turmoil for religion and science. The Protestant Reformation was sweeping across Europe, and the Catholic Church's response, the Counter Reformation, was in full swing. In 1543, Copernicus upended the classical world—and Church teaching—by proposing a heliocentric universe. Thirty years later, a brilliant light appeared in the sky, one so bright that it was visible even in daytime. The Danish astronomer Tycho Brahe found (using parallax measurements) that the light (now called Tycho's Supernova) originated from far beyond the planets. Contrary to Aristotelian doctrine, which held that the stellar sphere was eternally immutable, this was a new star! Inspired by Copernicus and Brahe, a brash Dominican priest named Giordano Bruno proposed a universe unimaginably larger and more splendid than the worlds envisioned by Aristotle, Ptolemy, or Copernicus: an infinite, centerless cosmos populated by innumerable sun-like stars, each with its own family of planets, some surely inhabited by intelligent beings. But Bruno's brazen attacks on Church doctrine (e.g., he disavowed the divinity of Christ) were his undoing. Caught in the violent crosswinds of sixteenth century religion and science, he was forced to flee Italy in 1576. He returned to Italy 15 years later to apply for a mathematics professorship at the University of Padua. But the position went to a younger man—Galileo Galilei—and shortly afterwards, Bruno was arrested by the Roman Inquisition and imprisoned on the charge of heresy. In the early morning hours of February 17, 1600, he was taken to a public marketplace and burned at the stake. (Thirty years later, Galileo faced the same inquisition.) Today, a statue of Giordano Bruno stands at the site of his execution, in the Campo de' Fiori (field of flowers) in Rome. Although he was executed for heresy, his statue is regarded today as a monument to free thought.

Four centuries after his death, Bruno's grand vision of the universe has withstood the test of time. Scientists have now verified the existence of nearly 2000 extrasolar planets, or exoplanets, within a small region of our own Milky Way Galaxy. The total number of planets within the galaxy is now estimated to be at least 100 billion, about 1 planet per star! It now seems certain that we will soon discover one—or many—inhabitable Earth-twins in our own cosmic backyard.

8.1 INTRODUCTION: THE "WAR ON MARS"

The year 2009—designated the International Year of Astronomy—marked the 400th anniversary of Galileo's first use of the telescope, and the equally important publication of Johannes Kepler's monumental work, *Astronomia Nova*, the book that liberated astronomers from classical Greek models of the solar system and universe. The stage was set earlier by the Danish nobleman Tycho Brahe, one of the most colorful characters in the history of science. Among other things, he lost part of his nose in a sword duel, kept a beer-swilling moose as a pet, and died under suspicious circumstances following a sumptuous banquet.*

* By one account, Tycho died of kidney failure aggravated by binge drinking at the banquet. By another account, he was poisoned, perhaps by Kepler himself, who had been denied full access to Brahe's data. To settle the question, his body has been exhumed twice, most recently in 2010. For further details, see "Murder! Intrigue! Astronomers?" by John Tierney, in the *New York Times*, November 29, 2010.

FIGURE 8.1 Tycho's large mural quadrant, erected at Uraniborg, had a radius of about 2 m. In the figure, the observer (far right) is peering through a sliding slit at a planet visible through the slot in the building's wall (upper left). An assistant (near right) records the time and a second assistant records data. The figures in the background, including Tycho, his dog, and his chemistry laboratory, are not drawn to scale. (From The Royal Library, National Library of Denmark, Copenhagen, Denmark, public domain.)

Despite his flamboyant lifestyle, Tycho was a master instrument designer and perhaps the greatest observational astronomer of the pre–telescope era. With lavish funding from the King of Denmark, he designed and equipped the finest observatory of his day, *Uraniborg*, on an island offshore of Copenhagen. For nearly two decades, Tycho used his massive quadrants and sextants (see Figure 8.1) to chart the skies with unprecedented accuracy. After losing the King's support, he abandoned Uraniborg and moved to Prague where he hired the young mathematician Johannes Kepler to help solve a nagging puzzle: Tycho's precise measurements of the position of Mars, carefully catalogued for over 20 years, were in stark disagreement with the circular-orbit models of both Ptolemy and Copernicus.

Following Tycho's death in 1601, Kepler gained unimpeded access to the vast trove of his astronomical data, and launched what he called his "war on Mars." He labored for nearly 4 years to reconcile Tycho's data with the Copernican model, but the disagreement proved to be too large: Tycho's measurements were accurate to $\pm 2'$ (2 arcminutes), but Kepler faced discrepancies as large as $\pm 8'$. Having absolute faith in Tycho's data, he made a bold and inspired guess, abandoning the circular orbits of Copernicus and Ptolemy in favor of *elliptical* orbits. To his delight, his hunch proved to be correct: theory and measurement now agreed within the limits of experimental uncertainty.

In 1609, Kepler published the *Astronomia Nova*, summarizing his calculations by proposing two laws governing the orbital motion of the planets. In 1619, he added a third law. Seven decades later, Kepler's laws guided Isaac Newton to the law of universal gravitation, which, besides describing the motion of planets and moons, correctly accounted for terrestrial gravity, ocean tides, the shape of the Earth, the precession of the equinoxes, and countless other phenomena. The law of universal gravitation allows us to determine the mass of the Earth, Sun, stars, and all other bodies that possess an orbiting satellite (e.g., a moon, planet, or companion star). In recent years, we have used it to discover dark (invisible) matter and hundreds of bodies orbiting stars other than our Sun—the *exoplanets* envisioned by Giordano Bruno. The law of universal gravitation is the master key that unlocks many of the mysteries of the universe.

8.2 KEPLER'S LAWS

Kepler knew that planets closer to the Sun moved faster than ones farther away. For example, Earth's average orbital speed is 30 km/s, whereas Venus' is 35 km/s and Mercury's is 48 km/s. Moreover, he knew that Mars did not move with constant speed: the closer it was to the Sun, the faster it traveled. He concluded that the planets were propelled by the Sun, not by some power source within the planets themselves (as was supposed by Ptolemy and Copernicus). In 1609, based on his exhaustive analysis of Tycho's data, Kepler proposed the following two "laws" of planetary motion:

1. The planets move in elliptical orbits, with the Sun at one focus.
2. The line drawn from the Sun to an orbiting planet sweeps out equal areas in equal times.

Ten years later, Kepler published his *Epitome Astronomiae Copernicanae* (summary of Copernican astronomy), in which he added a third law:

3. The square of a planet's period T is proportional to the cube of the semimajor axis a of its elliptical orbit.

Because of their central importance to the development of physics and astronomy, we will study these laws and their consequences with great care. Box 8.1 reviews the mathematics and geometry of the ellipse, while Figure 8.2 illustrates the second law. In the figure, the shaded regions S_1, S_2, and S_3 are the areas swept out by a line stretching from the Sun to the orbiting planet during three intervals of time. Kepler's second law states that for equal time intervals ($\Delta t_1 = \Delta t_2 = \Delta t_3$), $S_1 = S_2 = S_3$. Clearly, the planet is moving faster at t_1, when it is closer to the Sun, than at either t_2 or t_3.

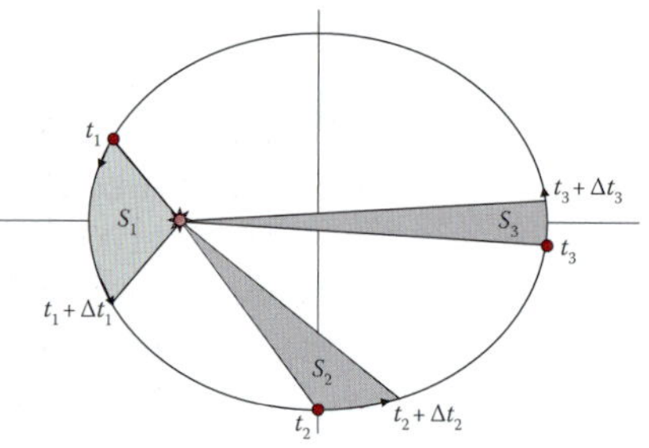

FIGURE 8.2 A planet (red dot) travels about the Sun in an elliptical orbit. S_1 is the area swept out in time Δt_1, and likewise for S_2 and S_3. Kepler's second law states that $S_1 = S_2 = S_3$ if $\Delta t_1 = \Delta t_2 = \Delta t_3$.

BOX 8.1 ALL ABOUT ELLIPSES

Here's how to construct an ellipse. Lay a sheet of paper on a flat surface. Pick any two points on the paper, and drive a thumbtack through each point. Tie the ends of a length of thread to the tacks, loop the thread around the tip of a pencil, and stretch the thread taut (see Figure B8.1a). Now move the pencil in a full circuit around the two tacks, keeping the thread tight and the pencil tip in contact with the paper. The tip will trace out an ellipse.

Mathematically, the locations of the tacks are called the *foci* of the ellipse (denoted F_1 and F_2 in Figure B8.1a), and the ellipse is defined as the set of points P such that $\overline{F_1P} + \overline{F_2P} = \text{const}$ (equal to the length of thread). Now draw a coordinate system with origin at the center of the ellipse, with F_1 and F_2 lying on the x-axis. The x- and y-axes divide the ellipse into four quadrants of equal area. The length (along the x-axis) of each quadrant is called the *semimajor axis* of the ellipse, and is denoted by a in the figure; the height (along the y-axis) of each section is called the *semiminor axis*, denoted by b. For a fixed length of thread, the ellipse becomes more circular as the two foci are moved closer together, and longer and skinnier—or more *eccentric*—as they are moved farther apart. The distance of each focus from the center of the ellipse is expressed as εa, where the *eccentricity* ε is a fraction between 0 (for a circle) and 1 (for a straight line). In Figure B8.1b, point P lies on the semiminor axis, so $\overline{F_1P} = \overline{F_2P} = a$. Applying the Pythagorean theorem to the triangle F_1PO, we obtain $a^2 = a^2\varepsilon^2 + b^2$, or

$$b = a\sqrt{1-\varepsilon^2}. \tag{B8.1}$$

The planets of our solar system move in elliptical orbits with the Sun occupying one focal point (F_1 in the figure). The closest distance to the Sun, called the planet's *perihelion*, is given by $r_P = a(1 - \varepsilon)$, and

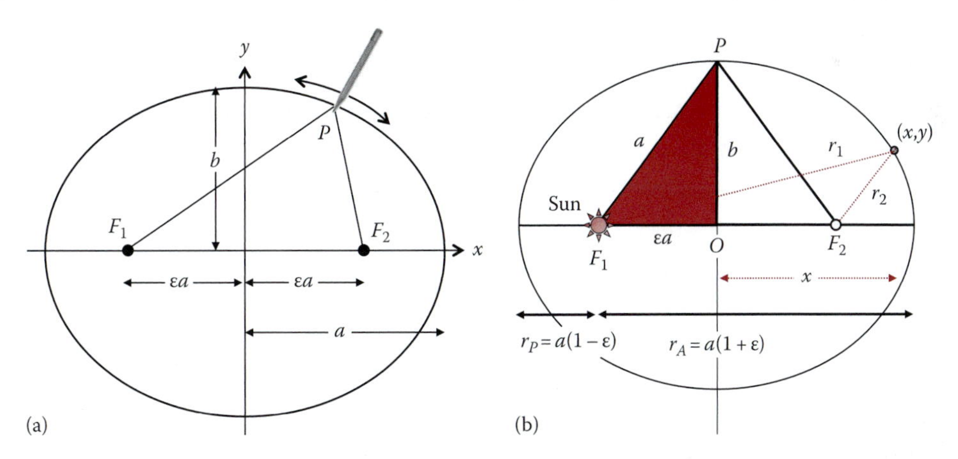

FIGURE B8.1 (a) An ellipse can be constructed by attaching a string between the focal points F_1 and F_2, and stretching it tight with the tip of a pencil (point P). Moving the pencil while keeping the string taut, the tip will trace out an ellipse. Since the length of the string is constant, $\overline{F_1P} + \overline{F_2P} = \text{const} = 2a$, where a is the semimajor axis of the ellipse. The distance of either focal point to the ellipse center is εa, so the perihelion distance is $r_P = a(1 - \varepsilon)$ and the aphelion distance is $r_A = a(1 + \varepsilon)$. (b) By the Pythagorean theorem, $b^2 + \varepsilon^2 a^2 = a^2$, so $b = a\sqrt{1-\varepsilon^2}$.

the farthest distance, called its *aphelion*, is $r_A = a(1 + \varepsilon)$. For the seven planets from Venus to Neptune, $\varepsilon < 0.1$. Using the binomial approximation, this implies that $b/a \geq 0.995$, i.e., the orbits are very close to perfect circles. Only because of Tycho's exacting experimental standards was Kepler able to determine that Mars' orbit was an ellipse.

Let's find an expression for the x- and y-coordinates of an arbitrary point P on the ellipse. Let $P = (x, y)$, and let $r_1 = \overline{F_1P}$, $r_2 = \overline{F_2P}$ be the distances from the foci to P. From Figure B8.1b, $r_1 = \sqrt{(x+\varepsilon a)^2 + y^2}$ and $r_2 = \sqrt{(x-\varepsilon a)^2 + y^2}$. Since $r_1 + r_2 = 2a$, then $r_1^2 = (2a - r_2)^2$, or

$$r_1^2 - r_2^2 = (x+\varepsilon a)^2 - (x-\varepsilon a)^2 = 4\varepsilon a x = 4(a^2 - a r_2).$$

Canceling the common factor $4a$, we obtain $a - \varepsilon x = r_2$. Squaring,

$$a^2 - 2\varepsilon a x + \varepsilon^2 x^2 = (x-\varepsilon a)^2 + y^2,$$

which simplifies to

$$x^2(1-\varepsilon^2) + y^2 = a^2(1-\varepsilon^2)$$

or, using Equation B8.1,

$$\frac{x^2}{a^2} + \frac{y^2}{b^2} = 1. \tag{B8.2}$$

This is the standard (and perhaps familiar) equation for an ellipse.

Let's find an approximate expression for $x(y)$ that is accurate for points near the aphelion, where $y \ll b$. Rearranging Equation B8.2, $x = a\sqrt{1 - y^2/b^2}$, or, using the binomial approximation,

$$x \simeq a\left(1 - \frac{1}{2}\frac{y^2}{b^2}\right) = a - \frac{1}{2}\frac{a}{b^2}y^2.$$

The *radius of curvature* of any function $y(x)$ at a point (x_0, y_0) is the radius of the circle that best matches the shape of the function in the immediate vicinity of (x_0, y_0) (see Figure B8.2a). At aphelion, a planet's coordinates are $(x_0, y_0) = (a, 0)$. The circle that best approximates the shape of the ellipse at this point has a radius r_c and is centered on the x-axis at $(x_c, 0)$, where $x_c + r_c = a$. The equation for points on the circle is, therefore, $(x - x_c)^2 + y^2 = r_c^2$. Using the binomial approximation again, with $y \ll r_c$, we obtain

$$x = x_c + \sqrt{r_c^2 - y^2} \simeq x_c + r_c\left(1 - \frac{1}{2}\frac{y^2}{r_c^2}\right) = a - \frac{1}{2}\frac{y^2}{r_c}.$$

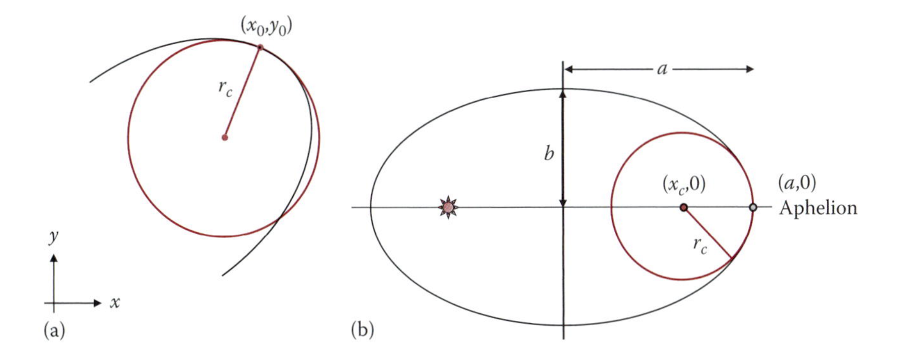

FIGURE B8.2 (a) The radius of curvature r_c of a function at a point (x_0, y_0) is the radius of the circle that best approximates the shape of the curve at the point (x_0, y_0). At the perihelion or aphelion of an ellipse, $r_c = b^2/a$.

Comparing the two equations for x, the radius of curvature of the orbit at aphelion is

$$r_c = \frac{b^2}{a}. \tag{B8.3}$$

The radius of curvature at perihelion is exactly the same as at aphelion. This expression for r_c is very handy, and we will use it repeatedly when analyzing planetary motion.

Exercise B8.1: Can you guess the radius of curvature at points $(x, y) = (0, \pm b)$, that is, where the planet's position is along the semiminor axis of the orbit?

Answer $r_c = a^2/b$

Kepler's third law is expressed mathematically as

$$\frac{T_i^2}{a_i^3} = \kappa_{Sun}$$

where

T_i is the orbital period of the ith planet of the solar system

a_i is the semimajor axis of its elliptical orbit

The proportionality constant κ_{Sun} has the same value for all bodies orbiting the Sun. For Earth, T_E = 1.00 year, and a_E = 1.00 AU, so κ_{Sun} = 1.00 year²/AU³. Figure 8.3 shows that the third law is in excellent agreement with modern data for the solar system.

Planet	Period T (Years)	Semimajor axis a (AU)	T^2/a^3
Mercury	.241	.387	1.00
Venus	.615	.723	1.00
Earth	**1.000**	**1.000**	**1.00**
Mars	1.881	1.524	1.00
Jupiter	11.86	5.204	.998
Saturn	29.46	9.582	.986
Uranus	84.01	19.20	.997
Neptune	164.8	30.05	1.00
(Pluto)	247.7	39.48	1.01

(a)

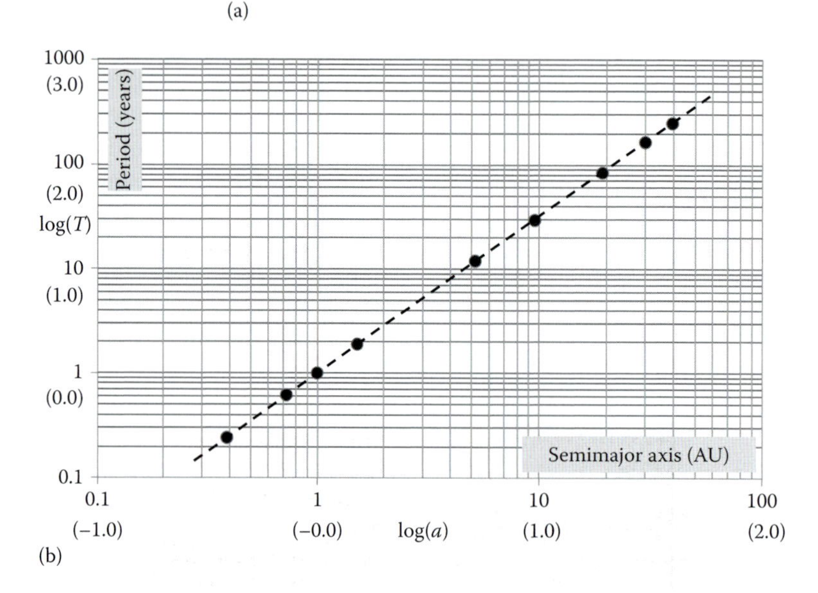

FIGURE 8.3 (a) The period T and semimajor axis a of each of the planets (plus Pluto) of our solar system. According to Kepler's third law, T^2/a^3 is the same for all. (b) A log–log plot of the data listed in part (a). If Kepler's third law is obeyed, the data should lie along a straight line. The small discrepancies shown in the table are caused by interactions between the planets.

EXERCISE 8.1

Figure 8.3b is a plot of $\log_{10}(T)$ vs. $\log_{10}(a)$ for the eight planets of the solar system and Pluto.

(a) Explain why the data fits a straight line.
(b) What is the slope of the straight line fitting the data? (Slope = $d(\log_{10} T)/d(\log_{10} a)$)
(c) If we instead expressed T in terms of seconds, and a in terms of meters, what would be the slope of the line fitting the data?

Answer (c) the same as (b)

8.3 VECTOR CROSS PRODUCT

It is useful to recast Kepler's second or "equal-area" law in the language of vectors. To do this, we need to introduce a new mathematical operation. The vector *cross product* of $\vec{a}$ and $\vec{b}$, written as $\vec{a} \times \vec{b}$, is defined to be a *vector* (1) with a positive magnitude $ab \sin\theta$, where θ is the angle between the two vectors and (2) directed perpendicular to both $\vec{a}$ and $\vec{b}$. To determine θ, slide the two vectors together so that their tails are touching (see Figure 8.4). To insure a positive magnitude, choose the angle θ to be the *smaller* of the two angles between the attached vectors ($0 \le \theta \le \pi$).

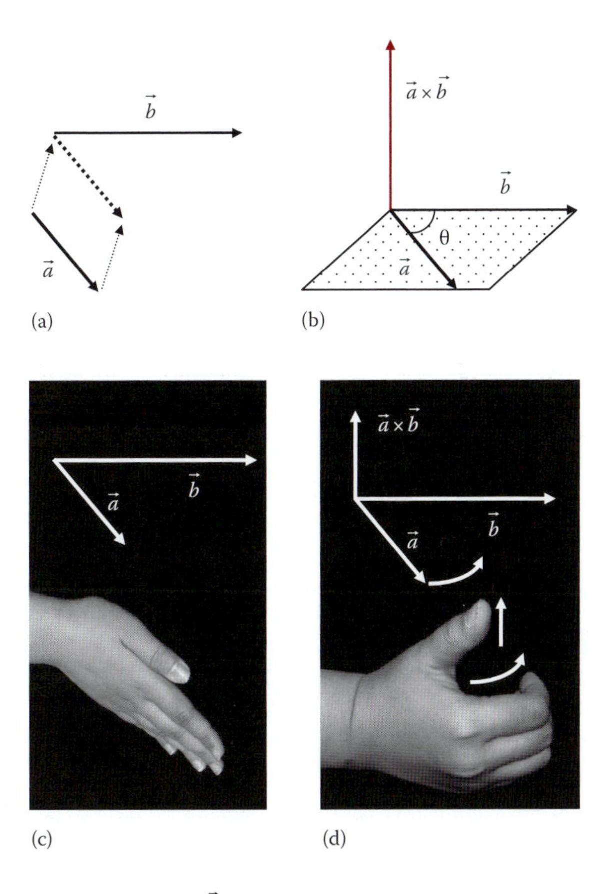

FIGURE 8.4 (a) To evaluate the cross product $\vec{a} \times \vec{b}$, first slide the two vectors together so that their tails are touching. (b) The attached vectors lie in a single plane (shaded) with an angle θ between them ($\theta \le 180°$). (c) The cross product is a vector directed perpendicularly to this plane. To determine the correct direction, point the fingers of your *right* hand in the direction of $\vec{a}$, and then curl them toward the direction of $\vec{b}$. (d) Now extend your thumb fully. It will point in the direction of $\vec{a} \times \vec{b}$. (Photos courtesy of Yue Du, Colgate '16.)

EXERCISE 8.2

Two vectors $\vec{a}$ and $\vec{b}$ are shown in the sketch to the right.

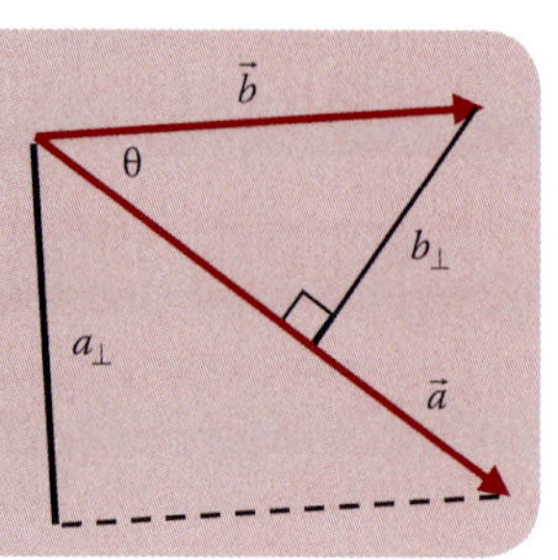

(a) Prove that $|\vec{a}\times\vec{b}| = ab_\perp$, where $b_\perp$ is the component of $\vec{b}$ perpendicular to $\vec{a}$.
(b) Similarly, show $|\vec{a}\times\vec{b}| = a_\perp b$, where $a_\perp$ is the component of $\vec{a}$ perpendicular to $\vec{b}$. These are very useful results.

The two vectors in Figure 8.4b lie in a single plane, so the cross product is perpendicular to this plane. But there are two perpendicular directions (pointing upward and downward in the figure). By convention, we pick the direction of $\vec{a}\times\vec{b}$ using the *right hand rule*. Flatten your *right* hand, and point your fingers in the direction of the first vector of the cross product (vector $\vec{a}$ in the case of $\vec{a}\times\vec{b}$). Now curl your fingers so that the fingertips point in the direction of $\vec{b}$. (Since your fingers cannot bend more than 180°, there's only one possible way to do this, so you may need to invert your hand.) Finally, with your curled fingers pointing along $\vec{b}$, extend your thumb fully. It will point in the direction of the cross product $\vec{a}\times\vec{b}$. This is the right hand rule (see Figure 8.4c and d). Make sure you use your *right* hand!

EXERCISE 8.3

Using the right hand rule, convince yourself that $\vec{b}\times\vec{a} = -(\vec{a}\times\vec{b})$, that is, the cross product $\vec{b}\times\vec{a}$ is *anti-parallel* to $\vec{a}\times\vec{b}$. Unlike the dot product $\vec{a}\cdot\vec{b}$, the order of the two vectors in the cross product matters.

EXERCISE 8.4

Use the right hand rule to find the following cross products of the unit vectors $\hat{i}$, $\hat{j}$, and $\hat{k}$.

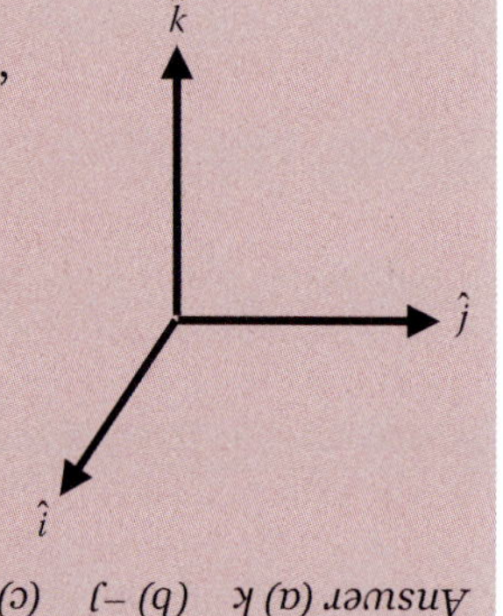

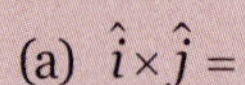
(a) $\hat{i}\times\hat{j} =$
(b) $\hat{i}\times\hat{k} =$
(c) $\hat{j}\times\hat{k} =$
(d) $\hat{k}\times\hat{k} =$
(e) Vectors $\vec{A}$ and $\vec{B}$ are parallel to each other. Evaluate $\vec{A}\times\vec{B}$.

Answer (a) $\hat{k}$ (b) $-\hat{j}$ (c) $\hat{i}$ (d) 0

8.4 APPLICATION: THE CROSS PRODUCT AND KEPLER'S SECOND LAW

Using the cross product, Kepler's equal-area law can be transformed into an important statement about the force $\vec{F}$ acting on an orbiting planet. Let $\vec{r}(t)$ be the vector pointing from the Sun to the planet at a time t, and let $\vec{v}(t)$ be the planet's velocity. Define $\theta(t)$ to be the angle between $\vec{r}$ and $\vec{v}$. In a short time interval Δt, the planet is displaced by $\vec{v}\Delta t$, and $\vec{r}(t)$ sweeps out the shaded triangular area ΔS shown in Figure 8.5. The altitude of the shaded triangle is $r\sin\theta$, and $v\Delta t$ is its base. Therefore,

$$\Delta S = \frac{1}{2} r\sin\theta \cdot v\Delta t.$$

Dividing by Δt and taking the limit as $\Delta t \to 0$,

$$\frac{dS}{dt} = \frac{1}{2} rv\sin\theta = \frac{1}{2}|\vec{r}\times\vec{v}|. \tag{8.1}$$

Kepler's equal area law states that dS/dt = const, which implies that the *magnitude* of $\vec{r}\times\vec{v}$ is a constant. Since the planet moves in a planar orbit, the *direction* of $\vec{r}\times\vec{v}$ (perpendicular to the plane) is also constant. Therefore, the vector $\vec{r}\times\vec{v} = \text{const}$, and its time derivative must equal 0. Differentiating the cross product,*

$$\begin{aligned} \frac{d}{dt}(\vec{r}\times\vec{v}) = 0 &= \frac{d\vec{r}}{dt}\times\vec{v} + \vec{r}\times\frac{d\vec{v}}{dt} \\ &= \vec{v}\times\vec{v} + \vec{r}\times\vec{a} \end{aligned} \tag{8.2}$$

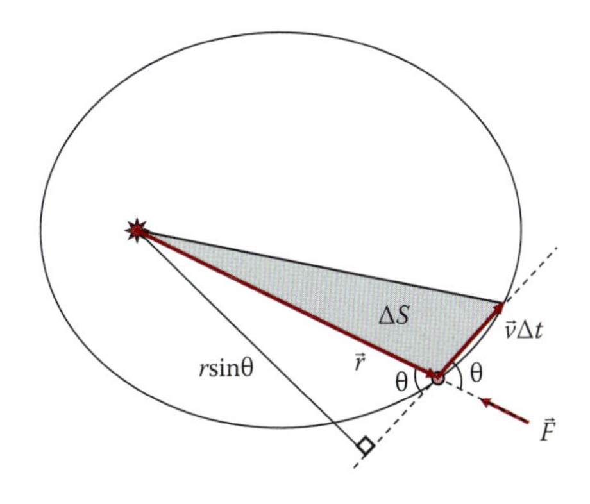

FIGURE 8.5 The shaded triangle is the area ΔS swept out by $\vec{r}(t)$ in the time interval Δt. The base of the triangle has a length $v\Delta t$, and the altitude is $r\sin\theta$. Therefore, $\Delta A = \frac{1}{2} rv\sin\theta \cdot \Delta t = \frac{1}{2}|\vec{r}\times\vec{v}|\Delta t$.

where $\vec{a}$ is the instantaneous acceleration of the planet. But $\vec{v}\times\vec{v}$ is always zero (as in Exercise 8.4e), and by Newton's second law, $\vec{a} = \vec{F}/m$, where $\vec{F}$ is the force on the planet and m is its mass Equation 8.2 can then be rewritten as

$$\vec{r}\times\vec{F} = 0. \tag{8.3}$$

Since neither $\vec{r}$ nor $\vec{F}$ is itself zero, and because the planet is continually deflected toward the interior of its orbit, Equation 8.3 requires that $\vec{F}$ is antiparallel to $\vec{r}$ ($\theta = 180°$ and $\sin\theta = 0$); that is, $\vec{F}$ always points from the planet directly back toward the central body (the Sun). We say that the force acting on the planet is a *central force*. This result is critically important for understanding planetary motion. We proved it using vector calculus, which was not available to either Kepler or Newton. Instead, Newton used a clever geometric argument to prove the converse: if the force on the planet is a central force, then Kepler's equal area law must be obeyed. See Box 8.2 for a description of Newton's derivation.

* Since $\vec{a}\times\vec{b} \neq \vec{b}\times\vec{a}$, the order of the two vectors cannot be changed when taking the derivative of the cross product. The usual product rule for differentiation becomes: $(d/dt)(\vec{a}\times\vec{b}) = (d\vec{a}/dt)\times\vec{b} + \vec{a}\times(d\vec{b}/dt)$.

BOX 8.2 NEWTON'S DEMONSTRATION OF THE EQUAL AREA LAW

Newton did not have vectors at his disposal, so he had to rely on geometry to analyze Kepler's second law. Instead of proving that the equal area law implies that the force on a planet is directed toward the Sun, he showed the converse: if the force is central (i.e., directed toward the Sun), the equal area law must be true. Figure B8.3a shows a planet P moving at constant velocity $\vec{v}$ with no forces acting on it. The points P, A, B, and C are the positions of the planet at times separated by equal intervals Δt, so $\overline{PA} = \overline{AB} = \overline{BC} = v\Delta t$. During the three time intervals, the areas swept out by the line extending from the Sun S to the planet are the areas of triangles PSA, ASB, and BSC. But each of these triangles has the

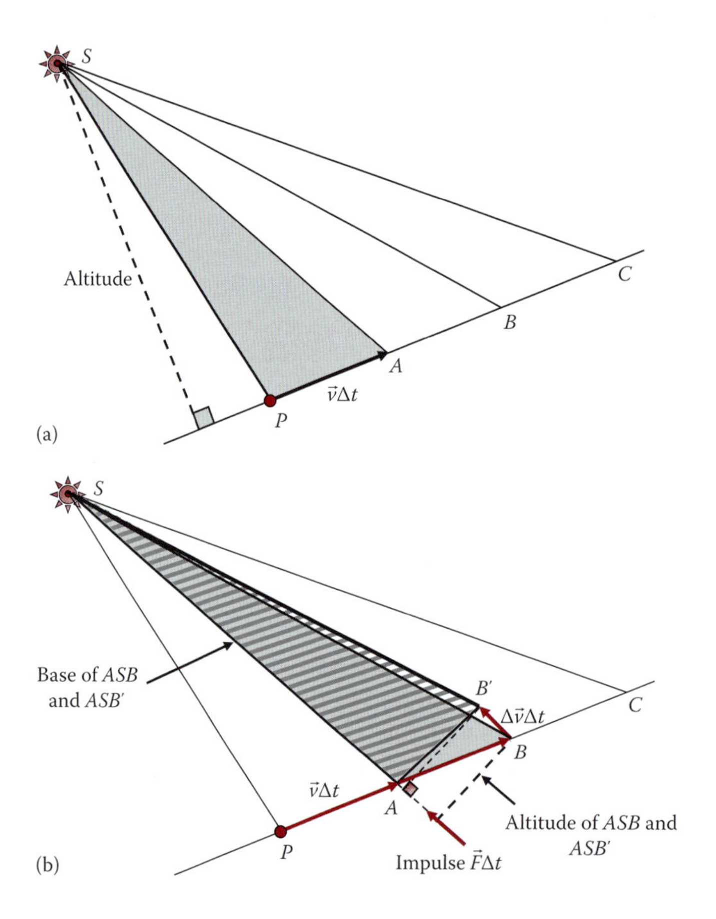

FIGURE B8.3 (a) A planet in a force-free region travels with constant velocity $\vec{v}$. Points A, B, and C are positions of the planet after successive intervals of time Δt. The triangles PSA, ASB, and BSC have equal length bases and share the same altitude, so they have the same area. (b) The attractive force of the Sun is modeled as an impulse $\vec{F}\Delta t$ directed toward the Sun, causing the planet to be deflected to point B' rather than B. Since $\overline{BB'}$ is parallel to $\overline{SA}$, the altitude of triangle ASB' is the same as the altitude of ASB. Likewise, triangles ASB' and ASB share the same base $\overline{SA}$. Therefore, the impulse does not change the area swept out.

same base length $v\Delta t$ and they share the same altitude. Therefore, in this simple force-free case, equal areas are swept out in equal times.

In Figure B8.3b, the Sun exerts a central force on planet *P*. Newton replaced the continuous force acting on the planet with a series of *impulsive* forces, short taps during which force is exerted for a very short time. One of these impulses is shown in the figure: when the planet is at point *A*, it experiences a short tap $\vec{F}$ for time Δt in the direction of the Sun. By Newton's second law, $\vec{F}\Delta t = m\vec{a}\Delta t = m\Delta\vec{v}$, so the planet acquires a velocity increment $\Delta\vec{v}$ toward the Sun. Without the impulse, the planet would have traveled in a straight line from *A* to *B*. With the impulse, it travels instead to point *B'*. Because $\vec{F}$ is directed toward the Sun, $\overline{BB'}$ is parallel to $\overline{SA}$, and triangles *ASB* and *ASB'* have the same altitude. They also have the same base ($\overline{AS}$) and so they have the same area. But we know that triangle *ASB* has the same area as triangle *PSA*, so area *ASB'* is equal to area *PSA*. The centrally directed impulse $\vec{F}\Delta t$ has no effect on the area swept out by the Sun–planet line: the equal area law holds.

Since the equal area law holds for each impulse, it must also hold for any number of impulses. To treat a continuously applied force, we must take the limit as $\Delta t \to 0$ and make the individual impulses and velocity increments $\Delta\vec{v}$ smaller but more frequent. This is how Newton proved that Kepler's equal area law was a consequence of a central force.

8.5 INVERSE-SQUARE DEPENDENCE OF THE PLANETARY FORCE

In the early 1680s, Newton's contemporaries Edmond Halley, Robert Hooke, and Christopher Wren (the founders of the Royal Society of London) met frequently in London to discuss the physics of planetary motion. They agreed with Kepler that the Sun governs the motion of the planets and suspected that the force exerted by the Sun on a planet diminishes as the inverse-square of the distance r between the two bodies ($F \propto 1/r^2$). But suspicion is not the same as proof. Wren offered his comrades a 40 shilling book as a prize* for proving that elliptical orbits arise from an inverse-square force. None of the three men succeeded. Halley visited Cambridge in August 1684, where Newton confided that he had already completed a proof, and promised to send the details in a few days. In fact, he did much better than that, although Halley had to wait 3 months for Newton's response. In November, Newton sent Halley a 10 page manuscript entitled *De Motu Corporum in Gyrum* (*On the Motion of Bodies in Orbit*), which contained the promised derivation, and, in addition, the roots of his subsequent masterpiece—the *Principia*.

It's straightforward to derive the inverse-square behavior of gravity for the special case of a planet undergoing *uniform circular motion*. Starting from the familiar expression for centripetal acceleration $a_c = v^2/r$, and recalling that $v = 2\pi r/T$, where T is the orbital period,

$$a_c = \frac{v^2}{r} = \frac{4\pi^2 r}{T^2}.$$

* This was a pretty attractive offer: 20 shillings, or £2 in 1680, is presently worth about £250 or U.S. $400.

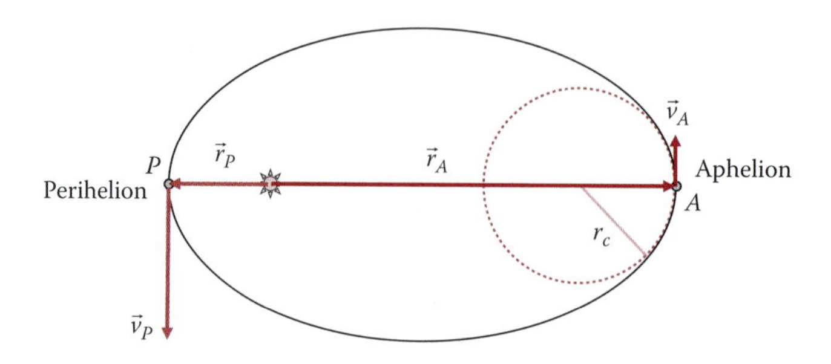

FIGURE 8.6 At its perihelion P or aphelion A, the planet's velocity $\vec{v}$ is perpendicular to its position $\vec{r}$. The radius of curvature of the orbit at either P or A is $r_c = b^2/a$, where a (b) is the semimajor (semiminor) axis of the ellipse.

For circular orbits, Kepler's third law is $T^2 = \kappa_{Sun} r^3$, from which the inverse-square dependence follows directly:

$$F = ma_c = m\frac{4\pi^2}{\kappa_{Sun}}\frac{r}{r^3} = m\frac{4\pi^2}{\kappa_{Sun}}\frac{1}{r^2}, \tag{8.4}$$

where m is the mass of the planet. Since κ_{Sun} is the same for all bodies in orbit about the Sun, it cannot depend on m, so a_c is independent of the planet's mass (just like the gravitational acceleration $\vec{a} = \vec{g}$ for projectiles on Earth).

It took the genius of Isaac Newton to prove that *elliptical* motion also follows from the inverse-square law. His proof is based on geometry and is much too complicated to reproduce here. Shortly after the *Principia* was published, the philosopher John Locke wrote to Newton, complained that he could not follow its lengthy geometric arguments, and asked for a simpler demonstration of the inverse-square law in elliptical orbits. Newton replied with the following illustration. Consider a planet at either its perihelion P or aphelion A, where its velocity $\vec{v}$ is perpendicular to its position $\vec{r}$ relative to the Sun (see Figure 8.6). Since $\vec{r} \times \vec{v} = \text{const}$ and $\sin\theta = 1$ at these two points, we know that $r_A v_A = r_P v_P$, or $v_A/v_P = r_P/r_A$. The acceleration of the planet at either point is given by v^2/r_c, where r_c is the *radius of curvature* at either point (see Box 8.1). Therefore,

$$\frac{F_A}{F_P} = \frac{mv_A^2/r_c}{mv_P^2/r_c} = \frac{v_A^2}{v_P^2} = \frac{r_P^2}{r_A^2}$$
$$= \frac{1/r_A^2}{1/r_P^2}.$$

At least for these two special points of the ellipse, $F \propto 1/r^2$.

Once we are satisfied that the planetary force is inverse-square, we can show that Kepler's third law must be true for elliptical orbits. Look once more at Figure 8.6. Using Equation 8.1 with $\theta = 90°$, the area swept out per unit time at the aphelion (A) is

$$\frac{dS}{dt} = \frac{1}{2}r_A v_A,$$

and because the force on the planet is a central force (Kepler's second law), dS/dt remains constant throughout the orbit. The total area within the ellipse* is $S = \pi ab$, where a and b are the semimajor and semiminor axes. In time T, the planet sweeps out the entire area of the ellipse, so

$$\frac{dS}{dt} = \frac{1}{2} r_A v_A = \frac{\pi ab}{T}.$$

Squaring and multiplying by 4

$$r_A^2 v_A^2 = \frac{4\pi^2 a^2 b^2}{T^2}.$$

The radius of curvature at A (and P) is given by $r_c = b^2/a$, so substituting $b^2 = r_c a$ into the above expression, the centripetal acceleration at aphelion is

$$\frac{v_A^2}{r_c} = \frac{4\pi^2 a^3}{r_A^2 T^2}.$$

From the inverse-square dependence of the force, the acceleration at A is proportional to $1/r_A^2$, so

$$\frac{4\pi^2 a^3}{r_A^2 T^2} \propto \frac{1}{r_A^2} \quad \text{or} \quad \frac{T^2}{a^3} = \text{const} \equiv \kappa_{Sun}. \tag{8.5}$$

This is Kepler's third law for elliptical orbits. Because the semiminor axis b cancels out of the derivation, the period T is independent of the eccentricity of the orbit and depends only on the semimajor axis a. Kepler based his third law on experimental data, while we have proven that it follows from the inverse-square force law.

EXERCISE 8.5

Two planets are in orbit about the same star. Planet A is in an elliptical orbit with aphelion 3 AU and perihelion 1 AU, whereas planet B has a circular orbit with radius R = 2 AU. Compare the periods of the two planets.

Answer $T_A/T_B = 1$

---

* If the equation for the area of an ellipse is unfamiliar to you, note that, for a circle, $a = b = r$, yielding $A = \pi r^2$.

8.6 NEWTON'S GRAND SYNTHESIS: UNIVERSAL GRAVITATION

Before Newton wrote the *Principia* in 1686, it was widely assumed that the forces governing celestial motion were distinct from those governing terrestrial motion. What causes the planets to move? Kepler suggested that they experienced a magnetic attraction to the Sun, while Descartes imagined that they were immersed in an invisible fluid that was churned into a giant whirlpool by the spinning Sun, sweeping the planets along with it. Newton took a different approach. Believing in the simplicity of Nature—that similar behaviors are likely to arise from the same cause—he concluded that the "celestial" force acting on the planets and the "terrestrial" force governing projectile motion on Earth, are one and the same thing—gravity! He boldly proposed that *all* bodies attract each other gravitationally according to the same inverse-square law, whether they be stars, planets, moons—or cannonballs. In the *Principia*, he developed his law of universal gravitation in exquisite mathematical detail, synthesizing the previously separate mechanics of Earth and sky.

How is projectile motion akin to orbital motion? In the first case, a body falls quickly to ground, whereas, in the second, it remains aloft indefinitely. Newton linked the two motions through a *gedanken* experiment. Imagine placing a powerful cannon atop a high mountain and aiming it horizontally (see Figure 8.7a). Neglecting air resistance, a cannonball fired at ordinary speeds drops to the Earth's surface along a parabolic arc. At higher speeds, it travels beyond the horizon, and the curvature of the Earth must be taken into account to predict where the ball will land. As the speed is increased still further, the trajectory evolves smoothly into orbital motion. Nevertheless, as long as the cannonball—now a satellite—remains near the Earth's surface, it undergoes the same acceleration $\vec{g}$ toward the surface as any other falling body. For an infinitesimal interval of time $\Delta t \to 0$, neither the magnitude nor the direction of the acceleration changes, and the familiar expression Equation 6.4 must still apply: $\vec{r}(\Delta t) = \vec{r}_o + \vec{v}_o \Delta t + \frac{1}{2}\vec{g}\Delta t^2$. Launched

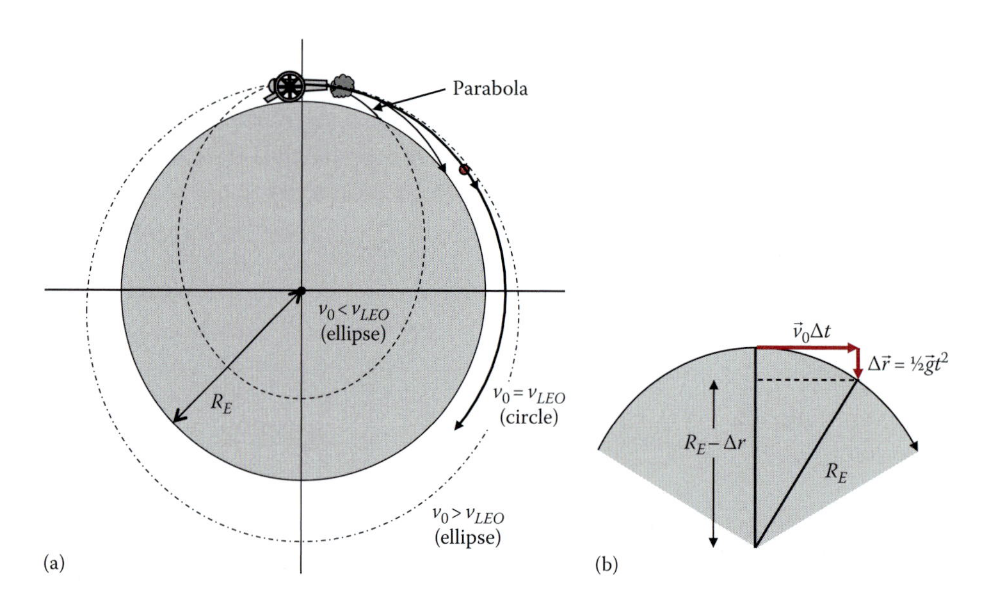

FIGURE 8.7 (a) Newton's cannon: the trajectory of the cannonball evolves from a parabolic arc ($v_o < v_{LEO}$) to a circular orbit ($v_o = v_{LEO}$) as the launch speed is increased. For $v_o > v_{LEO}$, the ball will follow an elliptical orbit, with the launch site as its perigee. (b) As long as the ball remains near the Earth's surface, its acceleration is directed toward the center of the planet and is equal to $\vec{g}$. In time Δt, it falls toward the Earth by $\Delta\vec{r} = \frac{1}{2}\vec{g}\Delta t^2$. To remain in a circular low Earth orbit (LEO), $(R_E - \Delta r)^2 + v_o^2 \Delta t^2 = R_E^2$, or $v_o = \sqrt{gR_E}$.

horizontally, the ball falls by a distance $\Delta\vec{r} = ½\vec{g}\Delta t^2$ while it moves horizontally by $\vec{v}_0\Delta t$ (Figure 8.7b). To remain aloft in a *circular* orbit—without gaining or losing altitude—its distance R to the center of the planet must not change. Using the Pythagorean theorem and Figure 8.7b, $(R_E - \Delta r)^2 + v_0^2\Delta t^2 = R_E^2$. Expanding the first term and simplifying, $v_0^2\Delta t^2 = 2R_E\Delta r - \Delta r^2$. In the limit $\Delta t \to 0$, $\Delta r \ll R_E$, so the last term vanishes and

$$\frac{v_0^2\Delta t^2}{2R_E} = \Delta r = \frac{1}{2}g\Delta t^2.$$

To enter a circular *low-Earth orbit* (LEO), the initial velocity must be

$$v_0 = \sqrt{gR_E} \equiv v_{LEO}. \tag{8.6}$$

At this speed, the cannonball/satellite will remain aloft indefinitely in a circular orbit skimming the planet's surface, *even though the body is always falling toward the center of the planet with acceleration* $\vec{g}$. Solving Equation 8.6 with R_E = 6370 km and g = 9.8 m/s², we find v_{LEO} = 7.9 km/s (18,000 mph!).

EXERCISE 8.6

Derive the result of Equation 8.6 by equating the satellite's centripetal acceleration to the acceleration of gravity g.

If we launch the ball horizontally with an initial speed $v_0 > v_{LEO}$, it will follow an elliptical orbit with the center of the planet as its focus, and the launch site will be the point of the orbit closest to the center of Earth—its *perigee*. If $v_0 < v_{LEO}$, and if the planet did not obstruct the motion, the ball would again execute an elliptical orbit, but now the launch site would be the orbit's *apogee*. The parabolic path of a projectile on Earth is just a small segment of a larger elliptical trajectory, as shown in Figure 8.7a. There is no fundamental physical difference between parabolic motion and orbital motion.

Of course Newton did not have artificial satellites or powerful artillery to test his hypothesis. Instead, he had the Moon. To associate terrestrial gravity with the inverse-square force acting on celestial bodies, he needed to prove that the gravitational force on Earth ($\vec{F} = m\vec{g}$) was also derivable from an inverse-square force law. How do you calculate the gravitational force between *extended* bodies (such as the Earth and Moon) rather than between idealized point masses? His solution, sometimes called "Newton's superb theorem," was a critical milestone along the path to the law of universal gravitation.

8.7 SHELL THEOREM, OR NEWTON'S "SUPERB THEOREM"

Imagine a point mass m in the vicinity of a large body such as Earth (see Figure 8.8). Divide the large body's total mass M_E into a large collection of identical infinitesimal masses dM_i. According to the hypothesis of universal gravitation, m is attracted to each of these small masses with a force that diminishes with the distance r_i squared: $d\vec{F}_i \propto 1/r_i^2$. The total force on m is the vector sum of all these forces: $\vec{F} = \sum_i \vec{F}_i$. In 1666, Newton guessed that if M_E were a spherically symmetric body, the force exerted on m would be the same as if the entire mass M_E were concentrated at its center. But it was not until he wrote the *Principia*, 20 years later, that he could prove rigorously that his guess was correct. His brilliant derivation is called the *shell theorem*, or Newton's superb theorem.* A calculus-based derivation will be presented in Chapter 10, after we have introduced the concept of potential energy. For now, we will simply state Newton's conclusions.

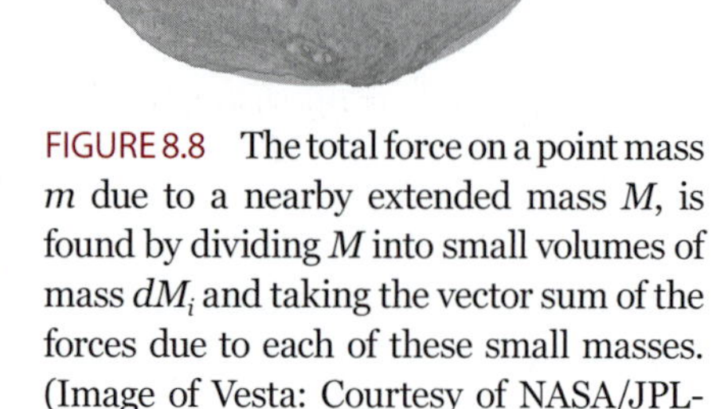

FIGURE 8.8 The total force on a point mass m due to a nearby extended mass M, is found by dividing M into small volumes of mass dM_i and taking the vector sum of the forces due to each of these small masses. (Image of Vesta: Courtesy of NASA/JPL-Caltech/MPS/DLR/IDA, Pasadena, CA.)

The shell theorem gives the force between a point mass m and a thin homogeneous spherical shell of total mass M and radius R (see Figure 8.9a). Let r be the distance between m and the center of the shell. Assuming that the inverse-square hypothesis is correct, the force exerted on m depends critically on whether m is inside or outside the shell's surface:

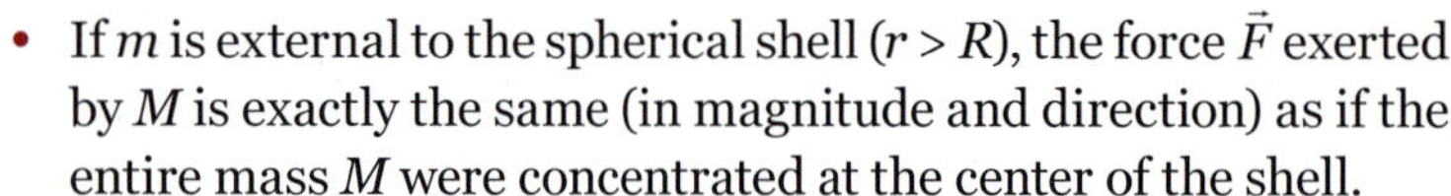

- If m is external to the spherical shell ($r > R$), the force $\vec{F}$ exerted by M is exactly the same (in magnitude and direction) as if the entire mass M were concentrated at the center of the shell.
- If m is inside the shell ($r < R$), $\vec{F}$ *is exactly zero*, no matter where m is placed within the shell.

A little thought suggests that these results are reasonable. For $r > R$, the spherical symmetry of M dictates that $\vec{F}$ must point toward the center of the shell, and r is an "average" distance between m and the various parts of the shell. For $r = 0$, m is equally attracted to all parts of the surrounding shell, so the total force $\vec{F} = 0$. For $0 < r < R$, m is attracted more strongly to mass on the near wall of the shell, but there is more

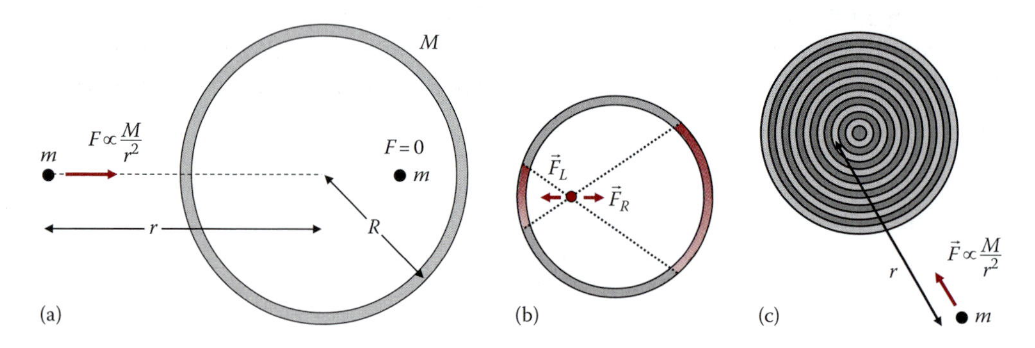

FIGURE 8.9 (a) The force on a point mass m, due to a spherical *shell* of mass M and radius R, depends on whether m lies inside or outside the shell. (b) When m is inside the shell, it is attracted to the near side of the shell by $\vec{F}_L$ and to the far side by $\vec{F}_R$. Since there is more mass to the right, $\vec{F}_L = -\vec{F}_R$, and the total force on m is zero. (c) A spherically symmetric body may be considered to be a collection of nested shells. If m lies outside the body, the force on it is the same as if the entire mass M were located at its center.

* Surprisingly, his proof is based on geometry, not calculus. As the mathematician J.E. Littlewood remarked, it "must have left its readers in helpless wonder."

mass on the far wall pulling m in the opposite direction. Overall, the two contributions to the total force cancel out (Figure 8.9b) and $\vec{F} = 0$ again.

The shell theorem can be generalized to real physical bodies other than shells. Earth may be pictured as a collection of nested spherical shells—an astronomical onion (Figure 8.9c). Although the density of the planet varies with depth, the density within each shell may be regarded as constant. The force on a body m situated *on or above the planet's surface* is the same as if the mass of each shell were at the Earth's center, or collectively, as if the entire mass of the planet M_E were located at its center.

EXERCISE 8.7

A body of mass m interacts gravitationally with two concentric spheres, each of mass M. Points a and b are, respectively, at distances $2r$ and r from the center of the spheres c. Let $F(a)$ be the force on m when it is located at point a, and similarly for $F(b)$ and $F(c)$.

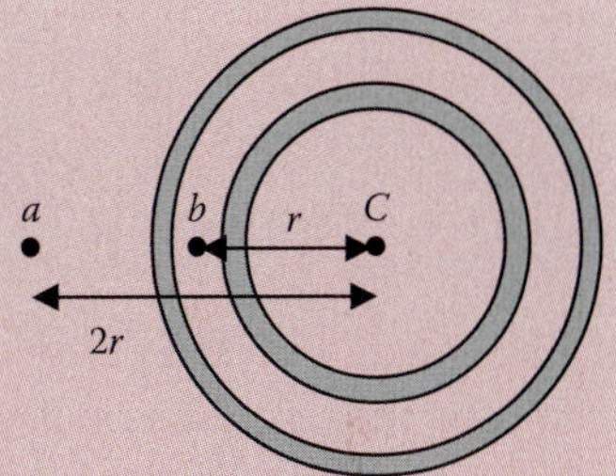

(a) Is $F(a)$ greater than, less than, or equal to $F(b)$?
(b) Evaluate the force $F(c)$.

Answer: (a) $F(a) = \frac{1}{2}F(b)$ (b) $F(c) = 0$

8.8 THE APPLE AND THE MOON

Armed with the shell theorem, Newton now set out to prove that the same inverse-square force accounts for projectile motion on Earth as well as orbital motion in space. According to the shell theorem, the force $\vec{F} = m\vec{g}$ may be considered as the inverse-square attraction of a body m to a mass M_E situated at the planet's center, a distance R_E below its surface: $g \propto 1/R_E^2$. Likewise, the acceleration of the moon a_M is due to its interaction with the *same mass* M_E but at a much greater distance R_{EM}.* From the time of Hipparchus, astronomers knew that the center-to-center distance R_{EM} between the Earth and the Moon is very nearly equal to $60R_E$. Therefore, according to the hypothesis of universal gravitation,

$$\frac{a_M}{g} = \frac{1/R_{EM}^2}{1/R_E^2} = \frac{1/(60R_E)^2}{1/R_E^2}$$

or

$$a_M = \frac{g}{60^2} = \frac{9.8}{3600} = 2.7 \times 10^{-3} \text{ m/s}^2.$$

* As a corollary of the shell theorem, Newton showed that the interaction of two spherical bodies was the same as if the mass of each body were concentrated at its center.

The Moon's centripetal acceleration can be calculated directly from its orbital period T and radius R_{EM}:

$$a_M = \frac{v_M^2}{R_{EM}} = \frac{4\pi^2 R_{EM}}{T^2}.$$

Using the known values $R_{EM} = 60R_E = 60 \times (6.4 \times 10^6\ \text{m}) = 3.85 \times 10^8\ \text{m}$ and $T = 27.3\ \text{days} = 2.36 \times 10^6\ \text{s}$, we find

$$a_M = \frac{4\pi^2(3.85\times10^8\ \text{m})}{(2.36\times10^6\ \text{s})^2} = 2.7\times10^{-3}\ \text{m/s}^2,$$

in agreement with the prediction of universal gravitation. The "celestial" force that guides the Moon in its orbit is merely Earth's gravity, reaching far above the planet's surface to control a body nearly half a million kilometers away. Newton finally had convincing evidence for universal gravitation.*

EXERCISE 8.8

A body in free fall on Earth drops 4.9 m in the first second after it is released. How far does the Earth's Moon fall toward Earth in 1 min?

Answer 4.9 m

* In 1666, when Newton began thinking about gravity, he lacked an accurate measure of the Earth's radius R_E. The best available estimate was $R_E \simeq 5500$ km, a 15% error that caused similar errors in R_{EM} and a_M. (The first accurate measure of R_E was performed by Jean Picard in 1671.) Judging that a 15% discrepancy between a_M and $g/3600$ was unacceptably large, he abandoned the inverse-square hypothesis for several years.

8.9 LAW OF UNIVERSAL GRAVITATION

Newton next looked to other systems to bolster his hypothesis of universal gravitation. In 1684, only 4 of Jupiter's 66 moons were known. These were the ones discovered by Galileo in 1610, and they were known to have nearly circular orbits with the semimajor axes and periods listed in Figure 8.10. As shown in the figure, $T^2/a^3 = \text{const} \equiv \kappa_{Jup}$ for the four moons, implying that the force exerted on them by Jupiter is inverse-square, just like the force exerted by the Sun on the planets. Five of Saturn's many moons had also been discovered, and Newton found that Kepler's third law held for them as well: $T^2/a^3 = \text{const} \equiv \kappa_{Sat}$. This common behavior was strong evidence that the forces on all planets and moons have the same inverse-square dependence. Kepler's third law holds for each of the three systems studied, but the constants are different in each case: $\kappa_{Sun} \neq \kappa_{Jup} \neq \kappa_{Sat}$. Apparently, the value of κ depends on the central body, and only on that body.

	$a \times 10^6$ (km)	T (days)	ε	T^2/a^3
Io	.4218	1.769	.0041	41.70
Europa	.6711	3.551	.0094	41.72
Ganymede	1.0704	7.155	.0011	41.74
Callisto	1.8872	16.69	.0074	41.44

(a)

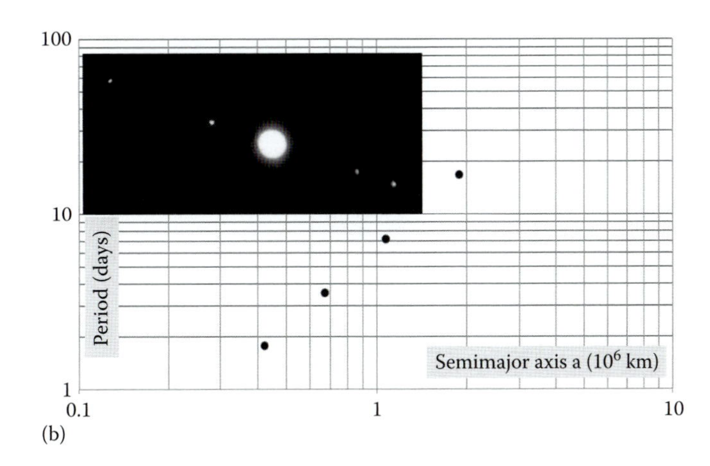

FIGURE 8.10 (a) The orbital semimajor axis a and period T of each of the Galilean moons of Jupiter. Newton's hypothesis of universal gravitation predicts that T^2/a^3 = const for the moons. (b) A log–log plot of T vs. a is a straight line, as anticipated. Inset: telescopic image of the four moons. Their orbits are in a plane nearly perpendicular to the page. (Image courtesy of Chanan Greenberg, www.greenhawkobservatory.com.)

EXERCISE 8.9

The semimajor axes, periods and eccentricities of the five moons of Saturn that were known in Newton's time are given below, in units of 10^6 km and (Earth) solar days.

(a) Show that Kepler's third law holds for the first four satellites listed.
(b) Find the period of the fifth moon, Dione.

	Discovered by	$a \times 106$ (km)	T (days)	ε
Titan	Huygens (1665)	1.222	15.95	0.0288
Iapetus	Cassini (1671)	3.561	79.33	0.0293
Rhea	Cassini (1672)	0.527	4.518	0.0010
Tethys	Cassini (1684)	0.295	1.888	0.0001
Dione		0.377	?	0.0022

Answer (b) 2.74 days

From Equation 8.4 the force exerted by the Sun on a planet of mass m_P is

$$F_{Planet} = m_P \frac{4\pi^2}{\kappa_{Sun}} \frac{1}{r^2},$$

where r is the distance between the two bodies. If the inverse-square force is universal, then the force on the Sun, due to the planet, must obey an equation of the same form:

$$F_{Sun} = M_{Sun} \frac{4\pi^2}{\kappa_P} \frac{1}{r^2},$$

where κ_P depends only on the properties of the planet. By Newton's third law (action–reaction), $F_{Sun} = F_{Planet}$, so $m_P/\kappa_{Sun} = M_{Sun}/\kappa_P$, or $m_P/M_{Sun} = \kappa_{Sun}/\kappa_P$. The simplest way to satisfy this relation is to require that $m_P \propto 1/\kappa_P$ and $M_{Sun} \propto 1/\kappa_{Sun}$, with the same proportionality constant in each case. Using this result,

$$F_{Sun} = F_P \propto \frac{M_{Sun} m_P}{r^2}$$

or

$$F_{Sun} = F_P = G \frac{M_{Sun} m_P}{r^2},$$

where the constant of proportionality G is called the *gravitational constant*, and must be determined experimentally. More generally, the force between any two bodies of mass m_1 and m_2 is

$$F_{12} = G \frac{m_1 m_2}{r^2}, \tag{8.7}$$

where r is the distance between the bodies. The direction of the force on body 1 is toward body 2, and vice versa. This is Newton's *law of universal gravitation*, applicable to bodies whose dimensions (e.g., radii) are small compared to the distance r between them or to bodies (such as stars and planets) that are spherically symmetric. Equation 8.7 is one of the most important and far-reaching equations in physics. It is "that mighty principle under the influence of which every star, planet, and satellite in the universe pursues its allotted course."*

For a body of mass m in a *circular* orbit about a central body of mass M, the force on m is

$$F = m\frac{v^2}{r} = m\frac{4\pi^2 r}{T^2} = G\frac{mM}{r^2} \quad \text{or} \quad \frac{T^2}{r^3} = \frac{4\pi^2}{GM}. \tag{8.8}$$

The constant κ in Kepler's third law is equal to $4\pi^2/GM$, where M is the mass of the central body.

---

* C. Vernon Boys, The Newtonian constant of gravitation, June 8, 1894, in *Proceedings of the Royal Institution of Great Britain*, Vol. XIV (1893–1895), pp. 353–377.

EXERCISE 8.10

A satellite is in low Earth orbit (LEO), just above the planet's atmosphere, with an orbital speed v_{LEO} and period of approximately 90 min. Now consider another satellite moving in a circular orbit with radius $r = 4R_E$.

(a) What is its acceleration in "g's"?
(b) What is its period in hours?
(c) What is its orbital speed as a fraction of v_{LEO}?
(d) What is the orbital radius of a *geosynchronous* satellite ($T = 24$ h)?

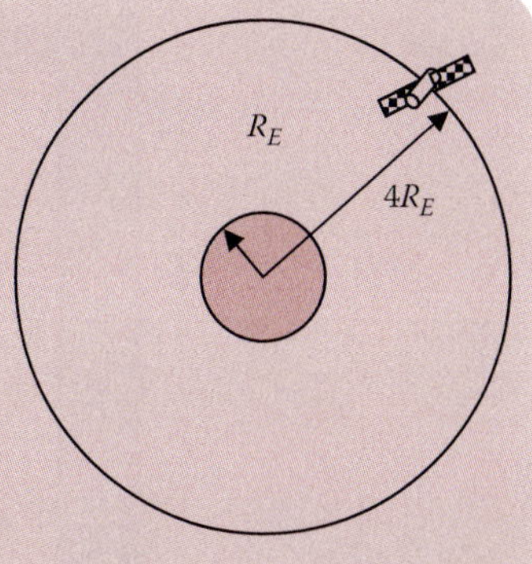

Answer (a) $g/16$ (b) 12 hrs (c) $\frac{1}{2}v_{LEO}$ (d) $6.35R_E$

For *elliptical* orbits, we have shown that T^2/a^3 is independent of the eccentricity (or semiminor axis b). Since a circle is just an ellipse with zero eccentricity, it follows that the period of an elliptical orbit is the same as the period of a circular orbit with $r = a$, given by Equation 8.8:

$$\frac{T^2}{a^3} = \frac{4\pi^2}{GM}. \tag{8.9}$$

Equation 8.9 says that *if we know the value of G*, we can find the mass M of any astronomical body that is host to an orbiting satellite (e.g., a planet or moon) with measurable period T and semimajor axis a. This includes the Sun, the planets from Earth to Neptune, and as you will see, many other stars and exotic objects far from our solar system—even black holes and entire galaxies! Mercury and Venus have no moons, and their masses were not known accurately until 1974, when the *Mariner 10* spacecraft executed close flybys of the two planets to sample their gravitational force directly. The value of G was not determined for more than a century after the publication of the *Principia*. Even so, Newton was able to calculate the *ratios* of the densities of the Sun, Jupiter, Saturn, Earth, and the Moon. The law of universal gravitation truly opened a new window on the universe.

EXERCISE 8.11

Use the data of Figure 8.10 and Exercise 8.9 to determine the ratio between Saturn's mass and Jupiter's mass. No fair looking it up!

Answer $M_S/M_J = 0.3$

8.10 WEIGHING THE EARTH AND SUN

For nearly 200 years after the publication of the *Principia*, physicists attempted to measure—not the gravitational constant G—but an equivalent quantity: the average density of Earth $\bar{\rho}_E$. Many clever experiments were performed, such as measuring the change of a pendulum's period as it was carried from sea level to

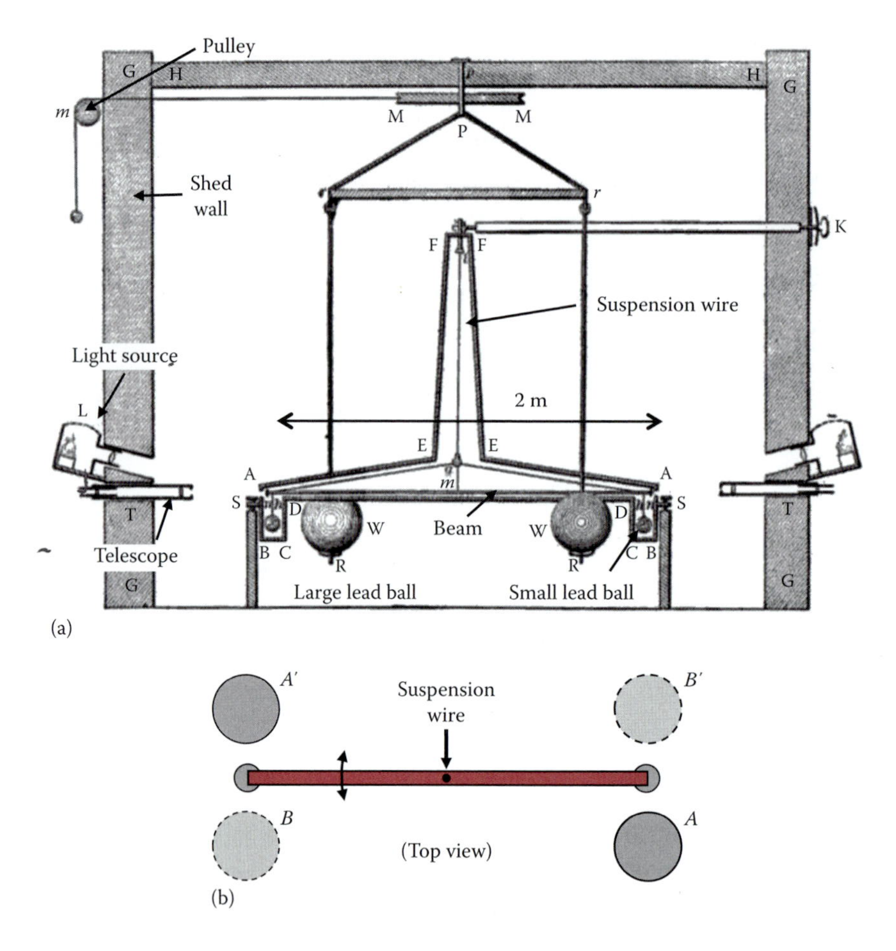

FIGURE 8.11 (a) Cavendish's torsion balance for measuring $\bar{\rho}_E$. The apparatus was enclosed within a sealed shed, and Cavendish measured the deflection of the beam using a telescope (at left) mounted in the shed wall. (b) Top view of apparatus. A rope and pulley were used to rotate the large lead balls from positions A and A' to B and B'. (From Cavendish experiment, public domain.)

the peak of a tall mountain, or as it was lowered into a deep mineshaft. The first accurate determination of $\bar{\rho}_E$ was carried out in 1798 by British physicist Henry Cavendish, who measured the exceedingly weak gravitational force between lead balls of known mass, and from that computed $\bar{\rho}_E$. To perform this challenging measurement, he constructed a sensitive apparatus called a *torsion balance* or *torsional pendulum*, shown in Figure 8.11.* Two lead balls of mass 0.8 kg were hung from the ends of a long wooden beam. The beam itself was suspended by a thin wire passing through its center, allowing it to rotate in the horizontal plane. Rotation twisted the suspension wire, which provided a tiny restoring force that induced simple harmonic

* The apparatus was originally designed by the Reverend John Michell, a geologist, who died before completing its construction. Cavendish's instrument conformed fairly closely to Michell's design.

motion with an oscillation period of about 7 min. (Measurement of the period allowed Cavendish to calculate the torsional "spring constant" of the wire.) With the beam at rest at its equilibrium position, two large lead balls of mass 160 kg were placed near the smaller hanging balls (at positions A and A' in Figure 8.11b), and the attractive force between the large and small balls caused the beam to rotate slightly. Then the large balls were swung to diametrically opposite positions (positions B and B' in Figure 8.11b), causing the beam to swing in the opposite direction. The total deflection (about 0.2°) was measured and the force between the balls was calculated.

Because of the torsion balance's extreme sensitivity, Cavendish took great care to shield it from air currents. The entire apparatus was sealed inside a small shed, and Cavendish carried out his measurements while standing outside the shed, using a telescope inserted through a peephole in the shed's wall. By comparing the weight of the lead spheres with the force between them, he determined a value for the Earth's density that is within 1% of the modern accepted value: $\bar{\rho}_E = 5515\ \mathrm{kg/m^3}$. From this measurement and the Earth's mean radius (R_E = 6371 km), the Earth's mass can be calculated: $M_E = 5.97 \times 10^{24}$ kg. In fact, Cavendish referred to his experiment as "weighing the Earth."

Of course, the gravitational constant G is of far greater fundamental importance than the Earth's density.* From the shell theorem and Equation 8.7, the force on a point mass m at the Earth's surface is

$$F = G\frac{mM_E}{R_E^2} = mg \quad \text{or} \quad g = \frac{GM_E}{R_E^2}. \tag{8.10}$$

Then, using modern values for g, R_E, and M_E, the gravitational constant is found to be

$$G = 6.67 \times 10^{-11}\ \mathrm{N\,m^2/kg^2}. \tag{8.11}$$

Knowing G, we can now determine the mass of other astronomical bodies.

EXERCISE 8.12

We are in a nearly circular orbit about the Sun, with a radius of 1.5×10^{11} m and a period of 1 year (3.15×10^7 s). Use this data and Equation 8.9 to find the mass of the Sun.

Answer $M_{Sun} = 2 \times 10^{30}$ kg

Suppose we wanted to determine the mass M' of the central star of another solar system. If we have measured the orbital period T' and semimajor axis a' of a planet in the alien system, we can calculate the star's mass by using Kepler's third law, just as you did in the above exercise. But rather than crunch all

* In an address to the Royal Institution of Great Britain in 1894, the physicist C. V. Boys remarked: "Owing to the universal character of the constant G, it seems to me to be descending from the sublime to the ridiculous to describe the object of this experiment as finding the mass of the Earth, or the mean density of the Earth."

the numbers again, it is easier to find M' in terms of M_{Sun}. Let T_E (=1 year) and a_E (=1 AU) be the period and size of the Earth's orbit. Then,

$$\frac{T_E^2}{a_E^3} = \frac{4\pi^2}{GM_{Sun}} \quad \text{and} \quad \frac{T'^2}{a'^3} = \frac{4\pi^2}{GM'} \quad \text{or} \quad \frac{(T'/T_E)^2}{(a'/a_E)^3} = \frac{1}{(M'/M_{Sun})}. \tag{8.12}$$

Since each of the terms in parentheses is dimensionless, it does not matter what units we use for T_E, T', a_E, and a', as long as we use them consistently. The units of years and AUs are particularly convenient for our calculation, and by using them we can reduce the work of finding M' considerably. To appreciate this, try the following exercise.

EXERCISE 8.13

If the alien planet's period is 2 years, and its orbital radius is 4 AU, what is the mass of the central star, expressed in terms of M_{Sun}? No fair using a calculator!

Answer $M' = 16M_{Sun}$

8.11 APPLICATION: BINARY STARS

In order to understand the physical processes occurring within a star, it is imperative to know the star's mass. Luckily, about 50% of the visible stars in the sky belong to multiple star systems: two or more gravitationally linked bodies orbiting a common center of mass (CM). *Visual binaries* are two-star systems where both stars can be resolved and tracked telescopically. For example, the "dog star" Sirius*—the brightest stellar object in the northern hemisphere—is a binary system with a very luminous primary star (Sirius A) and a faint secondary companion (Sirius B). Both M_A ($\approx 2\ M_{Sun}$) and M_B ($\approx 1\ M_{Sun}$) can be determined by measuring the period and semimajor axes of their elliptical orbits. But there is one hitch. The systems we treated earlier all involved a *stationary* central body M (e.g., the Sun) that was much more massive than the planets or moons in orbit around it (as in Equation 8.9). For two interacting bodies M and m, it is their *center of mass* that remains stationary. Kepler's third law, as expressed in Equations 8.8 and 8.9, holds only when M is coincident with the CM, that is, when $M \gg m$. This is not the case for binary stars.

Let's look first at the motion of two stars M_1 and M_2 that are moving in *circular* orbits about their common CM (see Figure 8.12a). Choosing the origin of coordinates to coincide with the system's CM,

$$M_1\vec{r}_1 + M_2\vec{r}_2 = 0,$$

and differentiating with respect to time,

$$M_1\vec{v}_1 + M_2\vec{v}_2 = 0,$$

* We met Sirius in Chapter 1. Its first sunrise appearance (in July) coincided with the flooding of the Nile and marked the start of the ancient Egyptian calendar.

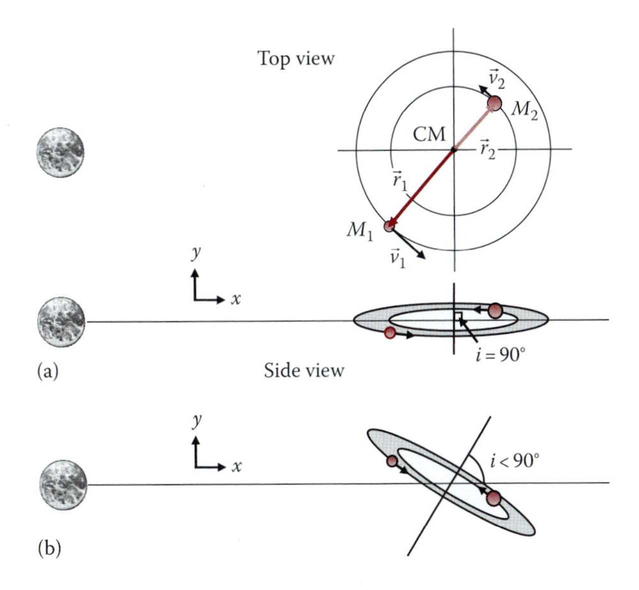

FIGURE 8.12 Two stars of a binary system are in circular orbit about their common CM. (a) The plane of their motion lies parallel to the line of sight from Earth (x-axis), with inclination angle $i = 90°$. (b) The orbital plane is inclined to the line of sight. Except for eclipsing binaries, i is unknown, adding an unknown factor $\sin i$ to the calculation of the stellar masses M_1 and M_2.

where $\vec{r}_i$ and $\vec{v}_i$ are the position and velocity of M_i with respect to the CM. To satisfy these equations, $\vec{r}_1$ and $\vec{v}_1$ must be antiparallel to $\vec{r}_2$ and $\vec{v}_2$, respectively, and their magnitudes must be related by

$$\frac{r_1}{r_2} = \frac{v_1}{v_2} = \frac{M_2}{M_1}.$$

The two stars are always in line with, and on opposite sides of, their CM. They share the same period $T_1 = T_2 \equiv T$, and they are always separated by the distance $r_1 + r_2$. Let's apply Newton's second law to M_1, which is moving with centripetal acceleration v_1^2/r_1:

$$M_1 \frac{v_1^2}{r_1} = M_1 \frac{4\pi^2 r_1}{T^2} = G \frac{M_1 M_2}{(r_1 + r_2)^2},$$

where we have eliminated the velocity using $v_1 = 2\pi r_1/T$. Simplifying and rearranging terms,

$$T^2 = \frac{4\pi^2}{G}(r_1 + r_2)^2 \frac{r_1}{M_2}.$$

In the same way, we can derive the equivalent expression for M_2:

$$T^2 = \frac{4\pi^2}{G}(r_1 + r_2)^2 \frac{r_2}{M_1}.$$

Now combine these two equations by multiplying both sides of the first equation by M_2, both sides of the second by M_1, and then adding the two expressions together:

$$(M_1 + M_2)T^2 = \frac{4\pi^2}{G}(r_1 + r_2)^2 \cdot (r_1 + r_2) = \frac{4\pi^2}{G}(r_1 + r_2)^3 \quad \text{or} \quad \frac{T^2}{(r_1 + r_2)^3} = \frac{4\pi^2}{G(M_1 + M_2)}. \tag{8.13}$$

Equation 8.13 is a generalization of Kepler's third law valid for *any* two bodies in circular orbit about their CM. Note that if $M_1 \gg M_2$, then $r_1 \ll r_2$ and the above expression reduces to Equation 8.8. What if the orbits are elliptical rather than circular? Then, the two ellipses share a common focus located at the CM, and Equation 8.13 is modified in the same manner as Equation 8.8 was adapted for elliptical orbits:

$$\frac{T^2}{(a_1 + a_2)^3} = \frac{4\pi^2}{G(M_1 + M_2)} \tag{8.14}$$

where a_1 and a_2 are the semimajor axes of the two orbits.

EXERCISE 8.14

Two stars, M_A and M_B, execute elliptical orbits with semimajor axes a_A and a_B. The figure shows the orbit of M_B. If $M_A = 2M_B$ (as in Sirius A and B), sketch the orbit of M_A. Where is M_A at the time shown in the figure?

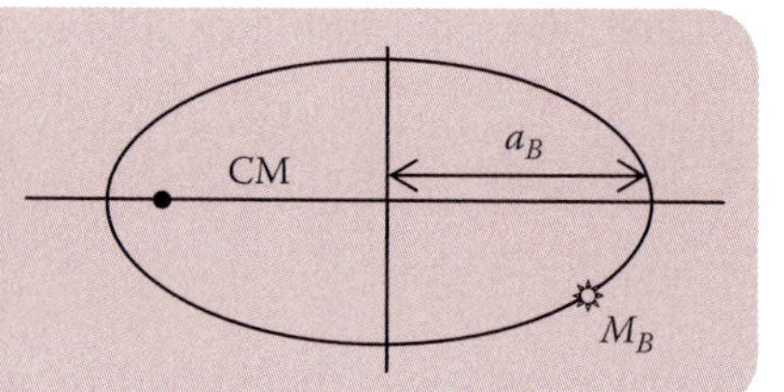

For visual binaries, we can *see* the two stars and can measure their period and orbital axes directly. More often, though, the stars are too far away or too close together to be resolved by even the largest telescopes. In spite of this, we can still use Equation 8.13 or 8.14 if we can measure the *Doppler shift* of the light emitted by each star as it traces out its orbit. Star systems where this is possible are called *spectroscopic binaries*. In the simplest case, shown in Figure 8.12a, the orbits of M_1 and M_2 are viewed edge on (orbit inclination $i = 90°$) by observers on Earth.* Let the orbits lie in the x–y plane of our coordinate system, and

* If $i \approx 90°$, the stars will eclipse each other, that is, one star will periodically pass in front of the other, temporarily blocking some or all of the more distant star's light from view. This is observed as a periodic reduction in the total luminosity of the system, and allows astronomers to determine the stars' radii.

define the x-axis to coincide with the line of sight from Earth to the binary system's CM. If the two stars are in uniform circular motion, the velocity of M_1 may be expressed as

$$\vec{v}_1(t) = v_1 \cos(\omega t + \theta_0)\hat{i} + v_1 \sin(\omega t + \theta_0)\hat{j}$$

and similarly for $\vec{v}_2(t)$. The orbital period $T = 2\pi/\omega$ can be measured directly from the periodicity of $\vec{v}_1(t)$. Recall from Equation 3.11 that the Doppler shift is proportional to the line-of-sight component of the velocity, which in this case is the x-component, so

$$\frac{\Delta\lambda_1(t)}{\lambda} = \frac{v_1 \cos(\omega t + \theta_0)}{c},$$

where c is the speed of light. An identical expression holds for $\Delta\lambda_2(t)$. Even though we cannot see the stars moving, we know for certain that their orbits are circular if their Doppler light curves are sinusoidal (see Figure 8.13). The peak values of $\Delta\lambda_1(t)$ and $\Delta\lambda_2(t)$ give us the speeds v_1 and v_2. Knowing v_i and T, we can calculate the orbital radii r_1 and r_2, and then use Equation 8.13 to find M_1 and M_2. (If $i \neq 90°$, as in Figure 8.12b, the analysis is more complicated, but lower limits to the mass can still be determined. See Problem 8.39.)

For example, Figure 8.14 shows Doppler data for the binary system Delta Circini (HD135240), a pair of massive stars about 1100 pc (3700 light years) from Earth—too far away for the stars to be resolved. (In the figure, the Doppler shifts have already been converted into radial velocities.) Each curve is sinusoidal, so

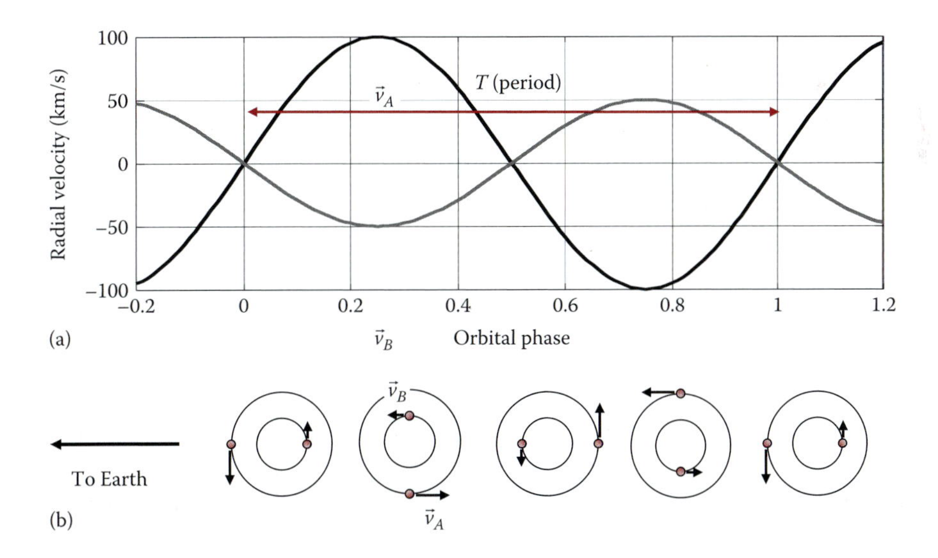

FIGURE 8.13 (a) When two stars of a binary system are in circular orbit, their Doppler shifts and radial velocities vary sinusoidally with time. In the figure, $v_A = 2v_B$, so $r_A = 2r_B$ and $M_A = ½M_B$. (b) The positions of the two stars at the peaks and zero crossings of the radial velocity curves.

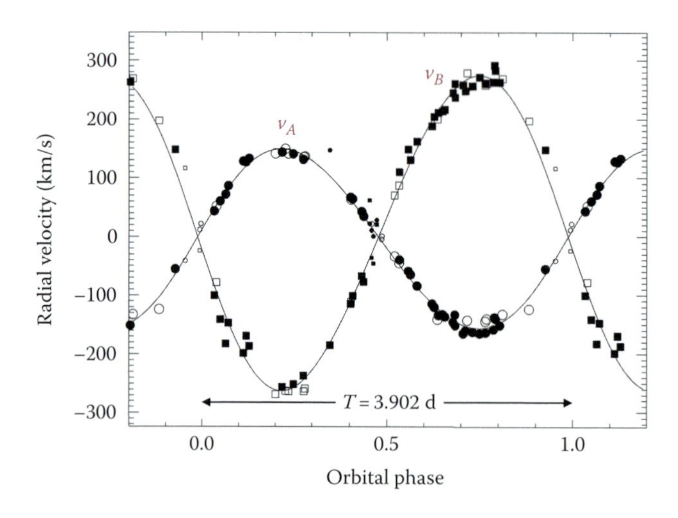

FIGURE 8.14 A real binary system: the Doppler-derived radial velocity curves of Delta Circini. Because $v_B/v_A = 1.73$, we know that $M_A/M_B = 1.73$. (From Penny, L.R. et al., *Astrophys. J.*, 548, 889, 2001, Figure 3, p. 893.)

we know that the orbits are circular (more properly, they are ellipses with small eccentricity). From the amplitude of each curve, calculated as $(v_{max} - v_{min})/2$, we find $v_1 = 149$ km/s and $v_2 = 257$ km/s, giving us the ratio $M_1/M_2 = v_2/v_1 = 1.73$. The period of either orbit is observed to be 3.902 days ($= 3.37 \times 10^5$ s $= 1.07 \times 10^{-2}$ years), allowing us to calculate the orbital radii:

$$r_1 = \frac{v_A T}{2\pi} = 7.99 \times 10^6 \text{ km} = 0.0533 \text{ AU}$$

and

$$r_2 = 1.73 r_A = 0.0922 \text{ AU}.$$

EXERCISE 8.15

Use Equation 8.13 and the numbers given above to find (a) the *total* mass of Delta Circini, expressed in terms of M_{Sun} and (b) the individual masses M_1 and M_2, once again in terms of M_{Sun}. (*Hint:* Take advantage of the strategy suggested in Equation 8.12.)

Answer (a) $M_1 + M_2 = 26.9M_{Sun}$ (b) $M_1 = 17.1M_{Sun}$, $M_2 = 9.9M_{Sun}$

8.12 APPLICATION: SEARCHING FOR EXOPLANETS

Are we alone in the universe? The scientific search for extraterrestrial intelligence (SETI) began with astrophysicist Frank Drake in the 1960s and was popularized by the Cornell astrophysicist Carl Sagan a decade later. But until 1995, there was no evidence for a planetary system about any Sun-like star other than our own.

Then, astronomers Michel Mayor and Didier Queloz at the Geneva Observatory in Switzerland discovered a Jupiter-sized planet circling the star 51 Pegasi, located about 50 light years from Earth. Since then, astronomers using Earth- and space-based telescopes have found nearly 2000 extrasolar planets, or *exoplanets*, in our Milky Way Galaxy, and the number of discoveries is growing steadily. The first planets were found using the Doppler technique discussed above. Since 2009, the orbiting *Kepler* space observatory has been detecting planets by searching for *transit* events: when a planet passes between its central star and the *Kepler* satellite, it blocks a small portion of the star's light, and the periodic dimming of the starlight reveals the planet's presence. In this section, we will discuss only the Doppler technique for finding exoplanets.

Consider the system shown in Figure 8.15a: a single planet of mass m is in orbit about a star of mass M. Just as for binary stars, the two bodies move in concert about their common CM. But in this case, only the star emits light. The presence of one or more planets is indicated by periodicity in the Doppler light curve of the star. For simplicity, we will assume that both trajectories are circular, with planetary orbital radius r and stellar orbital radius R. Because we expect planets to be much smaller than stars, we can safely assume that $m \ll M$, and consequently $R \ll r$. From the star's Doppler curve (Figure 8.15b), we can measure its orbital period T and orbital velocity V. For example, the historic Doppler measurements* of 51 Pegasi shown in the figure indicate that the star has an orbital speed $V = 53\,\text{m/s}$ and a period $T = 4.231\,\text{days}$. The star's orbital radius is therefore $R = VT/2\pi = 58.2\,\text{km}$. To find the *planet's* mass m and orbital radius r, we will use the generalized form of Kepler's third law, Equation 8.13:

$$\frac{T^2}{(R+r)^3} = \frac{4\pi^2}{G(M+m)}$$

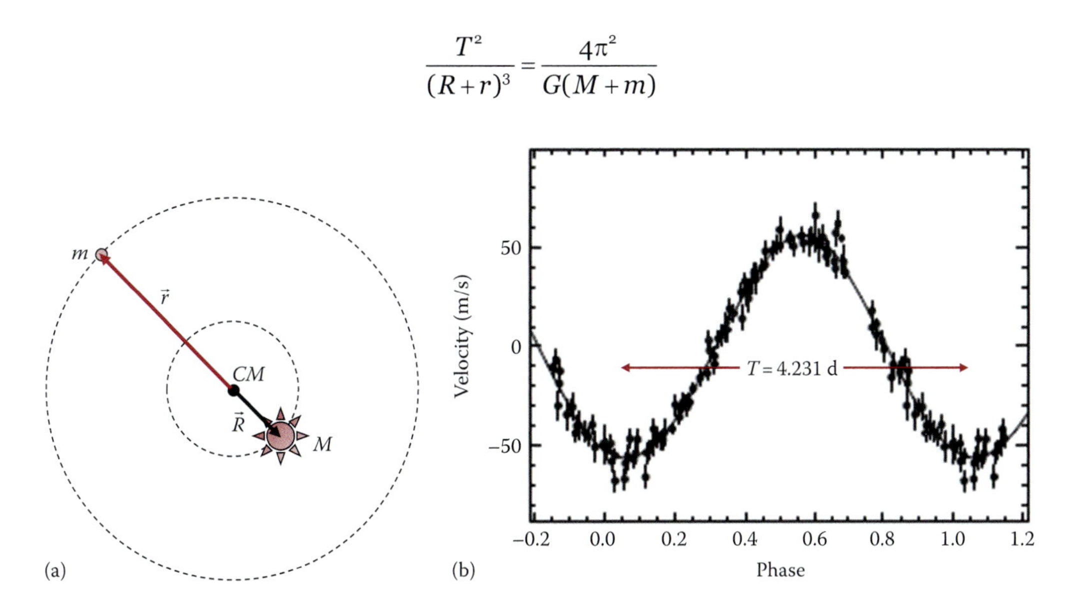

FIGURE 8.15 (a) An invisible exoplanet m and its parent star M are in circular orbit about their common CM. Since $M \gg m$, $R \ll r$. (b) The Doppler-derived radial velocity curve of the star 51 Pegasi. A phase change of 1.0 corresponds to a full period of the orbit. (Data from Marcy, G. and Butler, R., *Astrophys. J.*, 481, 926, 1997, Figure 3, p. 927.)

* Note the high precision needed for these measurements: $\Delta\lambda/\lambda = V/c < 2 \times 10^{-7}$.

which reduces to the simpler expression Equation 8.8 when $m \ll M$ and $R \ll r$:

$$\frac{T^2}{r^3} = \frac{4\pi^2}{GM}.$$

But there are two unknowns in this equation, M and r. Fortunately, based on decades of binary star studies, astronomers can confidently estimate the mass of a typical ("main sequence") star by measuring its luminosity and color. In the case of 51 Pegasi, $M = 1.11 \pm .06 M_{Sun}$. With this extra information, we can calculate the orbital radius r of the (invisible) planet and then determine the planet's mass from $m/M = R/r$. Try the following exercise.

EXERCISE 8.16

The orbital period of 51 Pegasi is 4.231 days = 1.183×10^{-2} years. (a) Use $M = 1.11 M_{Sun}$ and the strategy suggested by Equation 8.12 to determine the orbital radius (in AU) of the planet, called 51 Peg b. (b) Then find the mass m of the planet. Express your answer in terms of the mass of Jupiter ($M_J = 0.95 \times 10^{-3} M_{Sun}$). Do you think there could be life on this planet?

Answer: (a) 0.052 AU (b) $0.47 M_J$

8.13 APPLICATION: WHY STARS, PLANETS, AND (MOST) MOONS ARE ROUND

The Cavendish experiment vividly illustrates the weakness of the gravitational force between two lab-scale objects. Yet the same force between *astronomical* bodies can be strong enough to distort the bodies or even tear them apart. Gravitational interactions within an *isolated* planet or moon can be large enough to crush rocky material and change the body's overall shape. This explains why large astronomical objects are spherical, why smaller ones are not, and allows us to predict the maximum possible height of mountain ranges on rocky planets and moons.

Why are planets, stars, and most moons spherical? On the scale of billions of years (the age of the solar system), even "rigid" materials like rock can deform and flow at exceedingly slow rates: the higher the temperature, the faster the flow. The core region of a planet such as Earth is very hot and—as we will see—subject to huge compressive forces, so core planetary matter can flow and reshape the planet via its own internal gravitational interactions. Imagine two bodies, one of which is spherical (see Figure 8.16a). In the spherical body, the weight of the outer shells compresses the core *isotropically* (uniformly in all directions), so the body's spherical shape remains stable. For the nonspherical body (Figure 8.16b), inscribe the largest sphere possible within the body. Matter within this region squeezes the core isotropically. But the bulges outside the inscribed sphere add to these forces, squeezing the core more along the large dimension ℓ_1 than along the shorter dimension ℓ_2. Core material will slowly flow inward along ℓ_1, and leak outward along ℓ_2,

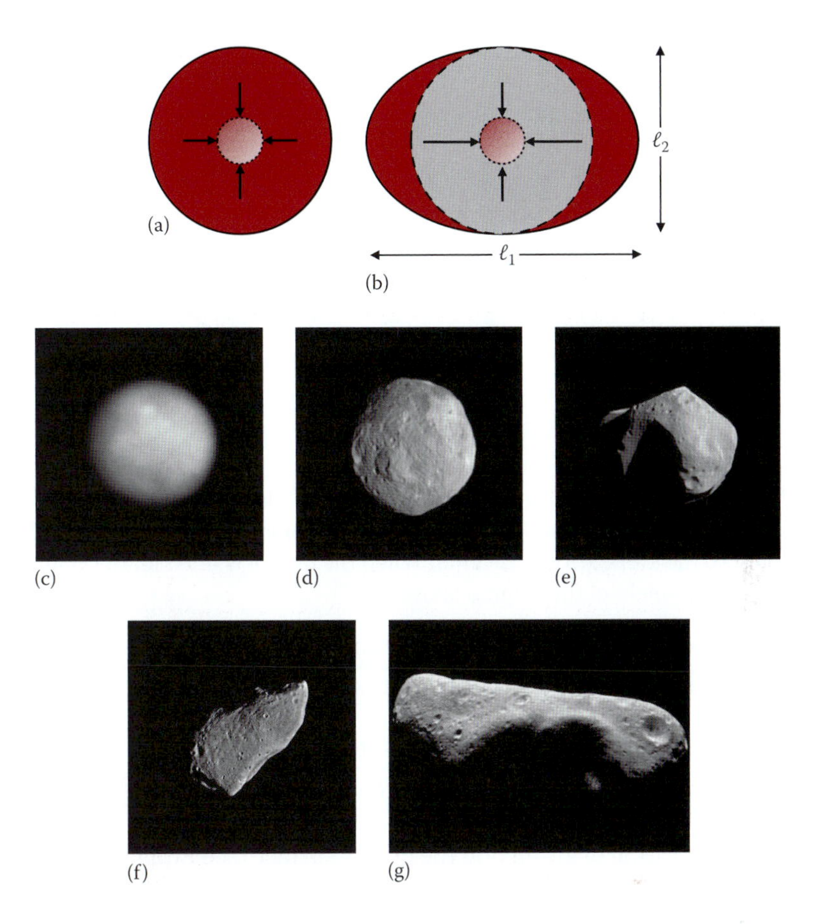

FIGURE 8.16 (a) The core of a spherically symmetric body is squeezed isotropically. (b) In a nonspherical body, the compressive pressure varies with direction, causing flow of material toward the core in the horizontal direction and out of the core along the vertical direction. (Asteroid shapes, Courtesy of NASA/JPL, Pasadena, CA.) (c) Asteroid Ceres has a mean diameter of 950 km, and is nearly spherical. (d) The smaller asteroid Vesta (mean diameter 520 km) is noticeably nonspherical. Asteroids (e) Mathilde, (f) Gaspra, and (g) Eros have maximum dimensions <70 km and are highly irregular in shape.

redistributing itself until a spherical shape is attained. Of course, if the body is fluid (as in stars or gas giant planets like Jupiter and Saturn), the flow occurs quickly.

Observations of asteroids and moons by the Hubble space telescope and interplanetary spacecraft indicate that objects with radii greater than about 250 km are spherical, whereas smaller bodies are often irregularly shaped. Consider a spherical body of uniform density ρ and radius R with total mass $M = \frac{4}{3}\pi R^3 \rho$ (see Figure 8.17). Let r denote the radial distance from the center of the body, and imagine drilling a slender mineshaft from the body's surface ($r = R$) to its center ($r = 0$). Let's find the gravitational force F exerted by M on a small mass m placed within the mineshaft at position r. The shell theorem tells us that the portion

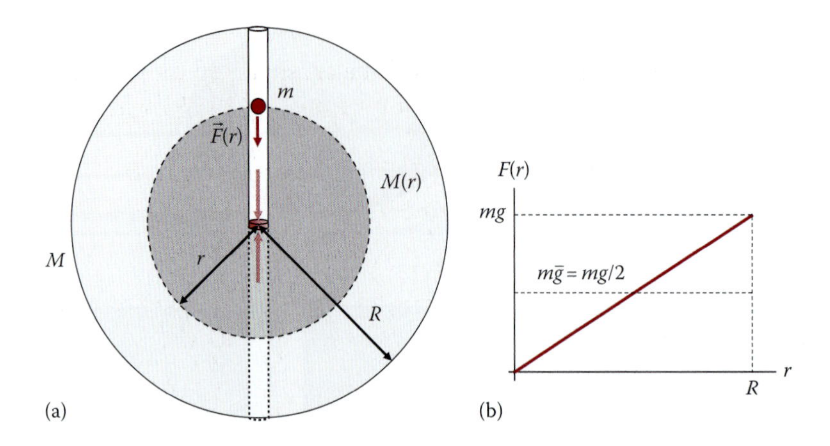

FIGURE 8.17 (a) A ball m is placed in a tunnel at a distance r from the planet's center. The force on the ball is proportional to $M(r)$, that portion of M that lies closer to the center of the planet. (b) The force on the ball decreases linearly with distance to the center, so the average force is ½mg. When the tunnel is filled in, the compressive force on a thin slab at the center is equal to the weight of the column above it.

of the sphere that contributes to the force on m is contained within r of the sphere's center. This mass is given by $M(r) = {}^4\!/_3\,\pi r^3 \rho = Mr^3/R^3$, so the force on m (its weight) is

$$F(r) = G\frac{M(r)m}{r^2} = \frac{GM}{R^2}\frac{r}{R}m = \frac{r}{R}mg,$$

where $g = GM/R^2$ is the acceleration of gravity on the body's surface (see Equation 8.10). At $r = 0$, $F = 0$, and at $r = R$, $F = mg$. The force on m, averaged over all locations $0 \le r \le R$, is therefore $\bar{F} = mg/2$, or equivalently, $\bar{g} \equiv \bar{F}/m = g/2$. If we fill the mineshaft from top to bottom with matter of density ρ, its total mass will be ρAR, where A is the cross-sectional area of the shaft, and the total weight of the column will be $F = \rho AR\bar{g} = \rho ARg/2$. Now imagine a thin slab of matter at the bottom of the shaft. This slab is compressed by the weight of the column above it, and by an equal and opposite force exerted by the matter below it. The weight F is the compressive force squeezing the slab, and the pressure P is defined as the compressive force per unit area: $P = F/A$, expressed in units of N/m² or pascals (Pa):

$$P = \frac{1}{2}\rho g R = \frac{2}{3}G\pi\rho^2 R^2 \equiv P_{max}. \tag{8.15}$$

EXERCISE 8.17

Prove the second identity in Equation 8.15 from the expressions for g and M.

The compressive force $P(r)$ at any point within the body is proportional to the weight of the column of matter above it, so $P(r)$ increases with increasing depth. Equation 8.15 is therefore the maximum pressure within the body. For Earth, $P_{max} \approx 170$ GPa. If the maximum pressure exceeds the core *compressive strength* P_C (set by the strength of chemical bonds), the core will be crushed and capable of flow (on geologic time scales). The condition for fluidity is $P_{max} \geq P_C$; or using Equation 8.15, $R \geq R_C$, where R_C is the critical radius:

$$R_C = \sqrt{\frac{3P_C}{2\pi G\rho^2}}. \tag{8.16}$$

Spherical moons and asteroids have a radius greater than R_C, whereas bodies with dimensions smaller than R_C are stable against core collapse and may remain irregularly shaped.

EXERCISE 8.18

The density of concrete is about 2400 kg/m^3 and its compressive strength is roughly 50 MPa (1 MPa = 10^6 N/m^2). Calculate the maximum radius of a nonspherical concrete body.

Answer R_C = 250 km

As you can see, the calculated value of R_C for concrete agrees rather well with astronomical observations.* Of course, moons, planets, and asteroids are not made out of concrete. Nevertheless, for the materials found in real astronomical bodies, ρ and P_C do not vary widely, and the value of R_C found above is representative of real bodies to within a factor of 2.

* For a different approach, see S.J. Heilig, Why are so many things in the solar system round, *The Physics Teacher* **48**, 377–380 (2010).

8.14 APPLICATION: THE SHAPE (OR FIGURE) OF THE EARTH

The Earth's spin changes the planet's shape from a perfect sphere to an oblate ellipsoid, slightly flattened at the poles and bulged at the equator. As a result, the measured gravitational acceleration g varies with latitude, an effect first observed by Jean Richer during his expedition to French Guiana in 1671 to determine the parallax of Mars and the distance to the Sun (see Sections 1.7 and 7.10). In the *Principia*, Newton used Richer's observations to test his hypothesis of universal gravitation. Following Newton, imagine drilling a vertical well of cross-sectional area A from the north pole to the center of the planet, and then connecting it to a horizontal well extending from the Earth's center to a point on the Equator (Figure 8.18a). The length of the vertical well is r_V and that of the horizontal well is r_H. Now fill the L-shaped channel with a fluid of uniform density ρ. According to the discussion above, the weight of the fluid in the vertical column is $W_V = m_V\bar{g} = ½\rho A r_V g$; likewise, the weight of the horizontal section is $W_H = m_H\bar{g} = ½\rho A r_H g$. If the planet were not spinning ($\omega = 0$), it would be spherical ($r_H = r_V = R_E$) and $W_H = W_V$. In rotation, the horizontal fluid is in uniform circular motion, which requires a net inward force to maintain its centripetal acceleration. The Earth provides this extra force by changing its shape ($r_H > r_V$) so $W_H > W_V$. In the following, we will assume that the change in shape is small, so $r_V - R_E \ll R_E$ and likewise for r_H.

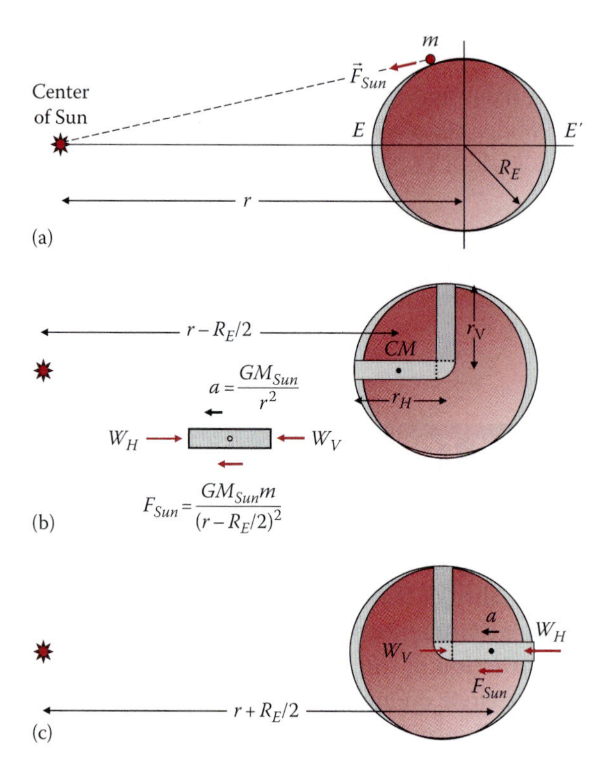

FIGURE 8.19 (a) The force on the mass m due to the Sun varies with its position on the surface of Earth. This variation in force causes the ocean tides. (b) To determine the height of the tide at point E, replace the surface ocean by a canal allowing water to flow between the north pole and point E on the equator. The FBD for the horizontal column of water m includes the attraction to the Sun and the weights of water in the horizontal and vertical legs of the canal. (c) To find the height of the tide at E', a canal is dug joining the pole to E'.

attraction toward the Sun is set equal to its value at the column's CM, a distance $r - R_E/2$ from the star. Applying Newton's second law to the horizontal fluid,

$$m_H a = m_H \frac{GM_{Sun}}{r^2} = W_V - W_H + \frac{GM_{Sun} m_H}{(r - R_E/2)^2}$$

or, rearranging terms,

$$W_H - W_V = \rho g A \frac{r_H - r_V}{2} = \frac{GM_{Sun} m_H}{(r - R_E/2)^2} - \frac{GM_{Sun} m_H}{r^2}.$$

Using the binomial approximation, $(1 + x)^n \simeq 1 + nx$ for $x \ll 1$,

$$\frac{1}{(r - R_E/2)^2} = \frac{1}{r^2}\left(1 - \frac{R_E}{2r}\right)^{-2} \simeq \frac{1}{r^2}\left(1 + \frac{R_E}{r}\right)$$

so that

$$\rho g A \frac{r_H - r_V}{2} \simeq \frac{GM_{Sun} m_H}{r^2}\left(1 + \frac{R_E}{r} - 1\right) = GM_{Sun} m_H \frac{R_E}{r^3}.$$

We can simplify this result using $g = GM_E/R_E^2$ and $m_H \simeq \rho A R_E$, to obtain

$$\frac{r_H - r_V}{R_E} \simeq 2\frac{M_{Sun}}{M_E}\frac{R_E^3}{r^3}.$$

To avoid complication, we have left out one detail: the Sun exerts a downward force on the vertical column of water, increasing W_V and changing the prefactor above from 2 to 3/2. The correct result is

$$\frac{r_H - r_V}{R_E} = \frac{3}{2}\frac{M_{Sun}}{M_E}\frac{R_E^3}{r^3}, \tag{8.18}$$

where $r_H - r_V$ is a measure of the height of the ocean tide at point E.

EXERCISE 8.20

(a) Plug in numbers for the Sun and the Earth, and find $r_H - r_V$. (b) Show that the tidal effect due to the Moon as about twice as large as that due to the Sun. When the Moon and Sun are aligned with Earth (full moon or new moon), the two effects add to create an especially high tide, called a spring tide. When they are in quadrature (first or third quarter moons), the tides are diminished (neap tide).

Answer (a) $h \approx 25$ cm

Finally, what happens at E', the point farthest from the Sun? Repeating the above analysis, let's drill an L-shaped canal from the North Pole to E', and note that the CM of the horizontal section is a distance $r + R_E/2$ from the center of the Sun. Applying Newton's second law to the horizontal fluid (Figure 8.19c),

$$m_H a = m_H \frac{GM_{Sun}}{r^2} = W_H - W_V + \frac{GM_{Sun} m_H}{(r + R_E/2)^2}$$

or

$$W_H - W_V = \rho g A \frac{r_H - r_V}{2} = \frac{GM_{Sun} m_H}{r^2} - \frac{GM_{Sun} m_H}{(r + R_E/2)^2}.$$

though the modern media are prone to crediting a single person with a great breakthrough, no one works in isolation, and the full story of discovery is often inconveniently complicated. Perhaps Newton said it best:

> If I have seen further, it is by standing on the shoulders of giants.

PROBLEMS

8.1 The table lists the semimajor axes and periods of 11 moons found in the solar system. Each moon orbits one of the planets Jupiter, Saturn, or Uranus.

Moon	*a* (km)	*T* (s)
A	3.774×10^5	2.365×10^5
B	1.070×10^6	6.182×10^5
C	4.363×10^5	7.522×10^5
D	1.222×10^6	1.378×10^6
E	4.216×10^5	1.528×10^5
F	2.947×10^5	1.631×10^5
G	1.909×10^5	2.177×10^5
H	1.880×10^6	1.442×10^6
I	2.660×10^5	3.580×10^5
J	5.270×10^5	3.904×10^5
K	6.709×10^5	3.068×10^5

(a) Separate the 11 moons into three groups, each group containing all of the moons orbiting the same planet.

(b) The heaviest planet is Jupiter and the lightest is Uranus. Match each group of moons with the appropriate planet.

(c) The *Galileo* orbiter circled Jupiter just above its surface. The planet has a radius of 6.9×10^4 km. Find the period of the spacecraft.

8.2 The star HD168443 has two heavy planets (called HD 168443b and c) orbiting about it. The orbits are elliptical, with periods and semimajor axes given in the following table.

Planet	*T* (days)	*a* (AU)
b	58.112	0.2931
c	1749.8	2.8373

(a) Prove that the data is consistent.

(b) Compare the mass of the central star with that of our Sun. Do this using ratios rather than by plugging in all the numbers.

8.3 In the 1980s, paleontologists reported fossil evidence of a 26 million year cycle of major extinctions of life on Earth. In response, astronomers suggested that the extinction events might be caused by "Nemesis," an unseen "death star" companion of the Sun that periodically disturbs the Oort cloud (a spherical shell of icy debris roughly 50,000 AU from the Sun), causing a spike in the number of comets that plunge into the inner solar system and impact Earth.

(a) Assuming the mass of Nemesis to be much less than M_{Sun}, find the semimajor axis of its orbit. Use the strategy of Section 8.10 to simplify the calculations.

(b) Recalculate the semimajor axis assuming $M_{Nem} \approx M_{Sun}$.

8.4 The radius and period of two moons in the solar system are listed below.

Moon	*R* (km)	*T* (days)
A	3.548×10^5	5.877
B	5.513×10^6	360.16

(a) Are these moons associated with the same planet?

(b) Is Saturn heavier or lighter than this (these) planet(s)?

8.5 The figure shows the radial velocities of a spectroscopic binary star system identified as HD 48099. The stars' period is 3.078 days = 2.66×10^5 s = 8.43×10^{-3} year, and the velocity of each star (measured by the Doppler shift) is plotted vs. time (Note: speeds are given in km/s).

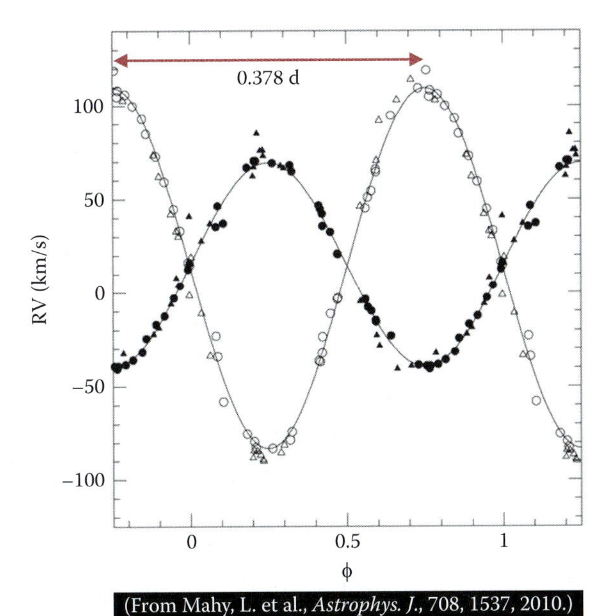

(From Mahy, L. et al., *Astrophys. J.*, 708, 1537, 2010.)

(a) Find the radii of the orbits (assumed circular). Express these distances in terms of AU (1 AU = 1.5×10^{11} m).

(b) Which curve in the above figure is associated with the heavier mass? Find the ratio of the two masses.

(c) Find the value of each mass in terms of solar masses ($M_{Sun} = 2 \times 10^{30}$ kg).

8.6 Vectors $\vec{A}$ and $\vec{B}$ are shown in the figure to the right. Let $\vec{C} = \vec{A} \times \vec{B}$ (not shown). The dotted lines are parallel and equally spaced.

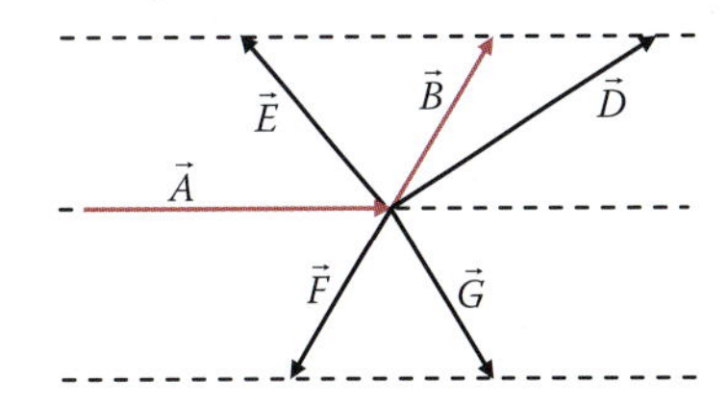

(a) In what direction is $\vec{C}$?

(b) Compare the magnitude and direction of $\vec{C}$ with those of $\vec{A} \times \vec{D}$, $\vec{A} \times \vec{E}$, $\vec{A} \times \vec{F}$, and $\vec{A} \times \vec{G}$.

(c) $\vec{B} \times \vec{F} = ?$

(d) If $|\vec{a} \times \vec{b}| = \vec{a} \cdot \vec{b}$, what is the angle between $\vec{a}$ and $\vec{b}$?

8.7 The figure shows a typical velocity curve (Doppler curve) of a star in close proximity to a massive planet. The period of the star's wobble is 3.097 days, and its velocity is 69.40 ± .45 m/s. By observing the star's luminosity and color spectrum, it is possible to estimate its mass: $M_{star} = 1.037 \pm .025\, M_{Sun}$. In the following, assume $M_{star} \gg m_{planet}$.

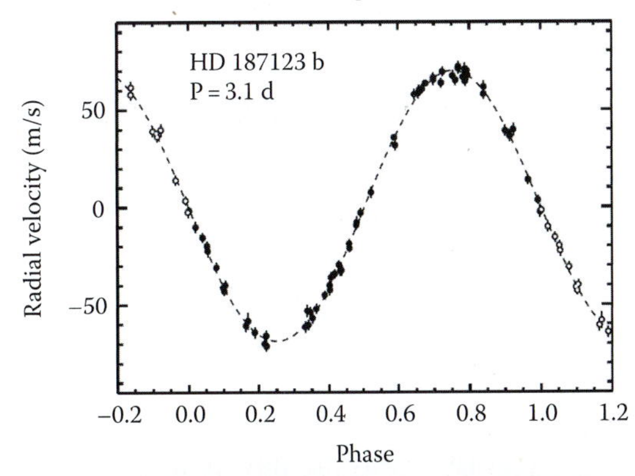

(From Wright, J. T. et al., *Astrophysical Journal*, 693, 1084, 2009.)

(a) What can you say about the shape of the planet's orbit?

(b) Find the orbital radius of the star. Compare it to the radius of our Sun, which is 7×10^8 m.

(c) Find the radius of the planet's orbit. Express your answer in terms of AU.

(d) Find the mass of the planet. The mass of Jupiter is about $10^{-3}\, M_{Sun}$. How massive is the planet relative to Jupiter?

8.8 Black holes must be observed indirectly, because no light can escape from them. One piece of evidence for their existence comes from binary star systems consisting of a visible star and an invisible black hole. The two bodies are in circular orbit about their common CM, as shown in the figure.

The black hole Cygnus X-1 was discovered by observing the motion of its companion star, which has a mass $m_s = 20\, M_{Sun}$. Doppler measurements fix the period of the star as 5.6 days (4.84×10^5 s) and its speed as 165 km/s. Assume the motion is circular.

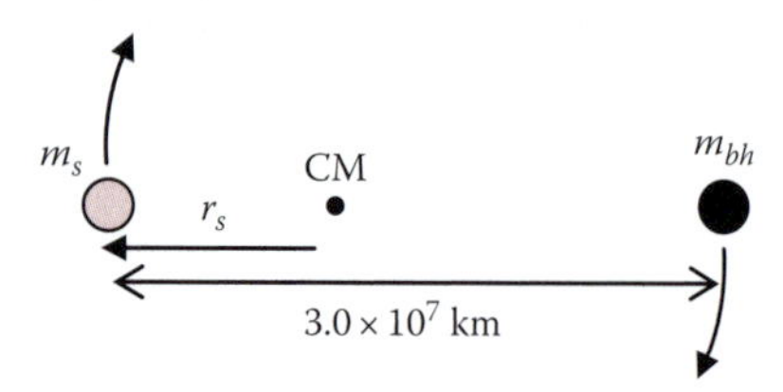

(a) Find the radius r_s of the visible star's orbit.

(b) Orbital theory tells us that the star and the black hole are separated by 3.0×10^7 km. Find the mass m_{bh} of the black hole, measured in solar masses.

Another way to detect a black hole is to observe its effect on other objects. A supermassive (but invisible) black hole is known to be in motion at a constant velocity $\vec{V} = V_{bh}\hat{i}$ parallel to the x-axis of our coordinate system. A stream of gas undergoes elastic collision with the black hole, with initial velocity $3\hat{i}$ Mm/s and final velocity $(2\hat{i} + 2\hat{j})$ Mm/s.

(c) Write equations in $(\hat{i},\hat{j})$ notation for $\vec{u}$ and $\vec{u}'$, the velocities of the gas relative to the black hole before and after colliding with the black hole.

(d) What is the speed V_{bh} of the black hole?

8.9 The Titius–Bode "law" is an empirical relationship dating from 1766, which, for unexplained reasons (or maybe just by coincidence), seems to predict the orbital radii of our solar system planets rather well. The law can be written as follows:

R(Mercury) = 0.4 AU
R(Venus) = $0.4 + 0.3 \times 2^0$ AU
R(Earth) = $0.4 + 0.3 \times 2^1$
R(Mars) = $0.4 + 0.3 \times 2^2$
R(???) = $0.4 + 0.3 \times 2^3$
R(Jupiter) = $0.4 + 0.3 \times 2^4$
R(Saturn) = $0.4 + 0.3 \times 2^5$
$R(n) = 0.4 + 0.3 \times 2^n$

(a) Uranus ($n = 6$) was discovered by William Hershel in 1781. Compare its measured semimajor axis of 19.2 AU (see Figure 8.3) with the value predicted by the Titius–Bode law. Agreement with the law helped convince astronomers that Uranus was indeed a planet, and added credence to the law itself.

(b) The gap at $n = 3$ vexed eighteenth century astronomers and motivated an international search for a "missing" planet nestled between the orbits of Mars and

Jupiter. In 1801, the Sicilian astronomer Giuseppe Piazzi discovered a tiny body with semimajor axis 2.77 AU. Does this fit with the "missing planet" hypothesis of the Titius–Bode law? Piazzi named the body Ceres, after the patron goddess of Sicily. It is the largest asteroid (or dwarf planet) of the solar system and contains about 1/3 of the total mass of the asteroid belt.

	HD 10180 Exoplanets		
n	Planet	R (AU)	P (days)
1	b	0.0223	1.178
2	c	0.0641	5.760
3	d	0.1286	16.357
4	e	0.2695	49.797
5	f	0.4924	122.72
6	g	1.422	602
7	h	3.40	2229

More generally, the Titius–Bode law may be expressed in exponential form as $R(n) = c_1 c_2^n$, where c_1 and c_2 are constants that depend on the mass of the central body, and n = 1, 2, 3, ... denotes the planets starting from the innermost one. (For our solar system, c_1 = 0.267 and c_2 = 1.56.) In 2011, Lovis et al. (*Astronomy and Astrophysics* **528**, A112) found evidence for seven exoplanets, labeled b–h, orbiting the Sun-like star HD 10180. Their calculated orbital parameters are shown in the table above.

(c) Make a plot of $R(n)$ vs. n, using a semi-log graph to determine if the data shown is in agreement with the exponential law given earlier. (If you are ambitious, you can also determine c_1 and c_2 from this graph.)

(d) In 2012, M. Tuomi (*Astronomy and Astrophysics* **543**, A52 (2012)) found evidence for two additional planets in the HD 10180 system, with semimajor axes 0.0904 AU (fitting between n = 2 and 3 above) and 0.330 AU (between n = 4 and 5 above). Replot your data (you must change the numbering to accommodate the two additional planets). Based on these results, how much confidence do you have that the Titius–Bode law is a consequence of real physical principles?

8.10 A spaceship is going from Earth to the Moon, along a path joining the centers of the two bodies. At what distance from the center of the Earth will the net gravitational force on the spaceship exerted by the Earth and Moon equal zero?

8.11 Suppose that the attractive interaction between a star of mass M and a planet of mass $m \ll M$ is of the form $F = KMm/r$, where K is the gravitational constant. What would be the relation between the planet's circular orbit radius and its period?

8.12 One World Trade Center, in New York City, which until 2001 was the tallest building in the Western Hemisphere. At the top of its spire, the gravitational acceleration is 0.017% less than its value at ground level. What is the height of the building? Why do you suppose this height was chosen? (*Hint:* 1 ft = 0.3048 m.)

8.13 A white dwarf is a collapsed star with a mass roughly the same as the Sun, and a radius roughly that of the Earth.

(a) Calculate the acceleration of gravity on the star's surface. If you are just strong enough to lift a mass of 50 kg on Earth, how much mass could you lift if you were standing on such a body? (Ignore the obvious environmental complications.)

(b) Find the speed and orbital period of a satellite in circular orbit just above the white dwarf's surface.

8.14 A satellite in circular geosynchronous orbit about Earth has an orbital period of 24 h.

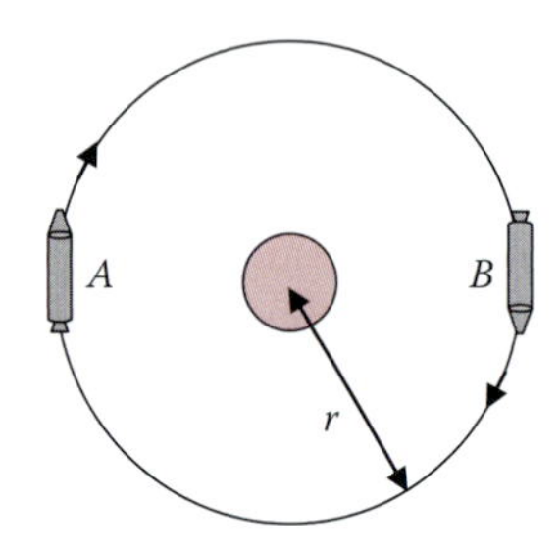

(a) Find the radius r of the satellite's orbit.

Now consider two such satellites traveling in the same sense in the same geosynchronous orbit, but located in diametrically opposed positions (see the figure). We want to alter the orbit of the spacecraft at A so that it overtakes the second spacecraft 10 days later at point B. To do this, we will fire spacecraft A's rocket thrusters for a short time, to change the *tangential* velocity of the spacecraft. After this short burn, spacecraft A will enter an elliptical orbit and will have a different altitude when it passes B.

(b) By what fraction do you want the orbital period to change?

(c) Should the spacecraft's velocity be increased or decreased at point A to effect this change in the period? (Hint: after the burn, do you want the satellite's perigee or apogee to be at point A?)

(d) What will be the semimajor axis of the new elliptical orbit?

(e) What will be the eccentricity of the new orbit?

8.15 The fastest possible rotation rate of a planet about its N–S axis is that for which the gravitational force just barely prevents matter from separating from the surface at the equator. Assume a spherical planet of uniform mass density ρ.

(a) Show that the shortest period of rotation is given by

$$T = \sqrt{\frac{3\pi}{G\rho}}$$

(b) Evaluate the rotation period assuming $\rho = 3.5\ \mathrm{g/cm^3}$, a typical value for planets, moons, and asteroids. No gravitationally-bound astronomical object has ever been observed with a period shorter than one determined in this way.

8.16 A binary star system consists of two stars, each with the same mass as the Sun, revolving about their CM. The distance between the stars is 4 AU. What is the period of rotation? (*Hint:* No calculator is needed for this problem.)

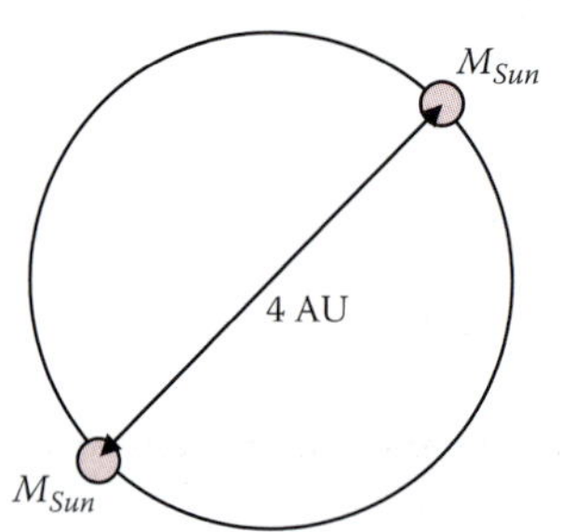

8.17 Light from a binary star system is found to be Doppler shifted with $\Delta\lambda/\lambda$ varying sinusoidally with time for each star. The maximum Doppler shifts are measured to be 6.67×10^{-5} for star A and 2.00×10^{-4} for star B. The period is 1.5 years.

(a) Which star is more massive?

(b) What is the ratio of the stellar masses?

(c) If the orbits of the stars are in a plane parallel to the line of sight (inclination angle $i = 90°$), find the separation of the two stars. Express your answer in AU.

(d) Calculate the individual stellar masses. Express your answer in terms of the solar mass.

8.18 Doppler shift measurements of the star Tau Boötes indicate the presence of a planet. As indicated in the figure, the period T of the star's motion is 3.313 days. By studying the color and luminosity of the star, astronomers have determined the star's mass to be $M_{star} = 1.33 \pm 0.11\ M_{Sun}$. The mass of the planet can safely be assumed to be much less than the mass of the star.

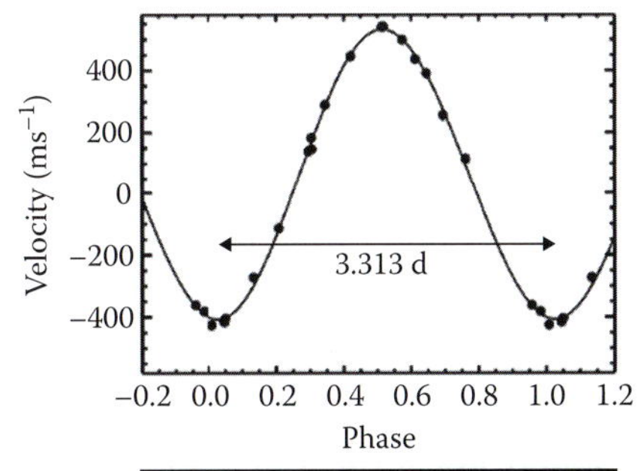

(From Butler, R.P. et al., *Astrophy. J.*, 474, L115, 1997.)

(a) What does the shape of the Doppler data indicate about the orbit of Tau Boötes?

(b) Using just the stellar mass and period, we can calculate the orbital velocity of the planet. *Prove* that and then calculate v_{planet}.

$$v_{planet} = \left(\frac{2\pi G M_{star}}{T}\right)^{1/3}$$

(c) In 2012, M. Lopez-Morales and her colleagues (*Astrophysical Letters* **753**: L25 (2012)) measured the radial velocity of the planet directly by observing the Doppler shift of light emitted by carbon monoxide (CO) molecules in the planet's atmosphere. They obtained a velocity $K_P = v_{planet} \sin i = 115$ km/s, where i is the inclination of the orbit illustrated in Figure 8.12a. Find the value of i in degrees.

(d) The orbit of the star is also inclined by the same angle i. As indicated in the figure, $K_{star} = v_{star} \sin i = 469$ m/s. Find the mass of the planet. What is the ratio of this mass to that of Jupiter?

8.19 On November 14, 2003, NASA-funded astronomers discovered a distant planetoid in orbit around the Sun. The object, dubbed "Sedna" for the Inuit goddess of the ocean, is the most distant object in the solar system to be discovered.

(c) If this tunnel were used to deliver mail, how long would it take for a letter to travel through the Earth? Compare this to the time for a satellite in LEO to travel halfway around the planet.

8.25 This is the same problem as the earlier one, except now the tunnel is not bored along a diameter, but along a chord through the Earth. Once again, assume a uniform Earth density and no friction.

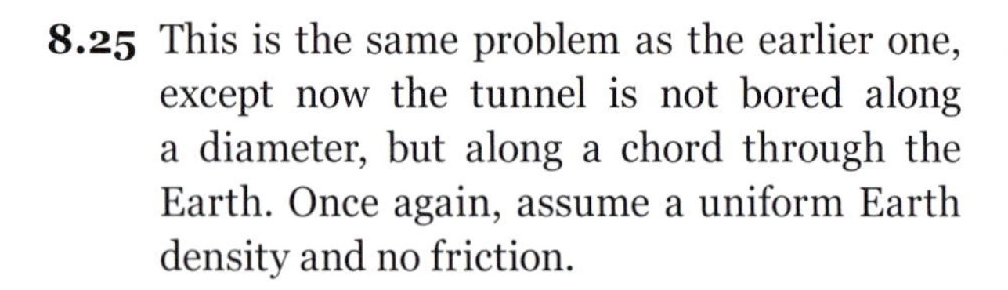

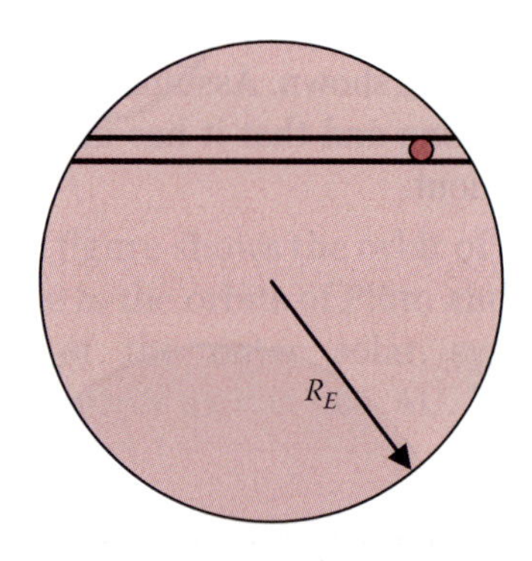

(a) Show that the motion is simple harmonic.

(b) Find the period.

(c) Will the object obtain the same maximum speed along the chord as it does along the diameter?

8.26 It is proposed to place a satellite in circular orbit 50,000 km from the center of the Earth. The orbit will be in the plane of the equator and is to be geosynchronous, that is, its period will be 1 day. The mass of the satellite will be 500 kg. In order to keep the satellite in this orbit, it is necessary for it to continuously fire its onboard rockets. In what direction should the rockets' thrust be exerted, and how large should this force be?

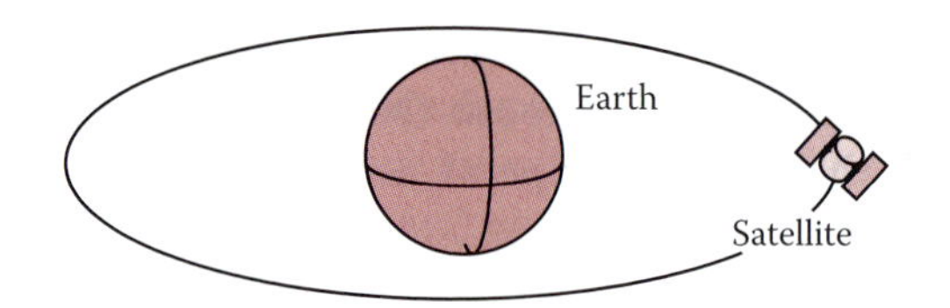

8.27 Upsilon Andromedae is a Sun-like star with three planets. The following table lists each planet's distance r from the star and its orbital period T. Assume circular orbits, and also assume that the star's mass is much greater that the mass of its planets.

Planet	r (AU)	T (days)
b	$0.0593 = 8.90 \times 10^{9}$ m	$4.617 = 3.989 \times 10^{5}$ s
c	$0.829 = 1.24 \times 10^{11}$ m	$241.5 = 2.087 \times 10^{7}$ s
d	$2.53 = 3.80 \times 10^{11}$ m	T_d

(a) Find T_d in either days or seconds.

(b) Find the mass of the star. (Express in kg or in terms of the mass of our Sun.)

(c) Now consider only the star and its innermost companion planet b. (Ignore planets c and d.) The orbital speed of the *star* is found to be 73.0 m/s. Find the mass of planet b. (Express your answer in kg or as a fraction of M_{Sun}.)

(d) If the radius of planet b is 6.0×10^{7} m, what is the value of "g" on its surface?

8.28 The table below is a compilation of data for many of the planets, moons, and rings in our solar system.

(a) Using a spreadsheet, set up a log–log graph as follows: On the x-axis, plot the planetary radii R_{planet} in Mm ($\times 10^{6}$ m). On the y-axis, plot r, the orbital radii (for moons) or ring radii.

Planet	R_{planet} ($\times 10^6$ m)	Moon/Ring	R_{moon} (km)	r ($\times 10^6$ m)
Earth	6.4	"Moon"	1740	384
Mars	3.4	Phobos	11.1	9.4
		Diemos	6.2	23.5
Jupiter	69.9	Metis	20	128
		Adrastea	12	129
		Almalthea	85	181
		Thebe	50	222
		Io	1820	422
		Europa	1570	671
		Ganymede	2630	1070
		Callisto	2400	1880
		Rings		123–181
Saturn	58.0	Pan	10	134
		Prometheus	50	139
		Mimas	200	186
		Enceladus	250	238
		Tethys	500	295
		Dione	560	377
		Rhea	765	527
		Titan	2500	1220
		Iapetus	735	3560
		Rings		67–140
Uranus	25.4	Cordelia	13	49.7
		Puck	81	86
		Miranda	235	130
		Ariel	579	191
		Umbriel	585	266
		Titania	789	436
		Oberon	761	584
		Rings		37–51
Neptune	24.6	Naiad	29	48
		Thalassa	40	50
		Larissa	94	73.5
		Proteus	209	118
		Triton	1350	355
		Nereid	170	5510
		Rings		42–63

Incomplete List of Moons and Rings of the Solar System

(b) Using three different colors or symbols, distinguish between moons with $R_{moon} > 250$ km, with $R_{moon} < 250$ km, and planetary rings.

(c) Now draw a line representing the Roche limit: $R_{roche} = 2.45 R_p$.

(d) What do you conclude from the graph?

8.29 Comet Shoemaker–Levy (SL9) was only the second comet known to orbit a planet. In July 1992, SL9 was in a highly elliptical orbit about Jupiter with a period of about 2 years. Its closest approach to Jupiter (perijove) was 6.4×10^{-4} AU. While passing through perijove, it split into pieces and lost sufficient energy by passage through Jupiter's atmosphere to put it on a collision course with the planet 2 years later. In July 1994, the fragments of Shoemaker–Levy crashed into the planet, creating a series of explosions that were easily observed from Earth.

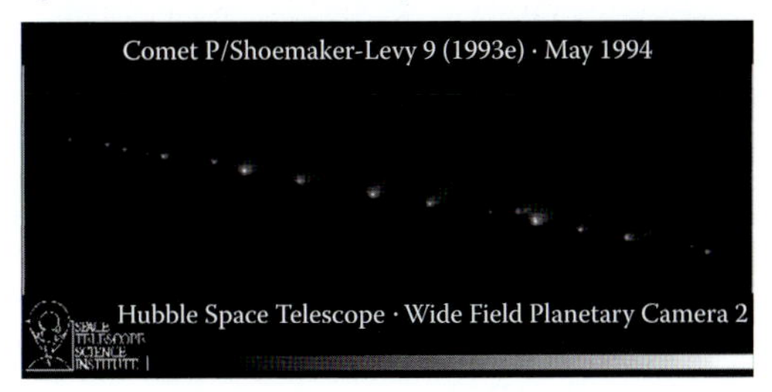

For the purposes of this question, ignore the motion of Earth and Jupiter. Jupiter's mass is 1.9×10^{27} kg $\approx 0.95 \times 10^{-3}\, M_{Sun}$, and its radius is 6.9×10^7 m.

(a) Why did the comet break into pieces in 1992?

(b) About how far from Jupiter was the comet in July 1993?

In March 1993, the comet pieces were strung out over an angular spread of 51 arcseconds (0.014°) when viewed from Earth, which corresponds to a physical length of 200,000 km. On the day before impact, the angular spread had grown to 20 arcminutes (0.33°), a physical length equal to 5×10^6 km.

(c) How far away from Earth was the comet in March 1993?

(d) Why did the fragment chain's length stretch so much from March 1993 to July 1994? (Hint: it has nothing to do with Jupiter's orbit.)

8.30 A binary star system has two stars whose masses are $m_1 = M_S$ and $m_2 = 4M_S$ ($M_S = 2 \times 10^{30}$ kg). The distance between them is 10 AU (1 AU = 1.5×10^{11} m) and they move in circular orbits. (It may be useful to remember that the Earth's orbital speed about the Sun is 30 km/s.)

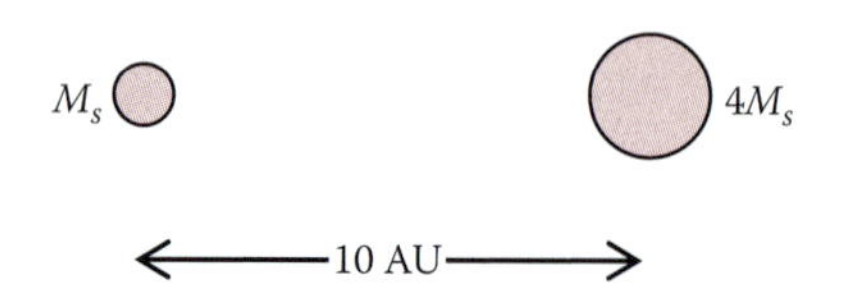

(a) How far from the more massive star is the CM? Indicate the location of the CM on the figure.

(b) What are the stars' velocities relative to the CM?

(c) What is the period of their orbit?

8.31 The space shuttle is in low earth orbit (LEO), circling the globe in about 84.6 min. (LEO implies that the shuttle's altitude h is very much less than R_{Earth}.)

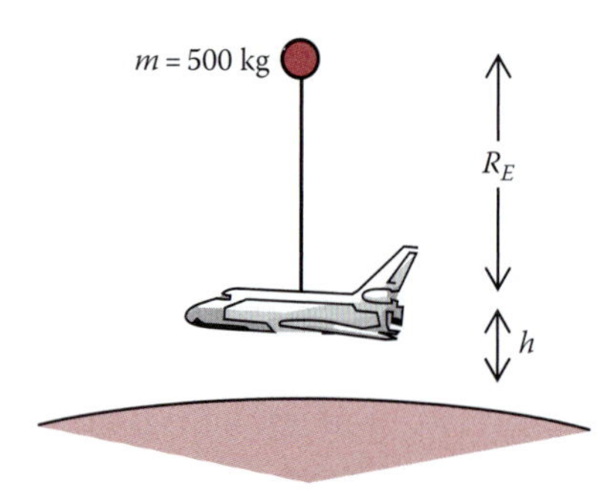

(a) What is its acceleration?

(b) Astronauts on the shuttle release a small 500 kg satellite that is tied to the shuttle with a wire of negligible mass (see the figure). The wire is extended until the satellite has an orbit radius of $2R_{Earth}$. What is the acceleration of the satellite?

(c) What is the tension in the wire? (*Hint:* The mass of the shuttle is 10^5 kg.)

(d) Because of the way it was deployed, the tethered satellite oscillates about its position directly above the shuttle. Find the period of this oscillation.

8.32 In December 2003, astronomers using the Parkes radio telescope in New South Wales, Australia (featured in the wonderful movie *The Dish*), discovered a binary star system consisting of two pulsars. The pulsars are called PSR J0737-3039 *A* and *B*. Pulsar *A* is easier to observe, and is found to have an orbital period T = 0.102 days and an orbital radius $r_A = 4.25 \times 10^8$ m.

The double pulsar is losing energy by radiating gravitational waves, making it a great experimental test bed for Einstein's general theory of relativity. By measuring the change in period, astronomers have determined that the total mass $M_A + M_B = 2.59M_{Sun}$.

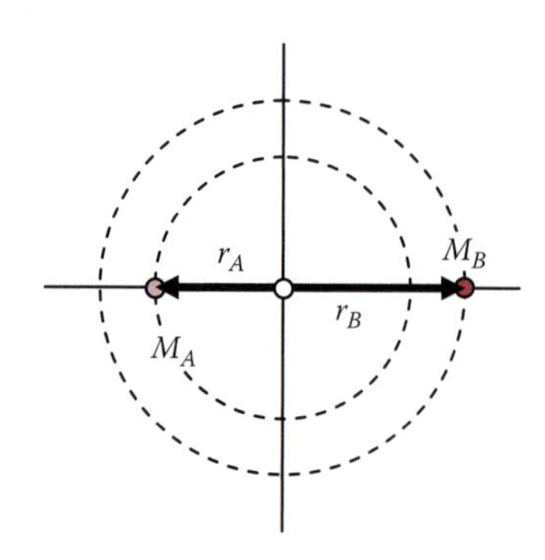

(a) What is the orbital speed of pulsar A?

(b) Find the total distance $r_A + r_B$.

(c) Find M_A and M_B. Express your answers in terms of M_{Sun}.

(d) Careful measurements reveal that the orbits of the two pulsars are not perfect circles, but are ellipses of eccentricity $\varepsilon = 0.088$. Let v_{max} be the highest speed of pulsar A in its orbit and v_{min} be its lowest speed. Find the ratio v_{max}/v_{min}.

8.33 *Kepler* is a space observatory launched by NASA in 2009 to search for extrasolar Earthlike planets that might support life. The spacecraft is in an "Earth-trailing" orbit about the Sun with its telescope pointed in a fixed direction away from the Sun. It detects planets by the "transit method," wherein an extrasolar planet passes in front of its star and blocks a small but detectable fraction of starlight. Kepler 5b is one of the first transiting exoplanets to be discovered by the spacecraft. It is in a circular orbit, with a period of 3.55 days, about a star with mass $M_{Star} = 1.37 M_{Sun}$ (determined by studying the star's luminosity and color). The radial velocity curve of the star is shown to the right. Its orbital velocity is 230 m/s.

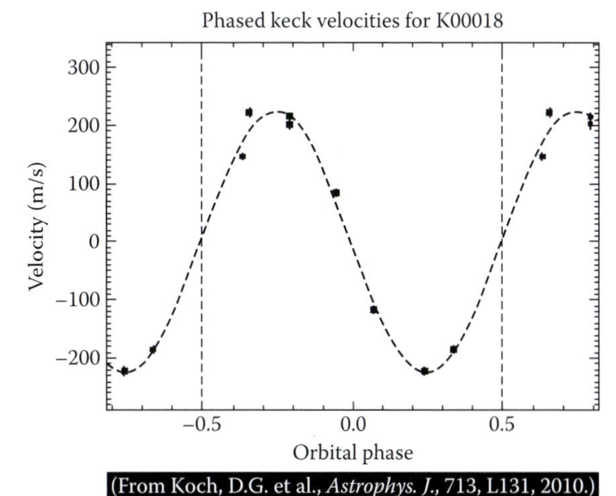

(From Koch, D.G. et al., *Astrophys. J.*, 713, L131, 2010.)

(a) Find the radius of the planet's orbit in AU.

(b) Find the orbital velocity of the planet.

(c) Determine the planet's mass and compare it to the mass of Jupiter. ($M_J = 10^{-3}\, M_{Sun}$)

When the planet crosses the face of the star, it blocks a small fraction of the starlight reaching *Kepler*'s camera. As shown in the second figure, the measured intensity of light decreases by a factor of 0.007 and the transit takes roughly 4.4 h from start to finish. Use this data to find

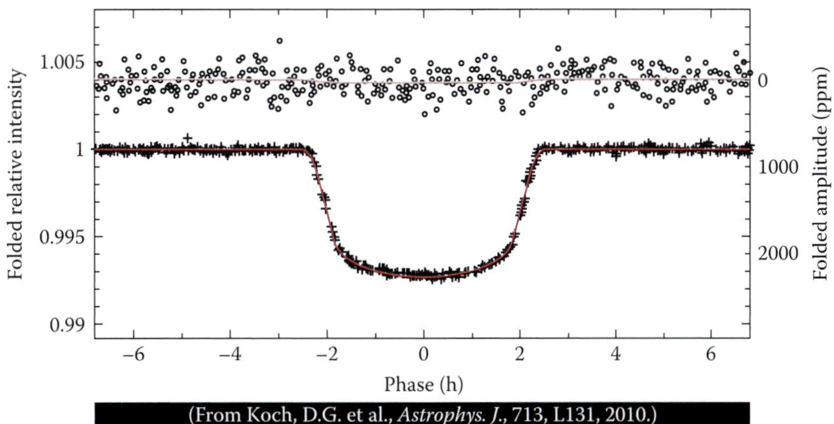

(From Koch, D.G. et al., *Astrophys. J.*, 713, L131, 2010.)

(d) The star's radius ($R_{Sun} = 7 \times 10^8$ m)

(e) The radius of the planet. Express this in terms of R_J.

The *Kepler* spacecraft is parked in a *heliocentric* (*not* a geocentric) orbit that is just slightly more eccentric than Earth's orbit. This was chosen to avoid disturbances from Earth and the Moon that would continuously change the orientation of the telescope. In 4 years' time, *Kepler* will lag behind Earth by a distance of about 0.5 AU.

(f) Find the orbital period of the spacecraft.

(g) Find the orbit's semimajor axis (in AU).

8.34 The *Mars Reconnaissance Orbiter* has been in orbit about Mars since 2006. Its mission is to take high-resolution images of the planet's surface to learn if water was ever present for long periods of time (long enough for life to evolve). The MRO is in a nearly circular orbit 300 km above the planet's surface. Mars has a mass of 6.42×10^{23} kg and a radius of 3380 km.

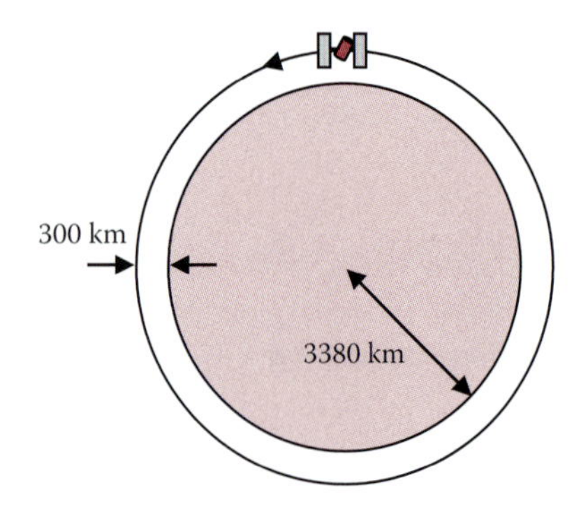

(a) Calculate the acceleration of gravity on the surface of Mars.

(b) What is the speed of the spacecraft in its orbit above Mars?

(c) Suppose NASA wanted to slow the spacecraft to half of the speed calculated in (b) while keeping its altitude the same (300 km). To do this, the MRO's on-board rockets would need to deliver a continuous thrust. Calculate the magnitude of this thrust force, using $M = 2000$ kg for the spacecraft's mass. (Ignore the mass of the exhausted fuel.)

(d) The figure shows a cartoon of the spacecraft with a protruding rocket engine nozzle. How would you orient the rocket to keep the spacecraft in its new, slower, orbit?

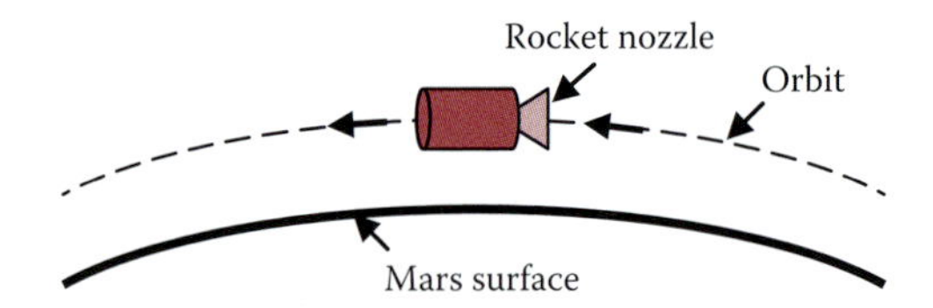

8.35 In May 2007, the star Gliese 436 was observed to jiggle with a period of 2.644 days. This suggested that there is a planet in orbit about the star. Assuming that this planet is in circular motion about the CM of the star–planet system, Doppler shift measurements of the starlight give a *star* velocity of 18.3 m/s.

(a) Find the radius of the star's orbit.

(b) Measurements of the star's luminosity and color allow astronomers to estimate the mass of the star: $M_{Gl} = 0.44 M_{Sun} = 8.8 \times 10^{29}$ kg. Find the radius r_p of the planet's orbit.

(c) Find the mass of the planet. Compare this to the mass of Earth ($M_{Earth} = 3 \times 10^{-6}\, M_{Sun}$).

(d) Whenever the planet passes between the star and Earth, the observed brightness of the star decreases. By studying the brightness vs. time, astronomers estimate that the radius of the planet $r = 4.10\ R_{Earth}$. Calculate the acceleration of gravity on the surface of the planet.

8.36 Pluto and its moons have been in the news recently. On July 11, 2005, Pluto's largest moon Charon briefly blocked the light from a distant star, allowing an accurate measurement of its radius. In May 15, 2005, astronomers working with the Hubble Space Telescope found two tiny additional moons, later named Hydra and Nix (see the following figure). Finally, on August 24, 2006, the International Astronomical Union demoted Pluto from "planet" status to "dwarf planet" status, leaving us with only 8 planets in our Solar System.

(Image courtesy of NASA, ESA, H. Weaver (JHU/APL), A. Stern (SwRI), and HST Pluto Companion Search Team.)

(a) Hydra and Nix are in near circular orbits about Pluto, with periods and orbital radii given in the following table.

Moon	*R* (km)	*T* (days)
Hydra	64,750	38
Nix		24.9

Fill in the missing information.

High resolution pictures from the Hubble Space Telescope show that Pluto and Charon are in circular orbits about their common CM, with a period $T = 6.387$ days ($=5.519 \times 10^5$ s), and a separation of 1.957×10^4 km.

(b) Find the total mass of Pluto and Charon. (Ignore the other moons.)

(c) Pluto's radius is $R_p = 1200$ km. Charon's radius was found to be $R_c = 600$ km. Assuming that the *densities* of the two bodies are the same, find the radius of Pluto's orbit.

(d) Astronomers estimate Hydra's radius to be 44 km and (assuming the same density as above) its mass to be 0.39×10^{18} kg. What is the acceleration of gravity ("g") on the surface of Hydra?

(e) A pendulum consisting of a 1 kg ball tied to a string of length 1 m is set in motion on the surface of Hydra. What is its period?

(f) Pluto is in an elliptical orbit around the Sun, with a perihelion (closest distance to the Sun) of 29.66 AU and an aphelion (farthest distance to the Sun) of 49.31 AU. Calculate the orbital period in years.

8.37 IM Pegasi is a bright binary star system in the constellation Pegasus. Doppler measurements of the light emitted by its two stars show that they are in circular orbits with period 24.65 days ($=6.75 \times 10^{-2}$ year).

The radial velocity curves shown in figure b below have amplitudes of 62.3 and 34.3 km/s.

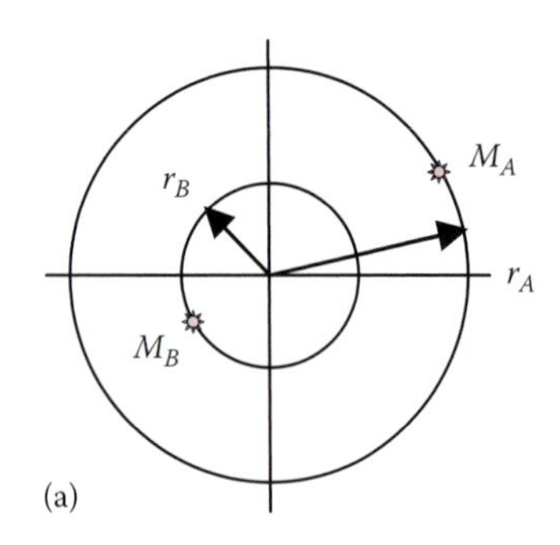

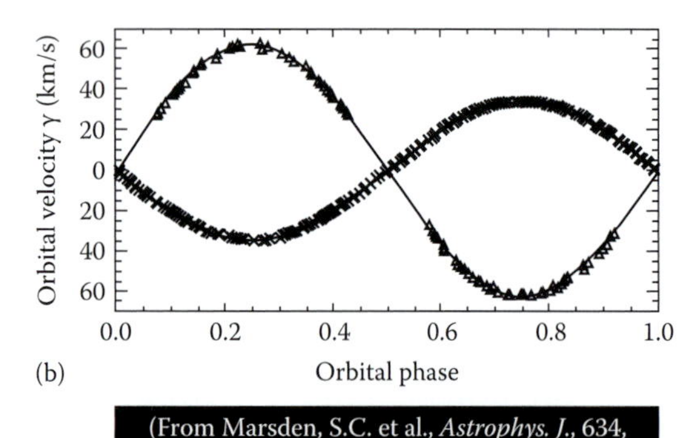

(From Marsden, S.C. et al., *Astrophys. J.*, 634, L173, 2005.)

(a) The two stars and their orbits are depicted in figure a. Which star is heavier? What is the ratio of their masses (M_A/M_B = ?)?

(b) The orbital radius is $r_A = 2.10 \times 10^{10}$ m (=0.140 AU). Find the masses M_A and M_B. (You may express these in terms of M_{Sun}.)

(c) Suppose instead that the smaller star is in an elliptical orbit, as shown in the next figure. The focus of the ellipse is at the point "*x*" in the figure. Describe the orbit of the heavier star. On this figure, sketch the orbit of the heavier star. Indicate its position when the lighter star is at the position shown.

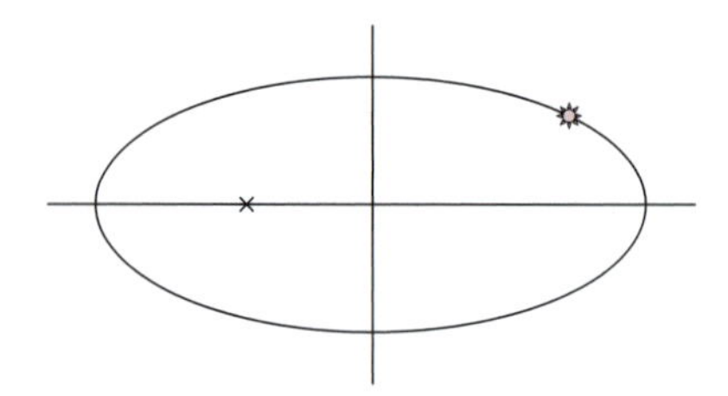

8.38 In 2012, the *Cassini* team of scientists found strong evidence for a subterranean ocean on Titan, the largest of Saturn's 53 known moons. The discovery was made by measuring Titan's gravitational force on the *Cassini* spacecraft during 6 close flybys of the moon in 2006–2011. Data analysis revealed that tidal forces due to Saturn distorted the thin crustal surface of Titan by about 10 m, far greater than possible for a solid sphere. For the purpose of this problem, imagine that Titan is a solid sphere covered by a liquid ocean of water (i.e., without an outer crust, figure b). The moon's mass is $M_T = 1.35 \times 10^{23}$ kg, and its radius is $R_T = 2575$ km. It orbits Saturn in 15.95 days with a semimajor axis equal to 1.22×10^6 km. Find the height of the tidal disturbance. (Your answer will be >10 m because Titan's icy outer crust (figure a) limits the allowed tidal distortion.)

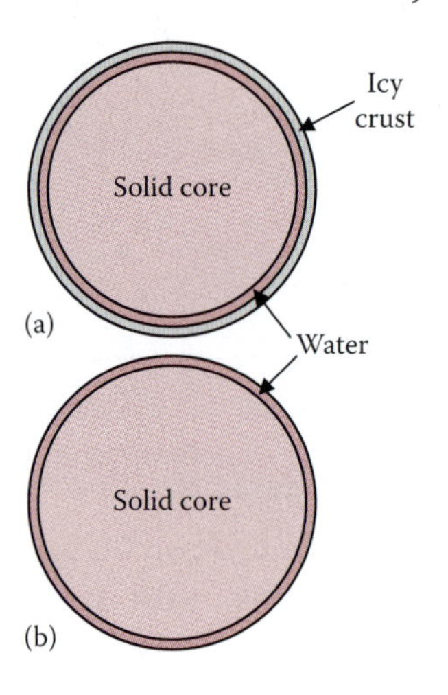

8.39 The derivation of exoplanet mass m presented in Section 8.12 assumes that the orbital plane of the planet is parallel to the line of sight from Earth. If the plane is inclined by an unknown angle i (as in Figure 8.12b), the Doppler technique can only determine $V\sin i$, where V is the speed of the star.

(a) Show that this allows us to calculate—not the planet mass—but the quantity $m\sin i$.

(b) Explain why $m\sin i$ is the *minimum* value of the actual planetary mass m.

8.40 Saturn has a mass 5.69×10^{26} kg and a *mean* radius 58.2×10^6 m. It rotates once every 0.444 days, and due to its spin, it is noticeably nonspherical when viewed through a telescope.

(Photo courtesy of NASA/JPL/STSI.)

(a) Calculate the gravitational acceleration g at the surface of the planet.

(b) Find the fractional difference $(a-b)/\bar{R}_S$ between the polar and equatorial radii of Saturn, and check your answer by direct measurements from the image shown.

8.41 A mountain of mass m may be modeled crudely as a cylindrical cone of constant density ρ, with height h and base radius r_b. Assume that the weight of the "mountain" $W=mg$ is distributed uniformly over the area of its base $A=\pi r_b^2$, so that the compressive pressure at base level is $P_{base}=W/A$. The mountain is situated on the surface of an astronomical body (a planet or moon) having a radius R and surface gravitational field g.

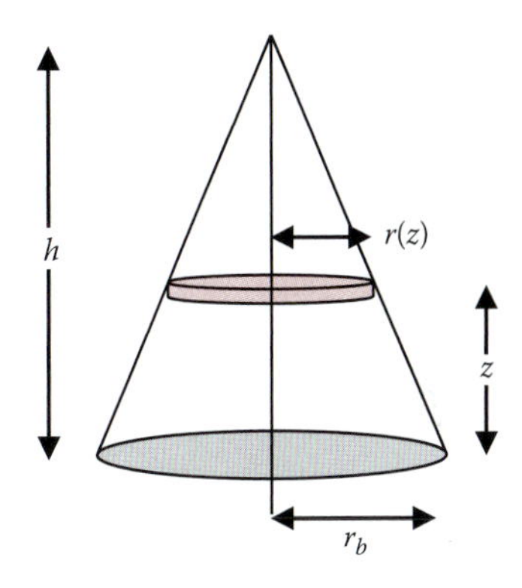

(a) Show that the mountain's mass is $m = \frac{1}{3}\rho A$. Do this by dividing the cone into a stack of thin disks of thickness dz and radius $r(z) = r_b(1 - z/h)$. See the figure.

(b) The maximum mountain height h_{max} is found by setting $P_{base}=P_C$, where P_C is the compressive strength of the material supporting the mountain. Using Equation 8.16, show that $h_{max} = 3R_C^2/2R$, where $R_C \approx 250\,\text{km}$ is the critical radius discussed in Section 8.13.

(c) Estimate the maximum height of a mountain on Earth. Compare your answer to the height of Mt. Everest (8.8 km) and Mauna Kea (10.2 km from ocean floor to peak). The radii of Mars and Venus are $R_{Mars}=0.53R_E$ and $R_{Venus}=0.95R_E$. Similarly, calculate h_{max} for Mars and Venus. The tallest mountain on Mars is

Olympus Mons (24 km), and the tallest mountain on Venus is Maxwell Montes (11 km). Compare your calculated heights to these figures.

(d) Why does the planet look round when viewed through a telescope? (*Hint:* Calculate h_{max}/R for an Earth-sized planet.)

8.42 The main asteroid belt is a band of rubble lying in the plane of the solar system between the orbits of Mars and Jupiter (a_{Mars} = 1.52 AU, a_{Jup} = 5.19 AU). Let T_{ast} (or T_J) be the orbital period of an asteroid (or Jupiter). When $T_J/T = n/m$, where n and m are small integers, an orbital *resonance* is said to exist and the asteroid is swept out of its orbit. Resonances create distinct circular gaps (called Kirkwood gaps, after Daniel Kirkwood, the American astronomer who discovered and explained them in 1866) in the asteroid belt at orbital radii 2.06, 2.5, 2.82, 2.95, and 3.27 AU.

(a) Find n and m for each of the five gaps listed above.

(b) The Cassini Division is the most prominent of the many gaps in Saturn's rings. Its inner edge lies 117,600 km from the center of the planet. Mimas, one of Saturn's >60 known moons, orbits the planet with a semimajor axis of 185,000 km. Show that the Cassini Division is likely due to an orbital resonance with Mimas.

Section III

CONSERVATION OF ENERGY

9

THE SECOND CONSERVATION LAW

Energy

*What we have discovered about energy is that we have a scheme with a sequence of rules. From each different set of rules we can calculate a number for each different kind of energy. When we add all the numbers together, from all the different forms of energy, it always gives the same total. But as far as we know there are no real units [of energy], no little ball-bearings. It is abstract, purely mathematical, that there is a number such that, whenever you calculate it, it does not change. I cannot interpret it any better than that.**

Richard P. Feynman

Even today, energy is an enigma. Ask for a definition of energy, and you are likely to "learn" that "Energy is the capacity to do work." Short of enlightening you, this merely raises a new question, "What is work?," which is equally difficult to answer. In his celebrated Lectures in Physics, the American physicist and Nobel laureate Richard Feynman illustrated the energy concept with a playful story about a cartoon character, a scamp named Dennis "the Menace." Dennis owns a set of 28 wooden blocks. Each evening, his mother gathers them up and puts them away. If she finds fewer than 28 blocks, she searches until she has located the missing ones, confident that there will always be exactly 28. One day, after looking high and low, she finds only 25 blocks, but notices that Dennis' toy box feels heavier than usual. Dennis has forbidden anyone from ever(!) opening his toy box, but Mom knows that each block weighs 3 oz, and the empty toy box weighs 16 oz, so she calculates and finds that

$$\text{Number of blocks seen} + \frac{(\text{mass of box}) - 16\text{ oz}}{3\text{ oz}} = 28.$$

On another day, she cannot account for the 28 blocks even after weighing the toy box. But she notices that the level of water in the bath tub is slightly higher than the usual 6 in., apparently because Dennis is dropping blocks into the bathwater. For each submerged block, the water level rises ¼ in. The water is too dirty for Mom

* From R. Feynman, *The Character of Physical Law*, The M.I.T. Press, Cambridge, MA, 1967.

to see the blocks directly, so she recalculates and finds to her immense satisfaction that

$$\text{Number of blocks seen} + \frac{(\text{mass of box}) - 16\text{ oz}}{3\text{ oz}} + \frac{(\text{height of water}) - 6\text{ in.}}{\tfrac{1}{4}\text{ in.}} = 28.$$

No matter how many ways Dennis finds to conceal his blocks, his mother is confident that she can derive a formula—however complicated—to account for all 28 blocks! The same thing is true for energy: each term of Mom's equation is analogous to a different form of energy. We are so confident that the sum of different forms of energy will remain constant (28 J, say), that if our formula does not give this result, we immediately conclude that there are other forms of energy—that we have not yet identified—which must be added to our energy formula. But unlike Dennis' toys, there are no physical "blocks" of energy, only terms of an equation that when summed, always give the same result. This is the essence of the law of energy conservation.

9.1 GALILEO AND THE ORIGINS OF THE ENERGY IDEA

The seeds of energy conservation were sown by Galileo in the early seventeenth century. In his final publication, *Discourses Concerning Two New Sciences* (1638),* he describes experiments conducted by rolling a ball along an inclined track. Picture the V-shaped track shown in Figure 9.1, where the left and right legs of the apparatus make equal angles with the horizontal ($\theta_L = \theta_R$). When the ball is released from point A, it accelerates down the left leg and decelerates up the right leg, coming to rest momentarily at point D. After correcting for the small effects of air drag and friction, D is found to be at the same height as A. This is not surprising, since motion up the right ramp is just a time-reversal of motion down the left. What *is* surprising is that we get the same result when $\theta_L \neq \theta_R$: the ball always comes to rest at a point D' having the same height as A, independent of the ramp angles! It is as if the ball remembers its initial state. But rather than ascribing memory to the ball, physicists prefer to say that something is *conserved*.

Galileo's results are fully consistent with Newton's Laws. Imagine a block sliding without friction down a straight segment of track inclined at angle θ (Figure 9.2a). Ignoring friction and air drag, the only component of force in the direction of travel (parallel to the track) is $F_{grav}\sin\theta = mg\sin\theta$. During a time interval $\Delta t = t_f - t_i$, the block moves between points i and f with constant acceleration $a = g\sin\theta$, or $a = (v_f - v_i)/\Delta t$ where v_i (or v_f) is its initial (or final) speed. The distance s separating i and f is related to the average velocity by $s = v_{avg}\Delta t = ½(v_f + v_i)\Delta t$. Multiplying the expressions for a and s, we obtain

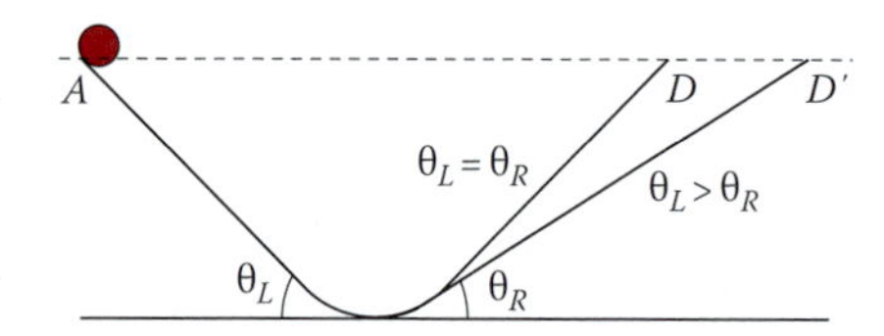

FIGURE 9.1 A ball, starting from rest at A, rolls down and up a bent track. Regardless of the track angles θ_L and θ_R, the ball always rises to the same height, stopping at D or D'.

$$a \cdot s = gs\sin\theta = \frac{1}{2}(v_f + v_i)(v_f - v_i) = \frac{1}{2}v_f^2 - \frac{1}{2}v_i^2.$$

* *Discourses* was written while Galileo was under house arrest in Florence. He was forbidden from publishing in Italy, and the book was smuggled out of Italy to be published in Holland.

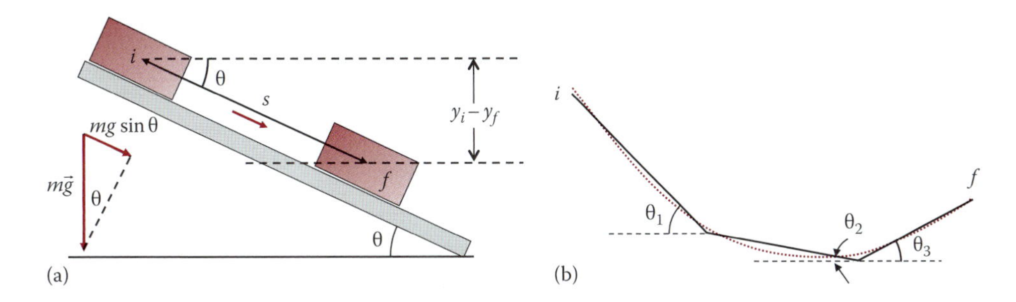

FIGURE 9.2 (a) A block slides without friction along an inclined track from i to f. The distance traveled is s, and the change in altitude is $y_i - y_f = s\sin\theta$. (b) A curved track is approximated by three straight line segments. The approximation to the original curved path improves as the number of straight segments is increased.

Since $s\sin\theta = y_i - y_f$, where y_i (or y_f) is the height of the initial (or final) point, the above equation may be rewritten as

$$\frac{1}{2}v_f^2 - \frac{1}{2}v_i^2 = g(y_i - y_f).$$

If the block is released from rest ($v_i = 0$) at height y_i, its speed depends only on the change in height; it will return to rest ($v_f = 0$) at the point f where $y_f = y_i$. This is just what Galileo observed. This result holds for all tracks, regardless of their shape, since a curved section of track may be considered to be a collection of many short straight sections tilted at different angles (Figure 9.2b).

Multiplying the above equation by the block's mass m, and rearranging terms, we find

$$\frac{1}{2}mv_f^2 + mgy_f = \frac{1}{2}mv_i^2 + mgy_i.$$

Since the points i and f were chosen at random, the above expression must be true for all i and j, or

$$K_{trans} + mgy = \text{constant}, \tag{9.1}$$

where the first term is the kinetic energy associated with the block's displacement, or *translational* kinetic energy: $K_{trans} = \frac{1}{2}mv^2$. No matter where the cart happens to be along the track, the two terms on the left side sum to the same number. They are like the terms derived by Dennis' mother to account for the total number of blocks.

Things get more complicated if we use a rolling ball (as did Galileo) rather than a sliding block. If we *measure* the ball's acceleration, we will find $a = \frac{5}{7}g\sin\theta$, where θ is the track angle. Then, proceeding just as we did above, $a \cdot s = \frac{1}{2}v_f^2 - \frac{1}{2}v_i^2 = \frac{5}{7}g(y_i - y_f)$, or

$$\frac{7}{5}\left(\frac{1}{2}mv_f^2 - \frac{1}{2}mv_i^2\right) = mg(y_i - y_f).$$

For a given change in height $(y_f - y_i)$, the change in translational kinetic energy of the ball is only 5/7 that of the block. The "missing" kinetic energy is associated with the ball's rotational motion: $K_{rot} = \frac{2}{5} K_{trans}$ (which we will prove in Chapter 11). Now, our energy equation has *three* terms,

$$K_{trans} + K_{rot} + mgy = \text{const.} \tag{9.2}$$

Note that we have not *defined* energy in either Equation 9.1 or 9.2. Instead, we have merely found three mathematical expressions which, when added together, always give the same result. If the ball (or block) experiences significant air drag or sliding friction as it moves, we will need to find additional terms to obtain a conserved quantity. Regardless of these complications, energy remains an exceedingly powerful concept simply because it is a *conserved.*

9.2 WORK AND KINETIC ENERGY

Our task for the remainder of this chapter is to figure out how to identify and express all of the terms—the forms of energy—that must be included in our conservation equation. We will proceed by focusing on the simplest case: the motion of one rigid body m under the influence of an applied force $\vec{F}$, which may arise from a single interaction or may be the vector sum of several applied forces: $\vec{F} = \vec{F}_1 + \vec{F}_2 + \cdots$. Let m be moving with an initial velocity $\vec{v}$. The body accelerates according to Newton's Second Law: $m\vec{a} = \vec{F}$, and the rate at which its kinetic energy K changes is

$$\frac{dK}{dt} = \frac{d}{dt}\left(\frac{1}{2}mv^2\right) = \frac{d}{dt}\left(\frac{1}{2}m\vec{v}\cdot\vec{v}\right),$$

where the dot product has been used to express v^2. Taking the derivative of the dot product,

$$\frac{dK}{dt} = \frac{1}{2}m\vec{a}\cdot\vec{v} + \frac{1}{2}m\vec{v}\cdot\vec{a} = m\vec{a}\cdot\vec{v}$$

or, using $\vec{F} = m\vec{a}$

$$\frac{dK}{dt} = \vec{F}\cdot\vec{v}.$$

The quantity $\vec{F}\cdot\vec{v}$ is the change of energy per unit time, and is called the *power* supplied by the force $\vec{F}$.

During a short time interval Δt, in which $\vec{F}$ and $\vec{v}$ do not change appreciably, the change of kinetic energy is

$$\Delta K = \frac{dK}{dt}\Delta t = \vec{F}\cdot\vec{v}\Delta t = \vec{F}\cdot\Delta\vec{r} \equiv W, \tag{9.3}$$

where $\Delta\vec{r} = \vec{v}\Delta t$ is the displacement of the body during the time Δt. The quantity $W \equiv \vec{F} \cdot \Delta\vec{r}$ is called the *work* done on the body by $\vec{F}$, and Equation 9.3 is called the *work-kinetic energy theorem*. Because the dot product is involved, only the force component *parallel to the displacement* contributes to the work. Equivalently, only the *displacement parallel to the force* figures into the work: $W = F_{\parallel}\Delta r = F\Delta r_{\parallel}$.

EXERCISE 9.1

A heavy box is pulled over a horizontal surface by a worker using a rope that makes an angle of 40° with respect to the surface.

Ignoring friction, three forces act on the block: $\vec{N}, m\vec{g}$, and the rope tension $\vec{F}_{ten} = 50$ N.

When the box has been displaced by 1 m, how much work is done on it by each force?

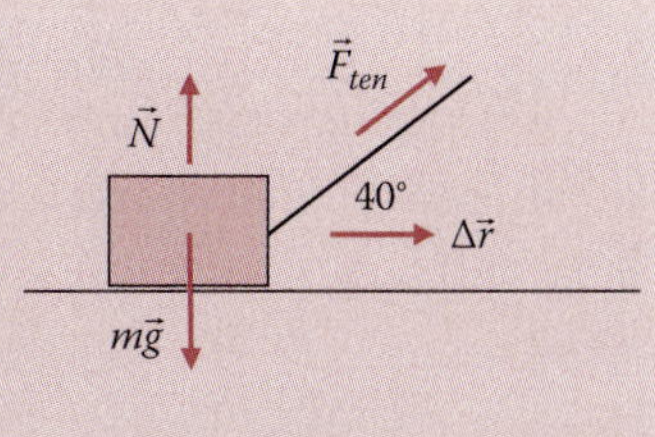

Answer: 0, 0, 38.3 J

Figure 9.3a shows the FBD of a block sliding without friction along an inclined plane. We have already analyzed this problem, but now let's recast our analysis using the language of work. The normal force $\vec{N}$ is perpendicular to $\Delta\vec{r}$, so it does no work on the block. Only the gravitational force $\vec{F}_{grav}$ has a component parallel to the displacement. Writing $\Delta\vec{r} = \Delta x\hat{i} + \Delta y\hat{j}$, the change in the block's kinetic energy when it slides from point *i* to point *f* is

$$\Delta K = W_{grav} = m\vec{g} \cdot \Delta\vec{r} = (-mg\hat{j}) \cdot (\Delta x\hat{i} + \Delta y\hat{j}) = -mg\Delta y$$

or (9.4)

$$\Delta K = \frac{1}{2}mv_f^2 - \frac{1}{2}mv_i^2 = mg(y_i - y_f).$$

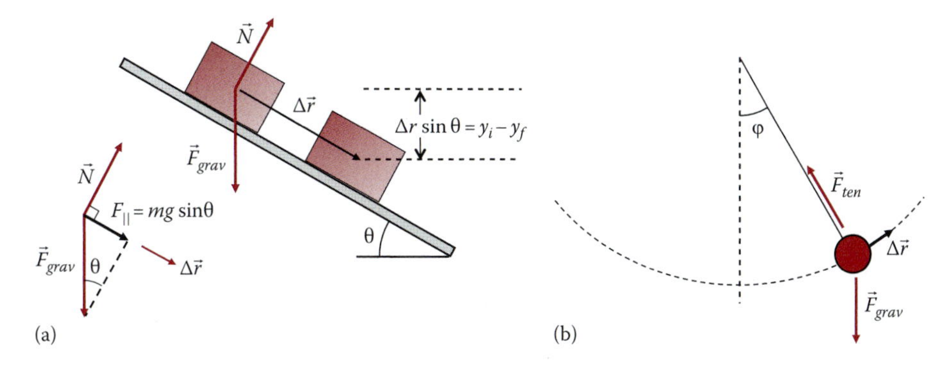

FIGURE 9.3 (a) The forces acting on a body sliding without friction on an inclined surface are $\vec{N}$ and $\vec{F}_{grav}$. The normal force $\vec{N}$ is perpendicular to $\Delta\vec{r}$ and does no work. (b) The tension force acting on a pendulum does no work because $\vec{F}_{ten} \perp \Delta\vec{r}$. In either (a) or (b), $W = mg(y_i - y_f)$.

This is identical to Equation 9.1. Notice that the track angle θ does not appear in this equation. Equation 9.4 is valid for any angle θ, even if the body is falling vertically (θ = 90°) in the absence of a track. The work done by the gravitational force $\vec{F}_{grav} = m\vec{g}$ depends *only on the change in altitude* Δy.

Equation 9.4 holds whenever the work done on the body is due solely to the gravitational force $\vec{F}_{grav} = m\vec{g}$. For example, consider the motion of a projectile in flight. Ignoring air resistance, the only force acting on the body is $\vec{F}_{grav}$, so $\Delta K = W_{grav} = m\vec{g}\cdot\Delta\vec{r} = -mg\Delta y$, the same as for the sliding block.

EXERCISE 9.2

A cannonball is fired from ground level at a 45° angle with respect to the horizontal. At its highest point, it is 250 m above the ground. Neglect air drag.

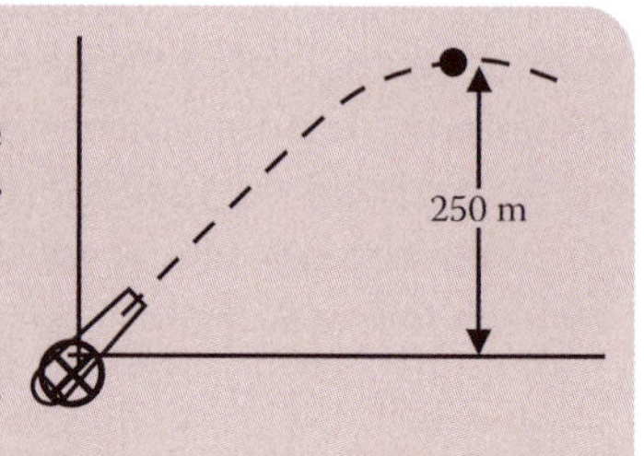

(a) Compare the cannonball's kinetic energy at its peak altitude with its initial kinetic energy, that is, K_{peak}/K_{init} = ?
(b) Find the initial speed of the ball.

Answer (a) 1/2 (b) ≈ 100 m/s

To further illustrate the utility of the work-kinetic energy approach, let's take a fresh look at the motion of a pendulum. In Chapter 7, we restricted the bob to small displacements about its equilibrium position. In this limit, the string tension F_{ten} is approximately constant, the motion is nearly horizontal, and the equation of motion (simple harmonic motion) is easily found. For larger displacements, these approximations are no longer valid, and a solution based solely on force and Newton's laws is quite difficult. Fortunately, the work–energy approach comes to our rescue.

The forces on the pendulum bob, shown in Figure 9.3b, are the string tension $\vec{F}_{ten}$ and the gravitational force $m\vec{g}$. As the deflection angle $\varphi(t)$ changes, $\vec{F}_{ten}$ changes in both magnitude and direction, yet it remains perpendicular to the instantaneous velocity $\vec{v}$ of the bob at all times: $\vec{F}_{ten}\cdot\Delta\vec{r} = 0$. The only force that does work on the bob is the gravitational force, so $\Delta K = W_{grav} = m\vec{g}\cdot\Delta\vec{r} = -mg\Delta y$, just as in the two cases considered above. The simple pendulum may look very different from a ball on a track, or a projectile in flight, but from a work-kinetic energy perspective, the physics is essentially the same.

EXERCISE 9.3

A 500 g ball is hung from a string of length ℓ = 1 m and released at an angle $\varphi_{max} = \pi/3$ (rad).

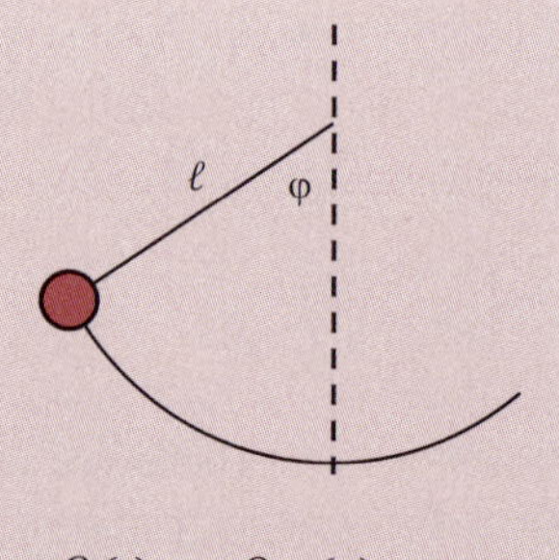

(a) At what point does the ball reach its highest speed?
(b) At this point, how far below its initial position is it?
(c) What is the speed of the ball at this point?

Answer (b) 0.5 m (c) 3.1 m/s

EXERCISE 9.4

A pendulum is mounted adjacent to a vertical wall. When the bob is released from point A, it swings through the center position B and comes to rest at C before reversing its motion. A peg is now inserted into the wall at D, directly above B. When the bob swings, the string catches on the peg and the bob is deflected along the arc $\widehat{BE}$.

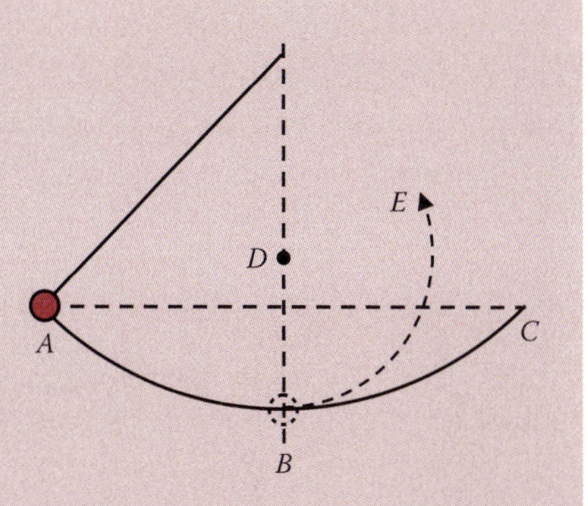

(a) What is the maximum height of the bob along arc $\widehat{BE}$?

(b) Is there a position D of the peg for which the bob will rise above its initial height at A?

This experiment was performed by Galileo and reported in *Discourses Concerning Two New Sciences*. He had no formal idea of energy at that time.

Answer (a) level with $\overline{AC}$ (b) No

The dimensions of kinetic energy are mass × velocity², and in SI units, K is measured in joules* (J): 1 J = 1 kg m²/s². The dimensions of work are force × distance, which are the same as mass × velocity² (check this), so 1 J = 1 kg m²/s² = 1 N m. All forms of energy, including potential energy (to be discussed shortly), rest mass energy, chemical energy, etc., can be expressed in terms of joules. The SI unit of *power* is the watt (W): 1 W = 1 J/s. For atomic- and nuclear-scale energies, the joule is inconveniently large and the electron volt (eV), megaelectron volt (MeV), or gigaelectron volt (GeV) is used instead: 1 eV = 1.60218... × 10^{-19} J. These units are discussed in greater detail in Box 9.2.

9.3 VARIABLE FORCES, WORK, AND THE LINE INTEGRAL

In the above examples, the only force that did work on the body was the constant gravitational force $m\vec{g}$. Imagine instead a force that changes with the position of the body: $\vec{F} = \vec{F}(\vec{r})$. To calculate W and ΔK, we must now generalize the definition of work given in Equation 9.3. Figure 9.4a depicts the curved path of a body moving between $\vec{r}_i$ and $\vec{r}_f$. To calculate the work done on the body, replace the actual path by a series of n straight line segments $d\vec{r}_1, d\vec{r}_2, \ldots, d\vec{r}_n$. As n increases and each segment shrinks, the approximation to the real path improves, becoming exact in the limit $n \to \infty$. The incremental amount of work done along the mth displacement is $dW_m = \vec{F}(\vec{r}_m)\cdot d\vec{r}_m$, where $\vec{F}(\vec{r}_m)$ is the average force exerted on the body during this displacement. The total work done on the body is given by the infinite sum

$$W = \Delta K = \lim_{n\to\infty}\sum_{m=1}^{n} \vec{F}(\vec{r}_i)\cdot d\vec{r}_i \equiv \int_{\vec{r}_i}^{\vec{r}_f} \vec{F}(\vec{r})\cdot d\vec{r}, \qquad (9.5)$$

* Named for the British physicist James Joule who determined the relationship between the unit of mechanical work (1 N m) and the unit of thermal energy (1 Cal). See Section 9.11.

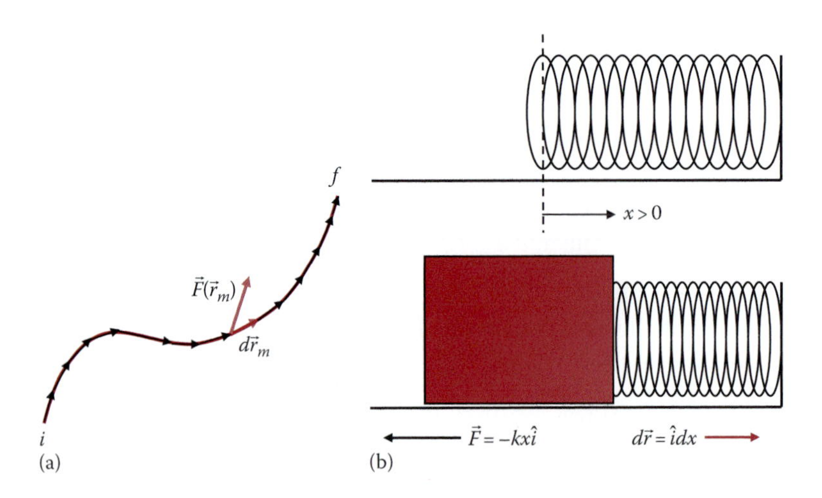

FIGURE 9.4 (a) A curved path is approximated by a series of infinitesimal displacements $\Delta\vec{r}_m$. The differential amount of work done during the displacement $d\vec{r}_m$ is $dW_m = \vec{F}(\vec{r}_m)\cdot d\vec{r}_m$, and the total work done from i to f is the sum, or integral, of the differential work. (b) A block moving in the positive x-direction compresses a spring. The work done in a displacement from $x\hat{\imath}$ to $(x+dx)\hat{\imath}$ is $dW = -kx\,dx$.

where we have replaced the infinite sum with an integral. This integral is called the *line integral* of $\vec{F}$ along the chosen path linking $\vec{r}_i$ to $\vec{r}_f$. Equation 9.5 is the generalized version of the work-kinetic energy theorem given earlier. If the force $\vec{F}$ is independent of position, it may be removed from the integral, and with $\int_i^f d\vec{r} = (\vec{r}_f - \vec{r}_i) = \Delta\vec{r}$, Equation 9.5 becomes identical to Equation 9.3.

EXERCISE 9.5

The force $F(x)$ acting on a 1 kg body is shown in the figure as a function of position x (m). The force points in the positive x-direction.

(a) Find the work done on the body as it moves from $x = 0$ to $x = 7$ m.
(b) If the body starts from rest, what is its speed when it arrives at $x = 7$ m?

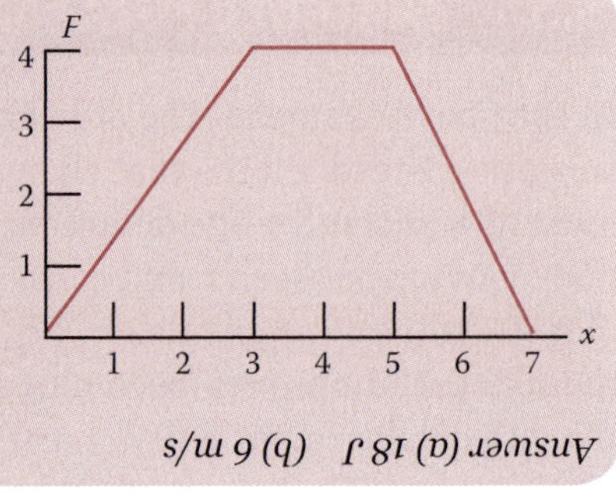

Answer (a) 18 J (b) 6 m/s

Picture a block attached to a spring resting on a frictionless horizontal surface. See Figure 9.4b. Let $\vec{r} = x\hat{\imath} = 0$ be the position of the block when the spring is fully relaxed, so x indicates the amount by which the spring is compressed ($x > 0$) or stretched ($x < 0$). Vectors are not required in this one-dimensional example, but let's retain them briefly to illustrate how the dot product in Equation 9.5 is handled. The force exerted on the block by the spring is $\vec{F}(x) = -kx\hat{\imath}$, where k is the spring constant. When the block is displaced incrementally from $x\hat{i}$ to $(x+dx)\hat{i}$, the work done on the block is $dW = \vec{F}(x)\cdot d\vec{r} = -kx\hat{i}\cdot\hat{i}\,dx = -kx\,dx$. By the work-kinetic energy theorem,

$$dW = dK = d\left(\frac{1}{2}mv^2\right) = -kx\,dx.$$

For a larger (nonincremental) displacement of the block between x_i and x_f, Equation 9.5 yields

$$\Delta K = \frac{1}{2}mv_f^2 - \frac{1}{2}mv_i^2 = -\int_{x_i}^{x_f} kx\,dx = \frac{1}{2}kx_i^2 - \frac{1}{2}kx_f^2 \quad \text{or} \quad \frac{1}{2}mv_f^2 + \frac{1}{2}kx_f^2 = \frac{1}{2}mv_i^2 + \frac{1}{2}kx_i^2. \tag{9.6}$$

Once again, we see that something is conserved: the "final" terms on the left side of the equation sum to the same value as the "initial" terms on the right side. Since x_i and x_f were chosen at random, Equation 9.6 can be expressed more generally as

$$\frac{1}{2}mv^2 + \frac{1}{2}kx^2 = \text{constant}. \tag{9.7}$$

EXERCISE 9.6

A block of mass m is held just above the top of a massless spring having a stiffness k. When the block is released, it compresses the spring and momentarily comes to rest when the spring is compressed by an amount $\Delta\ell$.

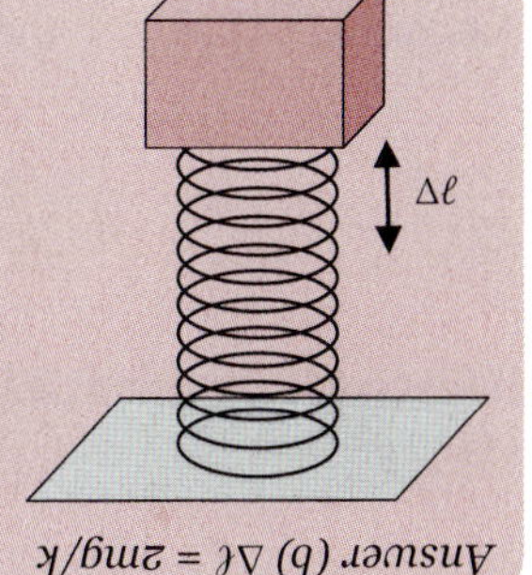

(a) Write an energy equation like Equations 9.1 and 9.7. (*Hint:* There are three terms.)

(b) Find an expression for $\Delta\ell$.

Answer: (b) $\Delta\ell = 2mg/k$

Clearly, energy is stored by a spring when it is stretched or squeezed. By using the work-kinetic energy theorem, we have identified a new term that must be included in our energy equation when we are dealing with springs or other elastic elements.

9.4 PATH INDEPENDENCE AND THE LINE INTEGRAL

Imagine that during spring break, you decide to hike up a nearby mountain (see Figure 9.5). There are two trails that lead to the peak, and from the parking lot (point i) to the mountaintop (point f) the altitude gain is 800 m. It takes n steps for you to reach the peak along path A, and your displacement on your mth step is $d\vec{r}_m = \hat{i}dx_m + \hat{j}dy_m + \hat{k}dz_m$ ($m = 1, 2,\ldots, n$). Your change in altitude on this step is $dz_m = \hat{k}\cdot d\vec{r}_m$, so your total change in altitude is

$$z_f - z_i = \sum_{m=1}^{n}\hat{k}\cdot d\vec{r}_m = 800\text{ m}.$$

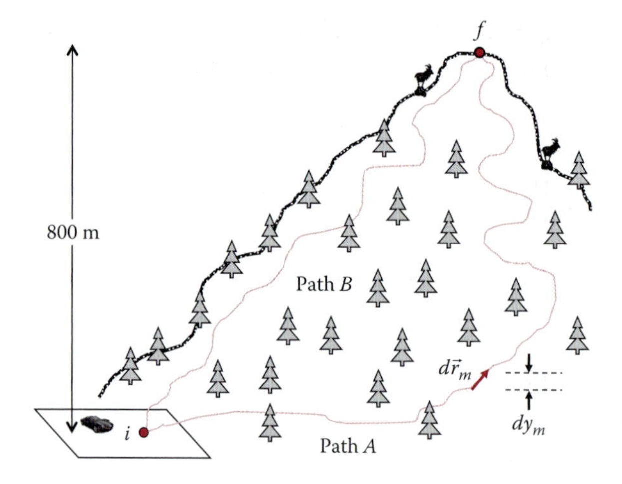

FIGURE 9.5 Two paths from the parking lot (point i) to the peak of the mountain (point f). For either path, the overall change in altitude is equal to 800 m. In a round trip to the peak and back, the overall change in altitude is 0.

Assuming that $n \gg 1$ and each step is short, the sum can be expressed as a line integral:

$$z_f - z_i = \int_{i,\text{path } A}^{f} \hat{k} \cdot d\vec{r} = 800 \text{ m}.$$

If you had taken path B instead of path A, you might have taken n' steps to reach the top, so

$$z_f - z_i = \sum_{m=1}^{n'} \hat{k} \cdot d\vec{r}_m \rightarrow \int_{i,\text{path } B}^{f} \hat{k} \cdot d\vec{r} = 800 \text{ m}.$$

The total change in height along either path is obviously the same. Having reached the peak following path A, you now descend the mountain along path B. Each *downward* step $d\vec{r}_m'$ you take is the vector opposite of the corresponding upward step, or $d\vec{r}_m' = -d\vec{r}_m$, so the line integral of descent will equal the negative of the line integral of ascent along the same path (note the switch in the limits of integration):

$$\int_{f,\text{path } B}^{i} \hat{k} \cdot d\vec{r} = -\int_{i,\text{path } B}^{f} \hat{k} \cdot d\vec{r} = z_i - z_f = -800 \text{ m}.$$

Your round trip change in altitude will (obviously) be zero, or in terms of line integrals,

$$\int_{i,\text{path } A}^{f} \hat{k} \cdot d\vec{r} + \int_{f,\text{path } B}^{i} \hat{k} \cdot d\vec{r} = 800 - 800 = 0.$$

Of course, this is true for *any* round trip, or *closed path*, starting and ending at the same point. Mathematically, the line integral of $\hat{k} \cdot d\vec{r}_m$ over a closed path is expressed more concisely as

$$\oint \hat{k} \cdot d\vec{r} = 0, \tag{9.8}$$

where the little circle added to the integral sign indicates that the line integral is to be evaluated over a *closed path*. Equation 9.8 implies that

$$\int_{i,\text{path } A}^{f} \hat{k} \cdot d\vec{r} = \int_{i,\text{path } B}^{f} \hat{k} \cdot d\vec{r}.$$

If this is true for *any* two paths linking any two points i and f, we say that the line integral $\int_i^f \hat{k} \cdot d\vec{r}$ is *path independent*. In the context of altitude gain, these results are easy to understand. In the next section, we will see that with minor alterations, the same mathematics is intimately connected to the principle of energy conservation.

9.5 CONSERVATIVE FORCES AND POTENTIAL ENERGY

Consider a body that starts out from point i with kinetic energy K_i. Under the influence of a force $\vec{F}(\vec{r})$, the body travels along a certain path to point f, and then returns to the starting point i along a different path, arriving with kinetic energy K_i'. See Figure 9.6a. If $K_i' = K_i$, we know that the total work done during the round trip is equal to zero, or

$$W = \Delta K = \oint \vec{F} \cdot d\vec{r} = 0. \tag{9.9}$$

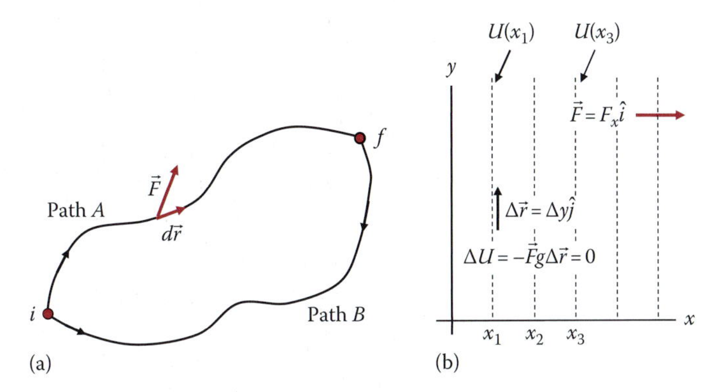

FIGURE 9.6 (a) If the force $\vec{F}(\vec{r})$ is conservative, the work done from i to f is independent of the path: $W_{i\to f}^{A} = W_{i\to f}^{B}$. The total work done going from i to f and then returning to i is 0: $W_{i\to f}^{A} + W_{f\to i}^{B} = 0$. (b) If the force $\vec{F}$ points in the x-direction, then the work done during a displacement perpendicular to $\hat{i}$ is zero, and $\Delta U = 0$. At all points $(x_1 y,z)$, $U(\vec{r}) = U(x_1)$, that is, U is independent of y and z.

This is very similar to Equation 9.8. Here, $\vec{F}$ replaces the unit vector $\hat{k}$ in the expression for the line integral, but otherwise the mathematics is unchanged. If Equation 9.9 holds for *any* closed path, we say that the force $\vec{F}$ acting on the body is a *conservative force*. And if the force is conservative, the work done on the body as it moves between points i and f is independent of the path taken:

$$W_{i\to f} = \int_{i,\text{path }A}^{f} \vec{F}\cdot d\vec{r} = \int_{i,\text{path }B}^{f} \vec{F}\cdot d\vec{r} = \int_{i,\text{path C}}^{f} \vec{F}\cdot d\vec{r} = \cdots = K_f - K_i. \quad (9.10)$$

EXERCISE 9.7

A body moves under the influence of force $\vec{F}(\vec{r})$ from point A to point B, or from A to C, along the paths shown. The work done by $\vec{F}$ along each path is indicated in the drawing. Assuming $\vec{F}$ is a conservative force, what is the work done if the body travels from B to C?

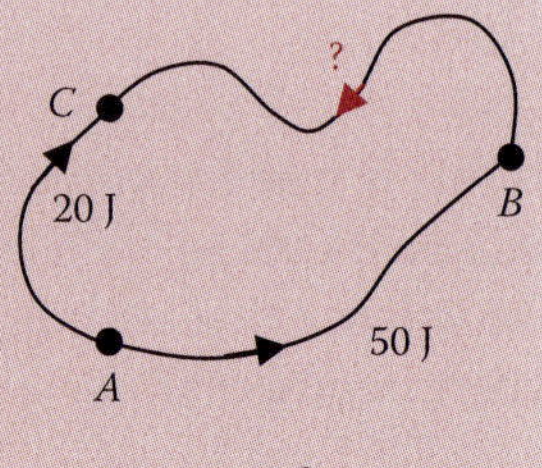

Answer −30 J

There is only one way for $W_{i\to f}$ to be the same for *all* paths linking the two end points i and f: the work integral cannot depend on the intermediate points along the path, and must be a function *of only the two end points*. In this case, we can express the work in terms of a new position-dependent function $U(\vec{r})$:

$$W_{i\to f} \equiv U(\vec{r}_i) - U(\vec{r}_f) = -\Delta U,$$

where $U(\vec{r})$ is called the *potential energy* function associated with the force $\vec{F}(\vec{r})$. Note that $W_{i\to f}$ does not determine either $U(\vec{r}_i)$ or $U(\vec{r}_f)$, but only determines the *difference* in the potential energy function at $\vec{r}_i$ and $\vec{r}_f$. We will examine this more closely in the next section.

The work-energy relation can now be restated (for conservative forces only):

$$\Delta K = K_f - K_i = W_{i\to f} = U_i - U_f = -\Delta U \quad \text{or} \quad K_f + U_f = K_i + U_i \equiv E_{mech}, \quad (9.11)$$

where $U_i = U(\vec{r}_i)$, etc., and the total *mechanical energy* E_{mech} is defined as the sum of the kinetic and potential energies. Summarizing, if the forces acting on a system are *conservative*, a potential energy function $U(\vec{r})$ can be defined for the system, and the mechanical energy $E_{mech} = K + U$ is conserved.

EXERCISE 9.8

Consider again the system shown in Exercise 9.7.

If the force is conservative, and if K_A (the kinetic energy at A) is 40 J while the potential energy U_A = 20 J, find

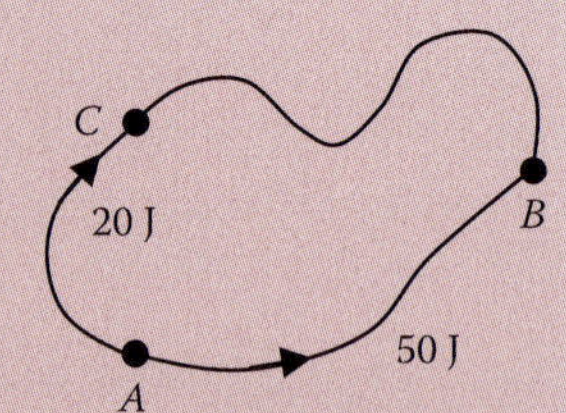

(a) K_B and U_B
(b) E_{mech} at point C

Answer (a) 90 J, −30 J (b) $E_{mech}(C) = E_{mech}(A) = 60$ J

Not all forces are conservative. Consider the motion of a block sliding *with friction* across a horizontal tabletop. The sliding friction force $\vec{F}_{fr}$ always opposes the motion, so when the block is displaced by $\Delta\vec{r} = \vec{r}_f - \vec{r}_i$, the work $W_{i\to f} = \vec{F}_{fr} \cdot \Delta\vec{r} < 0$. When the block returns from $\vec{r}_f$ to $\vec{r}_i$ (along any path), the friction force adjusts its direction so that it again opposes the motion and $W_{f\to i} < 0$. Therefore, for the round trip, $\oint \vec{F}_{fr} \cdot d\vec{r} < 0$. This violates the condition for a conservative force expressed by Equation 9.9, so no potential energy function can be written for friction nor for any other nonconservative force.

EXERCISE 9.9

Let $\vec{F}(\vec{r}) = \alpha z\hat{i}$, where α = const. $\vec{F}(\vec{r})$ points in the x-direction and increases in magnitude linearly with distance z from the x–y plane.

(a) What is the work done along path A?
(b) Along path B?
(c) Is $\vec{F}(\vec{r})$ a conservative force?

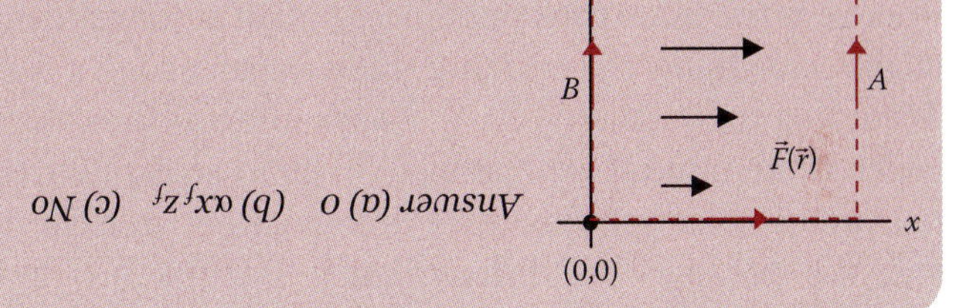

Answer: (a) 0 (b) $\alpha x_f z_f$ (c) No

9.6 FORCE AND POTENTIAL ENERGY

Potential energy is a powerful but profound concept. To use it effectively, we must be able to associate it with a concrete idea: force. Specifically, we need to know how to find $\vec{F}$ from $U(\vec{r})$, and *vice versa*. In a wide range of important physical situations, the force always points in the same direction. Let's define this to be the x-direction, so $\vec{F} = F_x\hat{i}$ and $\vec{F} \cdot d\vec{r} = F_x\,dx$. Equation 9.11 now reads

$$\Delta U = -W_{i\to f} = -\int_i^f F_x\,dx.$$

If the body's displacement $\Delta\vec{r}$ is perpendicular to the x-axis, the work done by F_x is zero, and $\Delta U = 0$. This means that all points (x, y, z) having the same value of x have the same potential energy, so $U(\vec{r})$ is independent of y and z: $U(\vec{r}) = U(x)$. See Figure 9.6b. Then,

$$\Delta U = \int_i^f dU = \int_i^f \frac{dU}{dx}\,dx = -\int_i^f F_x\,dx. \tag{9.12}$$

For conservative forces, ΔU is the same for all paths between points i and f, so we must have

$$F_x = -\frac{dU}{dx}. \tag{9.13}$$

*The force is the negative of the derivative of the potential function.**

Now let's ask the inverse question: If we know the force $\vec{F}(\vec{r})$, how do we determine the function $U(\vec{r})$? For the special case considered above, with $\vec{F} = F_x(\vec{r})\hat{i}$, we simply invert the above process, and integrate (i.e., take the antiderivative of) Equation 9.13 to obtain

$$U(\vec{r}) = -\int F_x\, dx + c, \tag{9.14}$$

where c is the usual constant of integration accompanying an indefinite integral. Equations 9.12 and 9.14 tell us the same thing: the work integral does not assign a definite value $U(\vec{r})$ to the potential energy at location $\vec{r}$. According to Equation 9.12, the work integral fixes the *difference* in potential energy between two points, which is the same as evaluating $U(\vec{r})$ up to an additive constant c, as in Equation 9.14. (Adding a constant c to U_f and U_i does not change the difference between the two potential energies.) Note also that c cannot influence the motion: when differentiating Equation 9.14 to find the force and the acceleration of the body, the constant c vanishes. We can freely add or subtract *any* constant to the potential $U(\vec{r})$ without affecting how the body moves. This freedom to choose c often allows us to simplify the solution of a problem.

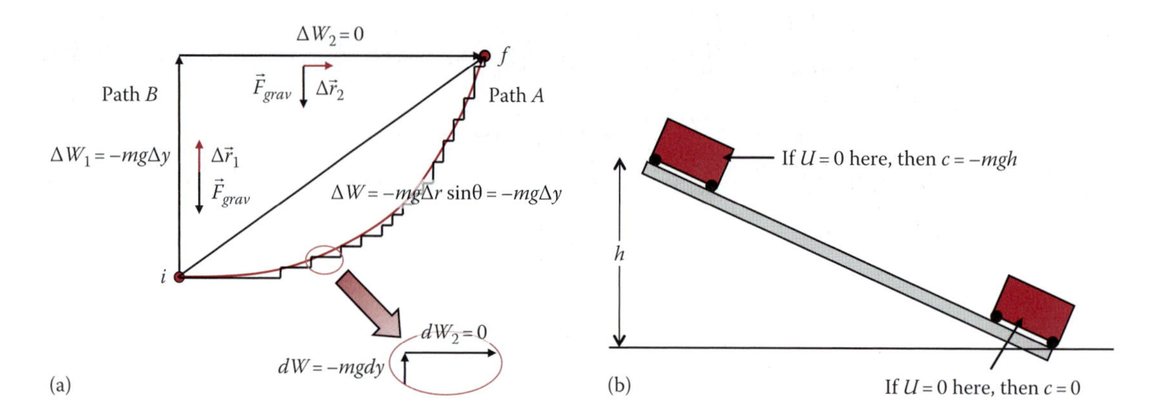

FIGURE 9.7 (a) A body moves from i to f along either a curved (A) or an L-shaped (B) path. The curved path can be approximated by a horizontal/vertical zigzag path and no work is done along the horizontal segments. The sum of the work done during the vertical segments is the same along either path. (b) The integration constant c may be chosen so that $U(y)$ is zero at either the top or bottom end of the ramp or at any other point.

* When $U(\vec{r})$ depends on y or z in addition to x, the full derivative appearing in Equation 9.13 is replaced by a partial derivative: $F_x = -\partial U/\partial x$, $F_y = -\partial U/\partial y$, etc. (If you aren't familiar with partial derivatives, don't worry. We will not mention them again in this course.)

Consider the potential energy associated with the gravitational force $\vec{F}_{grav} = F_y\hat{j} = m\vec{g} = -mg\hat{j}$. Figure 9.7a demonstrates that this is a conservative force: the work done by $\vec{F}_{grav}$ as m moves from i to f is given by $W_{i\to f} = \int_i^f \vec{F}_{grav}\cdot d\vec{r} = \int_i^f -mg\,dy = -mg(y_f - y_i)$, independent of the path, so from Equation 9.14,

$$U(y) = -\int F_y\,dy = -\int(-mg)\,dy + c = mgy + c.$$

We commonly say that mgy is the "potential energy of the body," but this is incorrect for two reasons. First, $U(y)$ arises from the gravitational force between the body m and the Earth, so it is a *shared* property of the Earth–body system. Second, from the inverse-square law of gravitation, $U(y)$ must be a function of the separation between the body and the center of the Earth. By choosing $c = 0$, we are implicitly defining the potential energy to be zero on the plane of our coordinate system where $y = 0$.

EXERCISE 9.10

Your instructor stands on top of a 1 m table while holding a 3 kg block at a height of 2 m above the tabletop. She defines her coordinate system such that $z = 0$ coincides with the height of the table, and declares that the potential energy of the block in her hands is 100 J.

(a) If she drops the block onto the table, what is its kinetic energy just before striking the surface? (Let $g = 10$ m/s².)
(b) Holding the block as before, she reaches over the edge of the table. What is the change in the block's potential energy?
(c) She releases the block. What are its kinetic and potential energies just before it strikes the floor?

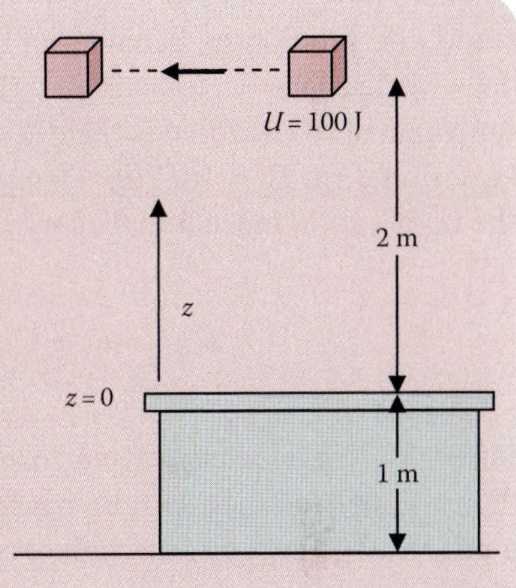

Answer: (a) 60 J (b) 0 (c) K = 90 J, U = 10 J (K + U = 100 J)

So what should we choose for c? Imagine a cart on an inclined track. Let the height of the track be h, and let $y = 0$ at the bottom of the track (see Figure 9.7b). If we set $c = 0$, so $U(y) = mgy$, we are choosing $U(0) = 0$ at the bottom of the track. This may be the most convenient choice for the zero of potential energy. Alternatively, we might set the potential energy to be zero at the top end of the track, $U(h) = 0$. In this case, $c = -mgh$ and $U(y) = mg(y - h)$. The important thing to remember is that our choice and the value we choose for c will not affect the motion of the cart.

For another example, consider a block attached to a spring. As discussed in Section 9.3, $F_x = -kx$, so

$$U(x) = -\int F_x\,dx = -\int(-kx)dx + c = \frac{1}{2}kx^2 + c,$$

where x is the amount by which the spring is squeezed or stretched. It seems natural to set the potential energy to be zero when the spring is completely relaxed. For this choice, $c = 0$, and $U(0) = 0$. Once again, the value of c fixes the location where $U = 0$. This location is ours to choose, and we can select the most convenient spot to match the problem at hand.

With the introduction of potential energy, we have a new term in our energy equation. But so far the analysis has been restricted to conservative forces, and even the humblest nonconservative force—friction—has been declared off-limits. Does this mean that energy is conserved only in unrealistic, idealized cases? Or, like Dennis' blocks, are there other places for energy to hide? We will see shortly.

9.7 A PICTURE WORTH A THOUSAND WORDS

Once we have found $U(\vec{r})$, we can follow the motion of the body *graphically*. This is a potent way to understand and predict the behavior of the system. Let's start with the case where gravity is the only force that does work: $U(\vec{r}) = U(y) = mgy + c$. This straight line function is plotted in Figure 9.8 for an arbitrarily chosen value of c. The force is the negative slope of $U(y)$, and is in the negative-y direction, in agreement with Equation 9.13: $F_y = -dU/dy = -mg$. If we know the kinetic energy K_0 of the body at some altitude y_0, then the body's total mechanical energy can be calculated:

$$E_{mech} = K_0 + U(y_0) = K_0 + mgy_0 + c.$$

Since $\vec{F}_{grav}$ is a conservative force, E_{mech} is constant, independent of position, and can be represented as a horizontal line on a graph of energy *vs.* altitude y. (Note that the horizontal axis of the graph corresponds to the *vertical* position of the body.) At altitude y, the kinetic energy $K(y) = E_{mech} - U(y)$ is equal to the separation between the lines drawn for E_{mech} and $U(y)$. This separation decreases for increasing y, indicating that the body decelerates as it rises. Because $K \propto v^2$, the kinetic energy cannot be negative, so the point where $U(y)$ and E_{mech} intersect (and $K = 0$) marks the maximum altitude y_{max} of the body. At y_{max}, the body reverses direction, from rising to falling, and y_{max} is commonly called a *turning point*. Graphs such as Figure 9.8 are called *potential energy diagrams*, and they encapsulate a great deal of information about the physics of a system. We will use them frequently to analyze the behavior of the systems we will study.

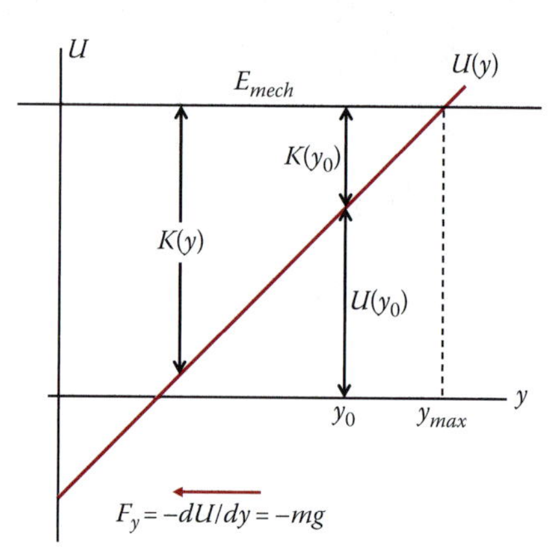

FIGURE 9.8 Potential energy diagram for $\vec{F} = -m\vec{g}$: $U(y) = mgy + c$. The kinetic energy at altitude y is found by taking the difference between E_{mech} and $U(y)$. The maximum altitude, or turning point y_{max}, is found where the lines for E_{mech} and $U(y)$ cross. The slope of $U(y)$ is the negative of the force.

EXERCISE 9.11

The potential $U(x)$ of a body is shown in the graph below. The body, initially at rest at $x = 2$ m, is given a gentle nudge to the left.

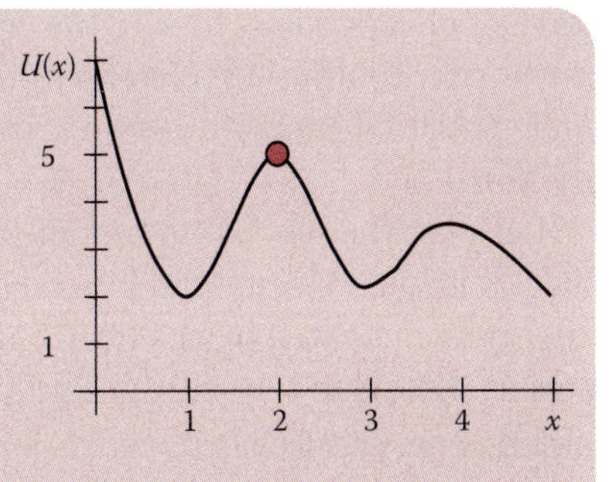

(a) Draw the line depicting E_{mech}.
(b) What is the kinetic energy at $x = 1$ m?
(c) In what direction is the force at $x = 0.5$ m?
(d) Where is the force equal to 0?
(e) What is the minimum value of x?

Answer (b) 3 J (c) ← (d) $x = 1, 2, 3, 4$ m (e) ≈ 0.25 m

9.8 POTENTIAL ENERGY AND SIMPLE HARMONIC MOTION

Oscillating systems are found in all scientific disciplines, and potential energy diagrams offer insight into their behavior. Let's return to our model oscillator, a block attached to an ideal spring sliding over a frictionless horizontal surface (Figure 9.9a). The potential energy is $U(x) = -\int -kx\,dx = \frac{1}{2}kx^2 + c$, where x is the block's displacement from equilibrium as well as the amount the spring is either stretched ($x < 0$) or compressed ($x > 0$). The corresponding potential energy diagram (with $c = 0$, shown in Figure 9.9b) is a parabola centered on the point $x = 0$. Note that the slope of $U(x)$ is zero at $x = 0$, indicating that the force is also zero at that point. For $x > 0$ (or $x < 0$), the slope is positive (or negative), confirming that $F_x = -dU/dx$ is a restoring

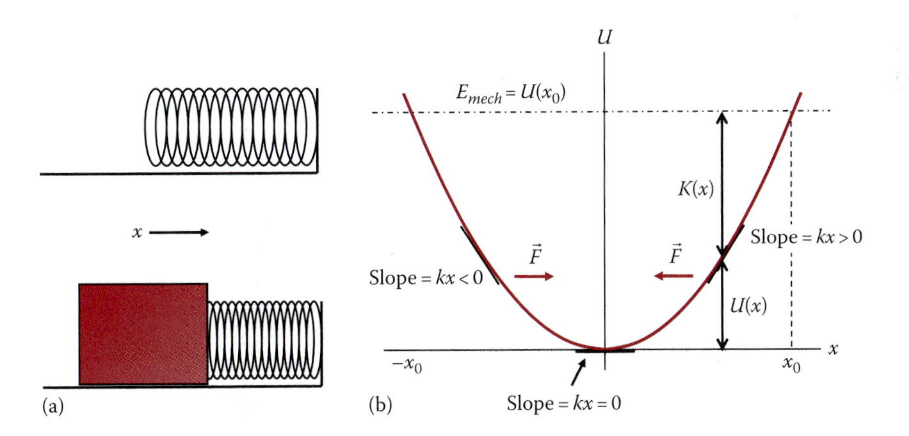

FIGURE 9.9 (a) A block attached to a spring moves on a horizontal frictionless surface. (b) The system's potential energy diagram $U(x) = (1/2)kx^2$. The turning points are located at $x = \pm x_0$. The force $F_x = -dU/dx$ always points toward the equilibrium point at $x = 0$ (where the slope $dU/dx = 0$).

force: it always points toward the equilibrium point $x = 0$. Finally, the curve $U(x)$ gets steeper as the displacement from equilibrium increases, just as the force increases linearly with distance from equilibrium. A *parabolic* potential energy function arises from a linear restoring force, and *is the signature of simple harmonic motion.*

Let's set the block into motion by pulling it aside by a distance x_0 and then releasing it from rest. The block's initial velocity and kinetic energy are zero, so its mechanical energy is $E_{mech} = U(x_0) = \frac{1}{2}kx_0^2$. This energy is indicated by the horizontal line in Figure 9.9b. After it is released, the block moves in the negative-x direction toward the equilibrium point at $x = 0$, where its potential energy is a minimum and its kinetic energy reaches its maximum value: $\frac{1}{2}mv_{max}^2 = \frac{1}{2}kx_0^2$. The block then continues beyond this point, and slows to zero velocity at $x = -x_0$, where $U(-x_0) = \frac{1}{2}kx_0^2 = E_{mech}$ and $K = 0$ once again. Reversing direction, it returns to $+x_0$, and repeats this cycle indefinitely. From the diagram, we can locate the turning points ($x = \pm x_0$) where $U(x)$ intersects with E_{mech}, the equilibrium position ($x = 0$) where $F_x = -dU/dx = 0$, and we can also find the speed of the block at any position, from $K = E_{mech} - U(x)$:

$$v = \sqrt{\frac{k(x_0^2 - x^2)}{m}}.$$

All of this can be extracted from the potential energy diagram with no direct reference to Newton's laws.

What if the surface is not horizontal, but tilted by an angle θ (Figure 9.10a)? Now gravity and the spring force do work on the block, so both contribute to the potential energy: $U(s) = \frac{1}{2}ks^2 + mgs\sin\theta + c$, where s is the compression of the spring and $s\sin\theta = \Delta y$ is the change in height of the block (Figure 9.10b). Note that $U(s)$ still has the quadratic form $U(s) = \alpha s^2 + \beta s + c$, so it is still a parabola, albeit not one centered on $x = 0$. Instead, it is centered on the new equilibrium point s_{eq}, located at the minimum of $U(s)$, or equivalently, the point where $F(s) = 0$. Using *no information other than the parabolic shape of the graph*, we can immediately infer that the block undergoes SHM centered on s_{eq}.

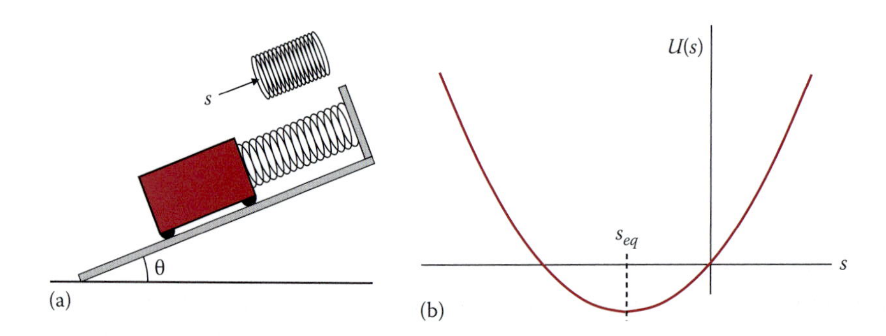

FIGURE 9.10 (a) If the block-spring oscillator surface is titled, the equilibrium point shifts to $s_{eq} < 0$, and the body undergoes SHM centered on s_{eq}. (b) The potential energy diagram for the tilted oscillator is still parabolic, the trademark of sinusoidal oscillation, but is offset from $s = 0$.

EXERCISE 9.12

A block of mass m = 0.5 kg is at rest, hanging vertically from a spring having a stiffness k = 20 N/m. The block is pulled downward by 10 cm and then released.

(a) Write an expression for $U(s)$, where s is the amount by which the spring is *compressed*.
(b) What is the equilibrium value of s?
(c) What can you say about the motion, based solely on the shape of the potential energy function $U(y)$?

Answer (a) $U(s) = \frac{1}{2}ks^2 + mgs + c$ (b) $s_{eq} \approx -25$ cm (c) *SHM of amplitude 10 cm about* s_{eq}

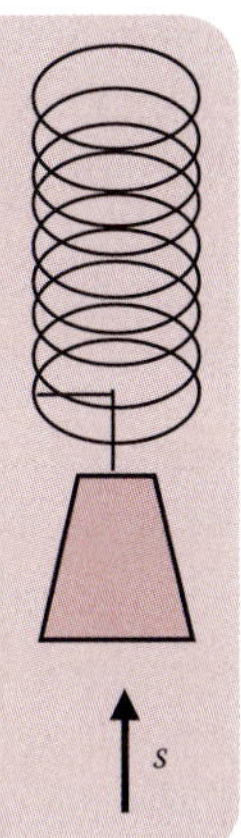

Finally, let's compare the present energy-based approach to SHM with the force-based approach taken in Chapter 7. Equation 7.12 gives the body's position as $x(t) = A\cos(\omega t + \theta_0)$, where $\omega = \sqrt{k/m}$, A is the amplitude of the motion, and θ_0 depends on the initial conditions. Differentiating to find the velocity, $v(t) = dx/dt = -\omega A \sin(\omega t + \theta_0)$. The kinetic and potential energies can now be expressed as functions of time:

$$K(t) = \frac{1}{2}mv^2 = \frac{1}{2}m\omega^2 A^2 \sin^2(\omega t + \theta_0)$$

and

$$U(t) = \frac{1}{2}kx^2 = \frac{1}{2}kA^2 \cos^2(\omega t + \theta_0).$$

Using $m\omega^2 = k$, the total mechanical energy is

$$\begin{aligned} K(t) + U(t) &= \frac{1}{2}kA^2(\sin^2(\omega t + \theta_0) + \cos^2(\omega t + \theta_0)) \\ &= \frac{1}{2}kA^2 = \text{constant} = E_{mech}. \end{aligned}$$

When $\cos(\omega t + \theta_0) = 1$, $\sin(\omega t + \theta_0) = 0$, and vice versa, the energy shuffles back and forth between kinetic and potential, while the total mechanical energy remains constant (see Figure 9.11). Note the striking contrast between the force approach and the energy approach: using force, we find the body's motion as a function of *time*; using potential energy, we describe the same motion vs. the body's *position*. Force is a vector, whereas energy is a scalar; scalars are easier to work with than vectors, and often the energy approach is easier, or the only tractable way to study a physical system.

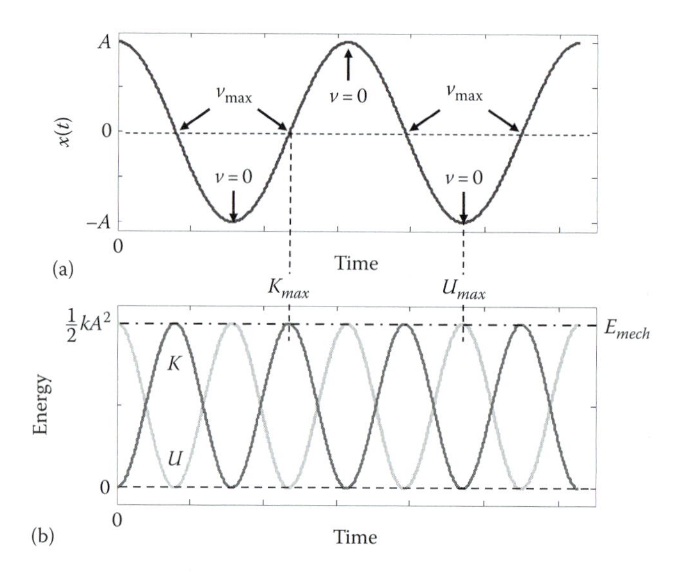

FIGURE 9.11 (a) Position vs. time for an oscillating body released from rest at point $x(0) = A$. (b) When the velocity $v(t) = \pm v_{max}$, the kinetic energy (dark gray curve) $K(t)$ is a maximum and the potential energy (light gray) $U(t) = 0$. When $v(t) = 0$, the opposite is true. At all times, $U(t) + V(t) = \text{const}$.

9.9 POTENTIAL ENERGY AND NONSINUSOIDAL OSCILLATIONS

When the restoring force is nonlinear, the potential energy is not a parabolic function, and the motion of the body is nonsinusoidal. Solving Newton's second law to find the position vs. time of the oscillating body may then be difficult or impossible, but with a potential energy diagram we can still learn a great deal about the motion. We can find the equilibrium point, the turning points, and the speed of the body as a function of position. We can also show that *all oscillating systems undergo simple harmonic motion in the limit of small amplitudes.* This is an exceedingly important result, because it allows us to derive the frequency of small oscillations in physical systems that are much more complicated and important than our ideal mass-spring system.

To illustrate this important point, let's take a closer look at the simple pendulum, a ball (mass m) attached to a string (length ℓ). See Figure 9.12a. The angular deflection of the string is φ, so the height of the ball above its lowest (equilibrium) position at $\varphi = 0$ is $y = \ell(1-\cos\varphi)$. Choosing $U(\varphi) = 0$ at $\varphi = 0$, the potential energy is $U(\varphi) = mgy = mgl(1 - \cos\varphi)$, which is decidedly *nonparabolic* in form (Figure 9.12b). Since $U(\varphi)$ is a continuous function at $\varphi = 0$, it can be approximated using a polynomial or "power series" expression:

$$U(\varphi) \simeq f_n(\varphi) = c_o + c_1\varphi + c_2\varphi^2 + c_3\varphi^3 + \cdots + c_n\varphi^n \tag{9.15}$$

where the coefficients $c_0, c_1, c_2,\ldots, c_n$ are constants to be determined. The polynomial $f_n(\varphi)$ is called a *Taylor series expansion* about the point $\varphi = 0$. Our task is to derive the values of $c_0, c_1, c_2,\ldots, c_n$ that

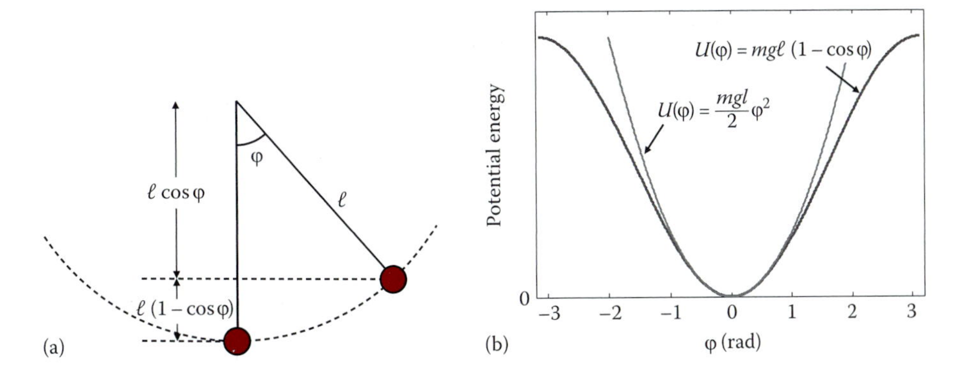

FIGURE 9.12 (a) A simple pendulum. The height y of the bob above its lowest point is given by $y = \ell(1-\cos\theta)$. (b) The potential energy function can be approximated by a parabola in the range $-1<\theta<1$ (rad). Therefore, oscillations with $\theta_{max}<1$ are nearly sinusoidal.

provide the best fit of $f_n(\varphi)$ to the real potential energy $U(\varphi)$. The optimum coefficients in Equation 9.15 are given by (see Box 9.1 for a complete derivation)

$$c_0 = U(0)$$

and for $m = 1, 2,\ldots, n$

$$c_m = \frac{1}{m!}\frac{d^m}{d\varphi^m}U(\varphi)\bigg|_{\varphi=0}$$

where the notation means that the derivatives are to be evaluated at $\varphi = 0$ and $m!$ ("m factorial") is defined as $m! = m\cdot(m-1)\cdot(m-2)\cdots 1$. (Don't be dismayed by the notation. Once you are familiar with the technique, these coefficients are not hard to calculate!) Using this recipe for the pendulum,

$$c_0 = U(0) = mgl(1-\cos 0) = mgl(1-1) = 0.$$

The next coefficient is equally easy to calculate:

$$c_1 = \frac{1}{1!}\frac{d}{d\varphi}mgl(1-\cos\varphi)\Big|_{\varphi=0} = mgl\sin\varphi\Big|_{\varphi=0} = 0.$$

When we are expanding the potential energy function about its equilibrium point, c_1 is *always equal to zero* because the first derivative of the function is proportional to the force, and the force is zero at the equilibrium point. This is a most important point, and the remainder of our discussion hinges upon it.

BOX 9.1 APPROXIMATING FUNCTIONS USING THE TAYLOR SERIES

It is often useful in physics to replace a function $y(x)$ with an approximate expression that closely matches the original function over a limited range of x. If this range of interest is centered on x_0, and if $y(x)$ is continuous there, then $y(x)$ can be approximated by a polynomial, or *power series*, in the form

$$f_n(x) = c_0 + c_1(x - x_0) + c_2(x - x_0)^2 + c_3(x - x_0)^3 + \cdots + c_n(x - x_0)^n \tag{B9.1}$$

where the coefficients $c_0, c_1, c_2, \ldots, c_n$ are chosen to optimize the agreement, or fit, between $f_n(x)$ and the original function $y(x)$. Intuitively, we expect that as n increases, the agreement between $f_n(x)$ and $y(x)$ will improve, and the range of x over which $f_n(x) \simeq y(x)$ will broaden. To illustrate, Figure B9.1 shows a function $y(x)$ in the range $0 \le x \le 2$. Let's find a power series approximation to $y(x)$ over a range of x centered on $x_0 = 1.4$. Crudely, we could replace $y(x)$ by a constant equal to its value at x_0. This *zeroth order* approximation is just the first term of the power series of Equation B9.1, $f_0(x) = c_0 = y(x_0)$, but it is a poor match to $y(x)$ except at the one point $x = x_0$. The *first order* expression $f_1(x) = c_0 + c_1(x - x_0)$ is better: By matching the value and the slope of $y(x)$ at x_0, $f_1(x)$ is reasonably accurate over the range $1.3 \le x \le 1.6$. A *second order* approximation $f_2(x)$ does even better: By matching the value, slope, and curvature of $y(x)$ at x_0, it yields a good fit over the wider range $0.8 \le x \le 1.9$. This suggests that, to find the optimum values for the coefficients $c_0, c_1, c_2, \ldots, c_n$, we should set $c_0 = y(x_0)$ as above, and for $n \ge 1$, equate the nth derivative of $y(x)$ with the nth derivative of $f_n(x)$ at the point x_0. Let's use this observation to find the coefficients c_n. Differentiating Equation B9.1 repeatedly, we find

$$f_n'(x) = c_1 + 2c_2(x - x_0) + 3c_3(x - x_0)^2 + \cdots + nc_n(x - x_0)^{n-1}$$

$$f_n''(x) = 2c_2 + 6c_3(x - x_0) + \cdots + n(n-1)c_n(x - x_0)^{n-2}$$

$$\vdots$$

$$f_n^{(n)}(x) = n(n-1)(n-2)\cdots 1c_n = n!c_n.$$

Next, evaluate each of the above derivatives at $x = x_0$,

$$f_n'(x_0) = c_1 \quad f_n''(x_0) = 2c_2 = 2!c_2$$

$$f_n'''(x_0) = 6c_3 = 3!c_3 \quad \ldots \quad f_n^{(n)}(x_0) = n!c_n,$$

where the factorial is defined as $n! = n(n - 1)(n - 2)\cdots 1$ (e.g., $4! = 4{\cdot}3{\cdot}2{\cdot}1 = 24$).

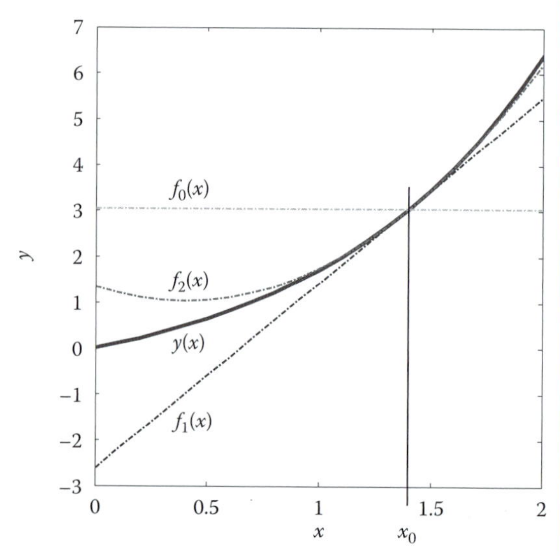

FIGURE B9.1 The nonpolynomial function $y(x)$ is approximated by a constant $f_0 = y(x_0)$, by a straight line $f_1(x) = f_0 + y'(x_0)(x - x_0)$, and by a parabola $f_2(x) = f_1(x) + \frac{1}{2}y''(x_0)(x - x_0)^2$. With each added term, the range about $x = x_0$ where the approximation is accurate grows larger.

For optimum fit, equate each of these derivatives with the corresponding derivative of the actual function $y(x)$ at the point x_0. For example, $f_n''(x_0) = 2!c_2 = (d^2y/dx^2)\big|_{x_0}$. Then, solving for the coefficients, we find

$$c_1 = \left.\frac{dy}{dx}\right|_{x_0} \quad c_2 = \left.\frac{1}{2!}\frac{d^2y}{dx^2}\right|_{x_0} \quad c_3 = \left.\frac{1}{3!}\frac{d^3y}{dx^3}\right|_{x_0} \quad \ldots \quad c_n = \left.\frac{1}{n!}\frac{d^ny}{dx^n}\right|_{x_0}.$$

Each of these expressions is a *constant*, proportional to a derivative of $y(x)$ evaluated at a fixed point x_0. The resulting polynomial using these coefficients is called the *Taylor series* expansion of $y(x)$ about x_0. Summarizing all of the above, the Taylor expansion can be expressed concisely as

$$y(x) \simeq f_n(x) = y(x_0) + \left.\frac{dy}{dx}\right|_{x_0}(x - x_0) + \left.\frac{1}{2!}\frac{d^2y}{dx^2}\right|_{x_0}(x - x_0)^2 + \cdots + \left.\frac{1}{n!}\frac{d^ny}{dx^n}\right|_{x_0}(x - x_0)^n. \quad \text{(B9.2)}$$

The number of terms we choose to keep depends on the accuracy and range required for the application we are working on.

In earlier chapters of this text, we used the small angle approximations $\sin\theta \simeq \theta$ and $\cos\theta \simeq 1$. It is easy to show that these are the lowest order terms of the Taylor series expansion for the sine and cosine. (Try it.) In addition, we have often used the binomial approximation $(1 + x)^n \simeq 1 + nx$. Writing this as a Taylor expansion, with $y(0) = (1 + 0)n = 1$,

$$\frac{d}{dx}(1+x)^n\Big|_0 = n(1+x)^{n-1}\Big|_0 = n \quad \text{and} \quad \frac{d^2}{dx^2}(1+x)^n\Big|_0 = n(n-1)(1+x)^{n-2}\Big|_0 = n(n-1)$$

so

$$(1+x)^n \simeq 1 + nx + \frac{1}{2!}n(n-1)x^2 + \frac{1}{3!}n(n-1)(n-2)x^3 + \cdots.$$

If $x \ll 1$, higher order terms are small relative to the first terms, and $y(x) \simeq 1 + nx$ is often sufficient.

Exercise B9.1: Find the Taylor series expansion about the point $x = 0$ for the exponential function $y(x) = e^x$. That is, find expressions for c_0 and c_n valid for all values of n. (Full disclosure: the function plotted in Figure B9.1 is $y(x) = e^x - 1$.) In many cases, $e^x \simeq 1 + x$ is a good enough approximation.

Answer: $c_0 = 1$, $c_n = 1/n!$

Not all of the coefficients are zero, of course. The next one is given by

$$c_2 = \frac{1}{2!}\frac{d^2}{d\varphi^2} mgl(1-\cos\varphi)\Big|_0 = \frac{mgl}{2}\cos\varphi\Big|_0 = \frac{mgl}{2},$$

and we leave it to you to derive the coefficients of higher order terms.*

EXERCISE 9.13

(a) Find the next two coefficients c_3 and c_4. (b) Satisfy yourself that the odd-numbered coefficients are all equal to 0, that is, $c_1 = c_3 = c_5 = \cdots = 0$.

Answer (a) $c_3 = 0$, $c_4 = -mgl/4! = -mgl/24$

The nonzero coefficients c_m become smaller as m increases, and for small values of φ, only the first few terms of the polynomial are necessary for a good approximation to $U(\varphi)$. For example, using the result of Exercise 9.11, $c_4/c_2 = 1/12$. If $\varphi_{max} < 0.5$ rad, the fourth order term $c_4\varphi^4$ makes at most a 2% $\left(\frac{1}{12}\times 0.5^2\right)$ correction to the potential energy. Higher order terms contribute even less. Hence, for the simple pendulum undergoing small deflection, the potential energy is approximately parabolic:

$$U(\varphi) \simeq c_2\varphi^2 = \frac{mgl}{2}\varphi^2$$

and the accuracy of the parabolic approximation improves as $|\varphi_{max}|$ gets smaller. As the amplitude of oscillation decreases, *the motion evolves into simple harmonic motion*: $\varphi(t) \simeq \varphi_{max}\cos(\omega t + \theta_0)$.

Moreover, this is true for *all* oscillating systems. Regardless of the exact force law obeyed by the oscillator, its potential energy can be approximated as

$$U(x) \simeq c_0 + c_2x^2 + (\text{higher order terms})$$

in the vicinity of its equilibrium point at $x = 0$. This follows because $c_1 = dU/dx = 0$ at the equilibrium point $x = 0$ and for small enough displacements x the higher order terms (c_3x^3, c_4x^4, ...) are much smaller† than second order term c_2x^2. By fitting a parabola to the potential energy function, we can also estimate the maximum amplitude for which SHM is a good approximation to the actual motion. Figure 9.12b suggests that the motion of a simple pendulum is nearly sinusoidal when $|\varphi_{max}| < 1$ (1 rad = 57.3°).

EXERCISE 9.14

The potential energy of a body of mass m is $U(x) = c_0 + c_2x^2 + c_3x^3$ in the vicinity of its equilibrium point.

(a) Show that $x = 0$ is the equilibrium point.
(b) Prove that the frequency of small oscillations is $\omega = \sqrt{2c_2/m}$.

* You may recognize that we are deriving the Taylor series expansion of the cosine: $\cos x = 1 - \frac{1}{2!}x^2 + \frac{1}{4!}x^4 + \cdots$

† The obvious exception is when $c_2 = 0$.

9.10 COLLISIONS AND ENERGY CONSERVATION

In 1668, the Royal Society of London issued a call for papers examining the physics of collisions. Christopher Wren and Christiaan Huygens described their studies of elastic collisions ($\Delta K_{total} = 0$), whereas John Wallis contributed his work on inelastic collisions. Together with Isaac Newton's later experiments, their work established the principle of momentum conservation, but also raised a puzzling question: Why was kinetic energy—or *vis viva* (live force), as it was then called—conserved in collisions between hard "elastic" objects, but not when softer bodies are used? An answer was proposed in 1686—not by Newton—but by another eminent physicist and mathematician, Gottfried Leibnitz, the coinventor of differential calculus: (In the following, read "kinetic energy" for "force.")

> [I maintain] that active forces are preserved in this world. The objection is that two soft or non-elastic bodies meeting together lose some of their force. I answer that this is not so. It is true that the wholes lose it in relation to their total movement, but the parts receive it, being internally agitated by the force of the meeting or shock... There is here no loss of forces, *but the same thing happens as takes place when big money is turned into small change.**

What a wonderful metaphor! In other words, some of the observable lab-scale motion of the colliding bodies (the "wholes") is converted into invisible microscopic motion of the "parts," meaning the atoms or molecules making up the two bodies. Here is a simple illustration.

Consider the interaction of the three carts shown in Figure 9.13a. The carts are all identical, each with mass m, and all roll without friction over a horizontal surface. Carts 2 and 3 are connected by an ideal spring, and are initially at rest. These two carts represent the internal "parts" of the larger, two-cart composite body. Cart 1, moving with speed v and kinetic energy $K = ½mv^2$, collides elastically with cart 2. As we learned in Chapter 5, cart 1 transfers all of its momentum to cart 2 at the moment of impact. So immediately after the collision, cart 1 is stationary, cart 2 is moving with speed v to the right, and cart 3 is still at rest (Figure 9.13b). Subsequently, the two coupled carts move in a complicated way, all the while conserving the initial momentum mv of cart 1. Imagine watching these two carts from a great distance: you are so far away that the two moving carts look like a single body of mass $2m$ moving to the right with constant velocity. This velocity is just their *CM* velocity: $V_{CM} = p_{tot}/2m = v/2$, that is, half the initial velocity of cart 1. From far away, the kinetic energy of the composite body appears to be $K' = ½2m(v/2)^2 = ¼mv^2 = ½K$. It looks like half of the initial kinetic energy has been lost, so the collision is inelastic. Now imagine a second observer, located at the *CM* of the two carts and moving with velocity V_{CM}. Immediately after the collision with cart 1, this observer measures the cart velocities to be $u_2' = v_2' - V_{CM} = v - v/2 = v/2$ and $u_3' = 0 - v/2 = -v/2$ (Figure 9.13c). To this observer, the two carts are undergoing pure SHM, with a total energy $½m(v/2)^2 + ½m(-v/2)^2 = ¼mv^2 = ½K$, exactly the quantity of kinetic energy that appeared to be lost in the collision! Overall, energy has been conserved, but some of the *observable* mechanical energy of cart 1 has been converted into *internal* energy—motion relative to the center of mass—that is unobservable

* From G. Leibnitz, *Philosophical Writings* (1686), edited by G.H.R. Parkinson, Dent, London, U.K., 1973, and quoted in J. Coopersmith, *Energy, the Subtle Concept,* Oxford University Press, 2010.

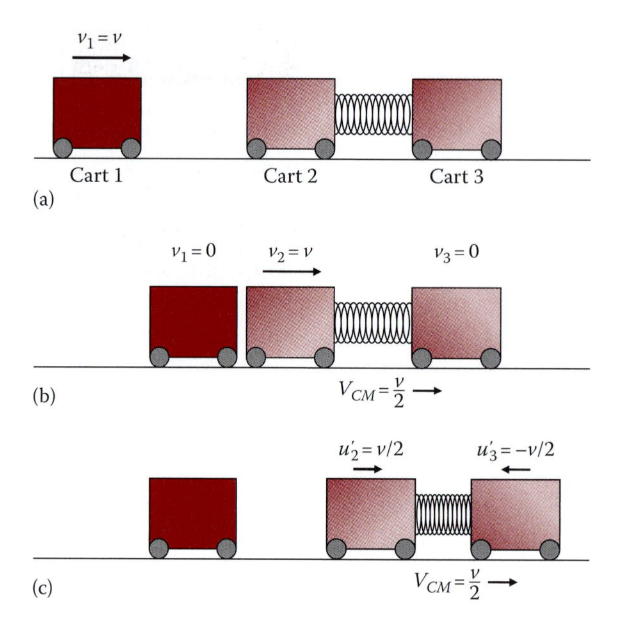

FIGURE 9.13 (a) Cart 1, moving with velocity v, approaches a stationary pair of identical carts coupled by a spring. (b) After colliding elastically with cart 2, cart 1 comes to a halt, transferring all of its momentum to cart 2. Cart 3 has not yet begun to move. The *CM* velocity of the two coupled carts is $v/2$. (c) For an observer in the *CM* frame, immediately after the collision carts 2 and 3 are moving toward each other with equal but opposite velocities $\pm v/2$.

from your remote observation point. The same thing occurs in collisions between real bodies. The observable kinetic energy is the energy associated with the body's *CM* motion, and the internal energy E_{int} is associated with the atomic-scale motion of molecules within the colliding bodies. This is how "big money is turned into small change." We must now find a way to include E_{int} in our energy equation.

9.11 NONCONSERVATIVE FORCES AND ENERGY

Up to now, we have studiously avoided nonconservative forces in our development of energy. This allowed us to introduce potential energy in a lucid way, but it fails to demonstrate the full scope of the energy principle. Nonconservative forces are the means by which mechanical energy is converted into other forms of energy.

A nonconservative force is one for which Equation 9.10 does not hold. The work done on a body moving between points i and f depends on the path taken, so $W_{i \to f} \neq U(\vec{r}_i) - U(\vec{r}_f)$. Friction is a nonconservative force. Since the friction force *always* opposes the displacement, $\vec{F}_{fr} \cdot d\vec{r} < 0$ always, so the work done over any closed path is always negative, $\oint \vec{F}_{fr} \cdot d\vec{r} < 0$. In the following discussion, we will use friction to illustrate the role of nonconservative forces in converting visible mechanical energy into other forms of energy that are not so easy to detect.

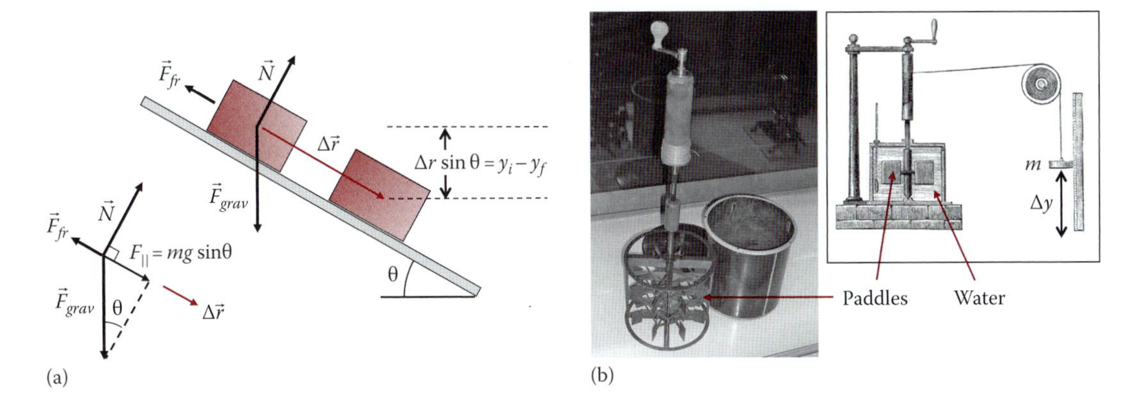

FIGURE 9.14 (a) The free body diagram for a block sliding *with* friction on a inclined surface. (b) The insulated bucket, or *calorimeter*, used by Joule to measure the mechanical equivalent of heat. Water in the bucket is agitated by the rotating paddlewheel, which is powered by a falling weight m. The mechanical energy transferred to the calorimeter is equal to $mg\Delta y$, and when compared to the rise in temperature, the relationship between mechanical energy and thermal energy (calories) is determined. (Images of Joule apparatus, public domain.)

Let's return to the familiar example of a body sliding down an inclined plane. Only this time, we will make it more realistic by including a friction force. Figure 9.14a is a copy of Figure 9.3a with the addition of $\vec{F}_{fr}$. Using the work–energy relation, the change in kinetic energy of the body as it slides a distance $\Delta\vec{r}$ down the inclined surface is

$$\Delta K = \vec{F} \cdot \Delta\vec{r} = \vec{F}_{grav} \cdot \Delta\vec{r} + \vec{F}_{fr} \cdot \Delta\vec{r}$$
$$= -mg\Delta y + \vec{F}_{fr} \cdot \Delta\vec{r}.$$

The gravitational work $-mg\Delta y$ depends only on the initial and final altitudes of the body, so this work term *is* path-independent and is still derivable from the potential energy function $U(y) = mgy + c$. The above equation can therefore be rewritten as

$$\Delta K = -\Delta U + \vec{F}_{fr} \cdot \Delta\vec{r} \quad \text{or} \quad \Delta K + \Delta U = \Delta E_{mech} = \vec{F}_{fr} \cdot \Delta\vec{r} \equiv W_{nc} \tag{9.16}$$

where W_{nc} is the work done by the nonconservative friction force. Since $\vec{F}_{fr} \cdot \Delta\vec{r} < 0$, the friction force reduces the system's mechanical energy. Before concluding that energy is not conserved, we might notice—using a sensitive thermometer—that the body is warmer than it was before it slid down the inclined plane. Can we associate this change of temperature with the loss of mechanical energy?

EXERCISE 9.15

A 1 kg block slides down a 30° inclined plane of height 0.5 m. A friction force F_{fr} = 1 N acts on the block. (Let g = 10 m/s^2.)

(a) How much work is done on the block by gravity as the body travels the length of the plane?
(b) How much mechanical energy is converted into "internal" energy?
(c) What is the change in potential energy?
(d) What is the change in kinetic energy?

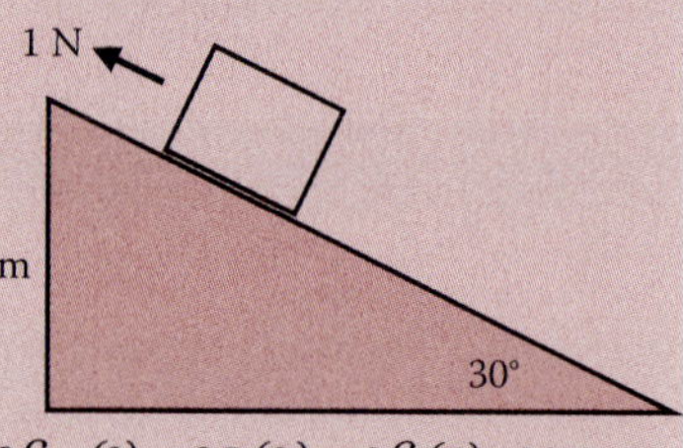

Answer (a) 5 J (b) 1 J (c) −5 J (d) +4 J

A definitive, experimentally verifiable answer to this question did not emerge until the mid-nineteenth century, more than 150 years after the publication of Newton's *Principia*. As is often the case, the story has a large cast of colorful characters. One of the leading roles was played by an American, Benjamin Thompson, whose loyalty to the British crown forced him to flee to England at the start of the American Revolution. Thompson, or Count Rumford, was a philanthropist, inventor, engineer, and notorious womanizer. In 1789, he was supervising the boring of cannon barrels in Munich. During the process, friction between the boring tool and the cannon barrel raised temperatures well above 100°C, requiring huge volumes of water to be used for cooling. At that time, it was supposed that heat was conveyed by an unseen fluid called "caloric," which passed from a hotter body to a cooler one when the two came into contact. Thompson observed that an unlimited amount of water could be boiled away as the boring continued, and concluded that this could not be due to the extraction of unlimited amounts of caloric from the boring bit, and endless flow of caloric to the cannon barrel and the water. Instead, he reasoned that the rise in *temperature must be associated with a form of microscopic motion induced by friction*. He communicated his observations to the Royal Society of London in 1798 in two papers that—even today—are fascinating to read for their scientific as well as historic content.* Forty years later, a young German physician, Julius Robert Mayer, signed on as ship's doctor for a voyage to Indonesia, where he observed that the blood in the veins of sailors he treated in the tropics was brighter red (indicating higher oxygen content) than it was in cooler climates. He attributed this difference to a lower rate of oxidation in warmer climates—a lower need for living bodies to stay warm by oxidation.† Upon returning to Germany, Mayer expanded his hypothesis into a sweeping statement of energy conservation:

> My position is perfectly definite. Gravitation, motion, heat, light, electricity, and chemical action are one and the same object in various forms of manifestation.

Mayer's papers were largely ignored, plunging him into deep despair and igniting a bitter dispute with supporters of the English physicist James Joule‡ (after whom the SI unit of energy is named). Both men set out to measure the *mechanical equivalent of heat*. This is the amount of mechanical work equivalent to 1 cal of

* *Philos. Trans. R. Soc. Lond.* **88**: 80–104 (1798) and *Phil. Trans. R. Soc. Lond.* **89**: 179–194 (1799).

† Modern medical practitioners would question Mayer's observations.

‡ Joule's father was a wealthy brewer, and Joule's Ale is still brewed in the United Kingdom.

thermal energy (the energy needed to raise the temperature of 1 cm^3 of water by 1°C). In exquisite experiments performed in 1844–1845, Joule measured the rise in temperature of a bucket of water churned by a paddlewheel driven by falling weights (see Figure 9.14b). The rise in temperature of the water was shown to be proportional to the reduction of the potential energy of the weights. Joule's result was within 1% of today's accepted value: 1 cal = 4.186 J. (Mayer's earlier result, published in 1842, was less accurate: 3.58 J/cal.) Following the death of two of his children in 1848, Mayer attempted suicide by leaping out of a third-story window (breaking both feet) and was admitted to a mental institution. In 1860, his groundbreaking work was finally recognized by leading physicists in Germany and England, and today he is rightfully regarded as the originator of the principle of energy conservation.

The experiments of Mayer and Joule linked a body's internal energy with its temperature. Nonconservative mechanical forces such as friction increase temperature—and internal energy—at the expense of mechanical energy: $W_{nc} = \Delta E_{mech} = -\Delta E_{int}$, so Equation 9.16 may be rewritten as $\Delta E_{mech} + \Delta E_{int} = 0$, or

$$E_{mech} + E_{int} = \text{constant}. \tag{9.17}$$

If we expand the definition of E_{int} to include *all* of the varieties of energy envisioned by Robert Mayer, plus others (magnetic, nuclear, etc.) we have subsequently discovered or have yet to discover, we are confident that Equation 9.17 remains valid. For each species of internal energy, another term must be added to our energy equation.

9.12 INELASTIC COLLISIONS AND WORK

At the beginning of this chapter, we remarked that it is often difficult to identify what we mean by the "work done" on a body. Here is an example of the problem. Imagine the totally inelastic collision between a stationary block of mass M and a high speed bullet of mass $m \ll M$ (see Figure 9.15a). After the collision, the bullet is lodged within the block, which subsequently slides over a friction-free surface with speed V. From momentum conservation, $(m + M)V = mv$, or $V \simeq (m/M)v$ (Figure 9.15b). The final kinetic energy is but a small fraction of the initial kinetic energy $\frac{1}{2}mv^2$:

$$K_f \simeq \frac{1}{2}MV^2 = \frac{m}{M}K_i,$$

but the block will feel warm to the touch. Clearly, much of the bullet's initial mechanical energy has been converted into internal energy.

By the work-kinetic energy theorem, $K_f = \bar{F}\Delta X$, where $\bar{F}$ is the average force acting on the block during the time interval between the initial impact of the bullet and when the bullet comes to rest within the block. ΔX is the displacement of the *block* during that time. By Newton's third law, the block exerts a reaction force $-\bar{F}$ on the bullet, reducing its kinetic energy to near zero (since $V \ll v$). Invoking the work-kinetic energy theorem again, $-\bar{F}\Delta x \simeq -K_i$, where Δx is the distance traveled by the *bullet* while it slows to rest within the block. Since $K_i/K_f = M/m \gg 1$, the two displacements are quite different: $\Delta x/\Delta X = M/m \gg 1$, and one must be careful about which distance to use when calculating work. Let's look first at the motion of the

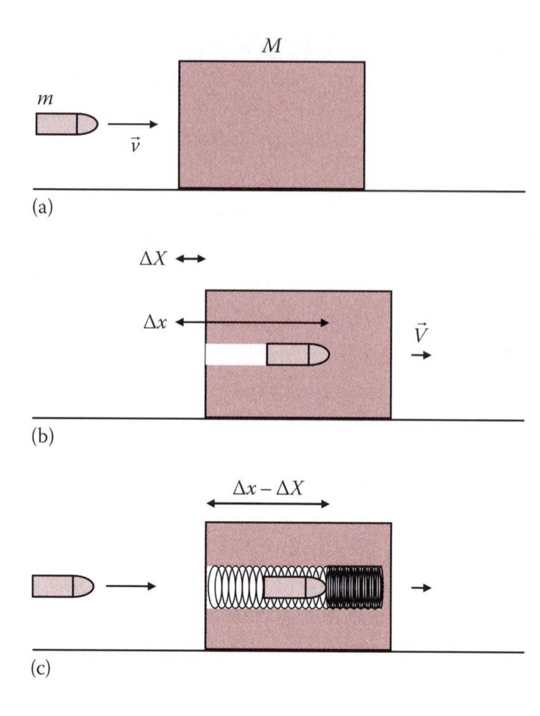

FIGURE 9.15 (a) A high velocity bullet m is fired into a stationary block M. (b) When the bullet is lodged within the block, conservation of momentum requires that the final velocity is $V = mv/(m + M) \simeq (m/M)v$. The displacement of the bullet Δx is much greater than the displacement of the block ΔX while the bullet is decelerating. (c) If the bullet is stopped by a spring, the amount of compression $s = \Delta x - \Delta X$ and the average force during this time is $(F_{final} + F_{initial})/2 = ½ks$.

block. K_f is the kinetic energy associated with the block's CM motion: $K_f = \Delta K_{CM}$, and ΔX is the displacement of the block's CM during the collision: $\Delta X = \Delta X_{CM}$, so

$$\bar{F}\Delta X_{CM} = \Delta K_{CM}. \tag{9.18}$$

Although we have derived this result for a highly specialized case, Equation 9.18 turns out to be generally true, and $\bar{F}\Delta X_{CM}$ is sometimes referred to as the "center of mass work."

The total "work" done on the block by the bullet is $\bar{F}\Delta x$, and this must be equal to the block's kinetic energy plus the internal energy associated with the heating and deformation of the block and bullet: $\bar{F}\Delta x = K_f + E_{int}$, or

$$E_{int} = \bar{F}(\Delta x - \Delta X).$$

For an explicit example, suppose the bullet comes to rest within the block by squeezing a spring of stiffness k (Figure 9.15c). Then $\Delta x - \Delta X \equiv s$ is the compression of the spring, and the average force is $\bar{F} = ½ks$, so $E_{int} = ½ks^2$, the usual equation for the energy stored within a compressed spring.

9.13 ENERGY AND SYSTEMS

In classical mechanics, we are generally interested in the behavior of a small set of interacting bodies which we will call a *system*. If these bodies interact only with one another (not with any external bodies), the system is said to be *isolated*. Within an isolated system, energy may exist in many forms: mechanical (macroscopic kinetic and potential energies) and internal (microscopic kinetic and potential energies associated with thermal motion; molecular, atomic and nuclear structure, etc.). As interactions take place, energy is exchanged between bodies, or is converted from one form to another, by a variety of processes: heat transfer, chemical or nuclear reactions, electric currents, electromagnetic or mechanical waves, or by mechanical work $W = \int \vec{F} \cdot d\vec{r}$. In the following discussion, we will use mechanical work exclusively to illustrate how energy is transformed. Conservative work converts kinetic into potential energy with no net change of mechanical energy: $\Delta E_{mech} = \Delta(K_{tot} + U_{tot}) = 0$. Nonconservative work transforms mechanical energy into internal energy, or vice versa, $\Delta E_{int} = -\Delta E_{mech}$. Overall, the total system energy remains constant: $\Delta(E_{mech} + E_{int}) = 0$. Earlier, when we were studying a block (or ball) on an inclined ramp, our system of interacting bodies (the block, ramp, and Earth) was treated as an isolated system. (Recall that we ignored interactions with the atmosphere, that is, air drag.) When we examined the mass-spring oscillator or the pendulum, our system was the mass and spring, or the pendulum bob and the Earth. In each case, the system was treated as an isolated system, and total energy was conserved.

If the system is *not* isolated, one or more external bodies may interact with it, and energy may be transferred into or out of the system. In Figure 9.16, a block m slides *with friction* over an inclined ramp. The block, ramp, and Earth form an isolated system with total energy $\frac{1}{2}mv^2 + mgy + E_{int} = \text{const}$ (Figure 9.16a). If the block is tied to a string, and is pulled up the ramp by the string tension $\vec{F}_{ten}$ (Figure 9.16b), the system is no longer isolated and its total energy will change. Newton's second law for the uphill acceleration is $ma = F_{ten} - mg\sin\theta - F_{fr}$. Multiplying by the block displacement Δs, and using $ma\Delta s = \Delta\left(\frac{1}{2}mv^2\right)$, we find

$$\Delta\left(\frac{1}{2}mv^2 + mgy\right) + F_{fr}\Delta s = F_{ten}\Delta s$$

or, since $F_{fr}\Delta s = -\vec{F}_{fr} \cdot \Delta\vec{r} = \Delta E_{int}$,

$$\Delta(E_{mech} + E_{int}) = F_{ten}\Delta s \equiv W_{ext}.$$

Because $\vec{F}_{ten}$ is applied by an external agent, we say that $\vec{F}_{ten}\Delta\vec{s} = W_{ext}$ is the *external* work done on the system to change its energy. This does not violate the principle of energy conservation. Suppose F_{ten} is maintained by a hanging mass m' (Figure 9.16c). If the system is expanded to include m', then $K_{tot} = \frac{1}{2}(m + m')v^2$ and $U_{tot} = mgy + m'gy'$, where y' is the altitude of m'. The total energy of the expanded system (m, m', the ramp and the Earth) is conserved.

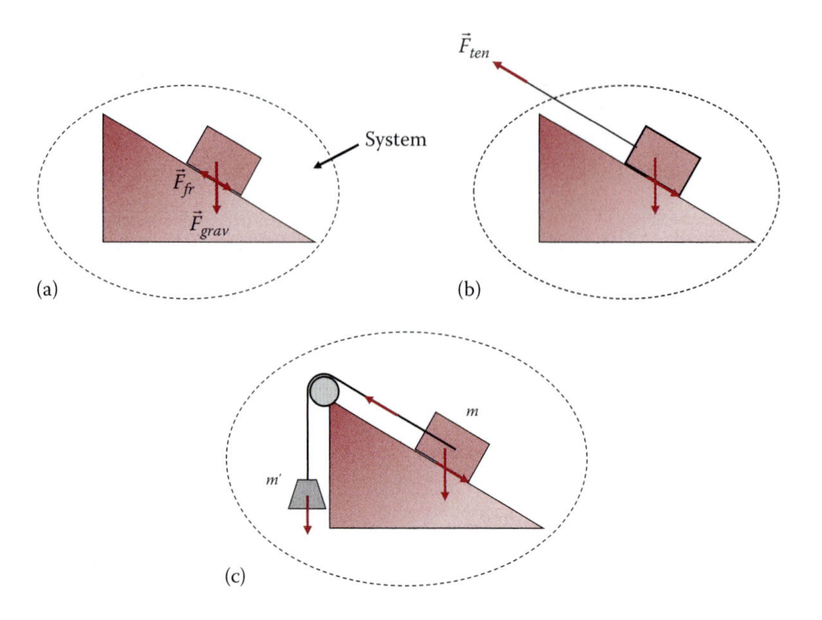

FIGURE 9.16 (a) The block, ramp, and Earth form an isolated system. (b) External work is done via the string tension force $\vec{F}_{ten}$. The system is not isolated. (c) $\vec{F}_{ten}$ is maintained by a hanging mass m'. If m' is included in the system, it is once again isolated.

EXERCISE 9.16

A 3 kg block rests at the bottom of a 30° ramp that is 1 m in length. The block is attached to a string that passes over a lightweight pulley and is tied to a hanging mass m. This mass is just large enough to pull the block up the ramp with little acceleration (so $K_{tot} \approx 0$). When the block is released, it experiences a sliding friction force $F_{fr} = 2$ N.

(a) Calculate m. (Use $g = 10$ m/s².)
(b) By how much is the potential energy of the block–Earth system increased when the block reaches the top of the ramp?
(c) By how much has the internal energy of the block–Earth system increased?
(d) Show that the total energy of the block–Earth-hanging mass is conserved.

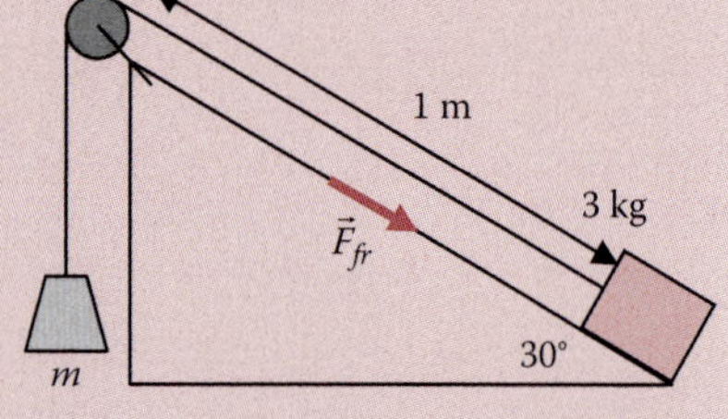

Answer (a) $m = 1.7$ kg (b) $\Delta U = 15$ J (c) $\Delta E_{int} = 2$ J

In ordinary matter, atomic nuclei are buried deep within their host atoms, and each nucleus (other than ¹H) is a near-perfect example of an isolated physical system: interactions with surrounding electrons are millions of times weaker than interactions between the protons and neutrons within the nucleus. Moreover, because of the equivalence between mass and energy, along with the development of modern mass spectrometry, we can now measure energy (mass) with astounding accuracy. This ability allows us to test the principle of energy conservation in minute detail, and—with the exception of some anxious moments in the 1930s—scientists are steadfastly convinced that the conservation law is universally valid. Let's take a look at those anxious moments.

9.14 APPLICATION: BETA DECAY AND THE DISCOVERY OF THE NEUTRINO

The early decades of the twentieth century saw an unprecedented surge in our understanding of the physical universe. Following the discovery of radioactivity in 1896, Planck proposed the *quantization* of energy in 1900; Einstein introduced his special theory of relativity in 1905, and in the late 1920s, quantum mechanics was developed. With each new development, the pillars of classical Newtonian physics seemed to crumble: space and time were not absolute, energy and mass were equivalent, particles behaved like waves and waves like particles, and certain (radioactive) atoms spontaneously changed their chemical identity. Would the most fundamental laws of physics—the conservation laws of momentum and energy—survive this onslaught of new discovery? The answer wasn't clear. One process in particular—beta decay—seemed to defy the conservation laws, triggering a crisis in the physics community.

Beta (β-) decay occurs when a radioactive nucleus changes its chemical identity by converting a neutron into a proton and ejecting a high-energy electron: $n \to p + e^-$. For example, the radioactive nucleus of isotope phosphorus-32 (^{32}P) contains 15 protons and 17 neutrons. It transforms by β-decay to a sulphur-32 (^{32}S) nucleus, with 16 protons and 16 neutrons: one of its 17 neutrons is converted into a proton, and an electron is ejected with kinetic energy K. Energy conservation and Einstein's mass–energy relation ($E = mc^2$) imply that the kinetic energy of the decay products is related to the change in mass:

$$E = M_P c^2 = (M_S + m_e)c^2 + K_e \quad \text{or} \quad K_e = \Delta M c^2 = [M_P - (M_S + m_e)]c^2, \tag{9.19}$$

where the subscripts stand for the sulfur (S) ion, the phosphorus (P) atom, and the electron (e). (The kinetic energy of the recoiling sulfur ion is negligible, a consequence of momentum conservation.) But β-decay experiments show that the electrons are not all emitted with the same kinetic energy! Instead, their energies are distributed over a continuous range, with a maximum equal to the value given in Equation 9.19: $K_e \le \Delta M c^2$. Worse, the electron momentum $\vec{p}_e$ is not antiparallel to the momentum of the recoiling sulfur ion, so the final total momentum $\vec{p}_S + \vec{p}_e \neq 0$ is not equal to the initial momentum of the phosphorus atom ($\vec{p}_P = 0$) (see Figure 9.17a). In 1930, it appeared that neither energy nor momentum is conserved in β-decay!

This observation was profoundly disturbing to physicists. Niels Bohr, the preeminent physicist of his day, was initially willing to accept that the conservation laws were violated in β-decay. In response, Wolfgang Pauli proposed a "desperate remedy," suggesting that the emitted electron was accompanied by a hitherto unknown and undetected particle of very low mass and zero electric charge. The conservation laws might be saved by including the energy and momentum carried away by this phantom particle—dubbed

BOX 9.2 ATOMIC AND NUCLEAR ENERGY UNITS

Albert Einstein's statement of mass–energy equivalence allows us to make a one-to-one correspondence between mass and energy: The rest mass of a body can be expressed—not in terms of kilograms—but as an equivalent amount of energy: $E = mc^2$. Since atoms and nuclei are so tiny, it is convenient to express their masses and equivalent energies in terms of appropriately scaled units. The *unified atomic mass unit* (u) is defined as one twelfth the mass of a carbon-12 atom,

$$1\,\text{u} = \frac{1}{12} m(^{12}\text{C}) = 1.660539... \times 10^{-27}\ \text{kg},$$

where the dots indicate additional significant figures. Clearly, 1 u is about the mass of a single proton or neutron. More exactly, $m_p = 1.007276...\text{u}$ and $m_n = 1.008665...\text{u}$. The atomic-scale unit of energy is called the *electron volt* (eV), and its value is based on the electric charge e of an electron or proton: $e = 1.602177... \times 10^{-19}$ coulombs:

$$1\,\text{eV} = 1.602177... \times 10^{-19}\ \text{J}.$$

The energy equivalent of 1 u is found by multiplying the mass in kg by $c^2 = (2.99792 \times 10^8\ \text{m/s})^2$,

$$E(1\,\text{u}) = 1.492413 \times 10^{-10}\ \text{J} = 931.491 \times 10^6\ \text{eV}$$
$$= 931.491\ \text{MeV} = 0.931491\ \text{GeV}.$$

In these units, the equivalent energy of a proton is $m_pc^2 = 938.268\,\text{MeV}$ and that of a neutron is $m_nc^2 = 939.562\,\text{MeV}$. For rough approximations, $m_pc^2 \approx m_nc^2 \approx 1\,\text{GeV}$.

The total energy of a free particle (i.e., in a force-free region of space) is given by $E = \gamma mc^2 = mc^2 + K$, where $\gamma = 1/\sqrt{1 - v^2/c^2}$ is the Lorentz factor and K is the particle's kinetic energy. The momentum $\vec{p} = \gamma m\vec{v}$ can be expressed in the same energy units by multiplying $\vec{p}$ by c. For example, if a proton is traveling at $\frac{4}{5}c$, $\gamma = 1/\sqrt{1 - \left(\frac{4}{5}\right)^2} = \frac{5}{3}$, and $pc = \frac{5}{3} m_p \times \frac{4}{5} c^2 = \frac{4}{3} m_p c^2 = 1251.02\,\text{MeV}$, while its kinetic energy is $K = E - m_pc^2 = \left(\frac{5}{3} - 1\right) m_p c^2 = 625.51\,\text{MeV}$. It takes a little while to get used to these energy units, but they are incredibly useful when dealing with atomic and subatomic processes.

Exercise B9.2: (a) The mass of an electron is $9.10938... \times 10^{-31}$ kg. Express its mass in appropriate energy units. (b) The neutrino is known to have a mass <2.2 eV. Compare its mass to that of an electron.

Answer (a) 0.510999 MeV (b) $\approx 4.3 \times 10^{-6}\, m_e$

Finally, let's use these results to find the energy and mass of the neutrino. By energy conservation, $E_\nu = 1.71 - 0.58 = 1.13\,\text{MeV}$, with an experimental accuracy of about ±10%. Since $p_\nu c = 1.17\,\text{MeV}$, the data suggests that $E_\nu = p_\nu c$, that is, the neutrino is a massless particle that, therefore, moves with speed c.* Pauli's desperate remedy was right. Energy and momentum are conserved in β-decay.

In 1956, Frederick Reines and Clyde Cowan reported *direct* detection of neutrinos by observing the *inverse* β-decay reaction $\bar{\nu} + p \rightarrow n + e^+$, where $\bar{\nu}$ denotes an antineutrino and e^+ is a positron. The antineutrinos were produced in copious amounts by reactors at nuclear weapons laboratories in Washington and South Carolina. Nearly 40 years later, Reines was awarded the Nobel Prize for this achievement. Paraphrasing Reines, "The poltergeist† of modern physics had been caught."

EXERCISE 9.17

True or false: In spontaneous radioactive decay, the total mass of the decay products is always less than the mass of the radioactive nucleus.

---

* Today, the electron neutrino and antineutrino are known to have nonzero rest mass energies $m_\nu c^2 < 2.2\,\text{eV}$.

† A poltergeist is a ghost that makes its presence known by making noise and moving objects about.

9.15 SUMMARY

> In the world of human thought generally, and in physical science particularly, the most important and fruitful concepts are those to which it is impossible to attach a well-defined meaning.
>
> **H.A. Kramers (1947)**

Energy conservation is the most important topic of this book. Unlike momentum and angular momentum (which we will soon study), energy plays a vital role in *all* fields of science (including the life sciences), and conservation of energy is the principle that unifies all of science. Given its ubiquitous importance, it is surprising that a clear definition of energy does not exist. We can define kinetic energy precisely, but what is the common attribute of other entities—mass, heat, compression, and position (potential energy)—that identifies them all as "energy?" In the early seventeenth century, Galileo glimpsed the idea of conservation of energy in his inclined track and pendula experiments. But the protean nature of energy eluded even the great Newton and was not fully appreciated until the mid-nineteenth century experiments of Robert Mayer and James Joule.

In this chapter, energy was introduced to Newton's force-centered mechanics via the work-kinetic energy theorem: $\Delta K = W_{i\rightarrow f} = \int_i^f \vec{F}\cdot d\vec{r}$. If the force $\vec{F}$ is a conservative force, the work done between any two points i and f is path independent, and a potential energy function $U(\vec{r})$ can be associated with $\vec{F}$: $U(\vec{r}_f) - U(\vec{r}_i) = \Delta U = -\int_i^f \vec{F}\cdot d\vec{r} = -W_{i\rightarrow f}$, so that $\Delta K + \Delta U = 0$, or $E_{mech} = K + U = \text{constant}$. If $U(\vec{r})$ is known, the force can be found by differentiation. In the simplest case where the potential energy is a function of only one coordinate, for example, $U(\vec{r}) = U(y)$, then $F_y = -dU/dy$. (Otherwise, we must use partial derivatives to express the force.) Potential energy is one of the most useful concepts in physics. A potential energy diagram allows us to analyze the motion of a body as a function of its position rather than of time. This is a powerful strategy and is often the only way to attack a problem.

In this chapter, we studied two conservative forces: the gravitational force $\vec{F}_{grav} = -mg\hat{j}$ and the spring force $\vec{F} = -kx\hat{i}$, and derived their respective potential energy functions: $U(y) = mgy + c$ and $U(x) = \frac{1}{2}kx^2 + c$, where the integration constant c can be chosen freely. Any location in space where $U(\vec{r})$ is a minimum is called an equilibrium point. In the vicinity of an equilibrium point, the potential energy can almost always be approximated by a parabolic function, $U(x) = c_0 + c_2x^2$, implying that for small oscillations about equilibrium, the motion is simple harmonic, or sinusoidal.

Nonconservative forces complicate matters by transforming mechanical energy into other forms of energy, which we lumped together and called "internal" energy. For example, friction between two sliding bodies warms the bodies, raising their internal energy at the expense of their visible mechanical energy. No potential energy function exists for these forces. Using nonmechanical instruments such as thermometers, electrical meters, and radiation detectors, we can account for other forms of energy, and in all cases, ranging from picosecond subatomic processes to the evolution of galaxies over billions of years, we find that total energy is conserved. With each new form of energy that we discover, we add a new and perhaps more esoteric term to the energy conservation equation.

But have we found all of the places that nature hides energy? Far from it! In 2011, three American-born physicists, Saul Perlmutter, Brian Schmidt, and Adam Riess shared the Nobel Prize for discovering that the Hubble expansion of the universe (Chapter 1) is *speeding up*; that is, distant galaxies are *accelerating* away from each other. It is as if space were filled with an unknown form of energy—*dark energy*—that overpowers the gravitational attraction galaxies feel for each other. We have no idea what dark energy is, but it accounts for 70% of everything there is in the universe. *Dark matter*, another mysterious substance, makes up another 25% of the contents of the universe. Ordinary baryonic matter (made of atoms and such), the stuff that we can see, makes up only 5% of the cosmos. Ironically, it seems like the bigger and better we build our telescopes, the less we know about the universe! Physicists place great faith in conservation laws. But science is not based on faith. To study nature with an open mind, we must be ready to abandon our most cherished beliefs if they are at odds with experimental evidence. In 1933, the neutrino saved the conservation laws, but there is always the possibility that they will again be challenged by new discoveries.

In his Nobel acceptance speech, Saul Perlmutter described the scientific exploration of the universe as an exhilarating activity joining humans "in teams from around the world, and in time across civilizations." By participating, he said, "We feel what it is like to be human."

> Perhaps the only thing better for a scientist than finding the crucial piece of a puzzle that completes a picture is finding a piece that doesn't fit at all, [which] tells us that there is a whole new part of the puzzle that we haven't even imagined yet, and the scene in the puzzle is bigger, richer than we ever thought.
>
> **Saul Perlmutter**
> *Stockholm, December, 2011*

PROBLEMS

9.1 A 1.0 kg mass on a horizontal frictionless surface is connected by a string to a hanging 2.0 kg mass. The mass is released.

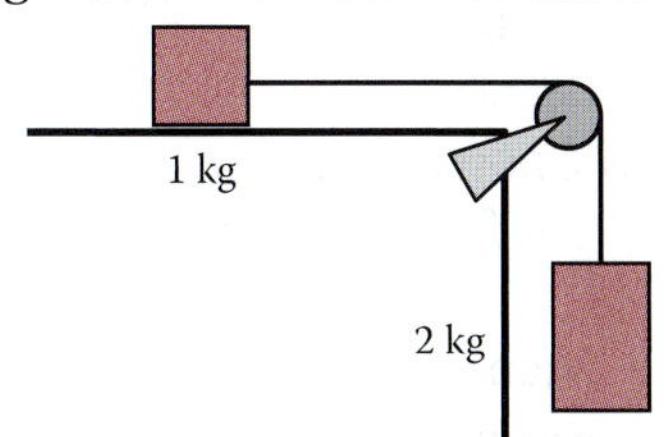

(a) When the bodies have moved 50 cm, what is the change in their potential energy?

(b) What is the speed of either body at this time?

9.2 A 2 kg projectile is launched from a "spring cannon" aimed 60° above the horizon. The initial stored energy in the spring is 400 J.

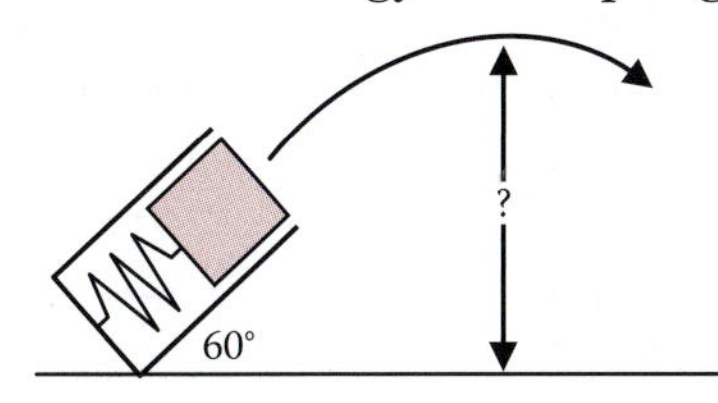

(a) What is the launch speed of the projectile?

(b) What is its maximum height? (Careful! Is the speed zero at this point?)

9.3 A block slides down the frictionless track shown in the figure. At point A, the block is not in contact with the track. What is the minimum height h_1 (in terms of r)?

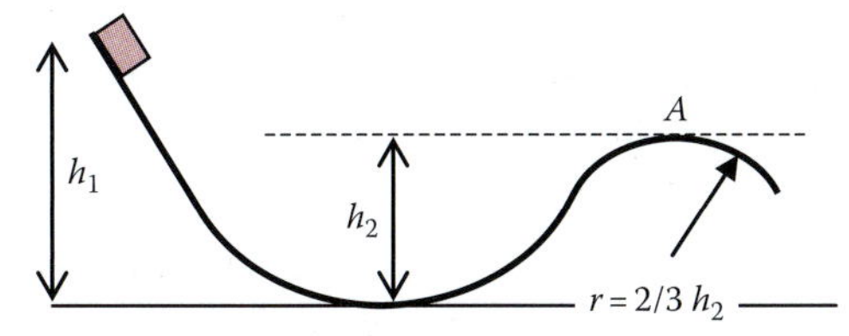

9.4 A block of mass $m_1 = 2.0$ kg slides along a frictionless table with a speed 10 m/s. Directly in front of it is a block of mass $m_2 = 5.0$ kg moving in the same direction with speed 3.0 m/s. A massless spring of spring constant $k = 1120$ N/m is attached to m_2. Find the maximum compression of the spring after the blocks collide. (*Hint:* What two quantities are conserved?)

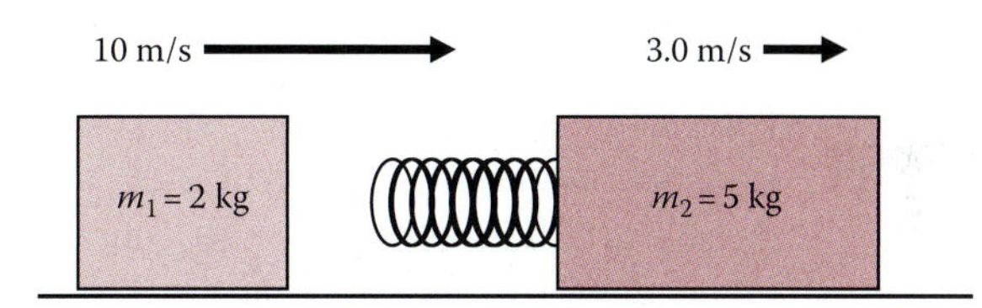

9.5 A ball of mass m is attached to the end of a lightweight rod of length L. The other end of the rod rotates on a frictionless bearing so that the ball can move in a vertical circle. The rod is released from the horizontal position shown with just enough of a velocity that it pivots around and just reaches the vertically upward position with zero velocity.

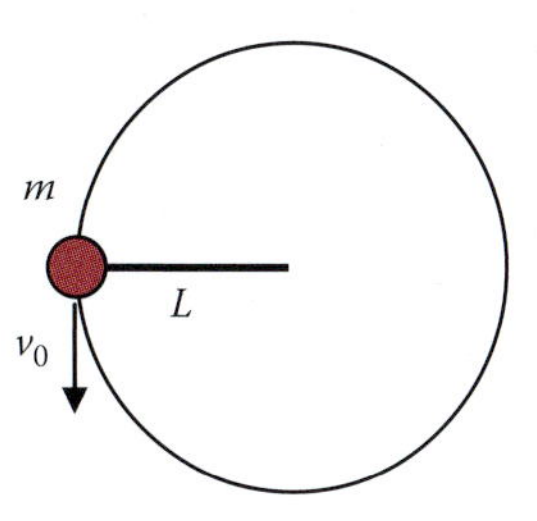

(a) What is the overall change in potential energy?

(b) What was the initial speed of the ball?

(c) What is the maximum speed of the ball?

9.6 On the Earth's surface, a pendulum and a mass-spring oscillator have the same period of oscillation. The pendulum consists of a 2 kg mass hanging from a cord of length ℓ = 1 m. The spring constant of the other system is k = 10 N/m.

(a) Find the mass m attached to the spring.

(b) Mass m is displaced 10 cm from its equilibrium position and then released. The pendulum bob is similarly displaced sideways by 10 cm and released. Which mass has the higher maximum velocity? Calculate this velocity.

(c) For small displacements, the pendulum bob undergoes SHM, that is, $x(t) = A\cos(\omega t)$. Choose one of the following answers for the potential energy function $U(x)$ (for small x).

- $U(x) = mgx^2/2\ell$
- $U(x) = -mg\ell^2/2x$.
- $U(x) = mgx$.
- $U(x) = mgx^2$
- None of the above

9.7 A massless spring of stiffness k = 15 N/m hangs vertically. Without stretching or compressing the spring, a body of mass m = 0.25 kg is attached to its free end and then released.

(a) How far below the initial position does the body descend? (Hint: Where is the equilibrium position?)

(b) What is the amplitude of the resulting oscillation?

(c) What is the period of the motion?

9.8 A particle of mass 2 kg moves along the x-axis through a region in which its potential energy $U(x)$ varies as shown in the figure. When the particle is at x = 3 m, its velocity is −2 m/s.

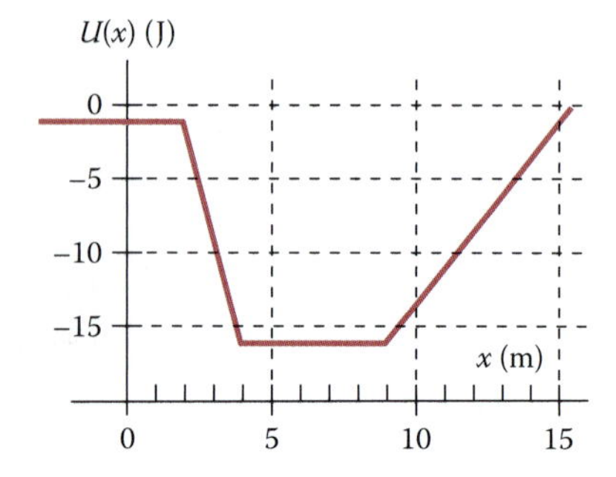

(a) What force acts on it at this position?

(b) Between what limits does the motion take place?

(c) How fast is it moving at x = 6 m?

9.9 The figure shows the potential energy of a 2 kg body as a function of position x.

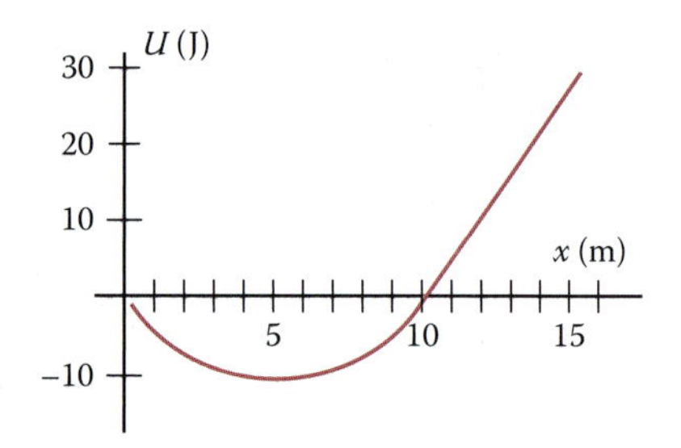

(a) Where is the force F = 0?

(b) Estimate the force at x = 12 m. What is the direction of the force?

(c) If the body is released from rest at x = 12 m, what is its maximum speed in the range $0 \le x \le 12$ m?

(d) When x = 5, the body is moving to the right with speed 6 m/s. What is the maximum possible value of x?

9.10 A body experiences a force $F(x) = 6 - 2x$ Newtons.

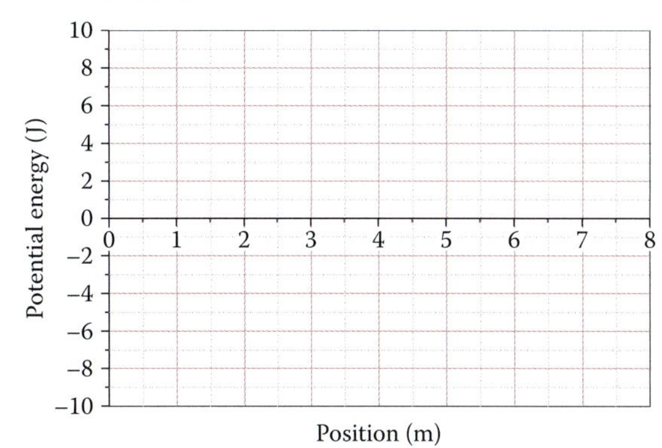

(a) Derive an expression for the potential energy function $U(x)$ for $0 < x < 7$ m. Let $U(0) = 1$ J.

(b) Plot your function on a graph like the one shown above.

(c) Check that your function $U(x)$ is consistent with the force. Where does the force equal zero? In what region is the force in the positive (negative) direction?

(d) If the body is released from rest at $x = 0$, what is its total mechanical energy? What is its maximum kinetic energy?

(e) Describe the motion.

9.11 A body of mass $m = 2$ kg moves along the x-axis through a region in which its potential energy $U(x)$ varies as shown in the figure. When the body is at $x = 10$ m, its velocity is -3 m/s.

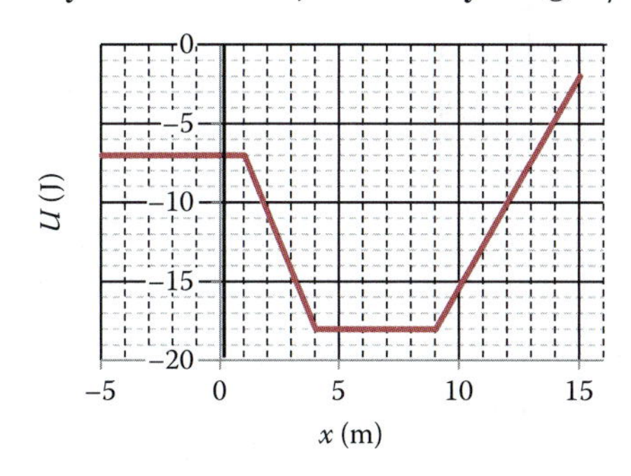

(a) Where does the force on m have its highest magnitude? How do you know?

(b) What force acts on m at $x = 12$ m?

(c) What is the total mechanical energy of the body?

(d) Between what limits is the motion confined?

(e) What is the maximum speed of the body?

9.12 The potential energy of a 2.0 kg body is shown in the figure.

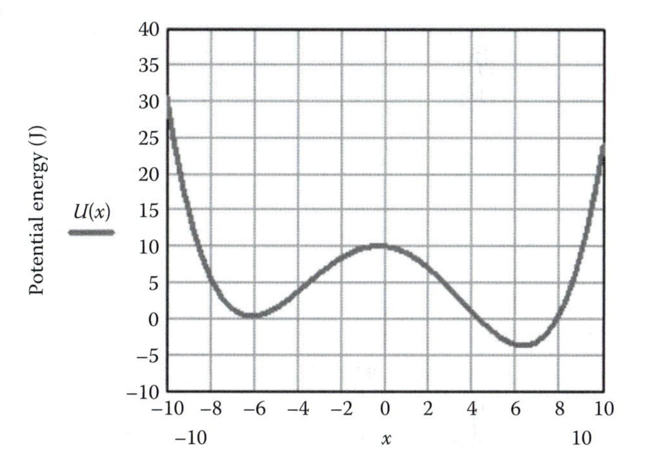

(a) What direction is the force on it at $x = 4$ m?

(b) If the body starts at $x = 8$ m with a velocity of 4 m/s to the left, approximately where will it come to rest?

(c) What would its velocity be when $x = 0$ m?

(d) If the body is released from rest at $x = -7$ m, describe its motion.

9.13 A 1.0 kg body moving in the x-direction has a potential energy $U(x) = x^2 - 4x$ J. This function is shown in the figure below.

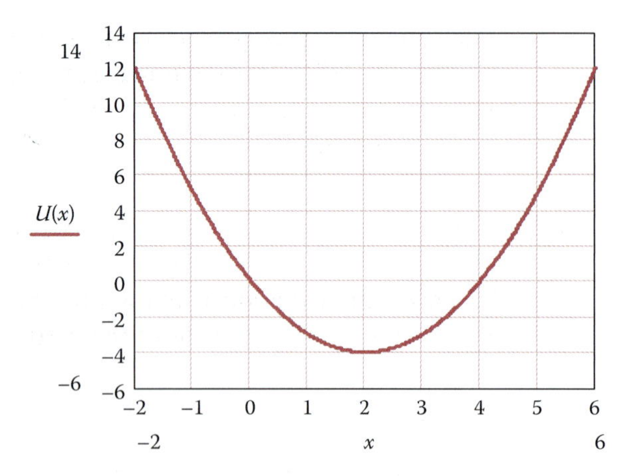

(a) Find the force on the body at $x = 0$. (Magnitude and direction.)

(b) If the body has a speed of 3.0 m/s at $x = 0$, what will be its maximum speed during its subsequent motion?

(c) If the body has a speed of 3.0 m/s at $x = 0$, estimate the maximum value of x that it reaches.

(d) Suppose the body starts from rest at $x = -1.5$ m. How much time will elapse before it reaches its maximum possible x position?

(e) Rewrite the potential function by "completing the square," that is, as $U(x) = (x - a)^2 + b$. Find a and b and compare your answers to the graph. Is this a simple harmonic oscillator? If so, what is the period of oscillation?

9.14 A rope of total mass 1.0 kg hangs symmetrically (half on either side) over a frictionless pulley at the top of a flagpole. The pole is 20 m high, and the length of the rope is also 20 m. The rope is given a tiny tug and rolls over the pulley and falls to the ground. What is the speed of the rope just before its bottom end reaches the ground? (The key to this problem is finding the initial and final potential energies of the rope. For each segment of the rope, $U = mgy_{CM} + c$, where y_{CM} is the height of that segment's center of mass.)

9.15 On a sunny day, a man is in deep thought while sitting on top of a hemispherical igloo of radius r. Due to a mild Earth tremor, he is given a slight nudge and begins to slide down the (frictionless) icy surface. Through what vertical distance does he slide before losing contact with the surface? (*Hint:* Draw an FBD for the man. When he loses contact with the surface, what is the value of the normal force?)

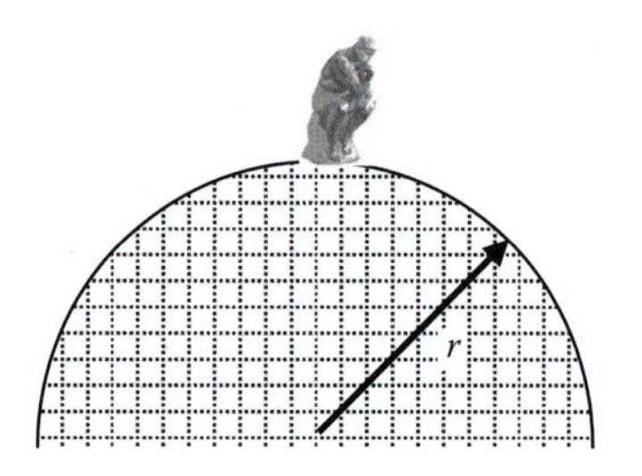

9.16 Starting from rest at point A, a block slides along the frictionless track shown in the figure.

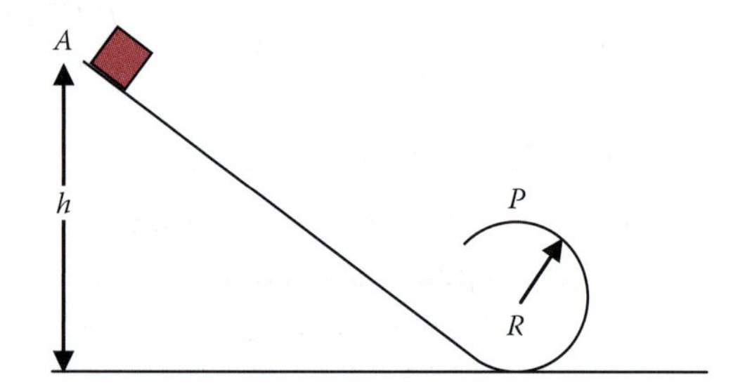

(a) What is the minimum height h such that the block reaches the top of the loop (point P) without losing contact with the track?

(b) For what initial height h would the normal force exerted by the track at point P be equal to the gravitational force $m\vec{g}$ on the body?

9.17 The potential energy of a body of mass $m = 2$ kg is shown in the graph below. The body is released from rest at $x = 5.5$ m.

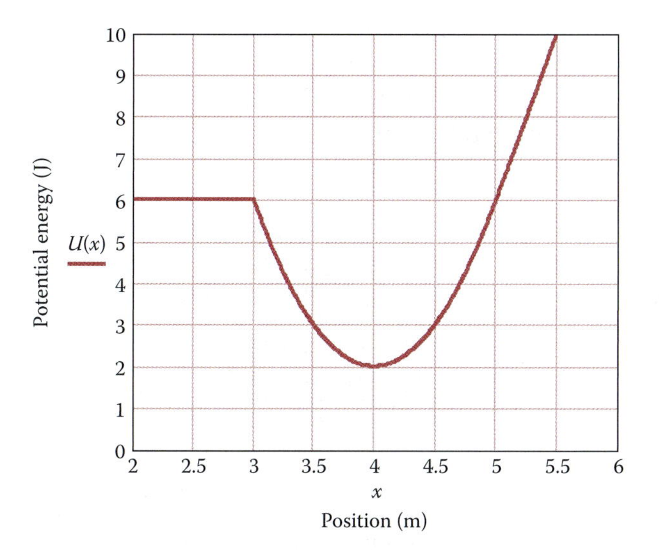

(a) What is the force (magnitude and direction) on the body at $x = 5.0$ m?

(b) What is the maximum speed?

(c) A scientist wants to confine the body to the region between $x = 3$ and $x = 5$ m. What kinetic energy should she give the body at $x = 4$ m?

(d) The body is released from rest at $x = 4.5$ m. Describe the resulting motion. How long does it take for the body to return to this position?

9.18 The potential energy of a body is given by $U(x) = -Ax + Bx^3$ J, where $A = 24$ J/m and $B = 2$ J/m^3. This function is shown by the solid line in the next figure over the range $0 < x < 5$ m.

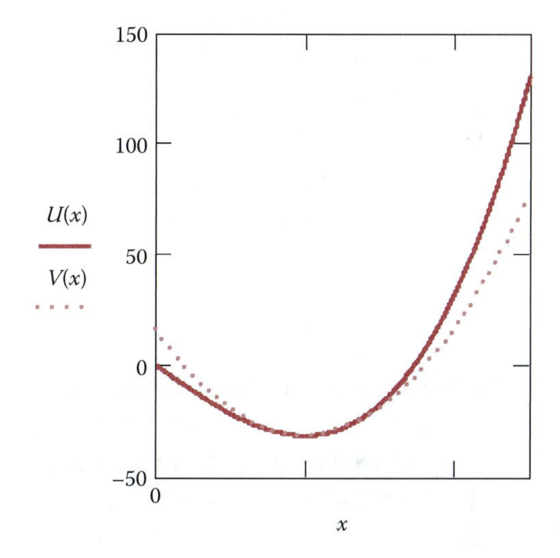

(a) Calculate the location of the equilibrium point x_0.

(b) Calculate the minimum value of the potential energy over the range shown in the figure.

(c) Expand $U(x)$ in a Taylor series about the equilibrium point, up to terms proportional to $(x - x_0)^2$. This expansion is shown as the dotted line in the figure.

(d) Examine the figure. Estimate the largest amplitude which would be approximately simple harmonic.

(e) If the body has a mass of 2 kg, what is the period of small amplitude oscillations of the body?

9.19 The Milky Way Galaxy consists of a dense central bulge plus a lower density thin disk. We are in the disk, about 8.5 kpc from the galactic center, in a near-circular orbit about the galactic center with a period of 240 Myr. In addition to this orbital motion, the Sun also oscillates perpendicular to the disk, attracted by the stars within the disk.

In this problem, we will ignore the orbital motion, and think only about the oscillatory motion.

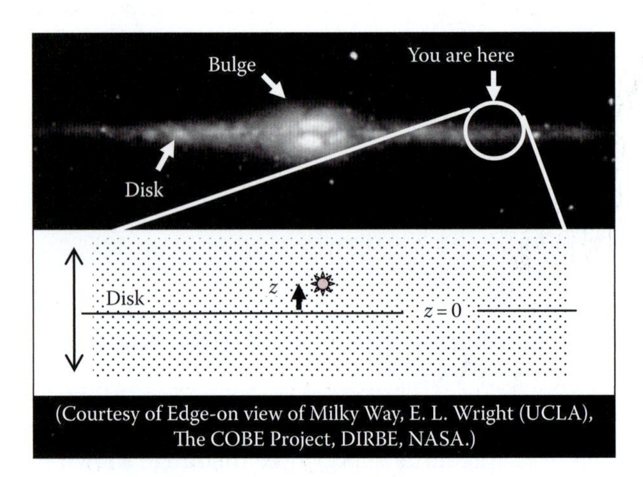

(Courtesy of Edge-on view of Milky Way, E. L. Wright (UCLA), The COBE Project, DIRBE, NASA.)

The potential energy of the Sun, based on observations of the local stellar density, is shown below as a function of distance z from the center of the galactic plane. The curve looks the same for negative values of z, that is, for positions below the central plane. (Note that we have chosen $U(z) = 0$ at $z = 0$.)

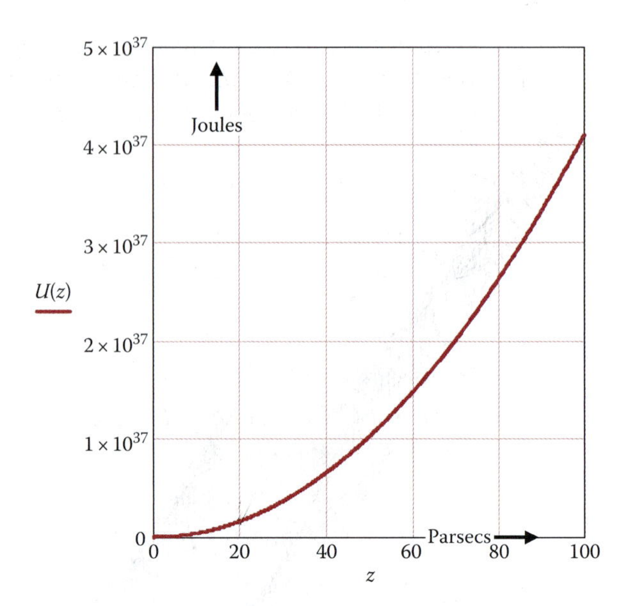

(a) The Sun ($M_S = 2 \times 10^{30}$ kg) is presently 20 pc above the central plane of the galactic disc and is moving away from the central plane. The amplitude of its oscillation is 100 pc. What is the Sun's present speed?

(b) Estimate the *acceleration* of the Sun when it is 90 pc from the central plane. (*Note:* 1 pc = 3.08×10^{16} m).

(c) The potential energy function shown is very nearly parabolic. Find the period of oscillation of the Sun about the central plane of the disk (1 Myr $\approx \pi \times 10^{13}$ s). (*Hint:* Don't forget to convert from parsecs to meters!)

(d) We can describe the oscillation by the sinusoidal function

$$z(t) = (100\,\text{pc})\cos(\omega t + \theta_0).$$

If we define the present time as $t = 0$, what is the value of θ_0? (Don't forget: The Sun is now 35 pc above the central plane and moving *away* from it.)

9.20 A system is described by a potential energy function $U(x) = (1/x^3) - (3/x)$ J. This function is shown in the figure below for $x \geq 0$. We will approximate $U(x)$ by a polynomial, for values of x close to the equilibrium position. That is

$$U(x) = U(x_0) + \left(\frac{dU}{dx}\right)_{x_0}(x - x_0) + \frac{1}{2}\left(\frac{d^2U}{dx^2}\right)_{x_0}(x - x_0)^2.$$

(a) Find $U(x_0)$, the potential energy at the equilibrium point x_0, where $F(x) = 0$.

(b) Calculate dU/dx and d^2U/dx^2 and evaluate these derivatives at x_0.

(c) Use your answers to (a) and (b) to write $U(x)$ as a second-order polynomial.

(d) In the above expression, $(d^2U/dx^2)_{x_0}$ is like an effective spring constant k. Why?

(e) If $U(x)$ is the potential energy of a body of mass $m = 0.667$ kg, what is the period T of small oscillations of m about the equilibrium point?

(f) Plot your quadratic function of part (c) in the graph below, and compare with the actual shape of $U(x)$. For what range of x do you think your answer to (e) is accurate?

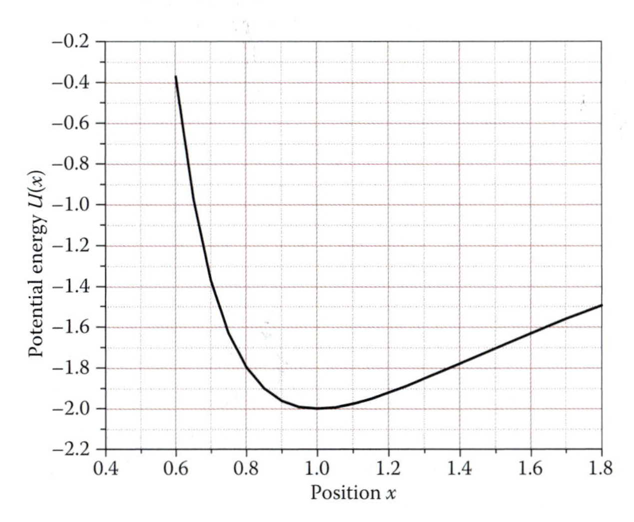

9.21 In a previous chapter, you considered the motion of a ball of mass m moving along a tunnel bored through the Earth.

(a) Prove once again that the force on the ball can be expressed as

$$F(r) = -mg\frac{r}{R_E}$$

where r is the distance to the center of the planet and R_E is its radius.

(b) If we choose the zero of potential energy to be at $r = 0$, what is the ball's potential energy when it is at the Earth's surface?

(c) If the ball is released from rest at the earth's surface, find its speed as it passes through the planet's center.

(d) We know that the ball undergoes simple harmonic motion $r(t) = R_E \cos \omega t$. Solve for ω and show that your answer is compatible with the velocity you found in part (c).

9.22 A body is subject to the force $\vec{F} = (ax - bx^3)\hat{i}$, where $a = 8$ N/m and $b = 2$ N/m^3.

(a) Derive an expression for the potential energy, choosing $U = 0$ at the origin.

(b) Find the positions x where $\vec{F} = 0$, and calculate $U(x)$ at these points. Then sketch the potential energy curve for $-4 < x < 4$ m.

(c) Use your curve to find the turning points of a body whose total energy is 4 J near the origin.

(d) A body near a *stable* equilibrium point will be attracted to that point, whereas a body near an *unstable* equilibrium point will be repelled from that point. Label each of the points found in (b) as either stable or unstable equilibria.

9.23 The figure shows the potential energy $U(x)$ for a particle of mass $m = 0.4$ kg.

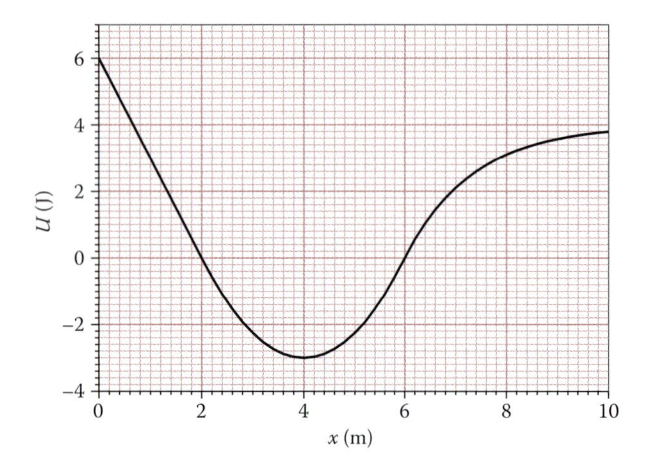

(a) Find the magnitude and direction of the force on the body at $x = 1$ m. Include units.

(b) If the body has a speed $v = 3$ m/s at $x = 2$ m, what is its maximum speed?

(c) In the interval $2 < x < 6$, the body undergoes simple harmonic oscillation. Estimate the period of oscillation.

9.24 You are trying to land your spacecraft on *Cloud City*, a distant space colony ruled by your old friend Lando. The mass of the disk-shaped colony $M = 1.6 \times 10^{15}$ kg and its radius $R = 10$ km. The mass of your spacecraft is $m = 500$ kg. The figure shows your potential energy vs. height h above the surface of the space colony. The potential energy is chosen to be 0 when $h \to \infty$. Use the graph to answer parts (a), (b), and (c).

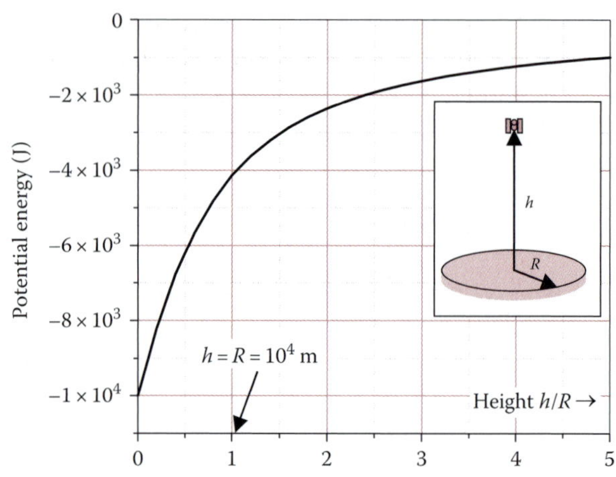

(a) What is the gravitational force (weight) on the rocket as it sits on the surface ($h = 0$) of the disk?

(b) If the rocket is launched from the surface ($h = 0$) with a vertical velocity $v = 10.0$ m/s, will it escape the colony? How do you know?

(c) If the craft escapes, what is its speed far from the colony ($h \to \infty$)? If it does not escape, what is the maximum value of h?

(d) The spacecraft's rocket engines use a fuel that has an exhaust velocity of 2500 m/s. What is the mass of fuel that must be burned to accelerate the craft to 10 m/s? Assume that the fuel is consumed in a short burst.

9.25 A block slides over a certain surface with a coefficient of friction μ_k. Its initial speed is v_0 and it slides until the frictional force brings it to a halt. The figure shows three situations: (a) the surface is horizontal, (b) the surface is tilted and the block slides up the slope or (c) down the slope. In which case is the increase in internal energy the greatest? The least?

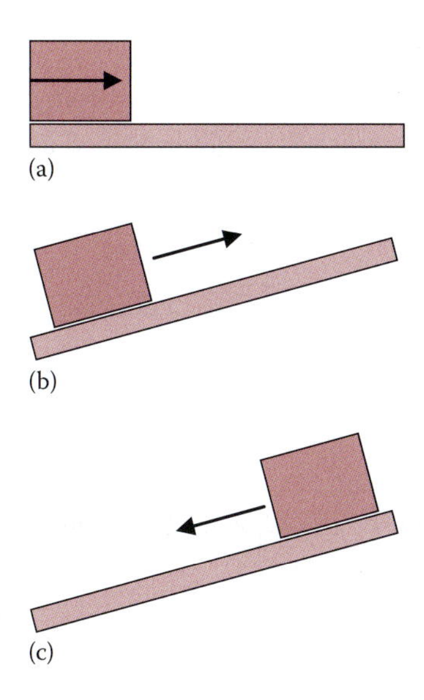

9.26 In 2015, NASA is scheduled to launch *Sunjammer*, a spacecraft that will be propelled by the radiation pressure of sunlight. Electromagnetic radiation is emitted in quantized bundles called *photons*. Each photon carries energy $E = hf = hc/\lambda$, where f and λ are its frequency and wavelength, and h is called Planck's constant: $h = 6.626 \times 10^{-34}$ J s. Photons act like massless particles and carry momentum p as well as energy: $p = E/c$. After launch into orbit, *Sunjammer* will deploy a large reflecting sail of area 1200 m^2 that will catch and reflect the Sun's light. Each photon reflected by 180° has its momentum reversed and imparts an additional momentum to the spacecraft $\Delta p = m\Delta v = 2hf/c$.

(Courtesy of Sunjammermission.com, courtesy L'Garde, Inc.)

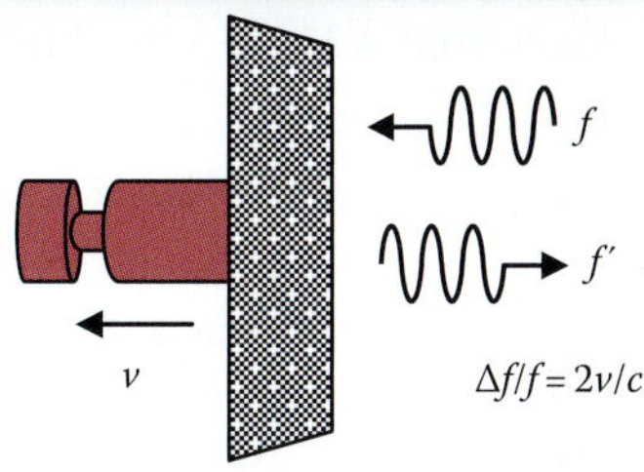

(a) At a distance of 1 AU, the intensity of the Sun's radiation is 1360 W/m^2, which means that about 2×10^{39} photons will strike *Sunjammer's* sail per second. This force exerted on the sail by the Sun's radiation is equal to the change of momentum per second. Show that this force is about 0.01 N.

(b) As photons are reflected from the space sail, the spacecraft gains kinetic energy and momentum. Consider the reflection of a single photon from a sail moving radially away from the Sun (see the figure) with speed v. The reflected photon is Doppler shifted by $\Delta f/f = -2v/c$ while the kinetic energy is increased by $\Delta K = \Delta(\frac{1}{2}mv^2) = mv\Delta v$, where m is the spacecraft mass. Show that energy is conserved.

9.27 The atomic weights of ^{32}P and ^{32}S are 31.973907 u and 31.972071 u, respectively. When ^{32}P β-decays to ^{32}S, $\Delta Mc^2 = K_e + E_\nu$, where K_e is the (relativistic) kinetic energy of the emitted electron and $E_\nu = p_\nu c$ is the energy of the neutrino. In the process, the ^{32}S atom recoils to the left, as shown in the figure.

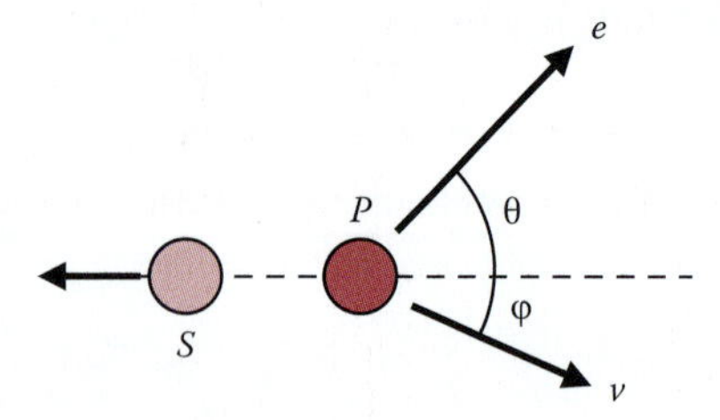

(a) Show that $\Delta Mc^2 = 1.710$ MeV.

In experimental studies of β-decay, the direction θ and magnitude of the electron momentum are measured along with the recoil momentum of the sulfur atom. The neutrino is not detected. In the following, we will express all momenta as energy pc. For one set of measurements, $p_e c = 0.645$ MeV and $\theta = 30°$. The recoil momentum of the sulfur is found to be 1.920 MeV. (*Note:* Its kinetic energy is negligible.)

(b) Find K_e. Assuming energy is conserved, find $p_\nu c$.

(c) Assuming momentum is conserved, find the angle φ.

9.28 Two carts, $m_1 = 1$ kg and $m_2 = 2$ kg, roll without friction along a straight horizontal track. Initially, cart 2 is stationary, and cart 1 is approaching it with speed 4 m/s. A spring of stiffness $k = 530$ N/m is mounted on m_2, so that m_1 compresses the spring as it approaches m_2. At some time, both carts are moving at the same speed.

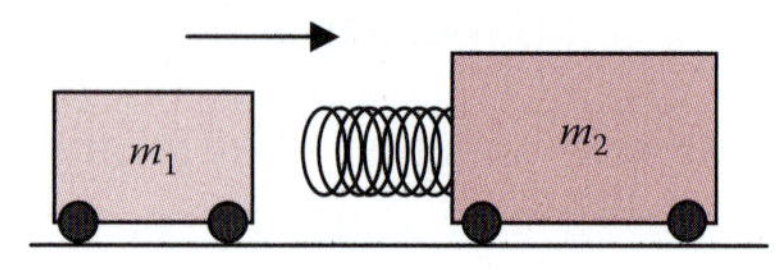

(a) Find this speed.

(b) Calculate the maximum compression of the spring.

(c) How long does it take to compress the spring by this amount? (*Hint:* Look at the motion relative to the *CM* of the two blocks, and review Section 7.10.)

9.29 A chain is held on a frictionless horizontal table with a small fraction (≪1) of its length hanging over the edge. The chain has a length L and a mass M. When released it slides off the table and falls to the floor, a distance $2L$ from the height of the table.

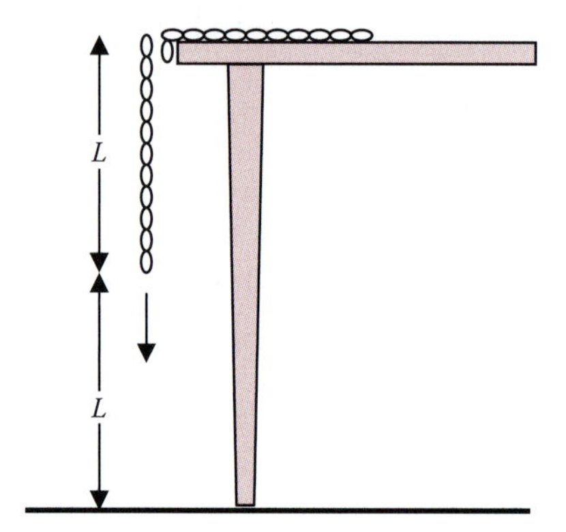

(a) What is the speed of the chain in the position shown, when its right edge has just slipped off the table?

(b) What is the speed of the lowest link when it strikes the floor?

(c) What is the speed of the highest link when it strikes the floor?

(d) This is a very rich mathematical problem. To see this, let y = the length of chain hanging over the table ($0 \le y \le L$). Show that the chain's acceleration is equal to

$$a = \frac{d^2y}{dt^2} = \frac{g}{L}y.$$

This looks very much like the equation for simple harmonic oscillation (just like the pendulum), except the minus sign is missing from the right hand side. The solution to this equation, for an initial velocity equal to zero, is

$$y(t) = y_o \cosh\left(\sqrt{\frac{g}{L}}t\right) \approx \frac{y_0}{2} e^{\sqrt{g/L}t}$$

where y_o is the length of chain initially hanging over the table edge, and the approximation $\cosh(x) \approx \frac{1}{2} e^x$ is good for large x. If $y_o = L/10$, *estimate* the time for the entire length of chain to slip off the table. (For numerical answers to parts (a)–(d), use $L = 0.5$ m.)

10

GRAVITATIONAL POTENTIAL ENERGY AND ORBITAL MOTION

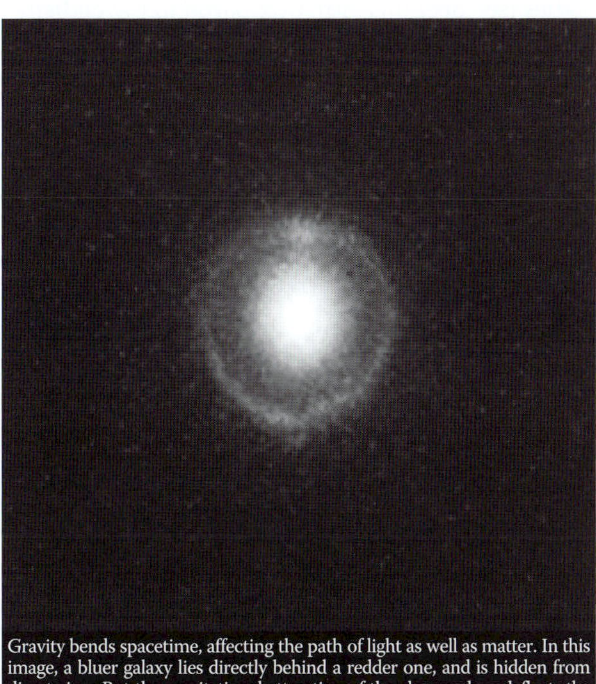

Gravity bends spacetime, affecting the path of light as well as matter. In this image, a bluer galaxy lies directly behind a redder one, and is hidden from direct view. But the gravitational attraction of the closer galaxy deflects the light from the more distant one, which we see as a ring. This is called gravitational *lensing*. By measuring the apparent size of the ring, astronomers can determine the mass of the foreground galaxy. (Courtesy of NASA, ESA, A. Bolton [Harvard-Smithsonian CfA], and the SLACS Team.)

Our knowledge of gravity is a road map by which we can navigate the marvels of heaven and Earth, the world of space and time, the valleys and waves in spacetime. It is a guide to traveling over land and sea, through air and space. To prospecting with gravimeter for *oil and mineral. To putting the tides to use. To launching rocket, satellite, and spaceship. To deciphering the doings of moon and planet, star and galaxy. To unpuzzling the power of the black hole and the dynamics of the universe.**

* From J.A. Wheeler, *A Journey into Gravity and Spacetime*, Scientific American Library, a division of HLHLP, New York, 1990.

John Archibald Wheeler

10.1 INTRODUCTION

In 1798, when Henry Cavendish measured the gravitational attraction between two lead spheres, he used a scaled-up version of the torsion pendulum developed earlier by Charles Coulomb to determine the electrical force between charged bodies (see Section 8.10). Compared to Cavendish's experiment, Coulomb's was a snap. The electrical forces he measured were gigantic relative to those measured by Cavendish: between two protons, the electric force is 10^{36} times greater than the gravitational force. But charge comes in two polarities (positive and negative), whereas mass—as far as we know—does not.† In macroscopic bodies, the number of positive charges is nearly equal to the number of negative charges, and this tendency toward electrical neutrality is more pronounced for larger bodies. As bodies grow larger, gravity ultimately wins out over the electrical force. It is gravity that governs the formation and structure of celestial bodies, guides their motion, and on the largest scale, orchestrates directs the evolution of the universe. An understanding of gravity—the feeblest of the four fundamental forces in nature—is the key to learning the history and ultimate fate of the cosmos.

† Physicists have learned to take nothing for granted. New experiments are underway to see if *anti*matter falls in the same direction and with the same acceleration g as ordinary matter.

10.2 WORK AND THE INVERSE SQUARE LAW

Let us begin by proving that the inverse-square gravitational force is a conservative force. Recall that the work done by a force $\vec{F}(\vec{r})$ on a body moving from $\vec{r}_i$ to $\vec{r}_f$ is given by the line integral

$$W_{i\to f}=\int_i^f dW=\int_i^f \vec{F}(\vec{r})\cdot d\vec{r}. \tag{9.5}$$

If $W_{i\to f}$ is independent of the path taken between $\vec{r}_i$ and $\vec{r}_f$, $\vec{F}(\vec{r})$ is a conservative force, and we can define a potential energy function $U(\vec{r})$ satisfying the following equation:

$$\Delta U=U(\vec{r}_f)-U(\vec{r}_i)=-W_{i\to f}=-\int_i^f \vec{F}(\vec{r})\cdot d\vec{r}. \tag{10.1}$$

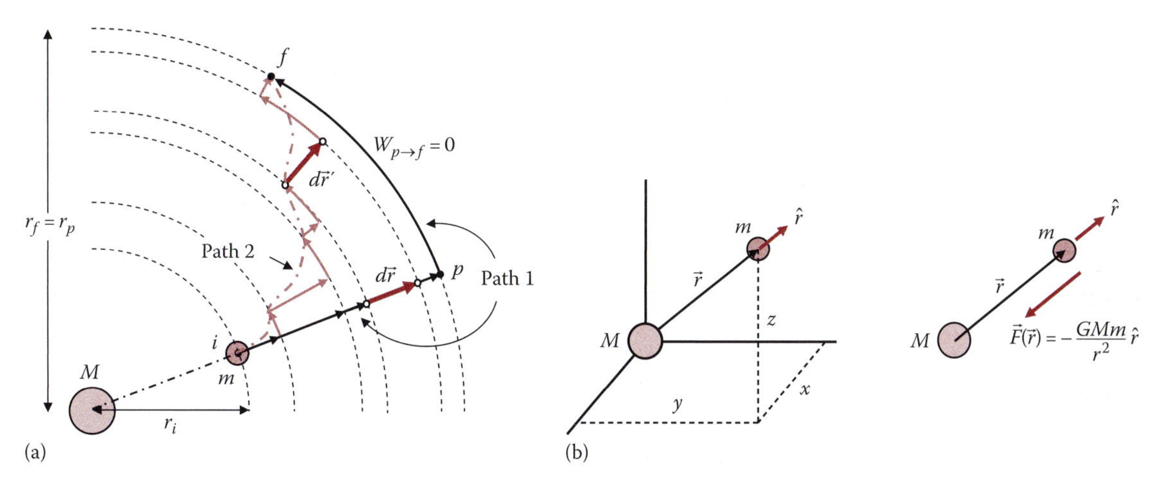

FIGURE 10.1 (a) Two paths are shown joining point i to point f. On Path 1 (black), body m first moves radially away from M to point p, then continues to f along a circular arc having M at its center. On Path 2 (red), the body follows a more complicated route that can be approximated as a sequence of radial and circular displacements. (b) The unit vector $\hat{r}$ is shown in x–y–z coordinates and without coordinates. The gravitational force on m is antiparallel to $\hat{r}$.

In Chapter 8, we learned that the magnitude of the gravitational force between two *point masses M* and *m*, separated by a distance r, is

$$F(r) = \frac{GMm}{r^2}.$$

This force is attractive and directed along the line joining the two bodies. Let's use these two features to prove that $\vec{F}(\vec{r})$ is conservative. Let M be located at the origin ($\vec{r} = 0$), and consider the gravitational work done on m as it moves from point i (at $\vec{r}_i$) to point f (at $\vec{r}_f$). Two possible paths are shown in Figure 10.1a. On Path 1, m first moves radially outward from i to p, and then continues on a circular arc of radius r_p joining points p and f. The work done along this trajectory is $W_{i\to f}^{Path1} = W_{i\to p} + W_{p\to f}$, or

$$W_{i\to f}^{Path1} = \int_i^p \vec{F}(\vec{r})\cdot d\vec{r} + \int_p^f \vec{F}(\vec{r})\cdot d\vec{r}.$$

But $\vec{F}(\vec{r}) \perp d\vec{r}$ along the circular arc from p to f, so the second integral vanishes ($W_{p\to f} = 0$), and $W_{i\to f}^{Path1} = W_{i\to p}$, the work done on m as it moves radially away from M.

Path 2 of Figure 10.1a represents an arbitrary route linking the same two points i and f. No matter how convoluted this path may be, it can be approximated by an alternating sequence of tiny radial and circular displacements. The approximation becomes exact as the number of incremental displacements approaches infinity. The work done along each circular arc of Path 2 is zero (just as $W_{p\to f} = 0$), so the total work done along Path 2 is just the sum of the work done along the radial displacements. Let's compare the work done along the radial segment $d\vec{r}'$ of Path 2 with that done along $d\vec{r}$ of Path 1. Since $d\vec{r}$ and $d\vec{r}'$ are the same distance from M, the *magnitude* of the force felt by m is the same on both segments: $F(\vec{r}') = F(\vec{r})$. The lengths of the two displacements are equal, $dr' = dr$, and in either case, the force is antiparallel to the displacement. Therefore, the differential work $dW' = \vec{F}' \cdot d\vec{r}' = -F'dr'$ is exactly equal to $dW = \vec{F} \cdot d\vec{r} = -Fdr$. For each radial segment of Path 2, there is a matching radial segment on Path 1, so $W_{i\to f}^{Path\,2} = W_{i\to f}^{Path\,1}$. This is true regardless of the shape of Path 2, proving that $W_{i\to f}$ is independent of the path. Therefore, *the inverse-square gravitational force is a conservative force.*

EXERCISE 10.1

A fictitious force exerted on a body by another body a distance r away has an inverse-square dependence, $F = A/r^2$, where A is a constant. When the body undergoes a displacement $d\vec{r}$, it feels a force that is always antiparallel to $d\vec{r}$. Is this a conservative force?

10.3 GRAVITATIONAL POTENTIAL ENERGY

We can now use Equation 10.1 to derive the gravitational potential energy function $U(\vec{r})$. First let's express $\vec{F}$ in vector form. Let $\vec{r} = x\hat{i} + y\hat{j} + z\hat{k}$ be the location of m relative to M, and define a *unit vector* $\hat{r}$ parallel to the position vector $\vec{r}$ (see Figure 10.1b):

$$\hat{r} = \frac{\vec{r}}{r} = \frac{x\hat{i} + y\hat{j} + z\hat{k}}{\sqrt{x^2 + y^2 + z^2}}.$$

In the following discussion, we will dispense with the coordinate axes and write all of our mathematics in terms of $\vec{r}$ and $\hat{r}$. Because m is attracted toward M, the force on m is in the $\hat{r}$ direction (i.e., antiparallel to $\hat{r}$), so the magnitude and direction of $\vec{F}$ can be expressed in vector form as

$$\vec{F}(\vec{r}) = -\frac{GMm}{r^2}\hat{r}. \tag{10.2}$$

Once again consider the work done along Path 1 in Figure 10.1a. Since $W_{p\to f} = 0$, Equation 10.1 tells us that $U(\vec{r}_p) = U(\vec{r}_f)$. More generally, $U(\vec{r})$ has the same value at all points equidistant from M. This set of points forms a circle in 2D, or a spherical shell in 3D, centered on M. In other words, $U(\vec{r})$ depends only on the *magnitude* of $\vec{r}$ and not on its direction: $U(\vec{r}) = U(r)$. To evaluate $W_{i\to f}$ along any path joining points i and f, it is sufficient to calculate the work done on a *radial* path that starts at a distance r_i from M and moves radially away to a distance r_f from M (e.g., points i and p in Figure 10.1a). From Equations 10.1 and 10.2, the change in potential energy is

$$U(r_f) - U(r_i) = -W_{i\to f} = -\int_{r_i}^{r_f} -\frac{GMm}{r^2}\hat{r} \cdot d\vec{r}.$$

When evaluating this line integral, there are a number of minus signs to keep straight, so let's proceed cautiously. If $r_f > r_i$, the vectors $\hat{r}$ and $d\vec{r}$ are parallel, $dr > 0$, and $\hat{r} \cdot d\vec{r} = |\hat{r}| \cdot |dr| \cos 0 = |dr| = dr$. If $r_f < r_i$, then $\hat{r}$ and $d\vec{r}$ are antiparallel, $dr < 0$, and $\hat{r} \cdot d\vec{r} = |\hat{r}| \cdot |dr| \cos \pi = -|dr| = dr$. So in either case, $\hat{r} \cdot d\vec{r} = dr$, and the line integral simplifies to an ordinary integral devoid of vectors:

$$U(r_f) - U(r_i) = -\int_{r_i}^{r_f} -\frac{GMm}{r^2} dr = GMm \int_{r_i}^{r_f} \frac{dr}{r^2} = -\frac{GMm}{r}\bigg|_{r_i}^{r_f}$$
$$= -\frac{GMm}{r_f} + \frac{GMm}{r_i}. \tag{10.3}$$

This expression gives us the *difference* in the potential energy when m is moved from $\vec{r}_i$ to $\vec{r}_f$ and is equivalent to writing the potential energy function as

$$U(r) = -\frac{GMm}{r} + c,$$

where c is a constant we are free to choose. Just as we saw in Chapter 9, c will not affect the motion of m in any way, and choosing c is equivalent to choosing the location where the potential energy is equal to zero. In the present case, it is intuitively sensible and convenient to set the gravitational potential energy to be zero when m is infinitely far from M, that is, $U(r = \infty) = 0$. This is the convention followed throughout physics. In the above equation, let $r_i = \infty$, and set $U(r_i) = U(\infty) = 0$ to obtain an expression for the gravitational potential energy function:

$$U(r) = -\frac{GMm}{r}, \tag{10.4}$$

where r is the distance between M and m. (Note that setting $U(\infty) = 0$ is equivalent to choosing $c = 0$.) This is an important result that we will use throughout the remainder of this course. As an additional bonus, Equation 10.4 can be used to derive Newton's shell theorem, which we promised to do in Chapter 8. See Box 10.1 for a detailed derivation.

The function $U(r)$ is plotted in Figure 10.2a. It is instructive to examine its shape in detail. First note that $U(r)$ is *negative* for all finite values of r, which may seem puzzling at first. But $U(r)$ is the *difference* in potential energy from when m and M are infinitely far apart ($r = \infty$); the closer the two bodies, the *lower* their potential energy. Since $U(\infty) = 0$, $U(r)$ must then be negative for all finite values of r. Second, $U(r)$ has curvature. How can this be reconciled with the familiar result $\Delta U(y) = mgy$ for a body at altitude y above the Earth's surface? To answer this question, let M_E and R_E be the mass and radius of the Earth. For all points r external to Earth ($r \geq R_E$), the shell theorem tells us to treat M_E as a point particle with its entire mass concentrated at the center of the planet. Equations 10.2 and 10.4 are thus valid for

BOX 10.1 DERIVATION OF THE SHELL THEOREM

In Chapter 8, we promised a calculus-based derivation of Newton's shell theorem, the crucial theoretical result needed to verify the law of universal gravitation. The shell theorem gives the gravitational force on a point mass m, due to a thin, spherically symmetric shell of mass M: (1) if m is *outside* the shell, the force is exactly the same as if all of M were concentrated at the shell's center and (2) if m is *inside* the shell, the force is exactly equal to zero. Newton's proof was based exclusively on geometric arguments and is quite challenging to follow. (It took him nearly 20 years to complete his derivation.) Even using integral calculus (coinvented by Newton) to sum the forces on m due to the various segments of the shell, the proof is not trivial. Using potential energy instead of force, the shell theorem is much easier to derive.

Suppose a body m interacts gravitationally with several other bodies M_1, M_2, M_3, ... The net force on m is the vector sum of the individual forces due to each mass: $\vec{F} = \vec{F}_1 + \vec{F}_2 + \vec{F}_3 + \cdots$. Just like forces, potential energy contributions can be added together. Let U_1 be the potential energy calculated as if m and M_1 were the only bodies present, and similarly for U_2, U_3,.... The total potential energy is given by $U(\vec{r}) = U_1(\vec{r}) + U_2(\vec{r}) + U_3(\vec{r}) + \cdots$. The crucial difference is that forces are vectors, whereas potentials are scalars, and adding scalars is a whole lot easier than adding vectors.

Imagine a thin spherical shell of radius R and total mass M, interacting gravitationally with a point mass m that is a distance r from the shell's center (see Figure B10.1a). Divide the shell into N tiny platelets,

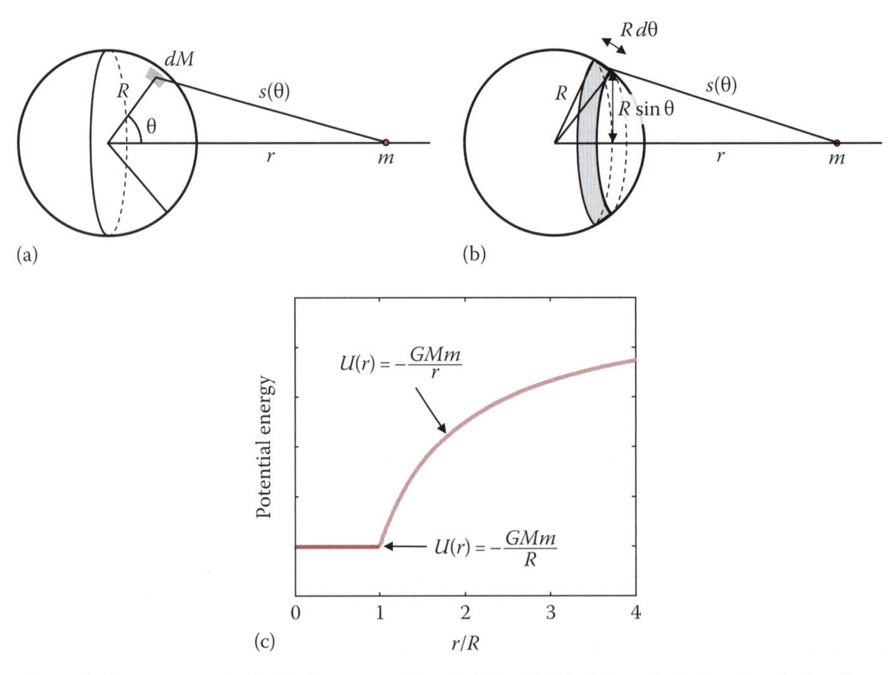

FIGURE B10.1 (a) To derive Newton's shell theorem, the shell is divided into infinitesimal platelets of area dA and mass dM. (b) The platelets are then grouped to form rings of radius $R\sin\theta$ and width $R\,d\theta$, like the one illustrated. All points on a ring are the same distance s from the test mass m. (c) After integration, the potential energy $U(r)$ takes the form shown. For $r < R$, $U(r)$ = const, and $F = 0$.

each of area dA containing a tiny bit of mass $dM = \sigma dA$, where σ ("sigma") is the areal mass density (mass per unit area) of the shell: $\sigma = M/4\pi R^2$. In the spirit of integral calculus, we are going to let $N \to \infty$ and dA, $dM \to 0$. From the discussion above, we know that the total potential $U(r)$ is equal to the sum—or integral—of the potential energy contributions from all of the differential masses dM. To simplify the calculation, group the platelets into thin rings such as the one illustrated in Figure B10.1b. The radius of the ring shown in the figure is $R\sin\theta$ and its circumference is $2\pi R\sin\theta$. Its width is $R\,d\theta$, so the ring's area is $dA = 2\pi R^2 \sin\theta d\theta$. All points on the ring lie at the same distance s from m, where s is found from the law of cosines: $s^2 = R^2 + r^2 - 2rR\cos\theta$. Therefore, the contribution of the ring to the potential energy of m is

$$dU = -\frac{GmdM}{s} = -\frac{2\pi R^2 \sigma Gm \sin\theta d\theta}{\sqrt{R^2 + r^2 - 2rR\cos\theta}}.$$

To find the total potential energy $U(r)$, we must sum the contributions from all of the rings making up the shell; that is, we will perform a *surface integral* over the shell. Although this looks like a daunting task, it is really quite easy! The only variable in the integrand is θ, and integration over the shell surface is simply integration over the range $0 \le \theta \le \pi$. Let $f(\theta) = R^2 + r^2 - 2rR\cos\theta$, so $df = 2rR\sin\theta\,d\theta$. The denominator in the above equation is the distance $f^{1/2}$, and in the numerator, $\sin\theta d\theta = df/2rR$, so

$$dU = -\frac{2\pi R^2 \sigma Gm}{2rR}\frac{df}{f^{1/2}}.$$

Simplifying and integrating,

$$U(r) = -\frac{\pi R \sigma Gm}{r}\int_{\theta=0}^{\pi}\frac{df}{f^{1/2}} = -\frac{\pi R\sigma Gm}{r}2f^{1/2}\Big|_{\theta=0}^{\pi}.$$

To finish the derivation, we must evaluate $f^{1/2}(\theta) = s(\theta)$ at the limits $\theta = 0$ and $\theta = \pi$. Using $\cos\pi = -1$ and $\cos 0 = 1$ we find

$$f^{1/2}\Big|_{\theta=0}^{\pi} = s(\theta=\pi) - s(\theta=0) = \sqrt{R^2 + r^2 + 2rR} - \sqrt{R^2 + r^2 - 2rR}.$$

But now we must be very careful, because we will get different answers depending on whether $r \ge R$ (when m is outside the shell) or $r < R$ (when m is inside the shell). The key thing to remember is that $s(\theta)$ is a distance, a positive number, so we must take the *positive* root of each radical above. For $r > R$, $\sqrt{R^2 + r^2 \pm 2rR} = r \pm R$, so $f^{1/2}\Big|_{\theta=0}^{\pi} = (r+R) - (r-R) = 2R$, and

$$U(r) = -\frac{4\pi R^2 \sigma Gm}{r} = -\frac{GMm}{r} \quad (r \ge R), \qquad \text{(B10.1)}$$

exactly as if the entire mass M were located at the center of the sphere. For $r < R$, the positive root of each radical is $\sqrt{R^2 + r^2 \pm 2rR} = R \pm r$ (notice the difference), so $f^{1/2}\Big|_{\theta=0}^{\pi} = (R+r) - (R-r) = 2r$ and

$$U(r) = -4\pi R\sigma Gm = -\frac{4\pi R^2 \sigma Gm}{R} = -\frac{GMm}{R} \quad (r < R), \tag{B10.2}$$

which is a *constant*, independent of r, and equal to the potential given by Equation B10.1 at $r = R$. These results are summarized in Figure B10.1c, and the shell theorem follows directly from them.

Exercise B10.1: Using $F(r) = -dU/dr$, show that Equations B10.1 and B10.2 lead to the conclusions stated in the shell theorem for $r \geq R$ and $r < R$.

Exercise B10.2: A point mass m lies on the axis of symmetry of a *hemi*spherical shell of radius R and mass M, just above its surface.

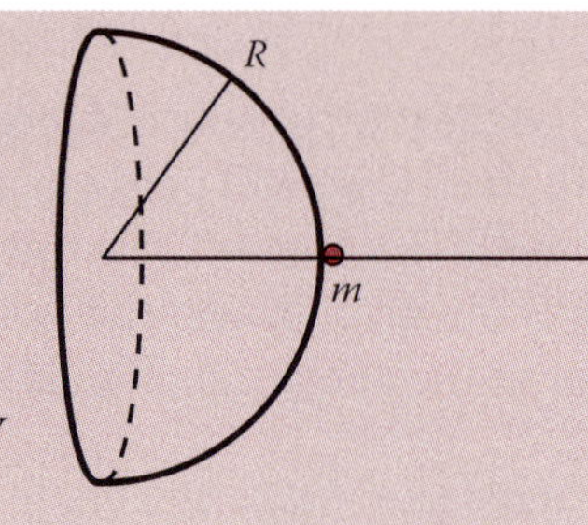

(a) Express the areal density σ in terms of M and R.
(b) Adapt the derivation of Equation B10.1 to find the potential energy of m in terms of m, M and R.

Answer: (a) $\sigma = M/2\pi R^2$ (b) $U = -\sqrt{2}GMm/R$

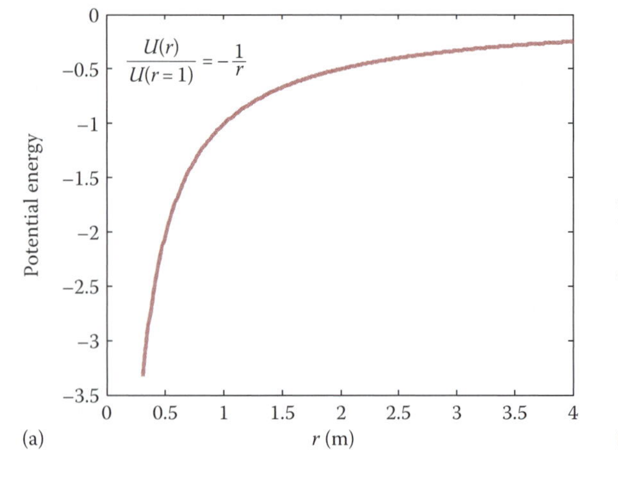

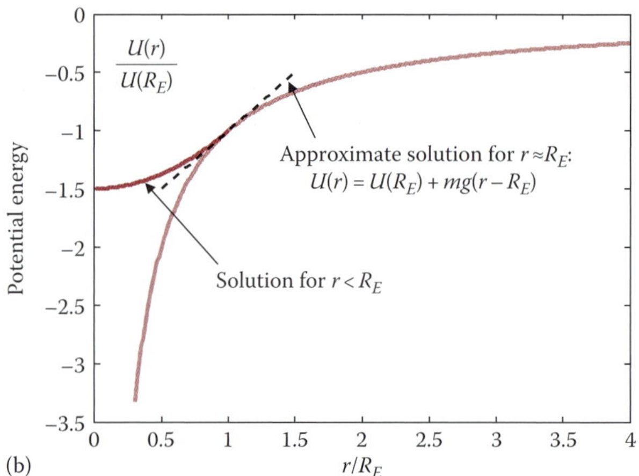

FIGURE 10.2 (a) The gravitational potential function $U(r)$ approaches 0 as $r \to \infty$ and $-\infty$ as $r \to 0$. For convenience, the potential is *normalized* to its value at $r = 1$ m, so the ordinate is expressed as $U(r)/GMm = -1/r$. (b) Here, the potential is normalized to its value at $r = R_E$. Near the Earth's surface ($r/R_E = 1$), $U(r)$ can be approximated by a (dotted black) straight line with slope mg.

all points with $r \geq R_E$. For small altitudes $y = r - R_E \ll R_E$, we can expand the potential in a Taylor series about $r = R_E$. From Box 9.1,

$$U(r) \simeq U(R_E) + \left.\frac{dU}{dr}\right|_{R_E} (r - R_E) + \frac{1}{2}\left.\frac{d^2U}{dr^2}\right|_{R_E} (r - R_E)^2 + \cdots$$

For small y, we need to retain only the first order term:

$$\begin{aligned} U(r) &\simeq U(R_E) - \left.\frac{d}{dr}\frac{GM_E m}{r}\right|_{R_E} (r - R_E) + \cdots \\ &\simeq U(R_E) + \frac{GM_E m}{R_E^2} y. \end{aligned}$$

In Chapter 8 (see Equation 8.10), we found that the gravitational acceleration near the Earth's surface is $g = GM_E/R_E^2$, so the above result simplifies to the familiar form

$$U(R_E + y) - U(R_E) = \Delta U \simeq mgy.$$

This *linear* approximation is accurate for small changes in altitude near the Earth's surface: $r \simeq R_E$ and $y \ll R_E$. For all other situations, the exact expression of Equation 10.4 must be used. The approximate solution is compared to the exact expression of Equation 10.4 in Figure 10.2b.

EXERCISE 10.2

In the above expression, we could choose $U(R_E) = 0$ and obtain the familiar expression $U(y) = mgy$ for points near the Earth's surface. In this case, what would be the value of $U(\infty)$? *Hint*: See Equation 10.3.

Answer $+GMm/R_E$

Note also that $U(r)$ approaches $-\infty$ as $r \to 0$. We can avoid this infinity by again invoking the shell theorem. Imagine a spherical body of uniform density $\rho = M/(4/3)\pi R^3$, where M and R are the body's mass and radius (see Figure 10.3a). The gravitational force on a point mass m embedded *within* the sphere (i.e., for $r < R$) is found from the shell theorem, $F(r) = GM(r)m/r^2$, where $M(r)$ is the mass of the sphere contained within

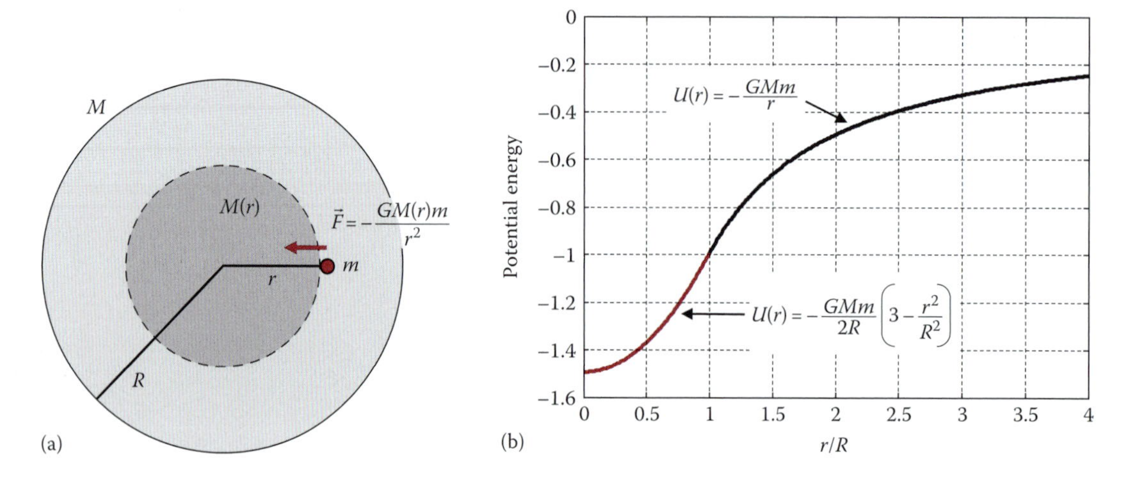

FIGURE 10.3 (a) A test mass m is embedded within a homogeneous sphere of radius R, at a distance r from its center. According to the shell theorem, only the mass $M(r)$ produces a nonzero force on m. (b) The solution for the potential energy for $r \geq R$ (black) and $r < R$ (red). The two solutions match up at $r = R$.

a distance r of its center: $M(r) = (4/3)\pi r^3 \rho = Mr^3/R^3$. The force on m is directed toward the center of the sphere, and is expressed in vector form as

$$\vec{F}(r) = -\frac{GM(r)m}{r^2}\hat{r} = -\frac{GMm}{R^3} r\hat{r}.$$

The potential energy function within the sphere is found by using Equation 10.1. Let's calculate the change in U as m is moved from a point on the surface of M ($r_i = R$) to an interior point ($r_f < R$):

$$U(r_f) - U(R) = -\int_R^{r_f} \vec{F}(\vec{r}) \cdot d\vec{r} = \frac{GMm}{R^3} \int_R^{r_f} r\hat{r} \cdot d\vec{r}$$

$$= \frac{GMm}{R^3} \int_R^{r_f} r dr = \frac{GMm}{2R^3}\left(r_f^2 - R^2\right).$$

The potential energy function must be a continuous function. Otherwise, the force $F = -dU/dr$ would be undefined at the discontinuity. From Equation 10.4, $U(R) = -GMm/R$, so substituting this into the above expression yields

$$U(r_f) = \frac{GMm}{2R^3}\left(r_f^2 - R^2\right) - \frac{GMm}{R} = \frac{GMm}{2R^3} r_f^2 - \frac{3}{2}\frac{GMm}{R}.$$

Dispensing with the subscript f, the complete solution for the potential energy, valid for $0 \le r \le \infty$, is

$$U(r) = \begin{cases} -\dfrac{GMm}{r}, & \text{for } r \ge R, \\ -\dfrac{GMm}{2R}\left(3 - \dfrac{r^2}{R^2}\right), & \text{for } r \le R. \end{cases} \tag{10.5}$$

This function is shown in Figure 10.3b as well as in Figure 10.2b. The negative infinity at $r = 0$ has been avoided and the expressions for the potential outside and inside the sphere's surface converge at $r = R$.

EXERCISE 10.3

(a) Verify from Equation 10.5 that the potential is continuous at $r = R$. (b) Also check to see that the force $F = -dU/dr$ derived from Equation 10.5 is continuous at $r = R$. (c) What is the value of $U(r = 0)$?

Answer (c) $U(0) = -(3/2)GMm/R$

10.4 APPLICATION: ESCAPE VELOCITY

Gravitational potential energy diagrams help us interpret a wide variety of important and interesting phenomena. Consider the launch of a spacecraft from Earth. What is the minimum launch speed needed if it is to escape the planet, that is, to increase its distance from Earth without limit? This is a critical engineering concern because it dictates the amount of fuel that the launch vehicle must carry. For simplicity, assume a stationary, nonspinning planet and ignore air resistance. In addition, suppose that the launch phase is short, so that the fuel used during liftoff is exhausted at low altitude ($y \ll R_E$). Ignoring this altitude, the total mechanical energy of the spacecraft at the end of the launch phase is

$$E = K_L + U(R_E) = \frac{1}{2}mv_L^2 - \frac{GM_E m}{R_E}, \tag{10.6}$$

where, m is the mass of the spacecraft and v_L is its speed when its engines are cut off.

From this time on, the spacecraft is in unpowered flight and E remains constant:

$$\frac{1}{2}mv^2 - \frac{GM_E m}{r} = E = \frac{1}{2}mv_L^2 - \frac{GM_E m}{R_E}.$$

As the spacecraft ascends, its potential energy $U(r)$ increases at the expense of its kinetic energy. It reaches a maximum altitude at the turning point r_{max} where $E = U(r_{max})$ and $v = 0$. This is illustrated in the potential energy diagram shown in Figure 10.4. The turning point is where the horizontal line representing E intersects the potential energy curve. If the launch speed is raised, the energy also increases (making E less negative), and the turning point r_{max} lies farther from the center of the planet. For the rocket to escape Earth, r_{max} must equal infinity, which requires $E \geq 0$. Solving Equation 10.6 for $E = 0$, the minimum launch speed needed for escape, called the *escape velocity* v_{esc}, is

$$v_{esc} = \sqrt{\frac{2GM_E}{R_E}} = \sqrt{2gR_E}, \tag{10.7}$$

where we have used $g = GM_E/R_E^2$.

EXERCISE 10.4

(a) Use $M_E = 5.97 \times 10^{24}$ kg and $R_E = 6380$ km, or $g = 9.8$ m/s², to find the escape velocity of a rocket from Earth. (b) For Mars, $M_M/M_E = 0.108$ and $R_M/R_E = 0.532$. Find v_{esc} from the Martian surface.

Answer (a) 11.2 km/s (b) 5.0 km/s

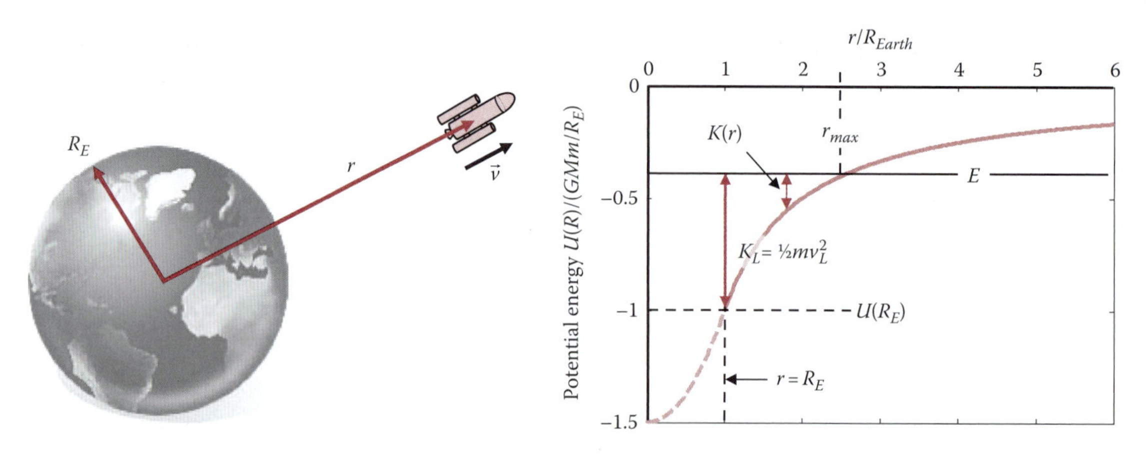

FIGURE 10.4 A rocket is launched from Earth with an initial kinetic energy K_L. Its highest altitude r_{max} is determined by the intersection of the total energy E with the potential energy $U(r)$. As K_L and E increase, the turning point r_{max} increases. When $E = 0$, r_{max} is at ∞, and the rocket is just capable of escaping from Earth.

Note that the rocket mass m does not enter into Equation 10.7. Consequently, the escape velocities of a spacecraft, a cannonball, a baseball, or even a molecule are *all the same*! This will be important in the discussion of planetary atmospheres below.

EXERCISE 10.5

Suppose the kinetic energy given to the spacecraft at launch is *twice* the minimum kinetic energy needed for escape from Earth. What will be the speed of the craft when it is far from Earth? (In the literature, this is called its *hyperbolic excess velocity*.)

Answer 11.2 km/s

10.5 APPLICATION: PLANETARY ATMOSPHERES

Remove the cap from a bottle of soda, and the carbon dioxide gas trapped above the liquid quickly escapes from the bottle. For a short while, it is replenished by gas evolving from the liquid, but soon the dissolved gas is depleted and the soda goes flat. Compare this to the Earth's atmosphere. It has been left open to space for 4.5 billion years, yet the planet has somehow managed to retain a life-supporting atmosphere of nitrogen, oxygen, and water vapor. Gravity acts as a leaky bottle cap for Earth. Molecules with speeds faster than the planet's escape velocity v_{esc} (= 11.2 km/s) leak away, or evaporate, whereas slower molecules remain trapped in the Earth's gravitational "bottle." This picture, proposed by James Jeans in 1925, accounts for the presence or absence of various molecules in the atmospheres of planets, moons, and other celestial bodies.

According to the *kinetic theory* of gases,* molecules move in random directions over a broad range of speeds characterized by the Maxwell–Boltzmann *distribution function* $P(v)$ shown in Figure 10.5a. The probability that a gas molecule has a speed in the range $(v, v + dv)$ is equal to the area $P(v)dv$ under the curve between these two limits (shown as the solid gray area in Figure 10.5a), and the total area under the curve is equal to 1. The most important result of the theory is that the average kinetic energy of a molecule is proportional to the temperature T of the gas (measured in Kelvin):

$$K_{avg} = \frac{1}{2}m(v^2)_{avg} = \frac{3}{2}k_B T,$$

or, in terms of a new variable v_{rms}, (10.8)

$$K_{avg} \equiv \frac{1}{2}mv_{rms}^2 = \frac{3}{2}k_B T,$$

where, m is the mass of the molecule and k_B is called Boltzmann's constant†: $k_B = 1.38 \times 10^{-23}$ J/K.

* Kinetic theory was developed by James Maxwell and Ludwig Boltzmann in the mid-nineteenth century.

† Boltzmann's constant is the ratio of two familiar constants: $k_B = R/N_A$, where $R = 8.314$ J/K is the ideal gas constant (as in $PV = nRT$), and N_A is Avogadro's constant: $N_A = 6.02 \times 10^{23}$.

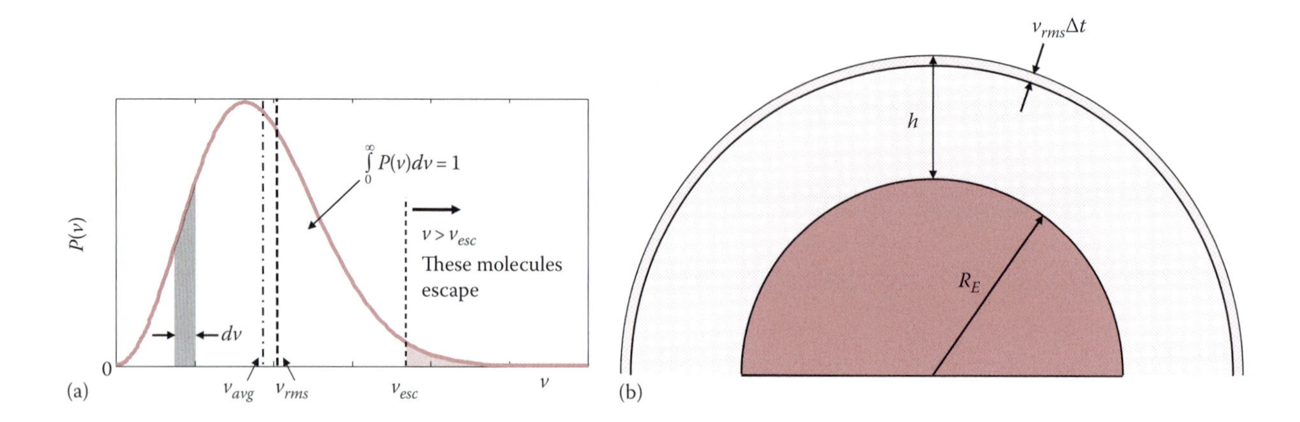

FIGURE 10.5 (a) The probability that a molecule has a velocity in the range (v, $v + dv$) is given by the solid gray area under the curve $P(v)$. The root-mean-square velocity v_{rms} lies slightly above the peak of the curve. Even if $v_{esc} > v_{rms}$, a small fraction of the molecules can escape the planet. (b) In time Δt, only those molecules that are within $v_{rms}\Delta t$ of the top edge of the atmosphere with speeds $v > v_{esc}$ can escape.

The new variable $v_{rms} = \sqrt{(v^2)_{avg}}$ is the "root-mean-square" speed; it is the square *root* of the *mean* (average) of the velocity *squared*. It is roughly equal to the average speed v_{avg} of the gas molecules. From Equation 10.8,

$$v_{rms} = \sqrt{\frac{3k_B T}{m}} \approx v_{avg}. \tag{10.9}$$

For example, a nitrogen molecule (N_2) has a mass $m = 28u = 28 \times 1.66 \times 10^{-27}$ kg $= 4.64 \times 10^{-26}$ kg. At room temperature ($T \approx 300$ K), a gas of nitrogen molecules has $v_{rms} \simeq 500$ m/s $= 0.5$ km/s. Although this is pretty fast (≈1100 mph!), it is still much slower than the Earth's escape velocity.

EXERCISE 10.6

Estimate v_{rms} for a hydrogen molecule (H_2) in a gas at temperature 600 K.

Answer 2.6 km/s

As illustrated in Figure 10.5a, the probability $P(v)$ decreases rapidly with increasing speed for $v > v_{rms}$. But even if v_{esc} is considerably greater than v_{rms}, there will be a tiny fraction of molecules in the gas

TABLE 10.1 Fraction of Molecules with Sufficient Kinetic Energy to Escape a Planet Decreases Very Rapidly as the Ratio v_{esc}/v_{rms} Increases

If v_{esc}/v_{rms} = ···	Fraction of Molecules with $v > v_{esc}$ is...
4.0	1.9×10^{-9}
4.5	4.0×10^{-12}
5.0	4.0×10^{-15}
5.5	1.8×10^{-18}
6.0	3.8×10^{-22}
6.5	3.7×10^{-26}

with sufficient kinetic energy to escape the planet. This fraction f is tabulated as a function of v_{esc}/v_{rms} in Table 10.1. As indicated, f depends *very* sensitively on the ratio v_{esc}/v_{rms}. Since v_{rms} depends inversely on the molecular mass, f must also be a sensitive function of mass. Different molecular compounds will have vastly different escape rates, and the ratio v_{esc}/v_{rms} determines which chemical species remain in the planet's atmosphere after billions of years of slow leakage into space.

EXERCISE 10.7

(a) Find the ratio of *rms* velocities for oxygen (O_2, $m = 32u$) and carbon dioxide (CO_2, $m = 44u$) at the same temperature. (b) If these two molecules are in the atmosphere of a small planet with $v_{esc}/v_{rms} = 5.0$ for O_2, estimate the ratio $f(CO_2)/f(O_2)$.

Answer (a) ≈ 1.2 (b) $\approx 10^{-7}$

Because of the strong dependence on v_{esc}/v_{rms}, other details of the escape process are less important, and a crude model is sufficient for predicting the chemical composition of a planetary atmosphere. Picture the atmosphere as a homogeneous layer of gas with a thickness (height) $h \ll R$, where R is the radius of the planet (see Figure 10.5b). Assume a uniform density n of molecules (number per unit volume) and a uniform temperature T throughout the atmosphere.* What fraction of these molecules escape into space during a short time interval Δt? To escape, a molecule must first reach the top of the atmosphere and then, via collisions with other molecules, acquire a speed $v \geq v_{esc}$. Only molecules within a distance $\approx v_{rms}\Delta t$ of the top can reach the top in time Δt. The fraction of molecules in the atmosphere satisfying this condition is $v_{rms}\Delta t/h$, and only a small fraction f of these molecules will have sufficient speed to escape. Therefore, the fraction of molecules escaping in time

* None of these assumptions is accurate, but—as we argued—the exact details are unimportant.

Δt is roughly $f \cdot v_{rms}\Delta t/h$. If the initial number of atmospheric molecules is N_0, then the number of molecules remaining after time Δt is

$$N_1 = N_0\left(1 - \frac{fv_{rms}\Delta t}{h}\right).$$

In the next time interval Δt, an equal fraction of these N_1 molecules escape into space and the number of molecules left behind after $2\Delta t$ is $N_2 = N_1(1 - fv_{rms}\Delta t/h) = N_0(1 - fv_{rms}\Delta t/h)^2$, and so on. Over a time t comparable to the age of the planet or moon, this escape process reoccurs $n = t/\Delta t$ times, and the number of molecules remaining after time t is

$$N_n = N_0\left(1 - \frac{fv_{rms}\Delta t}{h}\right)^n = N_0\left(1 - \frac{fv_{rms}t/h}{n}\right)^n,$$

or, using the binomial expansion $(1 + x)^n \approx 1 + nx$,*

$$N(t) \approx N_0\left(1 - \frac{fv_{rms}t}{h}\right).$$

To retain a significant atmosphere, $fv_{rms}t/h < 1$, or $f < h/v_{rms}t$.

Let's put in some numbers. The time t is approximately equal to the time since the formation of the solar system: $t \approx (4.5 \times 10^9 \text{ years}) \cdot (3.15 \times 10^7 \text{ s/years}) \simeq 1.5 \times 10^{17}$ s. A uniformly dense atmosphere of an Earthlike planet would be about 10 km thick (h = 10 km). So to retain an atmosphere of N_2 at a temperature of 300 K ($v_{rms} \simeq 500$ m/s), f must be less than about 10^{-16}. From Table 10.1, this requires $v_{esc} > 5.2v_{rms}$. A more rigorous analysis yields a more accurate criterion for retaining a significant fraction of a given molecular species in a planet's atmosphere,

$$v_{esc} \geq 6v_{rms}. \tag{10.10}$$

In real planetary atmospheres, the temperature is far from uniform, and we must use the temperature T_{exo} at the outer edge of the atmosphere (called the *exosphere*) to determine v_{rms} and the value of f. Although T_{exo} is difficult to measure, astronomers have fair estimates for the planets and some of the moons of our solar system. The average surface temperature on Earth is about 290 K, but its exosphere temperature is nearly 1000 K. For N_2 molecules in the Earth's exosphere, Equation 10.9 tells us that v_{rms} = 940 m/s or

* More properly, $e^{-x} = (1 - x/n)^n$ for $n \gg 1$, so the atmosphere leaks away *exponentially* with time.

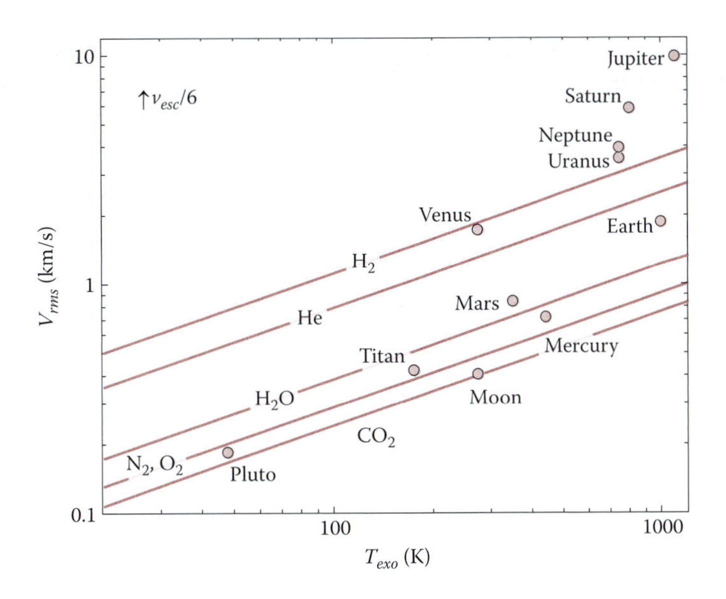

FIGURE 10.6 Each solid line is a plot of $v_{rms} = \sqrt{3k_BT_{exo}/m}$ as a function of the exosphere temperature T_{exo} for simple molecules of mass m. The coordinates of each point are (T_{exo},$v_{esc}/6$) for a particular planet or moon. If the point lies above the molecule's line, then $v_{esc}/6 > v_{rms}$, and the planet or moon will retain the molecule. (Data courtesy of D.C. Catling; Adapted from Catling, D.C. and Zahnle, K.J., *Scient. Am.*, 300(5), 36, May 2009.)

$v_{esc}/v_{rms} = 11.9$. Since 11.9 > 6, we can safely conclude that relatively little of Earth's N_2 has escaped into space during the past 4.5 billion years.

EXERCISE 10.8

Calculate v_{rms} for helium molecules ($m = 4u$) using $T_{exo} = 1000$ K. Compare your result with the Earth's escape velocity, and comment on the concentration of He (0.0005%) in the Earth's present atmosphere. Terrestrial helium is produced by α-decay of thorium and uranium within the Earth's crust, and is collected as a by-product of natural gas mining.

Answer $v_{rms} = 2.5$ km/s, $v_{esc}/v_{rms} = 4.5 < 6$

The solid lines in Figure 10.6 are plots of v_{rms} as a function of T_{exo}, calculated using Equation 10.9, for a number of common atmospheric molecules. Each solar system planet (along with two moons)* is represented by a point indicating its exosphere temperature and the value of $v_{esc}/6$ for that body. If this point lies above the line for a particular molecular species (e.g., H_2O), then $v_{esc} > 6v_{rms}$ and it is probable that if molecules of this chemical composition were originally present within the body, they still remain within its atmosphere.

---

* And one *dwarf planet*—Pluto!

10.6 ENERGY AND ORBITS

The principle of energy conservation offers an elegant way to analyze the trajectories of planets and other bodies in space. Imagine a body of mass m executing a circular orbit of radius r about a much more massive body M. Starting from Newton's second law and expressing the centripetal acceleration of the orbiting body as $a = v^2/r$, we can easily determine the body's kinetic energy K:

$$m\frac{v^2}{r} = \frac{GMm}{r^2},$$

so $v^2 = GM/r$ and

$$K = \frac{1}{2}mv^2 = \frac{1}{2}\frac{GMm}{r}. \tag{10.11}$$

Comparing Equations 10.4 and 10.11, $K = -(1/2)U$, and

$$E = K + U = \frac{1}{2}U = -\frac{GMm}{2r}. \tag{10.12}$$

EXERCISE 10.9

A satellite is in a circular orbit of radius r_o about Earth. Its potential energy is $U(r_o) = -5 \times 10^{10}$ J.

(a) What is its kinetic energy K?
(b) What is its total energy E_{mech}?
(c) If the satellite were initially at rest (not in orbit) at $r = \infty$ and allowed to fall toward Earth, what would its kinetic energy be at a distance r_o from the center of the planet?

Answer (a) 2.5×10^{10} J (b) -2.5×10^{10} J (c) 5×10^{10} J

As you know, the orbits of satellites, moons, planets, and interplanetary spacecraft are ellipses, not circles. We need to derive the counterpart to Equation 10.12 for elliptical orbits. Since the total energy is constant, it is sufficient to determine E at just one point along the orbit. Recall the description of ellipses presented in Box 8.1. At the *periapsis* of the orbit, the orbiting body is closest to the central body, and the distance between them is $r_p = (1 - \varepsilon)a$, where ε is the orbit eccentricity and a is the semimajor axis. At periapsis, the radius of curvature of the ellipse is $r_c = b^2/a$, where b is the semiminor axis of the ellipse, and we can regard the body at periapsis to be moving momentarily along a circular trajectory of radius r_c. Newton's second law can then be written

$$m\frac{v_p^2}{r_c} = \frac{GMm}{r_p^2},$$

so the kinetic energy at periapsis is

$$K_p = \frac{1}{2}mv_p^2 = \frac{1}{2}\frac{GMm}{r_p^2}r_c.$$

The potential energy at periapsis is $U_p = -GMm/r_p$, so the total energy of the orbiting body is

$$E = K_p + U_p = \frac{1}{2}\frac{GMm}{r_p^2}r_c - \frac{GMm}{r_p}.$$

Substituting for r_c and r_p, and using $b^2 = (1 - \varepsilon^2)a^2$,

$$E = \frac{GMm}{r_p}\left(\frac{b^2/a}{2r_p} - 1\right) = \frac{GMm}{(1-\varepsilon)a}\left(\frac{(1-\varepsilon^2)a}{2(1-\varepsilon)a} - 1\right).$$

Simplifying, we obtain the desired result:

$$E = -\frac{GMm}{2a}. \qquad (10.13)$$

EXERCISE 10.10

Derive Equation 10.13 from the expression preceding it.

The energy depends only on the length of the semimajor axis a; it is otherwise independent of the shape of the ellipse. For circular orbits, $a = b = r$, and Equation 10.13 reduces to Equation 10.12.

EXERCISE 10.11

When calculating planetary orbits, it is handy to know the value of GM_{Sun}/R_{ES}, where R_{ES} = 1 AU.

(a) Calculate this value and express your answer in J/kg.

A spacecraft is in a circular orbit about the Sun, with orbital radius 1 AU. Following a short rocket burst, its kinetic energy is boosted by 50%, and it enters an elliptical orbit.

(b) Compare its new energy E' with its original energy E. (c) What is the semimajor axis of this new orbit? (d) What is the period of this orbit?

Answer (a) 9×10^8 J/kg (b) $E' = (1/2)E$ (c) 2 AU (d) 2.83 years

As long as $E < 0$, the semimajor axis a remains finite, and the maximum separation between m and M occurs at the *apoapsis* $r_a = (1 + \varepsilon)a$. We say that m is *bound* to M and that its orbit is *closed*. As $E \to 0$, $a \to \infty$,

Except for the initial launch phase, Earth's influence can be safely ignored. To show this, let d_E be the distance from the spacecraft to Earth and d_S be the distance to the Sun (1 AU at launch). The force exerted by the Sun will be greater than that exerted by Earth when

$$\frac{GM_S}{d_S^2} > \frac{GM_E}{d_E^2},$$

or

$$\frac{d_E}{d_S} > \sqrt{\frac{M_E}{M_S}} = \sqrt{\frac{6\times10^{24}}{2\times10^{30}}} \approx 0.0017.$$

When the spacecraft is more than about 0.007 AU from Earth (a distance that is hardly visible on the scale of Figure 10.9a), the Sun's gravitational force is 16 times greater than Earth's, and from this time forward, the planet's gravitational tug on the spacecraft is negligible. A similar calculation can be done to compare the Martian and solar gravitational forces on the spacecraft near the end of the spaceflight. Since $M_M \simeq 0.1M_E$, the influence of Mars is unimportant until the spacecraft is within about 0.002 AU of the planet. Therefore, except at the very beginning and very end of its interplanetary journey, the spacecraft's trajectory is a nearly pure elliptical arc with the Sun at one focus.

Immediately after liftoff, the MSL and attached second-stage rocket are placed in a "parking" orbit about Earth before the rocket engine is reignited to boost the spacecraft's energy, allowing it to escape from Earth and enter its elliptical trajectory *en route* to Mars. (The rocket is then jettisoned.) After escape, the MSL's velocity relative to Earth is $\Delta\vec{v}_1$, so its velocity relative to the Sun is $\vec{v} = \vec{v}_E + \Delta\vec{v}_1$, where $\vec{v}_E$ is the orbital velocity of Earth. Since the spacecraft is not yet far from the planet, its distance from the Sun is still 1 AU when it enters its elliptical orbit (see Figure 10.9a). Therefore, its potential energy of interaction with the Sun is unchanged, and the change in total energy (referenced to the Sun) due to $\Delta\vec{v}_1$ is equal to the change in its kinetic energy:

$$\Delta E = \Delta K = \frac{1}{2}m(\vec{v}_E + \Delta\vec{v}_1)^2 - \frac{1}{2}mv_E^2 = m\left(\vec{v}_E \cdot \Delta\vec{v}_1 + \frac{1}{2}\Delta v_1^2\right), \quad (10.14)$$

where m is the spacecraft's mass.

From Equation 4.15, we can find the amount of rocket fuel needed to produce $\Delta\vec{v}_1$:

$$\Delta v_1 = v_{ex} \ln\frac{M_i}{M_f}, \quad (4.15)$$

where, v_{ex} is the propellant exhaust velocity (typically about 2 km/s) and M_i and M_f are the masses of the spacecraft before and after the fuel burn.

Let $M_i = M_p + m_{fuel}$, and $M_f = M_p$, the payload mass. Solving Equation 4.15 for the fuel mass m_{fuel}:

$$m_{fuel} = M_p \left(e^{\Delta v_1 / v_{ex}} - 1 \right).$$

For $\Delta v_1 / v_{ex} > 1$, the fuel required increases *exponentially* with Δv_1, so it is imperative to orient $\Delta \vec{v}_1$ in the direction which minimizes the fuel cost. Equation 10.14 shows that the gain in kinetic energy is greatest when $\Delta \vec{v}_1$ is parallel to $\vec{v}_E$ (see Figure 10.9b). With this orientation, $\vec{v} = \vec{v}_E + \Delta \vec{v}_1$ remains perpendicular to the spacecraft position $\vec{r} = \vec{r}_E$, so the burn location must be either the perihelion or aphelion of the interplanetary path. Since $\vec{v} > \vec{v}_E$, the speed is greater than it would be for a circular trajectory, so the burn must take place at perihelion and $r_p = R_{ES} = 1.0$ AU. Furthermore, because the orbit energy increases with increasing semimajor axis a, the lowest energy elliptical trajectory will be the one that just barely reaches the Martian orbit at its aphelion: $r_a = R_{MS} = (2a - r_p) = 1.5$ AU. This is exactly the geometry of the Hohmann transfer orbit, and so it makes sense that it is the most advantageous trajectory for interplanetary travel.*

EXERCISE 10.13

What if the spacecraft were traveling to an *inferior* planet, such as Venus? Then the spacecraft must be slowed from $\vec{v}_E$ to enter a Hohmann transfer orbit.

(a) What direction would you choose for $\Delta \vec{v}_1$ relative to $\vec{v}_E$?
(b) Where would the perihelion and aphelion of the transfer orbit be located?

EXERCISE 10.14

(a) How long does the trip to Mars take? (*Hint:* The trajectory is only half of a full ellipse.) (b) Figure 10.9a shows the location of Mars at the arrival of the spacecraft. Where on the diagram was the planet at launchtime? (Mars' orbital period is 1.88 years.)

Answer (a) 0.70 years or 255 days

Suppose the MSL spacecraft has escaped Earth and is coasting on a circular "Earth-leading" orbit of radius $R_{ES} = 1$ AU with speed $v_E = \sqrt{GM_S / R_{ES}} = 30$ km/s. Its total energy at this time is given by Equation 10.12, $E = -GM_S m / 2R_{ES}$, where m is its mass. A subsequent rocket burn then sends the spacecraft on its way to Mars. As explained above, the new rocket burn occurs at the *perihelion* of the trajectory, with $r_p = R_{ES} = 1$ AU

* Our discussion is only a plausibility argument, certainly not a proof. In fact, in his original 1925 publication (Reference 4), Walter Hohmann merely compared the fuel requirements of alternative trajectories with those of the orbit described above.

(distance from the Sun) and aphelion $r_a = 1.5R_{ES}$. The minimum energy needed to reach Mars is $E' = -GM_S m/2a$, with $2a = r_p + r_a = 2.5R_{ES}$. After the burn, the spacecraft's velocity is $\vec{v} = \vec{v}_E + \Delta\vec{v}_1$, and the total energy is

$$E' = -\frac{GM_S m}{R_{ES}} + \frac{1}{2}mv^2 = -\frac{GM_S m}{2.5R_{ES}}. \tag{10.15}$$

Solving for the kinetic energy, we obtain

$$\frac{1}{2}mv^2 = \frac{GM_S m}{R_{ES}}\left(1 - \frac{1}{2.5}\right) = \frac{3}{5}\frac{GM_S m}{R_{ES}},$$

or

$$v = \sqrt{\frac{6}{5}}\sqrt{\frac{GM_S}{R_{ES}}} = 32.9 \text{ km/s}.$$

Since $\vec{v}$, $\vec{v}_E$, and $\Delta\vec{v}_1$ are all parallel, $\Delta v_1 = v - v_E = 2.9$ km/s, and the spacecraft's rockets must consume enough fuel to change the velocity by this amount. This requires a significant amount of fuel. If $v_{ex} = 2.15$ km/s (hydrazine propellant), then $m_{fuel} \simeq 3M_P$. (Check this on your calculator). The actual MSL payload mass was nearly 4,000 kg, so 12,000 kg of fuel were required to inject the spacecraft into its interplanetary trajectory. The required fuel was carried and burned by the second stage Atlas rocket before it separated from the MSL.

EXERCISE 10.15

Suppose you choose the "simpler" transfer path shown below instead of the Hohmann orbit. After escaping from Earth, the spacecraft is moving along a circular orbit with radius 1.0 AU and speed 30 km/s, as shown. Its on-board rockets change the *direction* of travel by 90°, but not its speed or kinetic energy, so that the craft is now traveling *radially away* from the Sun with speed 30 km/s.

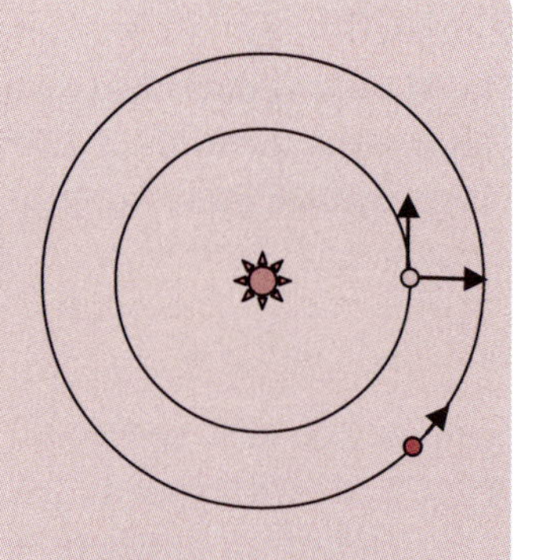

(a) Find the magnitude of Δv_1 needed for this maneuver.
(b) Compare the amount of fuel needed with the fuel required for the Hohmann orbit.

Answer: (a) $\Delta v_1 = 42.4$ km/s

Once the spacecraft reaches the vicinity of Mars, its velocity relative to the planet must be made small enough to either orbit the planet or descend safely to the Martian surface. The orbital speed of Mars is

$\sqrt{GM_S/R_{MS}} = 30/\sqrt{1.5} = 24.5$ km/s, and the speed v of the spacecraft when it reaches the planet can be found by once again invoking energy conservation as in Equation 10.15:

$$-\frac{GM_S m}{1.5R_{ES}} + \frac{1}{2}mv^2 = E' = -\frac{GM_S m}{2.5R_{ES}},$$

where we have used $R_{MS} = 1.5R_{ES}$ and $2a = 2.5R_{ES}$. Solving for the kinetic energy,

$$\frac{1}{2}mv^2 = \left(\frac{2}{3} - \frac{2}{5}\right)\frac{GM_S m}{R_{ES}},$$

or

$$v = \sqrt{\frac{8}{15}}\sqrt{\frac{GM_S}{R_{ES}}} = \sqrt{\frac{8}{15}} \times 30 \text{ km/s} = 21.9 \text{ km/s}.$$

To match the planet's orbital speed, the spacecraft's on-board thrusters must be reactivated to increase its speed by $\Delta v_2 = 24.5 - 21.9 = 2.6$ km/s. Overall, the sum $\Delta v_1 + \Delta v_2$ is a useful measure of the total propellant needed after launch to complete the mission.

On August 6, 2012, the MSL successfully landed on the surface of Mars, less that 2.5 km from its intended target. To execute this amazing feat, the spacecraft intercepted the Martian orbit *ahead* of the planet, traveling at 21.9 km/s relative to the Sun. In the reference frame of Mars, the MSL was then falling toward the planet at a speed of 2.6 km/s (see Figure 10.9c). It was then allowed to fall under the influence of Martian gravity. It first encountered the planet's atmosphere at an altitude of 120 km, roughly 3500 km from the center of the planet, where it was slowed initially by deploying a parachute. On-board retrorockets then brought the craft to a halt before lowering the rover *Curiosity* to the surface (see Problem 6.7).

EXERCISE 10.16

Far above the surface of Mars, the MSL was falling toward the planet with relative speed 2.6 km/s. What was its speed relative to Mars when it entered the Martian atmosphere, 3500 km from the center of the planet? (According to the NASA/JPL mission overview, the acceptable range of atmosphere entry speeds was $5.3 < v < 5.9$ km/s.) ($M_M = 0.642 \times 10^{24}$ kg)

Answer 5.6 km/s.

10.8 LITTLE NEWTONIAN COSMOLOGY

Cosmology is science on the grandest scale, where science and religion have overlapping—and often conflicting—interests. Cosmologists and clerics ask the same basic questions: When and how did the universe begin? When and how will it end? But when addressing these questions, scientists are confined to hypotheses that can be supported or refuted by hard physical evidence.* A proper treatment of modern cosmology requires concepts well beyond the scope of this text, yet it is still possible to fathom some basic understanding, and appreciate the stunning enigmas uncovered by recent research, using the Newtonian concepts you have already encountered.

On the supergalactic scale—where galaxies are just tiny dots—the universe is remarkably homogeneous. Wherever we point our telescopes, we find that the number of galaxies (as well as their properties) in a given volume of space is, on average, the same as in any other region of similar volume. No region of space is special. In cosmology, this observation is elevated to a foundational principle, the *cosmological* or *Copernican* principle. Just as Copernicus dislodged Earth from its special position at the center of the universe, cosmologists assert that our Milky Way galaxy is just an ordinary galaxy in an otherwise ho-hum region of space. The Copernican principle implies that astronomers in other galaxies draw the same conclusions about the universe as we do, and requires our cosmological theories to reflect this. For example, when we introduced the Hubble Law $\vec{v} = H_0\vec{r}$ in Chapter 3, we showed that observers in other galaxies see the same expansion law with the same value of H_0, in accord with the Copernican principle.

Hubble's Law—the experimental cornerstone of cosmology—states that the universe is expanding and distant galaxies in all directions are moving radially away from us. The Copernican principle adds that radial motion is the *only* possible motion. (Otherwise, the universe would look different to observers in other galaxies.) In Newtonian mechanics, this suggests that matter in the universe is distributed in a spherically symmetric pattern about us.† Consider a galaxy of mass m that is presently a distance r_0 from the Milky Way. Invoking the shell theorem, the gravitational mass $M(r_0)$ attracting this galaxy to us is the total mass enclosed within a sphere of radius r_0 centered on the Milky Way: $M(r_0) = (4/3)\pi r_0^3 \rho_0 \equiv M_0$ (see Figure 10.10a). In this expression, ρ_0 is the present mass density of the universe, which—according to the Copernican principle—must be the same throughout the universe. By Hubble's Law, galaxies within the sphere ($r < r_0$) are receding slower than m, while galaxies outside the sphere ($r > r_0$) are receding faster, so *nothing crosses the sphere's surface* as it expands outward. In other words, the mass M_0 acting gravitationally on m remains constant for all time. This is a familiar scenario, akin to the interaction between two point masses, and we can use the energy methods outlined earlier to study it.

In the following discussion, the subscript "0" refers to the *present* time t_0. The present location and velocity of galaxy m are denoted by $\vec{r}_0$ and $\vec{v}_0$, while ρ_0 and H_0 are the present density of the universe and the presently measured value of the Hubble constant. (In an expanding universe, the density must decrease with time, $\rho(t) < \rho_0$ for $t > t_0$.)

---

* Nevertheless, one of the first persons to propose an expanding universe was the Roman Catholic priest George Lemaître.

† This cannot be literally true. According to the Copernican principle, observers in a distant galaxy would draw the same conclusion of spherical symmetry centered on their galaxy, and two separated points cannot both be centrally located. This is a serious stumbling block in Newtonian cosmology, resolvable only by appealing to general relativity.

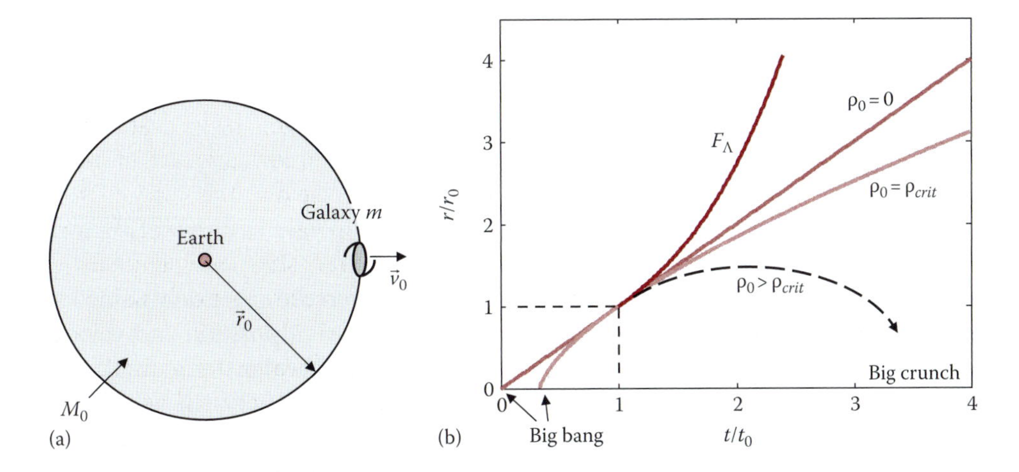

FIGURE 10.10 (a) A galaxy at a distance r_0 is gravitationally attracted toward the Milky Way by the total mass within a sphere of radius r_0. In Newtonian cosmology, this mass always slows the expansion rate of the universe. (b) The cosmic expansion for an "empty universe" $\rho = 0$; one with critical density $\rho = \rho_{crit}$; a closed universe $\rho > \rho_{crit}$; and one dominated by a repulsive force F_Λ derived from dark energy.

The total mechanical energy of galaxy m can be written in the usual way, $E = K + U$, or in terms of its present velocity and position relative to us,

$$E = \frac{1}{2}mv_0^2 - \frac{GM_0 m}{r_0}. \tag{10.16}$$

If $E < 0$, the Hubble expansion is bounded and the universe will eventually stop expanding. Solving Equation 10.16 for v_0, we find that for a bounded expansion,

$$v_0^2 < v_{esc}^2 = \frac{2GM_0}{r_0} = \frac{8}{3}\pi G\rho_0 r_0^2,$$

or, with $v_0 = H_0 r_0$ (the Hubble Law),

$$H_0^2 < \frac{8}{3}\pi G\rho_0.$$

Notice that r_0 has cancelled out of this equation, so that this result applies to *all* galaxies at all distances from us, that is, to the entire universe. For a *closed universe*, one that does not expand forever, the present density ρ_0 will be large enough to halt the expansion if it is greater than a critical density ρ_{crit}:

$$\rho_0 > \rho_{crit} \equiv \frac{3H_0^2}{8\pi G}. \tag{10.17}$$

EXERCISE 10.17

The present value of H_0 is $H_0 = 72$ km/s/Mpc $= 2.3 \times 10^{-18}$ s^{-1}.

(a) Calculate the critical density ρ_{crit} expressed in kg/m^3.
(b) How many hydrogen atoms per cubic meter does this represent? ($1u = 1.66 \times 10^{-27}$ kg.)

Answer (b) ≈6 atoms/m³

What does Newtonian physics predict for the evolution of the universe? Let's examine two special cases. For each, we will trace the position of a "test" galaxy m whose present position and velocity relative to us are $\vec{r}_0$ and $\vec{v}_0$. In the first case, called the "empty universe" model, the density $\rho_0 \ll \rho_{crit}$, allowing us to ignore the gravitational attraction between galaxies. In the absence of gravity, the velocity of galaxy m is constant, equal at all times to its present velocity, $v(t) = v(t_0) = v_0$. The galaxy's position relative to us at time t is $r(t) = v_0 t$, where t is the elapsed time since the "big bang," the time when the expansion began. At the present time, $r_0 = v_0 t_0 = H_0 r_0 t_0$, so the present age of the universe (in this model) is

$$t_0 = \frac{1}{H_0}.$$

EXERCISE 10.18

Find the age of the universe in this "empty universe" model. Express your answer in billions of years. (See Exercise 10.17 for the value of H_0.)

Using the above equations, the position of our test galaxy as a function of time can be written as

$$\frac{r}{r_0} = \frac{t}{t_0} = H_0 t. \qquad (10.18)$$

This is plotted as the straight line in Figure 10.10b. Clearly, an "empty universe" will expand forever at a constant rate.

In the second case, let $\rho_0 = \rho_{crit}$. Now the galaxy's speed is $v_0 = v_{esc}$, and its total mechanical energy $E = 0$. The total gravitating mass attracting the galaxy to us is $M_0 = (4/3)\pi r_0^3 \rho_{crit}$. Using Equation 10.17, this can be expressed conveniently as $2GM_0 = H_0^2 r_0^3$. From Equation 10.16 with $E = 0$, the galaxy m moves with relative velocity

$$v = \frac{dr}{dt} = \sqrt{\frac{2GM_0}{r}} = \sqrt{H_0^2 r_0^3}\, r^{-(1/2)}.$$

Rearranging, we obtain

$$r^{1/2} dr = H_0 r_0^{3/2} dt,$$

which can be integrated to obtain

$$\frac{r}{r_0} = \left(\frac{3}{2} H_0 t\right)^{2/3}. \qquad (10.19)$$

Due to the gravitational interaction between galaxies, the expansion is slowing down as time advances. (See Figure 10.10b.)

EXERCISE 10.19

(a) Do the integration and derive Equation 10.19. (b) Show that in the "critical density" model, the age of the universe is $t_0 = 2/(3H_0)$.

If $\rho_0 > \rho_{crit}$, the expansion will slow to a halt and then reverse itself, ending in a "big crunch" billions of years from now. But in any case, as long as $\rho_0 > 0$, the graph of $r(t)$ will lie below the straight line of the empty universe case, and the expansion rate will decrease with time.

But this is *not at all* what we observe! In stark contrast to these predictions, measurements taken during the past decade by two international teams* of astronomers indicate that the Hubble expansion is *speeding up*! A *repulsive* force $\vec{F}_\Lambda \propto mr\hat{r}$ of unknown origin is counteracting the attractive gravitational force on galaxies. This is not a small correction to our basic model: as the universe expands, its mass density $\rho(t)$ decreases ($\rho(t) \propto 1/r(t)^3$) while the repulsive force increases ($F_\Lambda \propto r$) and plays an ever more dominant role in driving the evolution of the cosmos. At some time in the future, assuming nothing changes from what we now observe, $\vec{F}_\Lambda$ will be the only cosmologically important force, and from then on the universe will expand at an exponential rate (shown in Figure 10.10b):

$$\frac{r}{r_0} = e^{H_0(t-t_0)}. \tag{10.20}$$

As bizarre, mysterious, and unphysical as it may seem, $\vec{F}_\Lambda$ has exceeded the gravitational force for nearly half the life of the universe, pumping huge amounts of "dark energy" into space to accelerate the Hubble expansion. Using $E = mc^2$, the equivalent mass density of this dark energy is presently 2.7 times greater than the density of "gravitational" matter. The latter includes ordinary baryonic matter (i.e., the ingredients of atoms and molecules) plus the equally mysterious dark matter that pervades space. We know nothing about dark energy or dark matter other than that they exist. They will be discussed in more detail in Chapter 14.

EXERCISE 10.20

In general, the Hubble "constant" is a function of time, and the symbol H_0 represents its present value (at time t_0). The Hubble Law at time t is $\vec{v}(t) = H(t)\vec{r}(t)$.

(a) Show that for the empty universe (Equation 10.18), $H(t) = H_0(t_0/t)$.
(b) Show that in the "dark energy" model of Equation 10.20, $H(t) = H_0$, that is, the Hubble constant is indeed constant with respect to time.

* The two teams were lead by Saul Perlmutter, Brian Schmidt, and Adam Riess, the trio who shared the Nobel Prize in physics in 2011 for their discovery of the accelerating expansion of the universe. See Section 9.15.

BOX 10.2 WHY DO THINGS FALL?

In Chapter 7, Box 7.1, we learned that time does not advance at the same rate for all observers. For example, two observers on opposite sides of the Earth might use their separate clocks to measure the times t_1 and t_2 when an orbiting satellite passes directly overhead. They then compute the time interval between these events to be $\Delta t = t_2 - t_1$. But an astronaut aboard the satellite would measure a different time interval $\Delta\tau$ between the same two events. The crucial difference is that $\Delta\tau$ is measured using a *single* onboard clock rather than separate clocks. $\Delta\tau$ is called *proper* time, or *wristwatch* time, and as you will see, it dictates how a body moves in a gravitational field.

Because of its *motion*, the astronaut's orbiting clock "ticks" at a slower rate than stationary clocks on the Earth's surface:

$$\Delta\tau = \Delta t\sqrt{\frac{1-v^2}{c^2}} \simeq \Delta t\left(1-\frac{1}{2}\frac{v^2}{c^2}\right), \tag{B10.3}$$

where we have used $v \ll c$, and $\Delta\tau$ (or Δt) is the time interval measured by the satellite (or Earth) clock. Conversely, because of its *altitude* z in Earth's gravitational field, the orbiting clock experiences a gravitational red shift that increases its tick rate: $\Delta\tau' = \Delta t(1 + gz/c^2)$. Adding the two effects, the time interval recorded by the satellite clock is related to the time interval measured by the surface clocks as

$$\Delta\tau \simeq \Delta t\left(1-\frac{1}{2}\frac{v^2}{c^2}+\frac{gz}{c^2}\right). \tag{B10.4}$$

This expression is valid for all clocks near the Earth's surface ($z \ll R_E$) moving with speed $v \ll c$.

Now imagine a body in a zero-gravity region of deep space. Let's allow the body to move only along the z-axis. At time $t_i = 0$, the body is at $z_0 = 0$, and at a later time $t_f = 2\Delta t$, the body has an altitude $z > 0$. Given these constraints, what is the body's trajectory $z(t)$ during the time interval $2\Delta t$? Of course, we already know the answer: it moves with uniform speed $\bar{v} = z/2\Delta t$. Therefore, the wristwatch time measured by a clock attached to the body is given by Equation B10.3:

$$\Delta\tau = 2\Delta t\left(1-\frac{\bar{v}^2}{2c^2}\right).$$

But let's consider other possibilities. Suppose the body travels half the time ($\Delta t_1 = \Delta t$) with speed $v_1 = \bar{v} + \Delta v_1$, and the other half the time ($\Delta t_2 = \Delta t$) with a different speed $v_2 = \bar{v} + \Delta v_2$ (see Figure B10.2a). Since it reaches in exactly the same time $t_f = 2\Delta t$, its average speed is still $\bar{v} = z/2\Delta t$, so $\Delta v_1 = -\Delta v_2 \equiv \Delta v$. In this case, the wristwatch time is

$$\Delta\tau' = \Delta t\left(1-\frac{(\bar{v}+\Delta v)^2}{2c^2}\right)+\Delta t\left(1-\frac{(\bar{v}-\Delta v)^2}{2c^2}\right) = 2\Delta t\left(1-\frac{\bar{v}^2}{2c^2}-\frac{\Delta v^2}{2c^2}\right).$$

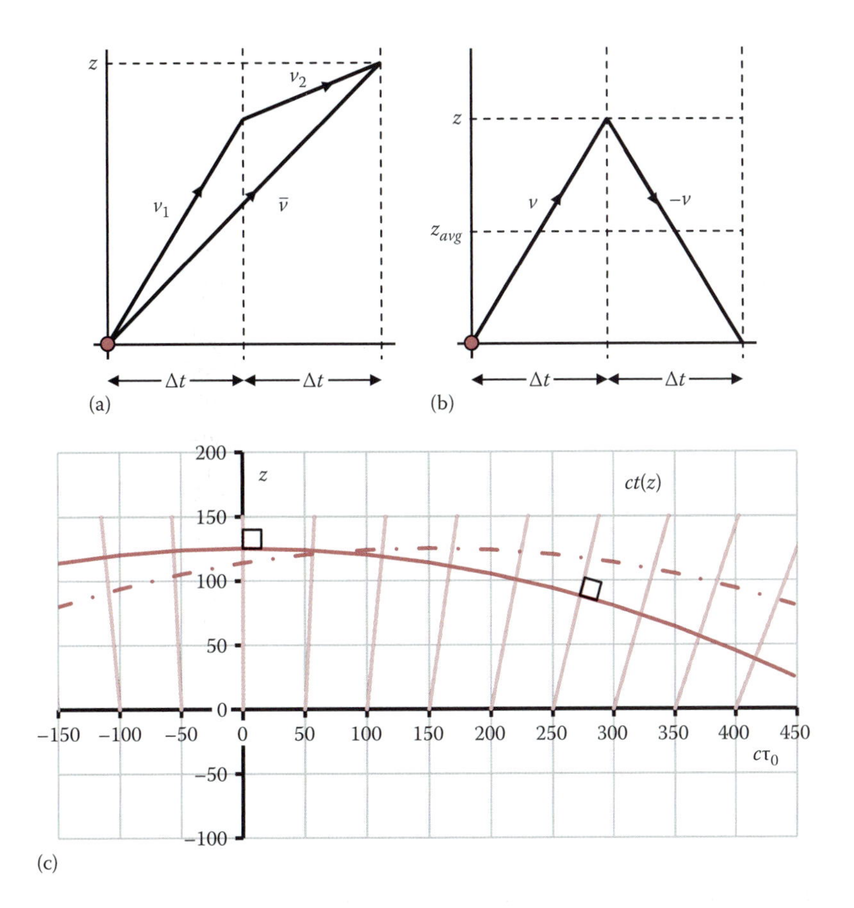

FIGURE B10.2 (a) A body in zero gravity starts from $z_0 = 0$ and arrives at a point $z > 0$ in time $2\Delta t$. The constant velocity trajectory yields the maximum value of $\Delta\tau$. (b) One trajectory for a body in a gravitational field $\bar{g} = g\hat{k}$. The body rises to height z before returning to its starting point $z_0 = 0$. (c) $c\tau(z)$ vs. z and $c\tau_0$ (radial lines). To emphasize the curvature of time, c is given the value 100 m/s (solid red line). Trajectory of a body in free fall, plotted as height z vs. $c\tau_0$. The body reaches its maximum height at $\tau(z) = 0$. At all times, the trajectory is perpendicular to $c\tau(z)$ (dash-dot red line). The same body with an initial velocity $v(0) > 0$ so that it reaches its maximum altitude at $c\tau_0 = 150$ m. The angle between the trajectory and $c\tau(z)$ remains constant throughout the motion.

Notice that $\Delta\tau'$ has its maximum value when $\Delta v = 0$. This result is true no matter how we slice up the time interval ($\Delta t_1 \neq \Delta t_2 \neq \Delta t_3 \neq \cdots$), and it remains true even if we allow motion in three dimensions: motion in the x- and y-directions merely increases the speed, which further reduces $\Delta\tau$ as prescribed by Equation B10.3. *The proper time is maximized when the body moves with constant velocity.*

What happens in a gravitational field $\vec{g} = g\hat{k}$? According to the equivalence principle, a free falling reference frame is an inertial frame in which the local effects of gravity disappear. So we might expect that the body will again move along the trajectory that maximizes its proper time. Imagine the body to be located at $z_0 = 0$ at time $t_i = 0$, and again at $z_0 = 0$ at a later time $t_f = 2\Delta t$. No forces other than gravity influence the motion. If the body were to remain stationary at $z(t) = 0$ throughout the time interval $2\Delta t$, then $v = 0$ and by Equation B10.4 $\Delta\tau = 2\Delta t$. But we can increase $\Delta\tau$ by allowing the body to move upwards. The higher the body goes, the faster its proper time advances. On the other hand, the higher it goes, the faster it moves, which slows the proper time rate. There must be an optimum height z_{max} for which $\Delta\tau$ is a maximum (see Figure B10.2b). Let's find that height.

Once again split the time interval Δt into two equal segments, and let the body rise from $z_0 = 0$ to height z at constant velocity $v = z/\Delta t$ during the first segment; then let it fall back to $z_0 = 0$ with $v = -z/\Delta t$ during the second segment. Using the average height $z_{avg} = z/2$ in Equation B10.4, we easily obtain

$$\Delta\tau = 2\Delta t\left(1 - \frac{v^2}{2c^2} + \frac{gz}{2c^2}\right) = 2\Delta t\left(1 - \frac{z^2}{2c^2\Delta t^2} + \frac{gz}{2c^2}\right).$$

To maximize $\Delta\tau$, take the derivative with respect to z and set it to zero:

$$\frac{d}{dz}\Delta\tau = 2\Delta t\left(-\frac{2z}{2c^2\Delta t^2} + \frac{g}{2c^2}\right) = 0,$$

or

$$z = \frac{1}{2}g\Delta t^2.$$

To maximize the proper time subject to the constraints specified, the body should follow a trajectory with maximum altitude $(1/2)g\Delta t^2$, in agreement with the equations of projectile motion from Chapter 6. Of course, the true path of the ball is parabolic, but it is fascinating to see how one feature of the trajectory follows from a totally different point of view: the maximization of proper time.

Exercise B10.3: Actually, we've only shown that $\Delta\tau$ is an extremum. For $z = (1/2)g\Delta t^2$, prove that $d^2\Delta\tau/dz^2 < 0$, so the extremum is indeed a maximum value.

Imagine a squad of observers stationed at various altitudes z. Each observer is equipped with a clock, and all clocks are synchronized to read 0 at the instant the body reaches its maximum height z_{max}. Let $\tau(z)$ be the time indicated by a clock at altitude z. According to Equation B10.4 with $v = 0$, $\tau(z) = \tau_0(1 + gz/c^2)$, where τ_0 is the time indicated by the clock at ground level ($z = 0$). In other words, $\tau(z)$ is the reading by a clock at altitude z that is simultaneous with a reading τ_0 by a clock at ground level. In Figure B10.2c, $c\tau(z)$ is plotted as a function of z for several values of $c\tau_0$. (Both axes have the dimensions of length.) The figure is drawn with $c = 100$ m/s, to magnify the slopes of the lines by a factor of 3×10^6. Also shown is the trajectory of a vertically falling body, $z = z_{max} - (1/2)g\tau_0^2$, as measured by an observer at ground level. Notice that the trajectory is *everywhere perpendicular* to the lines of simultaneity $\tau(z)$. If the body reaches its maximum altitude before or after $\tau(z) = 0$, the trajectory is shifted to the left or right, and the angle between it and $\tau(z)$ is no longer 90°. Nevertheless, this angle *remains constant* as the body falls. This is also shown in the figure.

What's going on? If there were no gravitational field, then $g = 0$ and the lines of simultaneity would be vertical: $\tau(z) = \tau_0$, as in Figure B10.2a. The trajectory of a body moving with constant velocity makes a constant angle with these lines. Likewise, when $g \neq 0$ the trajectory makes a constant angle with $\tau(z)$, but these lines are no longer parallel; in fact, they converge at $z = -c^2/g$ (≈ 1 ly [light-year]), indicating that at the Earth's surface, time is curved with a radius of curvature equal to 1 ly. The falling body moves along a "straight path" through this curved spacetime. In the jargon of relativity, this path is called a *geodesic*. It is akin to the great circle paths followed by aircraft on intercontinental flights.

These results are valid for motion in a "weak" gravitational field within a "small" region of space. Bodies fall to Earth, and the planets orbit the Sun, because they are following the straightest paths through curved time. For strong gravity (e.g., near a black hole) or for trajectories of interstellar length, the curvature of space must be considered along with the curvature of time.

In 1919, British astronomers led by Sir Arthur Eddington set out to measure the bending of starlight as it passed close to the Sun on its way to Earth. The angle of deflection predicted by the relativistic curvature of spacetime is twice the "Newtonian" prediction that supposes light is attracted to the Sun in the same way as gravitating mass. In May of that year, Eddington and his collaborators formed two teams that traveled to western Africa and Brazil to observe the starlight during a total eclipse of the Sun. Despite cloudy skies in Africa and faulty equipment in Brazil, the two teams were able to confirm the relativistic prediction. On November 7, 1919, almost exactly one year after the end of World War I, the *London Times* reported their scientific results in a headline article, "*Revolution in Science, New Theory of the Universe, Newtonian Ideas Overthrown.*" In a single day, Albert Einstein became an international celebrity.*

---

* For an interesting modern account of Eddington's experiment, see D. Kennefick, Testing relativity from the 1919 eclipse—A question of bias, *Physics Today* **62**, 37–42 (March 2009).

(c) It is desired to launch the spacecraft from the surface of the shell into a circular orbit of *altitude* R. What would be the *total* energy when the spacecraft is in this orbit? Indicate this energy by a line on the potential energy diagram.

(d) In this circular orbit, what would be the spacecraft's kinetic energy?

(e) What is the escape velocity from the surface of the shell?

(f) A 10 kg meteorite is approaching the planet. When it is very far away ($r/R \gg 1$), its speed is negligible. With what speed does it crash into the surface?

10.5 You are piloting a space capsule in a circular orbit 500 km above the Earth's surface. A stricken spacecraft lies 1000 km ahead of you in the same circular orbit. Your mission is to overtake it after one full orbit of your own space capsule, that is, you are to meet up at your present location.

(a) What is your present speed? What is your present orbital period?

To catch up to the disabled spacecraft, you fire your onboard rockets in a single short burn, changing your velocity by a small amount $\Delta\vec{v}$.

(b) In which direction should $\Delta\vec{v}$ point? Choose from the three choices (a)–(c) shown in the figure.

(c) Regardless of your choice, is your initial position (shown in the figure) the *perigee* of your new orbit, the *apogee*, or *neither*?

(d) How long will it take to meet up with the disabled craft?

(e) What is the lowest altitude of your new orbit?

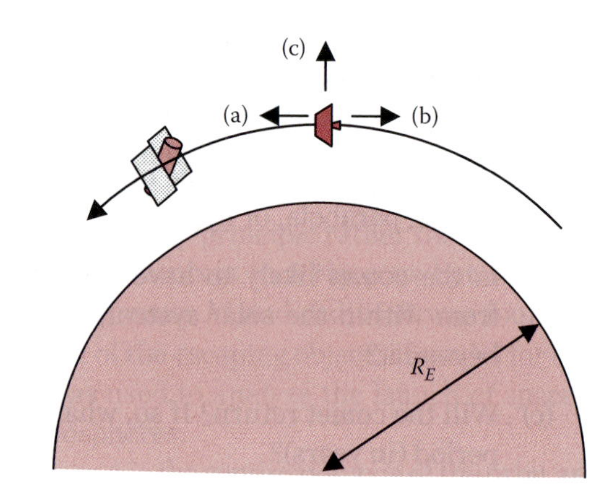

10.6 In February 2000, days before landing on asteroid 433 Eros, the *NEAR* spacecraft was "parked" in a circular orbit about Eros with radius $r = 5.0 \times 10^4$ m and period $T = 1.0 \times 10^5$ s (≈ 30 h).

(a) What was the acceleration of the spacecraft in this orbit?

(b) If we model Eros as a uniform density sphere of radius $R = r/6 = 8.33$ km and density $\rho \approx 3000$ kg/m^3, what is the acceleration of gravity "g" on its surface?

(c) Suppose you have landed on Eros. Could you throw a baseball fast enough to escape the gravitational pull of the asteroid? Show by explicit calculation. (*Hint:* 100 mph ≈ 44.5 m/s.)

(d) One way for the orbiter to descend to the surface is to fire its on-board rockets to first bring the spacecraft to a standstill ($v = 0$), and then allow it to fall freely. If the rocket fuel's exhaust speed is

1.6 km/s, and *NEAR*'s mass is 800 kg, what is the mass of the fuel needed for this maneuver?

(e) Is there a way to land NEAR on the surface of Eros using less fuel that the amount calculated above? If so, explain clearly how you would do this. If not, explain clearly why not. Include a figure in your answer. No calculations are required.

10.7 Two identical Earth satellites A and B are to be launched into nearly circular orbits about the Earth's center. Satellite A will orbit at an altitude of R_E and satellite B will orbit at an altitude of $3R_E$ ($R_E = 6400$ km).

(a) What is the ratio of the potential energies for satellite B and A?

(b) What is the ratio of the orbital kinetic energy of B to that of A? What is the ratio of their orbital speeds?

(c) In which case is the total energy greater?

(d) Does it take more energy to get satellite A to its orbital altitude (with zero velocity) than to put it in orbit once it is there? Answer the same question for B.

10.8 In 1864, Jules Vernes published his science fiction classic, *Journey to the Center of the Earth*. Let us follow Professor Hartwigg and his team of explorers down a vertical tunnel to the netherworld. The figure shows the potential energy $U(r)$ for a mass $m = 1.0$ kg vs. its distance from the planet's center ($r = 0$). R is the Earth's radius, so $r/R = 1$ at the Earth's surface. Far from the planet ($r/R \gg 1$), $U(r)$ approaches 0.

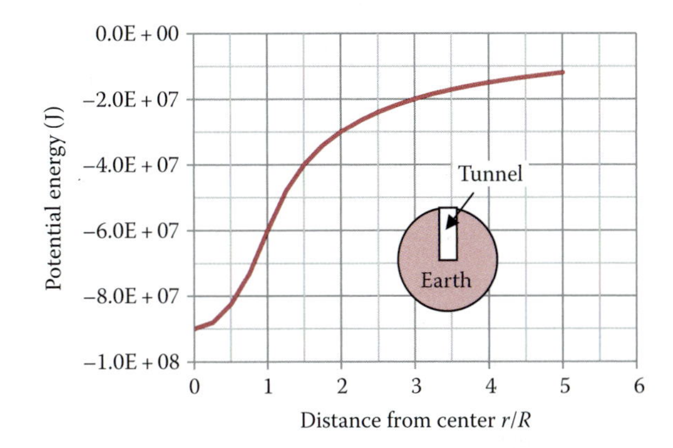

(a) Where is the force on m the greatest?

(b) If you dropped m from the Earth's surface, what would be its speed at the Earth's center?

(c) What is the escape velocity of m if it were launched from the center of Earth? What would be the escape velocity of a 2.0 kg mass launched from the same place?

(d) If m (1.0 kg) were in a *circular orbit* around Earth with $r/R = 1.5$ (i.e., at an altitude $0.5R$), what would be its kinetic energy?

10.9 We wish to send a 500 kg spacecraft to Venus via a minimum energy transfer orbit, that is, an elliptical orbit that is tangential to the nearly circular orbits of Earth and Venus. We know that Venus' orbital radius is 0.72 AU. Ignore the gravitational forces of Earth and Venus on the craft.

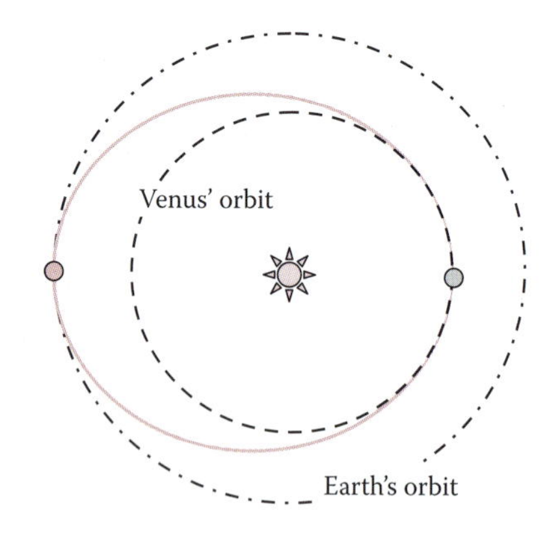

(a) What is the value of the semimajor axis of the ellipse?

(b) What is the total energy of the spacecraft?

(c) What velocity should the spacecraft have after escape from Earth?

(d) With what speed will it reach Venus?

(e) How long will it take the spacecraft to reach Venus?

(f) When the spacecraft reaches Venus, what velocity change will be necessary for it to orbit the Sun in the same circular orbit as Venus?

10.10 We want to discard a defunct 800 kg satellite. One possibility is to "drop" the satellite into the Sun. To do this, we would reduce the satellite's speed to 0 (from 30 km/s) and let it "fall" a distance 1 AU into the Sun.

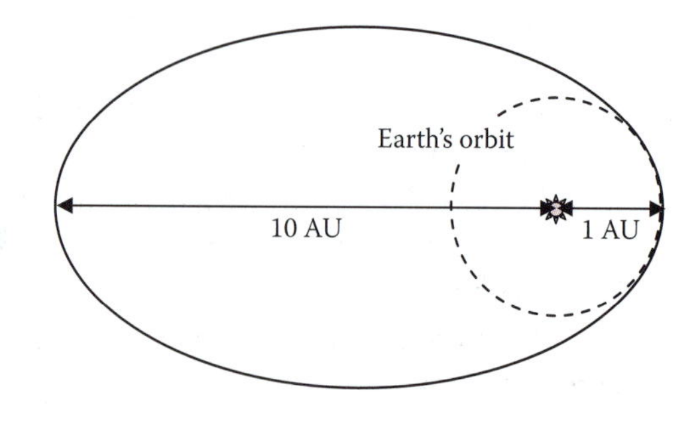

(a) Ignoring Earth's gravitational field, what impulse (momentum change) should we give the satellite? (Impulse is related to the amount of fuel used.)

Another way to discard the satellite is first to increase its speed such that it enters an elliptical orbit with perihelion at the firing point (1 AU) and aphelion at 10 AU. (See the figure.)

(b) What is the total energy of the spacecraft after the firing?

(c) What is the speed of the satellite at perihelion?

(d) Assuming we started with an orbital velocity of 30 km/s, what is the net impulse needed to boost the spacecraft into the elliptical orbit?

(e) At the aphelion, the spacecraft's thrusters fire to reduce the velocity to 0. Subsequently, the spacecraft falls to the Sun. What is the total impulse given to the satellite (perihelion plus aphelion)? How does this compare to the impulse in part (a)?

(f) Instead of dropping the spacecraft to the Sun, we might decide to send it out of the solar system, never to return. Assuming the spacecraft is initially in an Earthlike orbit, what impulse would be needed to do this? How does this compare with parts (a) and (e)?

(g) Which of the three schemes uses the least amount of rocket propellant?

10.11 Your space capsule is in a circular orbit about Earth at an altitude of 3200 km. At position *A* shown in the figure, you fire your retrorockets, slowing the spacecraft to put it into an elliptical orbit. You reenter the Earth's atmosphere at position B (the *perigee* of your new orbit), and deorbit by deploying a parachute that safely brings the spacecraft to rest on the Earth's surface.

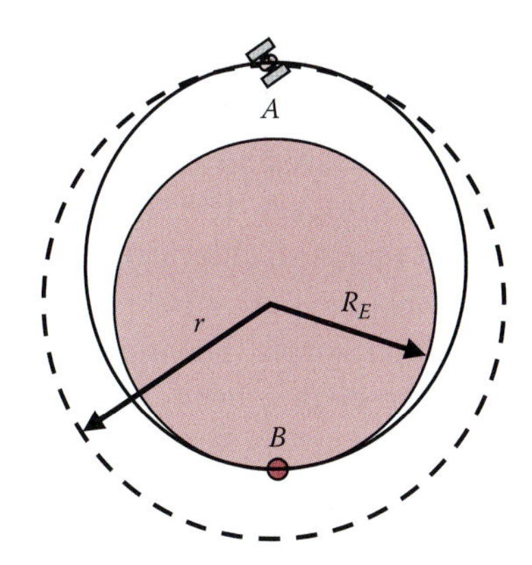

(a) What is the value of the semimajor axis a of the elliptical reentry orbit?

(b) What was your speed immediately after firing the rockets?

(c) Find the time between the rocket firing and reentry. Neglect the thickness of the Earth's atmosphere. (*Hint:* The period of a LEO satellite is 85 min.)

(d) What is your speed upon reentry (point B, where $r \approx R_E$)?

10.12 *Breaking news story*! Cameras on NASA's Solar Observatory Satellite (SOS) have spotted a previously undetected Earth twin in a circular orbit of radius 1 AU about the Sun. Viewed from Earth, it is directly behind the Sun, and since its period is exactly the same as Earth's, it is invisible from our planet. Scientists are eager to explore "Planet X" and plan to launch a spacecraft with an orbital period of 1.5 years, at which time Earth and Planet X will have exchanged places, and the spacecraft will meet up with X at the present position of Earth.

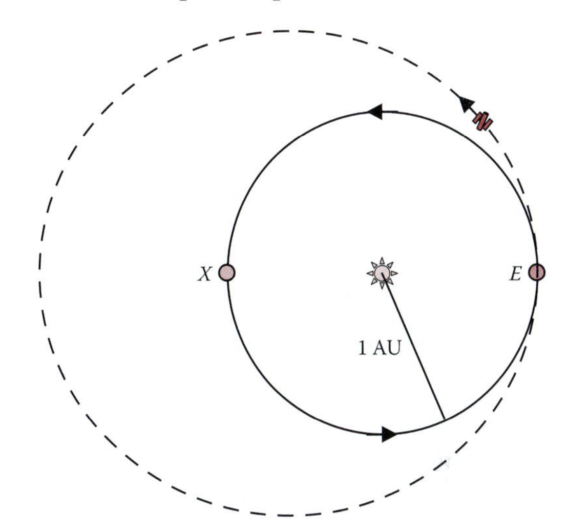

(a) Calculate the semimajor axis a (in AU) of the spacecraft's elliptical orbit.

(b) What must be the spacecraft's speed after escape from Earth? Neglect the Earth's gravitational pull on the satellite.

(c) What is the spacecraft's speed at aphelion (its farthest distance from the Sun)?

(d) Another plan is to intercept Planet X in half year, at the present position of Earth, by launching the spacecraft into an orbit with semimajor axis 0.63 AU and speed 19 km/s after launch. Which of the two schemes uses less rocket fuel?

10.13 *Eris* is a Pluto-sized dwarf planet that is in a distant elliptical orbit about our Sun. It was discovered in 2005 by astronomers at the Palomar Observatory in California. At the present time, it is very close to aphelion, 97.56 AU from the Sun, and is moving with speed 2.26 km/s.

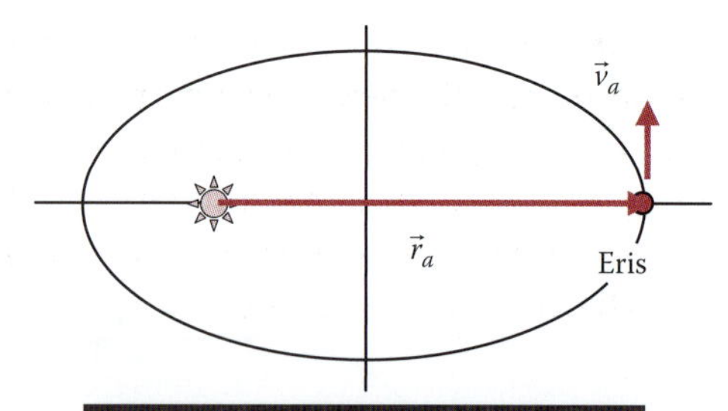

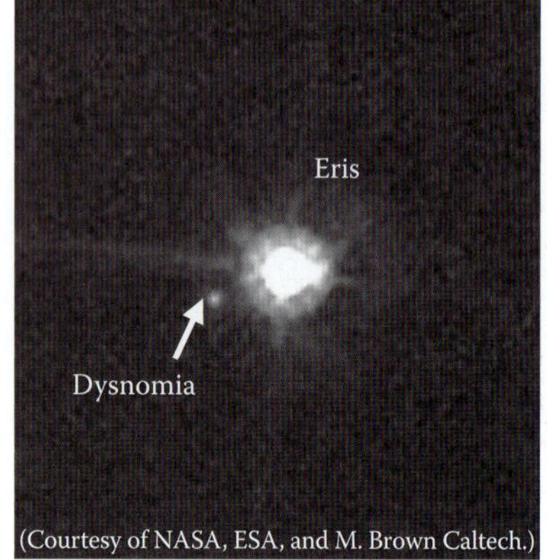

(Courtesy of NASA, ESA, and M. Brown Caltech.)

(a) Calculate the semimajor axis a of Eris' orbit.

(b) Calculate the period of the planet's orbit about the Sun. Express your answer in years.

Eris has a small moon, called *Dysnomia*, which is in a circular orbit about Eris. By comparing the relative brightness of Eris and Dysnomia, the Palomar astronomers estimate the radius of Dysnomia to be 50 km. Comparison to other small moons suggests that the surface gravitational acceleration $g = 0.014\ \text{m/s}^2$.

(c) Assuming that the moon is a solid sphere of uniform density ρ, find the density. What do you think is the chemical composition of Dysnomia?

(d) A professional baseball pitcher can throw a ball at a speed of 100 mph (44 m/s). If you were standing on the surface of Dysnomia, could you throw a baseball with enough speed so that it never returns to the moon? Support your answer by a calculation.

10.14 On October 9, 2009, NASA's LCROSS mission crash-landed a spacecraft near the south pole of the Moon, in order to excavate a crater and search for the presence of water. Prior to its impact, the spacecraft (mass = 2900 kg) was in an elliptical orbit with semimajor axis $a = 4.69 \times 10^8$ m perpendicular to the Moon's orbit. The impact occurred at the perigee of that orbit, a distance 3.85×10^8 m from Earth. (*Note:* In the figure, the Moon's nearly circular orbit is shown edge on, and appears as a narrow ellipse.)

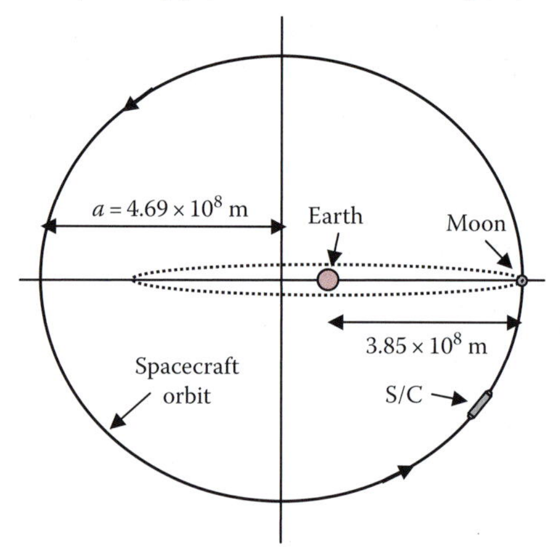

(a) At what point on the orbit of the spacecraft is its speed a minimum?

(b) Calculate the period (days) of the spacecraft's orbit.

(c) What is the spacecraft's speed when it is near perigee, but still far enough from the Moon so that the Moon's gravitational attraction can be neglected?

(d) Closer to the Moon, its gravitational attraction must be considered while Earth's can be neglected. Find the speed of the spacecraft when it impacts the surface of the Moon. ($M_M = 7.35 \times 10^{22}$ kg, $R_M = 1.74 \times 10^6$ m.)

(e) Do you expect your answer to (d) to be greater than, less than, or equal to the escape velocity from the Moon? Explain your choice clearly. (You should not need to evaluate v_{esc} to answer this question.)

10.15 NASA has found a potentially profitable mineral resource in asteroids. The asteroid *16 Psyche* has enough iron–nickel ore to supply Earth's needs for several million years! The plan is to build a fully automated mining facility on the asteroid, and eject 1000 kg (1 metric ton) packets of ore from the surface of the asteroid for refining. Your physics class has been assigned the task of designing and developing the launch device: *a giant spring.*

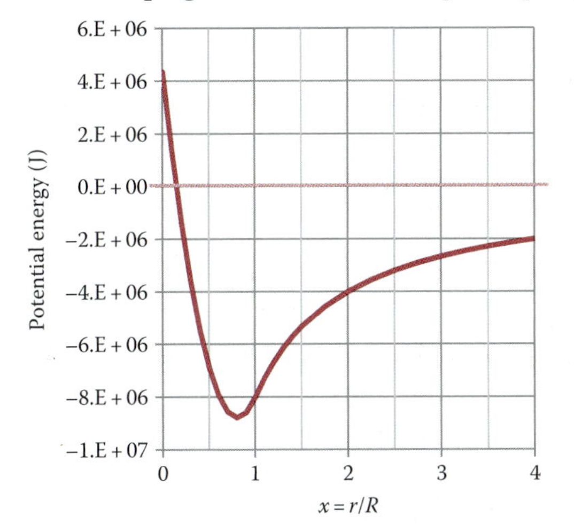

The asteroid has a mass $M = 2.27 \times 10^{19}$ kg and an average radius $R = 130$ km.

The plans call for boring a tunnel from the asteroid's surface to its center. The spring fits within the tunnel. The graph shows the total potential energy (gravity plus spring) for a 1000 kg mass as a function of $x = r/R$, where r is the distance to the center of the asteroid. The potential energy of the system has been chosen to be zero at $r = \infty$.

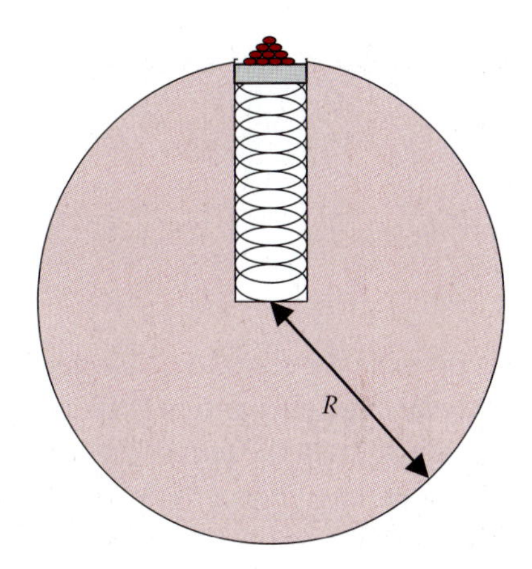

(a) From what point r (or x) would you release the ore packet in order to achieve escape velocity? Indicate this position on the graph.

(b) If the spring is fully compressed, so that the load of ore is released from rest at r =0, what is its maximum speed? Indicate on the PE diagram the position where $v = v_{max}$.

(c) From the graph, estimate the maximum force exerted on the ore load. (Caution: when $\Delta x = 1$, $\Delta r = R$.)

10.16 The table gives the mass M, the radius R, and the exosphere temperature T_{exo} of the solar system planets and the Earth's moon. Calculate the escape velocity for each planet or moon. Then find v_{rms} for H_2, He, N_2, O_2, H_2O, and CO_2 for each planet or moon, and indicate if the atom or molecule might be found in its atmosphere. (*Hint:* This is an ideal spreadsheet exercise!)

Object	M/M_{Earth}	R/R_{Earth}	T_{exo} (K)
Mercury	0.055	0.38	440
Venus	0.815	0.95	280
Earth	1.000	1.00	1000
Moon	0.012	0.27	270
Mars	0.107	0.53	350
Jupiter	318	10.77	1000
Saturn	95	9.01	800
Uranus	14.5	3.93	750
Neptune	17.2	3.87	750
Pluto	0.1	0.46	60

10.17 It is possible for stars to be ejected from the galaxy by close encounters ("collisions") with other stars. Treat the Sun as having a circular orbit about the center of the Milky Way Galaxy with an orbital radius $R_{Sun} = 8$ kpc ($=2.5 \times 10^{20}$ m) and an orbital speed 230 km/s. Treat the galaxy as a point mass $M_{Gal} = 1.0 \times 10^{11} M_{Sun}$ located at the center of the Sun's orbit. (For computational purposes, it may be useful to know that $GM_{Gal}M_{Sun}/R_{Sun} = 1.07 \times 10^{41}$ J.)

(Courtesy of ESO, European Space Agency, Spiral galaxy M83.)

(a) What is the minimum mechanical energy that must be added to the Sun in an encounter with another star to give the Sun an unbounded orbit?

(b) An encounter with a star gives the Sun a sudden increase in velocity, boosting it into an elliptical orbit with a semimajor axis of 40 kpc. What was the speed of the Sun just after the encounter?

(c) In its new orbit, what is the Sun's speed at *apogalacticon*?

(d) Qualitatively, what would happen to the other star in the encounter described in parts (a) and (b), and why? What would you expect the "end" state of the Milky Way Galaxy to be, far into the future?

(e) In 2005, Brown et al. discovered a star in the Milky Way halo, 110 kpc from the galactic center, moving with speed 710 km/s. Is this a runaway star? Assuming that the star was ejected by interaction with a massive black hole near the center of the galaxy, estimate when the interaction took place? (See Brown W.R. et al., *ApJ Lett.* **622**, L33 (2005).)

10.18 A planet of mass m is in a circular orbit with speed v and radius $R = 1$ AU about a star of mass $M = M_{Sun} = 2 \times 10^{30}$ kg. The star suddenly erupts explosively, leaving behind a smaller star of mass $M' = (3/4)M_{Sun}$. The explosion does not affect the instantaneous velocity of the planet.

(a) Find the total energy of the planet after the explosion in terms of G, M_{Sun}, m, and R.

(b) Show that the final orbit of the planet is an ellipse.

(c) Find the orbit's semimajor axis a and the distance to the star at perihelion and aphelion.

(d) How would your answers change if the final mass of the star were $M' = M_{Sun}/2$?

10.19 In the last leg of its trip to Saturn, the spacecraft *Cassini* executed an elliptical arc with a perihelion at the orbit of Venus (R_{VS} = 0.72 AU) and aphelion at the orbit of Saturn (R_{SS} = 9.54 AU). Let $m_{Cassini}$ = 5000 kg.

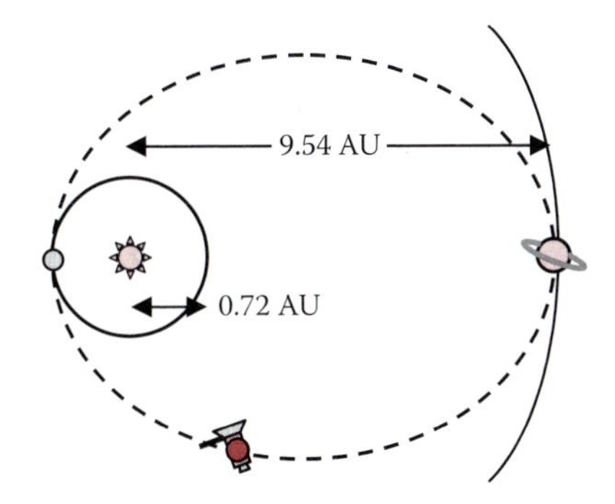

(a) Find the total energy of *Cassini* in its elliptical path.

(b) Find the spacecraft's speed at aphelion.

(c) What is the semiminor axis of the ellipse?

(d) How long did it take to go from Venus to Saturn?

10.20 A 1000 kg satellite is sent into a circular low Earth orbit of altitude $h \ll R_E$. Its orbital velocity is 7.9 km/s. Ground control plans to send this satellite into a geosynchronous orbit, that is, an orbit with a period of one day (so the satellite always sees the same part of the Earth).

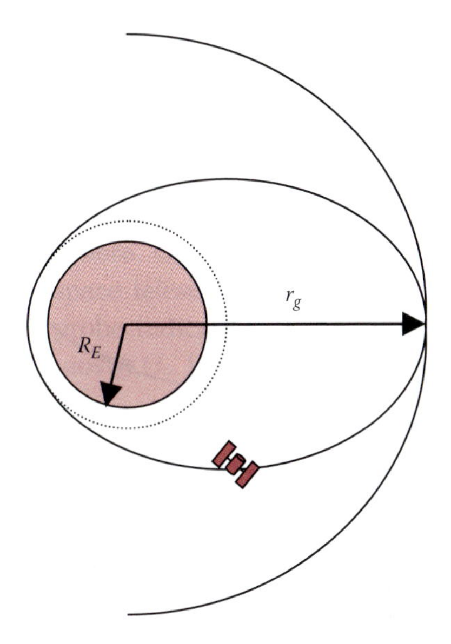

(a) What is the radius r_g of a circular geosynchronous orbit? Express your results in terms of R_E.

(b) When the satellite is in LEO, its thrusters are fired tangentially to its current trajectory, so that its new orbit is elliptical with apogee at r_g. Find the velocity immediately after the rocket burst that enabled this transfer orbit.

(c) Once the satellite reaches r_g, a computer malfunction brings it to a complete stop, and the satellite falls. With what kinetic energy does it impact the Earth?

(d) What might be the purpose of a satellite in geosynchronous orbit?

10.21 A satellite is in an elliptical orbit around the Earth. The semimajor axis of the ellipse is 7.2×10^7 m, and at perigee the craft is 9.7×10^6 m ($=r_p$) from the center of the planet. The eccentricity of the orbit is $\varepsilon = \sqrt{3}/2$, so the semiminor axis $b = a/2$.

(d) Calculate Δv_2, the magnitude of the velocity change at the intersection point. Compare the fuel requirements for this orbit with the requirements of a Hohmann transfer orbit.

10.29 Consider the force on a galaxy of mass m, now situated at a distance r_0 from the Milky Way (see Figure 10.10a). At the present time, the repulsive "dark energy" force F_Λ on the galaxy is 2.7 times stronger than the attractive gravitational force F_g. According to the discussion in Section 10.8, $F_g = GM_0 m/r_0^2$, where $M_0 = (4/3)\pi\rho_0 r_0^3$ and ρ_0 is the present mass density. The magnitude of the repulsive force on the galaxy can be expressed as $F_\Lambda = Amr_0$, where A is a constant.

(a) How far away was the galaxy when $F_g = F_\Lambda$? Express this distance as a fraction of r_0. (*Hint:* Does M_0 change with time?)

(b) Let $H(t) \approx H_0 = 2.3 \times 10^{-18}\ \text{s}^{-1}$ ($1/H_0 = 13.8$ Gyears). Use the critical density model, Equation 10.19, to estimate how long ago the two forces were equal in magnitude.

Section IV

CONSERVATION OF ANGULAR MOMENTUM

spinning with angular speed ω, Δm_i undergoes circular motion about the axis with speed $v_i = \omega r_i$. The total rotational kinetic energy of the pulley is therefore*

$$K_{rot} = \sum_{i=1}^{N} \frac{1}{2} \Delta m_i v_i^2 = \frac{1}{2} \left(\sum_{i=1}^{N} \Delta m_i r_i^2 \right) \omega^2 \equiv \frac{1}{2} I \omega^2, \qquad (11.1)$$

where we have defined the *moment of inertia* I about the pulley's axis of rotation as

$$I \equiv \sum_{i=1}^{N} \Delta m_i r_i^2.$$

In the limit as $N \to \infty$ and $\Delta m_i \to 0$, the summation can be replaced by an integral:

$$I = \int r^2 dm, \qquad (11.2)$$

where r is the radial distance of the differential mass dm from the axis of rotation. Since translational kinetic energy is $K_{trans} = ½ MV_{CM}^2$ and rotational kinetic energy is $K_{rot} = ½ I\omega^2$, the moment of inertia may be thought of as the rotational analog of mass.

The hanging body M and the pulley are linked by the string, so their motions are not independent. When the pulley rotates one full turn (2π radians), a length $2\pi R_P$ of string unwinds from its rim, lowering M by the same amount. More generally, when the pulley rotates through an angle $\Delta\theta$, M falls by a distance $\Delta y = -R_P \Delta\theta$. The speed of the falling body is related to the angular speed of the pulley:

$$V_{CM} = \frac{dy}{dt} = -R_P \frac{d\theta}{dt} = -R_P \omega,$$

so the rotational energy is $K_{rot} = \frac{1}{2} I\omega^2 = \frac{1}{2} I \cdot V_{CM}^2 / R_P^2$. Our energy expression can be rewritten as

$$E = Mgh = \frac{1}{2}\left(M + \frac{I}{R_P^2} \right) V_{CM}^2 + Mgy.$$

* In this and the next chapter, we will frequently use the symbol $\sum$ as a convenient shorthand for the summation of a very large number of terms: $\sum_{i=1}^{N} x_i = x_1 + x_2 + \cdots + x_N$.

Solving for the velocity term,

$$V_{CM}^2 = \frac{Mg(h-y)}{\frac{1}{2}\left(M + I/R_P^2\right)}.$$

A larger I results in a smaller V_{CM}. To calculate V_{CM}, we must learn how to find the moment of inertia of the rotating body.

11.3 CALCULATING THE MOMENT OF INERTIA

The moment of inertia of a body depends on the body's mass and shape, as well as the location and orientation of the axis about which it is rotating. For example, the radial distance r in Equation 11.2 will be different for different axes of rotation. Most often we will be interested in the rotation of a highly symmetric body (e.g., a cylindrical pulley or a spherical ball) about an axis of symmetry passing through the body's center of mass. In such cases, the integration of Equation 11.2 can often be done without too much mathematical pain. Picture a thin circular hoop of mass M and radius R rotating about an axis perpendicular to the plane of the hoop and passing through its center (see Figure 11.2a). All points dm on the hoop are at the same distance R from this axis, so

$$I_{hoop} = \int r^2 dm = R^2 \int dm = MR^2.$$

That was easy! Now let's use this result to tackle something a bit harder: a uniform *solid disk* with mass M and radius R, rotating about the same axis as the hoop. To perform the integral of Equation 11.2, divide the disk into a collection of thin nested circular hoops. Figure 11.2b shows one of these hoops with radius r and thickness dr. The area of this hoop is its length (circumference) $2\pi r$ times its width (thickness) dr, $dA = 2\pi r\, dr$. Its mass dm is related to the total mass M and total area $A = \pi R^2$ by

$$\frac{dm}{M} = \frac{dA}{A} = \frac{2r\,dr}{R^2}.$$

Using our previous calculation for the thin hoop, the moment of inertia of this thin ring is $dI = r^2 dm$. To find the total moment of inertia of the disk, we must sum, or integrate, over the entire collection of hoops making up the disk:

$$I_{disk} = \int r^2 dm = \int_{r=0}^{R} r^2 M \frac{2r\,dr}{R^2} = \frac{2M}{R^2}\int_0^R r^3 dr = \frac{2M}{R^2}\frac{R^4}{4}$$
$$= \frac{1}{2}MR^2.$$

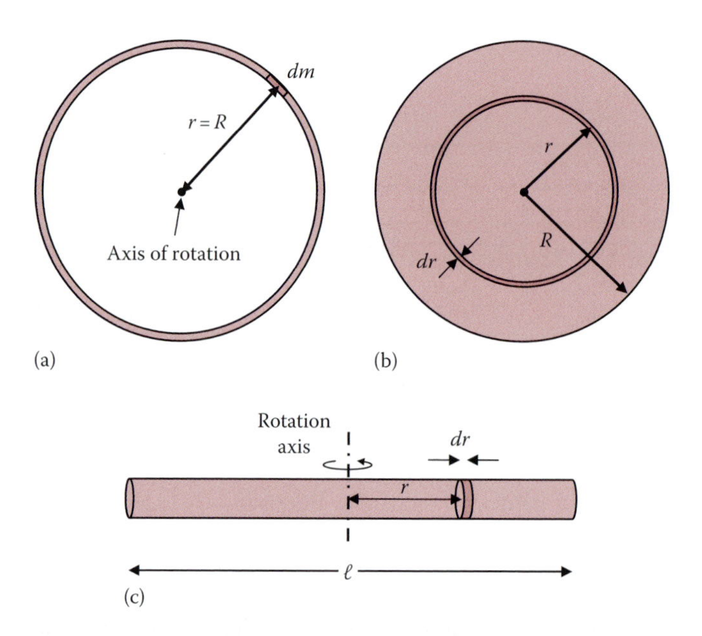

FIGURE 11.2 (a) A thin *hoop* or hollow cylinder has all of its mass M concentrated at a distance R from its central axis of symmetry. (b) A solid *disk* or cylinder may be thought of as a set of nested hoops of thickness dr. (c) To find the moment of inertia of a long rod, divide it into short segments of length dr.

EXERCISE 11.1

(a) Why is the coefficient of MR^2 smaller for the disk (=½) than for the hoop (=1)? (b) What is the moment of inertia of a *solid cylinder* of mass M, radius R, and length ℓ about the same axis of rotation considered for the solid disk?

Answer (b) $\frac{1}{2}MR^2$

For a third example, picture a thin rod of mass M and length ℓ, rotating about an axis perpendicular to the rod and passing through its center (see Figure 11.2c). In this case, divide the rod into short segments of length dr. The mass dm of a single segment is given by $dm/M = dr/\ell$. If this tiny segment is located a distance r from the rotation axis, its contribution to the moment of inertia is $dI = r^2\,dm = (M/\ell)r^2\,dr$. The moment of inertia of the entire rod is

$$I_{rod} = \int r^2 dm = \frac{M}{\ell}\int_{r=-\ell/2}^{\ell/2} r^2 dr = \frac{1}{3}\frac{M}{\ell}r^3\Big|_{-\ell/2}^{\ell/2} = \frac{1}{12}M\ell^2.$$

EXERCISE 11.2

(a) Calculate I for a thin rod of length ℓ and mass M rotating about an axis through *one end* of the rod. (In the figure, the axis is perpendicular to the page.) (b) This axis is a distance $d = \ell/2$ from the center of mass. We know that the moment of inertia for an axis through the *CM*, perpendicular to the rod, is $I_{CM} = \frac{1}{12}M\ell^2$. Show that the moment calculated in (a) can be expressed as $I = I_{CM} + Md^2$. This is one example of useful relation called the *parallel axis theorem*. See Box 11.1.

Answer (a) $I = \frac{1}{3}M\ell^2$

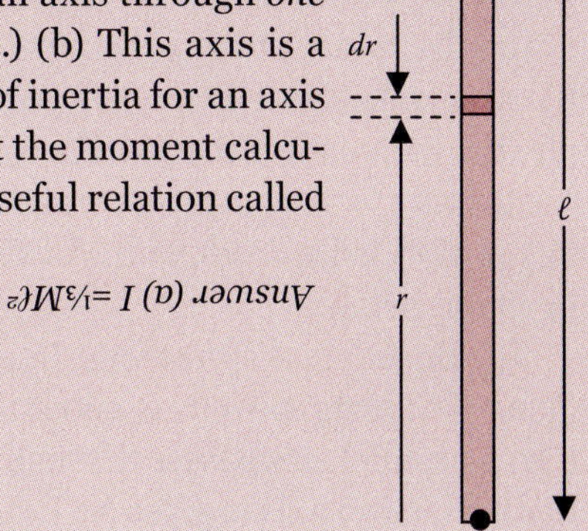

Figure 11.3 lists the moments of inertia of several common shapes having high symmetry. Box 11.1 explains how the more complicated expressions in the figure were derived.

Let's use the result of Exercise 11.2 to solve a favorite problem of physics instructors. Imagine a meter stick ($\ell = 1$ m) of mass M that is initially standing vertically on the floor. The lower end of the stick is attached to a hinge that is fixed to the floor and allows the stick to pivot (see Figure 11.4a or Exercise 11.2). The stick's initial kinetic energy is zero, so the total energy is equal to the initial potential energy U. To calculate U, isolate a short segment of the stick, of length dy, located a distance y above the floor (Figure 11.4b). The mass of this segment is $dm = M\,dy/\ell$ and its potential energy relative to the floor is $dU = gy\,dm$. The total potential energy is found by integrating dU over the length of the stick:

$$U(=E) = \int_0^\ell gy\,dm = \frac{gM}{\ell}\int_0^\ell y\,dy = \frac{gM}{\ell}\cdot\frac{\ell^2}{2} = \frac{1}{2}Mg\ell,$$

as if all of the stick's mass were located at its *CM*. Now suppose the stick is given a gentle sideways nudge that causes it to swivel and fall to the floor. What is the velocity of the moving (unhinged) end just before it strikes the floor? At impact, the potential energy is 0, so the stick's kinetic energy is $K = E = \frac{1}{2}Mg\ell$. The stick is in pure rotation about the hinge, so $K = K_{rot} = \frac{1}{2}I\omega^2$. Solving for ω, with $I = \frac{1}{3}M\ell^2$, we obtain $\omega = \sqrt{3g/\ell} = 5.4\ \text{s}^{-1}$. The speed of the far end of the stick is $v = \omega\ell = 5.4$ m/s. (It may surprise you to learn that, just before impact, the acceleration at the far end of the stick is greater than g! See Problem 11.12.)

This method of solution can be used when a body moves in pure rotation, so $K = K_{rot} = \frac{1}{2}I\omega^2$. In other cases, the expression for the kinetic energy will be more complicated. In the following section, we will derive a general result to cover these cases.

BOX 11.1 THREE STRATEGIES FOR CALCULATING MOMENTS OF INERTIA

Figure 11.3 summarizes the moments of inertia you will need in this course. But rather than just memorizing the results, take a moment to examine them carefully. There are obvious relationships between the moments about different axes of the same body or between moments of different bodies. For example, the moments of inertia of a solid cylinder about the perpendicular axes a and b (in Figure 11.3) are related as $I_a = \frac{1}{2}I_b + \frac{1}{12}M\ell^2$, and $\frac{1}{12}M\ell^2$ is the moment of inertia I_a of a *long thin rod*. Where does the factor of ½ come from, and why does the thin rod moment appear? The answers to these questions can be found below.

The parallel axis theorem: Imagine an arbitrarily shaped *thin plate* of mass M situated in the x–y plane. The plate is rotating about the z-axis, which pierces the x–y plane at a distance R from the body's *CM*. A parallel axis passes through the *CM* (see Figure B11.1a). The moment of inertia of the body relative to rotation (z-) axis is given by Equation 11.2,

$$I_z = \int r^2 dm = \int (\vec{R} + \vec{r}\,')^2 dm,$$

where, $\vec{R}$ is the position of the *CM* relative to the rotation axis and $\vec{r}\,'$ is the location of dm relative to the *CM*: $\vec{r} = \vec{R} + \vec{r}\,'$.

Using the dot product to expand the integral,

$$I_z = \int (\vec{R} + \vec{r}\,')\cdot(\vec{R} + \vec{r}\,')dm = \int R^2 dm + 2\vec{R}\cdot\int \vec{r}\,' dm + \int r'^2 dm.$$

The second term is exactly zero, because the *CM* is located at $\vec{r}\,' = 0$. Since R is a constant, the first term is equal to MR^2, while the third term is the moment of inertia about the parallel axis through the *CM*: $I_{CM} = \int r'^2 dm$. Therefore,

$$I_z = I_{CM} + MR^2. \tag{B11.1}$$

This is called the *parallel-axis theorem*. If you know the moment of inertia about an axis through the *CM* perpendicular to the plate (I_{CM}), you can easily calculate I about any other axis parallel to it.

The "stretch" theorem: Equation B11.1 is not restricted to thin plates. Let's stretch the plate out along the z-axis, as shown in Figure B11.1b. This makes no difference to the calculation of I_z, as long as the x–y cross-section and the total mass remain the same as before. To see this, choose the z-axis as the axis of rotation. Then, $I_z = \int r^2 dm = \int (x^2 + y^2)dm$. When we stretch the body to a length ℓ, each mass element dm is elongated into a column with the same x and y values as dm had in the thin plate. The coordinate z does not enter into the calculation for I_z. Mathematically, the integral for I_z is exactly the same as if all of the mass were concentrated in the x–y plane. This is the *stretch theorem*. If you know the moment of inertia of a body relative to a certain axis of rotation, stretching the body along that axis does not affect its moment of inertia. For example, the moment of inertia of a thin hoop along its axis of cylindrical

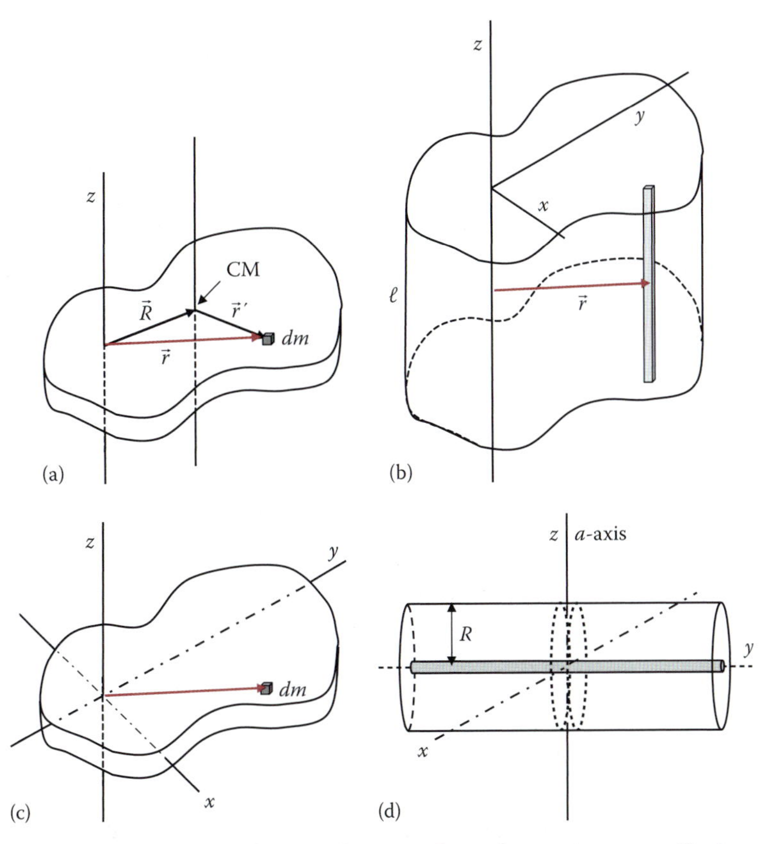

FIGURE B11.1 (a) A thin plate lies in the x-y plane, and rotates about the z-axis, perpendicular to the plate. The position of dm relative to the axis is $\vec{r} = \vec{R} + \vec{r}\,'$, where $\vec{R}$ is the vector from the z-axis to the CM, and $\vec{r}\,'$ is the position of dm relative to the CM. (b) Stretching the plate along the z-axis does not change I_z. (c) For a thin plate, the moments of inertia relative to the x- and y-axes are related to I_z as $I_x + I_y = I_z$. (d) To calculate I_z for a cylinder, start with a thin rod and expand the radius to R. The integral $\int x^2 dm$ for the cylinder has the same value as for a thin disk lying in the x-z plane.

symmetry is $I_b = MR^2$. If we stretch the hoop into a long hollow cylinder, the moment of inertia about the same axis (axis b in Figure 11.3) does not change.

Perpendicular axis theorem: Let's return to the thin plate lying in the x-y plane, rotating about the z-axis (Figure B11.1c). The moment of inertia about this axis is $I_z = \int (x^2 + y^2)dm$. Similarly, the moment of inertia about the x-axis is $I_x = \int (y^2 + z^2)dm$, but since the body is a plate with negligible thickness ($z \simeq 0$), $I_x = \int y^2 dm$. Likewise, $I_y = \int x^2 dm$. Therefore,

$$I_x + I_y = I_z$$

or, in cases where $I_x = I_y$, (B11.2)

$$I_x (= I_y) = \frac{1}{2} I_z .$$

This is the *perpendicular axis theorem* (applying to thin plates only). For example, the moment of inertia of a thin solid disk (a cylinder with length $\ell = 0$) about axis b (shown in Figure 11.3) is $I_b = ½MR^2$. The moment about axis a, perpendicular to b, is simply $I_a = ½I_b = ¼MR^2$.

Finally, let's put these ideas together to tackle something more complicated: the moment of inertia about axis a of a long solid cylinder of radius R and length ℓ. As before, designate this axis as the z-axis of our coordinate system. See Figure B11.1d. We know the moment of inertia of a *thin rod* of the same mass and length: $I_a^{rod} = \int y^2 dm = \frac{1}{12}M\ell^2$. Now let the rod swell into a cylinder of radius R, that is, stretch its radius but leave its length unchanged. The moment of inertia about the a-axis (the z-axis) is given by

$$I_a^{cyl} = \int (x^2 + y^2)dm = \int x^2 dm + \int y^2 dm$$
$$= \int x^2 dm + I_a^{rod}.$$

Because of the cylindrical symmetry of the body, $\int x^2 dm = \int z^2 dm$. In addition, $\int (x^2 + z^2)dm = I_b^{cyl}$, so $\int x^2 dm = ½I_b^{cyl} = ¼MR^2$. The total moment of inertia of the cylinder about the a axis is therefore

$$I_a^{cyl} = \frac{1}{2}I_b^{cyl} + I_b^{rod}$$
$$= \frac{1}{4}MR^2 + \frac{1}{12}M\ell^2$$

in agreement with Figure 11.3. The same strategy works for the cylindrical shell, the long annulus, and the parallelepiped in Figure 11.3. By exploiting these three rules, you can often sidestep difficult integrations when calculating moments of inertia. Perhaps now you can see how the more complicated results of Figure 11.3 were derived.

Exercise B11.1: (a) A solid parallelepiped, such as the one shown in Figure 11.3, has a square cross-section with $b = c$, and length $a > b,c$. The moment of inertia about an axis through its center, parallel to the body's length, is $I_b = \frac{1}{12}(b^2 + c^2)$. Use the theorems above to show that the moment of inertia about a perpendicular axis a through the body's center of mass is $I_a = \frac{1}{12}(a^2 + b^2)$, as given in the figure. (b) What is the moment of inertia about an axis a' parallel to a, but located at one end of the bar (a distance $a/2$ from axis a)?

Answer (b) $I_{a'} = \frac{1}{12}(4a^2 + b^2)$

11.4 TOTAL KINETIC ENERGY OF A RIGID BODY

A rigid body can have both translational and rotational motion: picture a rolling ball or a spinning planet in orbit about the Sun. In the most general terms, the body is a collection of small bits m_i (ultimately molecules) with a total mass M and a center of mass $\vec{R}_{CM}(t)$. The instantaneous position $\vec{r}_i(t)$ of the ith molecule

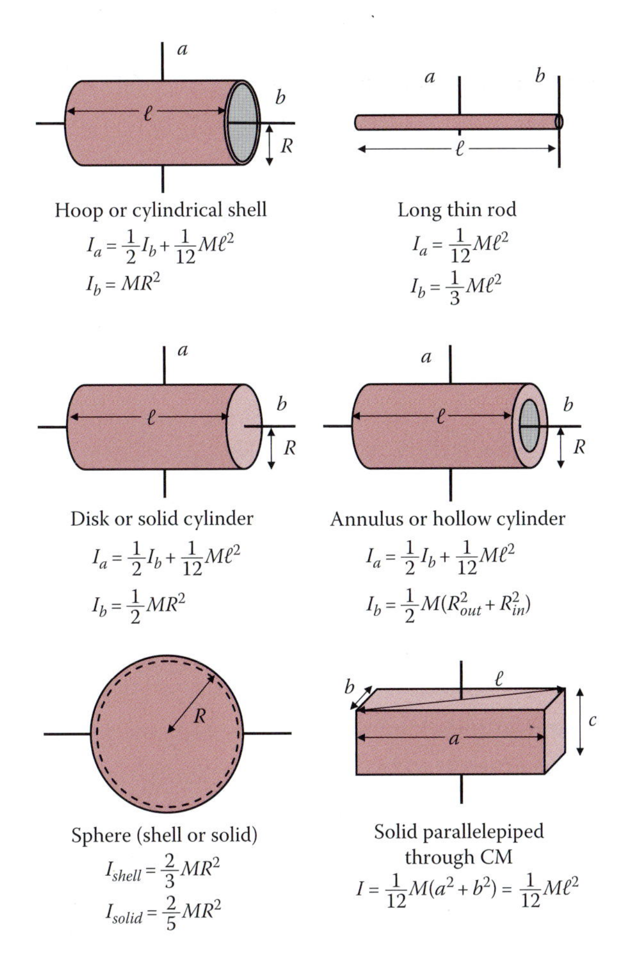

FIGURE 11.3 A summary of moments of inertia for common shapes. In all cases shown but one (I_b of the rod), the axes of rotation pass through the *CM* of the body.

can be expressed using *CM* coordinates as $\vec{r}_i(t) = \vec{R}_{CM}(t) + \vec{r}_i'(t)$, where $\vec{r}_i'$ is the location of m_i relative to the *CM*. As the body moves, each molecule moves with the *CM*, and also moves about the *CM*, by whatever modes of motion are available to it. The velocity of m_i is $\vec{v}_i(t) = d\vec{r}_i/dt = \vec{V}_{CM} + \vec{u}_i(t)$, where $\vec{u}_i = d\vec{r}_i'/dt$ is the velocity relative to the *CM*. Therefore, the total kinetic energy of the body is

$$K_{tot} = \frac{1}{2}\sum_{i=1}^{N} m_i v_i^2 = \frac{1}{2}\sum_{i=1}^{N} m_i(\vec{V}_{CM} + \vec{u}_i)\cdot(\vec{V}_{CM} + \vec{u}_i)$$
$$= \frac{1}{2}MV_{CM}^2 + \vec{V}_{CM}\cdot\sum_{i=1}^{N} m_i\vec{u}_i + \frac{1}{2}\sum_{i=1}^{N} m_i u_i^2.$$

The first term of this expression is just the kinetic energy associated with center of mass motion. The third term is the molecular kinetic energy relative to the *CM*, and the middle term is exactly zero.

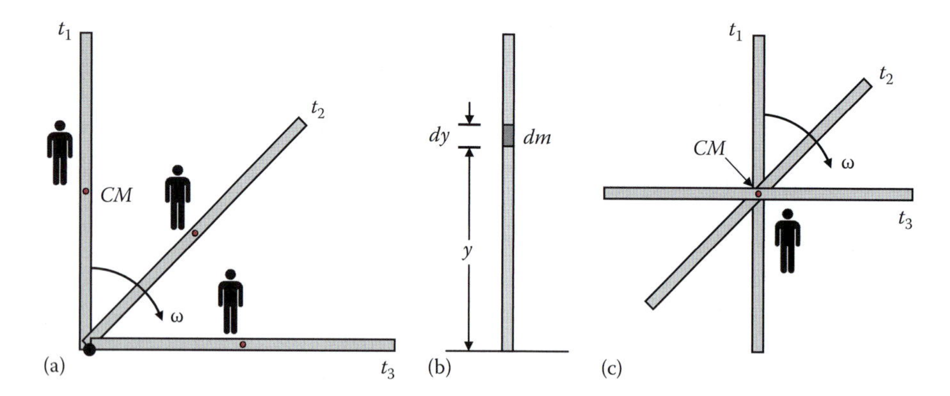

FIGURE 11.4 (a) A falling stick pivots one-quarter turn about a hinge mounted on the floor. (b) To calculate the initial potential energy, divide the stick into short segments of length dy and mass $dm = M\,dy/\ell$. (c) An observer moving with the *CM* sees the falling stick making a quarter turn before striking the floor at time t_3.

> **EXERCISE 11.3**
>
> Show that $\sum_i m_i \vec{u}_i = 0$. (You might want to refer to Section 5.4.)

We are left with a very important result having broad validity: the kinetic energy of a body (or, more generally, a collection of bodies) divides neatly into two parts:

$$K_{tot} = \frac{1}{2} M V_{CM}^2 + K_{rel}, \tag{11.3}$$

where $K_{rel} = \frac{1}{2}\sum_{i=1}^{N} m_i u_i^2$ is the kinetic energy that an observer traveling with the *CM* would measure. Picture a moving box filled with gas molecules in random motion. The first term in Equation 11.3 refers to the collective translational motion of the box and its gaseous contents, while the last term refers to the random molecular motion of the molecules within the box. An observer inside the box would "see" only the second part. For *rigid* bodies, where each molecule is clamped to its nearest neighbors, the only *macroscopic* mode of relative motion is rotation,* so

$$K_{tot} = \frac{1}{2} M V_{CM}^2 + \frac{1}{2} I \omega^2. \tag{11.4}$$

A word of caution: In Equation 11.4, I must be measured about an axis of rotation that passes through the body's *CM*. For example, let's revisit the previous example of the falling stick attached to a hinge at one

* In rigid bodies, molecules also vibrate about their equilibrium positions. The energy of this motion is included as *internal* energy rather than as *mechanical* energy.

end. To an observer standing on the floor, the rod executes one-quarter turn before striking the floor, as in Figure 11.4a. To an observer moving with the *CM* of the stick, it does the same thing in the same time (Figure 11.4c). Both observers therefore measure the same value of ω for the stick's angular velocity. Initially, $E = U_i = ½Mg\ell$. At the instant of impact with the floor, $U_f = 0$ and $E = K_{tot}$, so by conservation of energy,

$$E = \frac{1}{2}Mg\ell = \frac{1}{2}MV_{CM}^2 + \frac{1}{2}I\omega^2,$$

where $V_{CM} = \omega\ell/2$ and I is the moment of inertia *measured relative to the stick's CM*: $I = {}^1\!/_{12}M\ell^2$.

$$E = \frac{1}{2}Mg\ell = \frac{1}{2}M\frac{\omega^2\ell^2}{4} + \frac{1}{2}\frac{M\ell^2}{12}\omega^2 = \frac{1}{2}\frac{M\ell^2\omega^2}{3}.$$

Simplifying and solving for ω, we find $\omega = \sqrt{3g/\ell}$, just as we found before. The two methods of solution are equivalent. The important point to remember is, in either strategy, we must use the moment of inertia calculated relative to the appropriate axis: the *CM* (center of the stick) in the present method or the fixed axis of rotation (the end of the stick) in the previous one.

Rolling bodies move by translation as well as rotation. Consider, for example, a round object (a hoop, cylinder, or sphere) of mass M and radius R resting atop an inclined plane of height h. When released, the body rolls without slipping down the ramp. What is its *CM* speed V_{CM} at the bottom of the incline? At the top of the ramp, the body's kinetic energy is equal to 0. The change in potential energy as the body rolls from the top of the ramp to a point with altitude y is $\Delta U = Mg(y - h) < 0$. Since the body does not slip as it rolls, it experiences static friction only (which causes it to roll), but there is no loss of mechanical energy due to kinetic friction between the body and the ramp at the point of contact. Therefore, $\Delta E_{mech} = 0$, so $\Delta K_{tot} = -\Delta U$, or

$$\frac{1}{2}MV_{CM}^2 + \frac{1}{2}I\omega^2 = Mg(h - y),$$

where I is measured about an axis passing through the *CM*. The no-slip condition also imposes a condition between the body's *CM* speed and its rotational velocity ω. If there is no slippage between the body and the surface of the ramp, then the body's *CM* moves a distance $s = 2\pi R$ for each complete revolution. More generally, when the body rotates by an angle $\Delta\theta$, its *CM* moves by $\Delta s = R\Delta\theta$. Dividing by the time interval Δt and taking the limit $\Delta t \to 0$, $V_{CM} = ds/dt = R\,d\theta/dt = \omega R$. This result insures that, at the point of contact, the rim of the wheel is not moving relative to the ramp surface. This is illustrated in Figure 11.5b. Using $\omega = V_{CM}/R$ in the above energy equation, we obtain

$$Mg(h - y) = \frac{1}{2}\left(M + \frac{I}{R^2}\right)V_{CM}^2.$$

If the rolling body is a solid sphere, its moment of inertia about its *CM* is $I_{sphere} = {}^2\!/_5 MR^2$. At the bottom of the ramp ($y = 0$), the sphere's speed is easily found: $V_{CM}^2 = 2gh/(7/5) = {}^{10}\!/_7\,gh$.

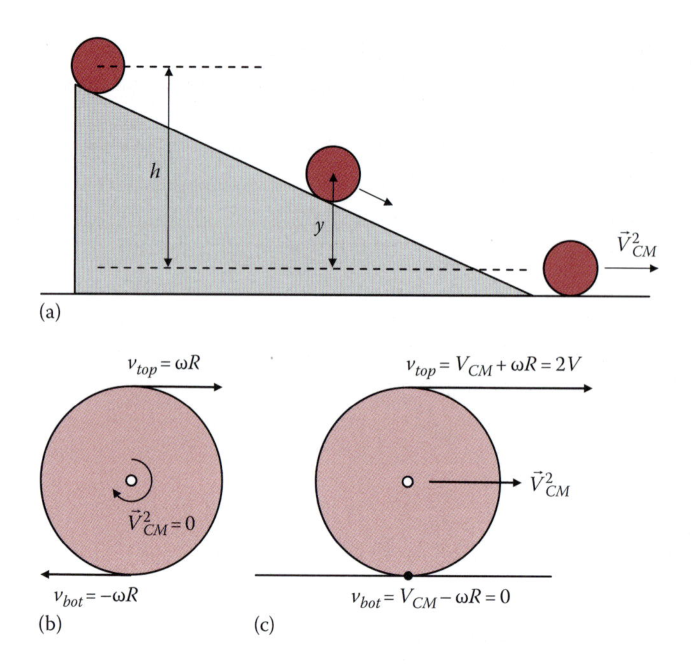

FIGURE 11.5 (a) A body rolls without slipping down an inclined plane. Its kinetic energy is due to the *CM* motion ½ MV^2_{CM} plus its rotational motion ½ $I\omega^2$. (b) Each point on the circumference of a body of radius R is moving with speed ωR relative to the *CM*. (c) When the body is rolling without slipping, the velocity of the body at the point of contact with the stationary surface is zero; the velocity at the top of the body, relative to the surface, is $v_{top} = 2V_{CM} = 2\omega R$.

EXERCISE 11.4

Here is a favorite lecture demonstration. A ball, a hoop, and a solid cylinder decide to race down an inclined plane. Their masses and radii are equal, and each object rolls without slipping. (a) Which object finishes first? Second? Last? (b) If $R_{hoop} > R_{cylinder} > R_{sphere}$, what would be the outcome of the race? (c) If, instead, $M_{hoop} < M_{cylinder} < M_{sphere}$, how would this influence the race?

Answer (a) The sphere wins, and the hoop comes in last (b) and (c) Will have the same result

EXERCISE 11.5

This is another great demonstration. Your instructor is about to eat lunch. He has two cans of soup to choose from. But the cans are unlabeled, and all he knows is that one of them contains (thin, watery) chicken broth and the other one contains (thick, hearty) bean-with-bacon soup. He wants the bean soup for lunch. How can he pick the right can, simply by rolling the two down a ramp?

Answer (Not telling)

11.5 NEW CONSERVATION LAW

One day you arrive a little late for physics class, and (no surprise!) the only empty seats are in the front row. As you quietly settle in, you notice that your instructor is sitting cross-legged—lotus style—on a small rotatable platform. Nearby, there is a bicycle wheel with a short handle attached to its hub (see Figure 11.6a). Your instructor calls for an assistant, looking straight at you! Given your late arrival and the shaky grade you received on the last quiz, you suspect it would be wise to volunteer. Your task is simply to hand the bicycle wheel to your seated instructor and then carefully observe what happens.

The teacher raises the wheel above her head and holds it so that the handle (which coincides with the wheel's axis of rotation) is vertical. She reaches up and sets the wheel spinning at 2.5 revolutions per second (rps) in a counterclockwise sense (CCW), as viewed from above the wheel. This causes her to rotate in the opposite, clockwise (CW) sense, with a rotation rate of 0.5 rps. Then she tilts the wheel's axis away from the vertical direction, which reduces her own rotation rate. When the handle is pointed horizontally, her rotation rate is zero, and when the handle points downward, she rotates at 0.5 rps in the opposite, CCW sense. Then, when she returns the wheel to its original upward orientation, her own spin reverses and returns to

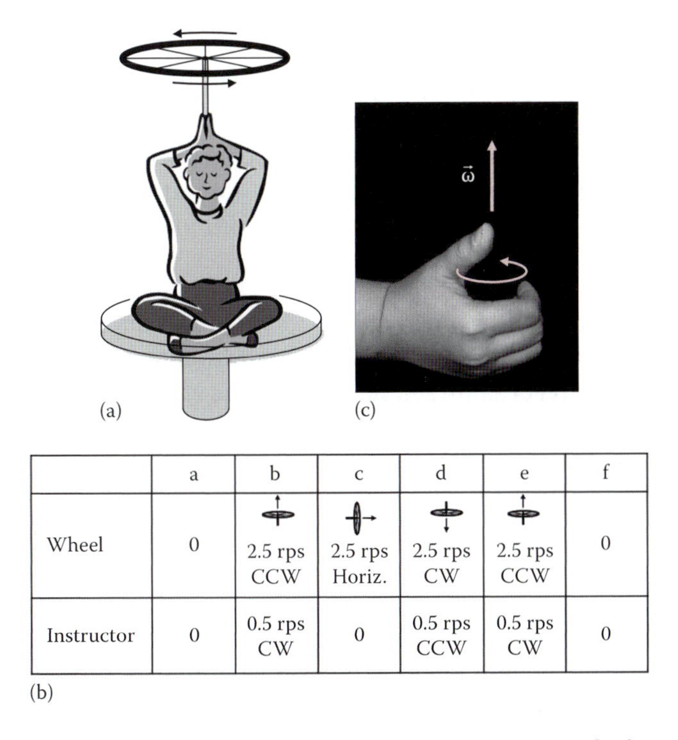

	a	b	c	d	e	f
Wheel	0	2.5 rps CCW	2.5 rps Horiz.	2.5 rps CW	2.5 rps CCW	0
Instructor	0	0.5 rps CW	0	0.5 rps CCW	0.5 rps CW	0

(b)

FIGURE 11.6 (a) The lecture demonstration: the instructor sits on a rotating platform holding a bicycle wheel. (b) A table of observations listing the rotation rate and direction of the wheel and instructor at six moments during the demonstration. (c) The direction of $\vec{\omega}$ is perpendicular to the plane of rotation. If you curl the fingers of your right hand in the sense of the rotation, your extended thumb will point in the direction of $\vec{\omega}$.

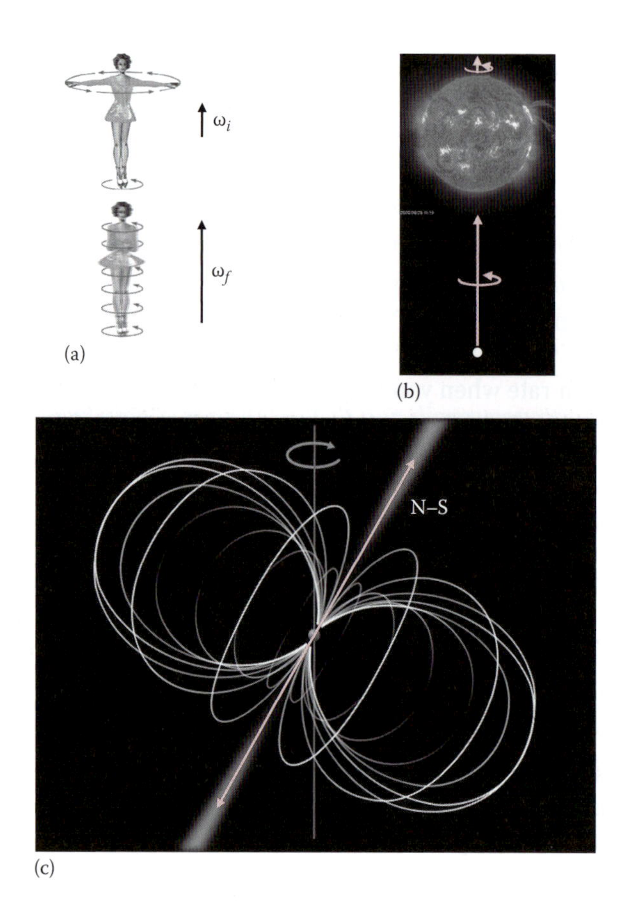

FIGURE 11.7 (a) By reducing her moment of inertia, an ice skater conserves angular momentum by increasing her spin rate. (b) The same is true when a star collapses to form a neutron star. In this case, the spin rate can increase by a factor $>10^6$. (c) A pulsar's axis of rotation (↑) is not aligned with its N–S magnetic axis. As the star rotates, its radiation pattern (white) rotates with the star. (Courtesy of NASA and ESA, SOHO/EIT Consortium, an international cooperative project of NASA and the European Space Agency.)

EXERCISE 11.9

Estimate the moment of inertia of you plus the two dumbbells when your arms are fully extended, and when they are folded across your chest. At arm's length, the weights are about 1 m from your central axis. (Assume that you are about the same size as your instructor. See Exercise 11.6.)

Answer $I_i \approx 3$ kg m², $I_f \approx 1$ kg m²

The conservation of spin angular momentum requires $I_i\vec{\omega}_i = I_f\vec{\omega}_f$, where i and f stand for "initial" and "final." The initial state is when you are spinning with your arms outstretched. In the above exercise, you found that $I_i \approx 3I_f$, so $\vec{\omega}_f \approx 3\vec{\omega}_i$. (You may be surprised that the 1 kg dumbbells affect the motion so strongly, since your own mass is much greater than 1 kg. Can you explain why the hand weights have such a large influence?)

The same phenomenon occurs in astronomical systems. When a massive dying star ($8 < M/M_{Sun} < 20$) runs out of nuclear fuel, its stellar core cools and collapses under its own weight. The core is squeezed so tightly by gravity that, at its center, electrons and protons are crushed together to form neutrons. The core is effectively transformed into a colossal nucleus (held together by gravity!) with a density exceeding 10^{17} kg/m^3. The surrounding layers of the star crash into this incompressible ball at high speed ($\approx 0.15c$), and then rebound to collide head-on with atoms falling from the outermost layers of the star. These latter collisions heat the outer layers to fantastically high temperatures, inducing nuclear fusion while blasting them outward into space. The collapse and subsequent explosion is called a *supernova*, and is responsible for producing all elements of the periodic table heavier than iron, including the radioactive isotopes that were essential for warming the early Earth and cradling life on our planet. In its wake, the supernova leaves its compressed core, a *neutron star* with mass $1\text{–}3M_{Sun}$ and radius of about 10 km.*

If the star is originally rotating, its collapsed core will spin very much faster in order to conserve spin angular momentum (Figure 11.7b). For example, consider the collapse of the Sun to form a neutron star. The Sun rotates once every 26 days and has a radius $R_i = 7 \times 10^8$ m. Its density is far from uniform, and as a result, its moment of inertia is roughly $I_i = \frac{1}{16}MR_i^2$ (rather than $\frac{2}{5}MR^2$ for a homogeneous sphere). A neutron star with $M = M_{Sun}$ will have a radius $R_f \simeq 10\,\text{km} \ll R_i$ and moment of inertia $I_f \approx \frac{1}{3}MR_f^2$. Conservation of spin angular momentum requires that $I_i\omega_i = I_f\omega_f$. Assuming no loss of mass (which is admittedly unrealistic), the ratio between the initial and final rotation periods ($T = 2\pi/\omega$) of the star is given by

$$\frac{T_f}{T_i} \left(= \frac{\omega_i}{\omega_f} \right) = \frac{I_f}{I_i} = \frac{16}{3}\frac{R_f^2}{R_i^2}$$

or

$$T_f = \frac{16}{3}\left(\frac{10^4 \text{ m}}{7 \times 10^8 \text{ m}}\right)^2 \times 26 \text{ days} \simeq 2.8 \times 10^{-8} \text{ days.}$$

Since 1 day = 8.64×10^4 s, $T_f \simeq 2.4$ ms. In this simplified scenario, a collapsed Sun will become a neutron star, rotating about 400 times per second.

The pulsars discovered by Jocelyn Bell are neutron stars that emit electromagnetic energy by a process called synchrotron radiation. Like Earth, a pulsar has magnetic north and south poles that are not aligned

* This scenario of neutron star formation was cautiously proposed in 1934 by Walter Baade and Fritz Zwicky, less than 2 years after the discovery of neutrons by James Chadwick.

an Earth-sized exoplanet. With appropriate understatement, the plaque's three designers concluded: "We do not know if the message will ever be found and decoded; but its inclusion ...seems to us a hopeful symbol of a vigorous civilization on Earth." If *Pioneer*'s plaque *is* discovered, the "little green men" who find it will share a secret of unprecedented significance: they will know that they are not alone.*

* In 1977, two *Voyager* spacecraft were launched, each carrying a "golden record" with characteristic images and sounds from Earth. Among the sounds is a Chuck Berry recording of "Johnny B. Goode." The record must be played on a 16⅔ rpm turntable, and is thoughtfully packaged with a suitable stylus. What impression will Chuck Berry make on college-age aliens? Will an 16⅔ rpm LP convince aliens that ours is an advanced civilization?

PROBLEMS

11.1 Find the moment of inertia of a thick ring of mass M, inner and outer radii R_1 and R_2, and thickness t. (This is called an *annulus*.) The axis of rotation passes through the ring's center perpendicular to its flat surfaces. Do this as follows:

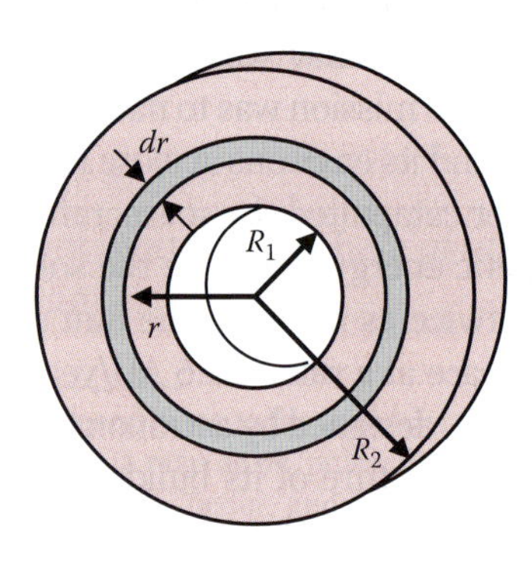

(a) Divide the ring into very thin interlocking ringlets, and consider a single ringlet of radius r and width dr. If the density of the object is ρ, what is the mass dm of this ringlet? What is its moment of inertia dI?

(b) Now sum (i.e., integrate) the moments of inertia of all the ringlets ranging in radius from R_1 to R_2.

(c) Derive an expression for the mass M in terms of the density ρ, R_1, R_2, and t.

(d) Combine your results in parts c and d to derive the standard expression for the moment of inertia of the annulus: $I = \frac{1}{2}M(R_1^2 + R_2^2)$. Check your results against other shapes: a hoop is a thin ring with $R_1 \approx R_2$; a solid disk has $R_1 = 0$.

(e) If $t \ll R_2$, find the moment of inertia about an axis passing through the center of the ring, *perpendicular* to the axis used above.

11.2 The moment of inertia of the Earth about its polar axis is $I_E = 0.331 M_E R_E^2$, whereas the moment of inertia of a solid sphere of uniform density is $I = \frac{2}{5} MR^2$. What does this tell you about the composition of the Earth?

11.3 (a) Find the moment of inertia of a thin rectangular slab about an axis through its center of mass that is perpendicular to the plane of the slab. The slab has a width a and length b.

(b) Find the moment of inertia about a parallel axis passing through one corner of the slab. (See the figure below.)

(c) Find the moment of inertia about a third axis lying in the plane of the slab and passing through its *CM*. See the figure. (*Hint*: Use the stretch theorem.)

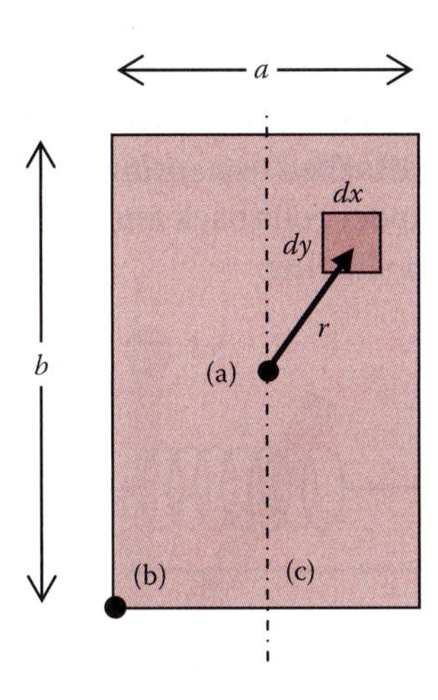

11.4 A disk of radius R and thickness t has a density that increases linearly from the center outward, varying as $\rho = \rho_0 r/R$, where r is the distance from the axis of the disk.

(a) Find the total mass M of the disk.

(b) Derive an expression for the moment of inertia I about the disk's axis of symmetry, in terms of M and R. Compare your result to that for a solid disk of uniform density, and for a hoop, each having the same mass M and radius R.

11.5 A thin spherical shell (like a ping-pong ball) of mass dm and radius r has a moment of inertia $dI = \frac{2}{3} r^2 dm$ about an axis passing through its center. Given this information, derive the moment of inertia of a *solid* sphere by treating the sphere as a collection of nested shells:

(a) Let the shell have radius r and thickness dr. Show that its mass is $dm = 4\pi r^2 \rho\, dr$, where ρ is the density. Using the above expression, show that the moment of inertia dI of the shell is

$$dI = \frac{8\pi}{3} \rho r^4 dr.$$

(b) To find the moment of inertia I of a solid sphere of radius R and total mass M, integrate dI over a collection of shells with $0 < r < R$. Express your answer in terms of R and the total mass M.

11.6 The moment of inertia of an oxygen molecule about an axis passing through its center of mass and perpendicular to the line joining the masses is measured (by microwave spectroscopy) to be 1.95×10^{-46} kg m^2.

(a) What is the center-to-center distance between the atoms? Express your answer in picometers.

(b) The calculated atomic radius of an oxygen *atom* is about 50 pm. In part (a), why is it ok to treat the two atoms of the molecule as point masses? (*Hint*: Think about the structure of an atom.)

11.7 A soup can filled with chicken broth rolls without slipping down an inclined plane. The height of the ramp is 0.25 m.

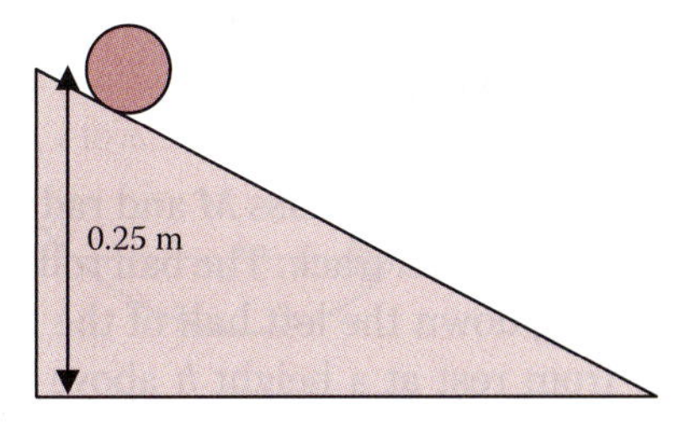

(a) Compare its motion to that of an empty can rolling down the same slope. Which can reaches the bottom first? (*Hint*: At first, the broth does not rotate at the same rate as the can.)

(b) What is the speed of the empty can at the bottom of the slope? (You may treat the can as a hollow cylinder.)

(d) Compare the ball to a sphere of uniform density. Choose one of the following responses.

(i) The density ρ is uniform throughout the disk.

(ii) The density ρ is greater near the surface of the ball.

(iii) The density ρ is greater near the center of the ball.

11.15 A woman sits in the exact center of a long canoe. The moment of inertia of the boat about a vertical axis through its center is 30 kg m^2 and that of the woman is 1.0 kg m^2. She is initially facing east. If she carefully stands and turns 180° so that she faces west, through what angle does the canoe turn? Describe the motion of the boat after she sits back down. If she is sitting in the front seat when she turns, would the canoe rotate through an angle greater than, less than, or the same as your first answer above?

11.16 A student is standing on a rotatable platform holding a bicycle wheel. He has a mass of 80 kg and can be approximated as a cylinder of radius 20 cm. The platform has a mass of 20 kg and the same radius. The bicycle wheel has a diameter of 80 cm and a mass of 4 kg.

(a) Holding the wheel with its axle directed vertically, he sets the wheel spinning. How many rotations of the wheel occur before he has turned one full circle on the platform?

(b) The student stops the wheel from turning by applying a gentle frictional force to the outside surface of the wheel. What happens to his rotational motion? What would happen if he instead applied a strong force, stopping the wheel suddenly?

11.17 You are standing on a rotatable platform that has a mass of 20 kg and a radius of 20 cm. Your mass is 80 kg and you can be approximated as a uniform cylinder of radius 20 cm. Initially, you and the platform are stationary. You wish to make an "about face" turn, so that you are facing the opposite direction. If you were standing on the floor, you could do this in 2 s.

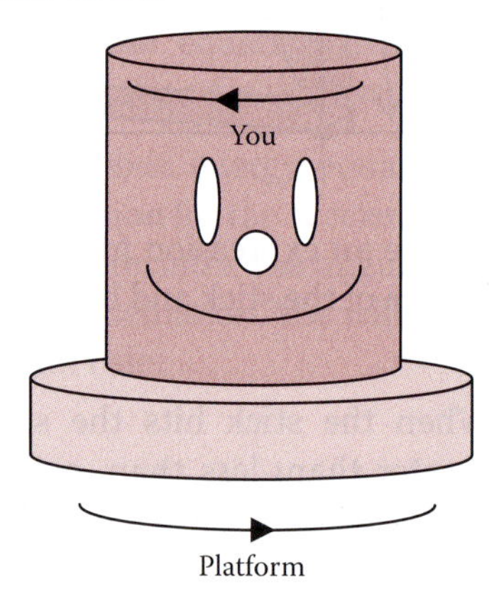

(a) What happens when you try to turn around? Why?

(b) How many revolutions will the platform have executed when you have turned by 180°?

(c) How long will it take you to do this while standing on the platform?

(d) If you held your arms straight out rather than at your sides, would it take a longer or shorter time to make the "about face" turn?

11.18 Two children, each of mass M, sit at opposite ends of a narrow board with length L and mass M (the same as the mass of each child). The board is mounted on a vertical axle through its center and is free to rotate in a horizontal plane without friction.

(a) What is the moment of inertia of the board plus the children about a vertical axis that coincides with the axle?

(b) What is the spin angular momentum of the system if it is rotating with an angular speed ω_0 in a clockwise direction as seen from above? (Include the direction of $\vec{L}$.)

(c) While the system is rotating, the two children pull themselves toward the center of the board until they are half as far from the center as before. What is the resulting angular speed?

(d) What is the change in kinetic energy of the system as a result of the children changing their positions? How is energy conserved during this motion?

11.19 An electric motor spins a solid disk with a moment of inertia $I_d = 2 \times 10^{-3}$ kg m^2 about its central axis. The device, called a *reaction wheel*, is used to change the orientation of spacecraft without the need for rocket thrusters. At position *A*, the reaction wheel axis is aligned with the central axis of the spacecraft. The craft has a moment of inertia $I_p = 120$ kg m^2 about this axis.

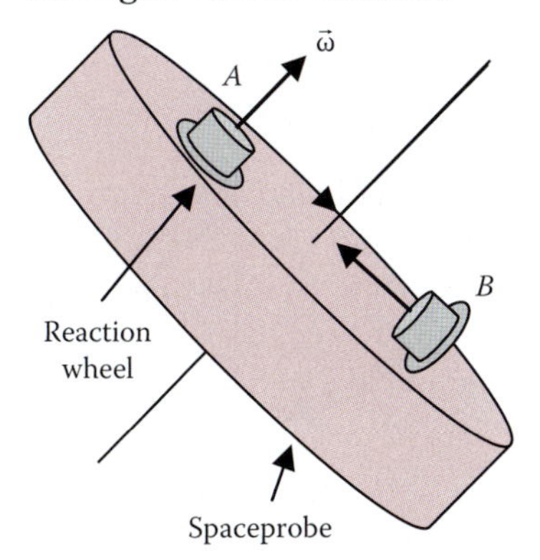

(a) If the reaction wheel is set spinning at 5000 rpm, find the time needed to rotate the spacecraft through 30°.

(b) If the reaction wheel is mounted at position *B*, with its rotation axis perpendicular to the central axis of the craft, will the spacecraft rotate when the wheel is set spinning? If so, in what direction?

11.20 The Martian "day" is almost exactly the same length as a day on Earth: $T_{Mars} = 1.03T_E$. In recent years, spacecraft landing on Mars have found a small seasonal variation in the length of the day, $\Delta T_{Mars} = 0.5$ ms, which can be attributed to the melting of the polar ice caps on the planet. According to the model, ice situated near the planet's axis of rotation periodically melts and is redistributed in a thin atmospheric shell surrounding the planet.

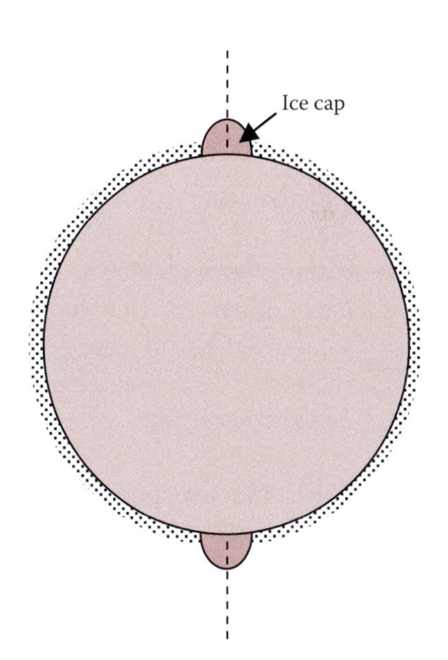

The mass of Mars is 6.24×10^{23} kg and the planet's radius is 3.38×10^6 m.

(a) Is ΔT_{Mars} positive or negative; that is, when the ice caps melt, does the Martian day get longer or shorter?

(b) Derive an equation for the fractional change $\Delta T_{Mars}/T_{Mars}$ in terms of the fractional change in the moment of inertia of the planet.

(c) Find the mass of the ice caps.

What if Mars had a subsurface ocean that decoupled the core of the planet from its surface crust? Then, the melting of the ice caps would affect only the rotation of the surface.

(d) If the planet were covered by a thin crust of density $\rho = 1$ g/cm^3 and thickness 100 km, the mass of this crust would be 1.4×10^{22} kg and its moment of inertia would be 1.1×10^{35} kg m^2. Predict ΔT_{Mars}, the change in time for 1 complete rotation of the Martian surface, for this model. (Based on the changing rotation rate of Saturn's largest moon Titan, astronomers suspect that it harbors a subsurface ocean.)

(e) Imagine that the crust and core of a planet (or Moon) are rotating at slightly different rates which gradually approach each other. As this happens, will the total mechanical energy of the system increase, decrease, or stay the same? Explain carefully.

11.21 To solve the overpopulation problem on Earth, the United Nations decides to hollow out the planet to make a thin spherical shell of uniform density, with outside radius R and shell thickness $0.1R$. The total mass of the planet remains unchanged.

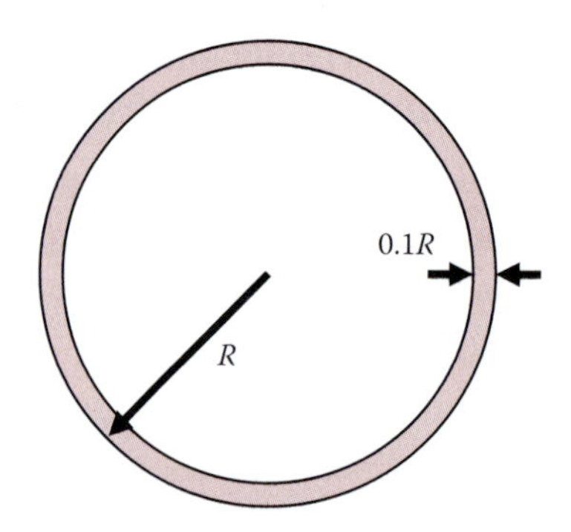

(a) Assuming that the density of the shell is equal to the average density of the present Earth, find the new R in terms of the present radius of the planet R_E.

(b) What is the new value of "g" on the surface of the hollowed out planet?

(c) How long is one "day" on the new planet?

11.22 If the polar icecaps of the Earth were to melt and the water returned to the oceans, the ocean level would rise by about 30 m. Assume that the ice caps lie on or close to the axis of rotation, so that initially they do not contribute significantly to the planet's moment of inertia.

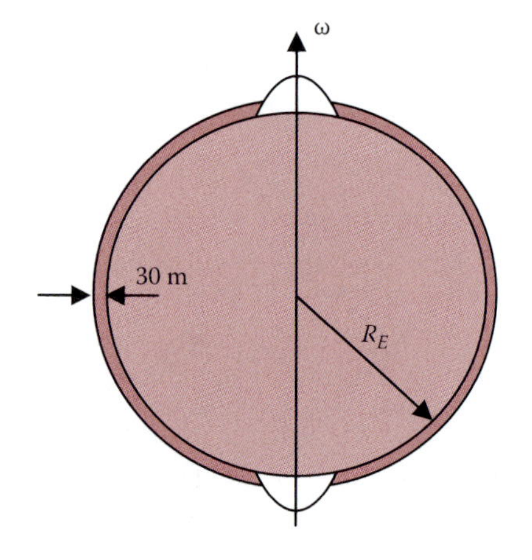

(a) The water from the ice caps forms a 30 m shell around the Earth. Find an expression for the moment of inertia of this shell. ($\rho_{H_2O} = 1000\ kg/m^3$.)

(b) Compare this to the total moment of inertia of Earth. Assume a uniform density sphere.

(c) Estimate the change in the length of an Earth day. (*Hint*: It is small but measurable.)

11.23 Picture the gravitational collapse of a uniform spherical ball of gas to form the Sun. If the initial radius of the gas ball was 10^{13} km, and it was rotating at a rate of 1 revolution per 10^6 years, what would be the period of rotation of the Sun, if it were a sphere of uniform density and radius 7×10^8 m? (Assume no loss of mass.)

11.24 The Sun's present radius is 7×10^8 m and its present period of rotation is 26 days.

(a) What would its period be if it expands to a red giant with a radius of 1 AU (engulfing the Earth)?

(b) Prove that the kinetic energy of rotation can be written as $K_{rot} = L^2_{spin}/2I$, where I is the moment of inertia of the spinning body.

(c) Would the Sun's kinetic energy increase or decrease as it expanded?

11.25 The Crab Nebula, shown in the figure, is the remnant of a historic supernova that was observed by Chinese, Japanese, and Arab astronomers in 1054 AD. The glowing filaments of the Crab are expanding outward at 1500 km/s, and *they are accelerating*. To power this acceleration, an energy source delivering 5×10^{31} W (J/s) is required. At the heart of the nebula is a pulsar with a 33 ms rotation period. Careful measurements performed at the Arecibo radio telescope in Puerto Rico indicate that the pulsar's period T is increasing at a rate $dT/dt = 4.21 \times 10^{-13}$, that is, its rotation rate is decaying. Treat the pulsar as a neutron star with mass $M = 1.4M_{Sun}$ and radius $R = 10$ km.

(a) Show that the decay of the rotation rate accounts for the energy output needed by the nebula.

(b) Pulsars have densities rivaling nuclear densities, yet they are still held together by gravity. Surface matter on the star's equator is in uniform circular motion with centripetal acceleration $\omega^2 R = 4\pi^2 R/T^2$. What is the minimum value of T if a pulsar having the mass and radius given above is to remain stable?

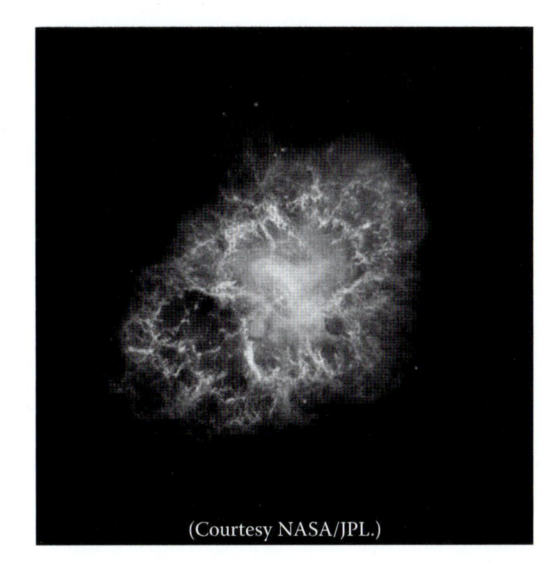

(Courtesy NASA/JPL.)

12

ANGULAR MOMENTUM AND ITS CONSERVATION

The Great Comet of 1680 over Rotterdam, by Lieve Verschuier, shows the comet as it appeared in the evening sky in December 1680. The viewers are sighting the comet with cross-staffs, navigation instruments for measuring angles, probably to determine the comet's altitude and angular extent of its tail. (From the public domain.)

Throughout history, comets were popularly regarded as harbingers of disaster. They warned of God's displeasure with man, and foretold the death of a king, or plague, war, and famine. Before the Principia (1687), scientific understanding of comets was sorely lacking: While planets were known to travel in elliptical orbits, comets were thought to move in straight lines at constant speed. (This view was shared by Galileo, Kepler, Huygens, Cassini, and even Newton!) In November, 1680, a bright comet appeared in the morning sky shortly before dawn. Day by day, it crept closer to the Sun before vanishing in its glare. Two weeks later, the comet reappeared shortly after dusk in the evening sky, now a brilliant spectacle with a gigantic tail. It transfixed Isaac Newton as well as the astronomers John Flamsteed and Edmond Halley. What was its nature? Where did it come from? Newton initially held the traditional view, that the morning and evening comets were different bodies. (They certainly looked very different!). Flamsteed argued for a single comet that made a U-turn before reaching the Sun. In the Principia, Newton adopted the one-comet hypothesis and analyzed its motion using the inverse-square law. The success of his analysis was a triumph for his theory of universal gravitation, and may have been what persuaded Halley of Newton's genius. Halley became Newton's staunchest supporter and benefactor, reviewing, editing, and underwriting the publication expenses of the Principia.

In 1687, Halley wrote to Newton and suggested that his theory and computational techniques (including calculus) be applied to other comets. Halley was especially intrigued by the bright comet of 1682. His calculations suggested that its orbit was elliptical, that is, periodic, and he suspected that the same comet had appeared earlier, in 1532 and 1607. By 1705, Halley had calculated the orbital parameters for 24 comets, including those of 1680 and 1682. Now convinced that the 1682 comet ("Halley's comet") was periodic, he boldly predicted its return in late 1758 or early 1759. Its appearance at the predicted time would provide a rigorous test of Newton's theory of gravitation.

Fifteen years after Halley's death, his prediction was refined by the French mathematician Alexis-Claude Clairaut and two young assistants, Joseph-Jerome LaLande and Mme. Nicole-Reine Lepaute, who computed the effect of Jupiter and Saturn on the comet's period. Throughout 1758, the trio raced to pinpoint the date of the comet's arrival. (A "prediction" made after the event would hardly be noteworthy.) In November, just 6 weeks before the comet was spotted, they publicized their results: Halley's comet would pass through perihelion in April, 1759. In fact, it did so just one month earlier. Employing Newton's theory of gravitation, Halley (and Clairaut) had achieved a stunning success: using data taken in 1682, they had foretold—with 0.1% accuracy—a celestial event occurring 75 years later.

In 1908, the British astronomer and seismologist Herbert H. Turner characterized the importance of Halley's achievement in the following way:

> *There can be no more complete or more sensational proof of a scientific law than to predict events by means of it. Halley was deservedly the first to perform this great office for Newton's Law of Gravitation.... Newton's great discovery of the Law of Gravitation, and the story of Halley's comet, thus form an integral part of the most important event in the whole history of science.**

* H.H. Turner, Halley's Comet, an evening discourse to the British Association, Dublin, September 4, 1908, Oxford University Press, cited by D.W. Hughes in The history of Halley's comet, *Philosophical Transactions of the Royal Society London, Series A*, 323, 349–367 (1987).

12.1 INTRODUCTION

In Chapter 11, we studied spinning bodies and found that they possess rotational kinetic energy plus another quantity that we loosely called "spin angular momentum": $\vec{L}_{spin} = I\vec{\omega}$. Using a rotatable platform and a bicycle wheel, we demonstrated that for isolated bodies, $\vec{L}_{spin}$ is conserved. But spin is only one example of angular momentum. We will now broaden the definition of $\vec{L}$, and show that all bodies, non-rotating ones, even bodies moving along a straight line, may possess angular momentum. Moreover, we will see that the total angular momentum of an isolated system does not change, giving us a new conservation law to add to those of momentum and energy. Conservation of angular momentum is the crux of Kepler's second law (the equal area law), and we will use it to find the shape of planetary orbits and to predict the trajectories of asteroids, comets, and spacecraft. Like momentum and energy, angular momentum is a powerful tool for studying motion, and it will be employed in Chapters 13 and 14 to study the behavior of rigid bodies in simple lab-scale situations as well as in phenomena that challenge our understanding of the cosmos.

12.2 BROADER DEFINITION OF ANGULAR MOMENTUM

Imagine that you are at point A, observing the motion of a small body. Relative to A, the body is located at position $\vec{r}$ and moving with momentum $\vec{p}$ (see Figure 12.1a). The angular momentum $\vec{L}$ of the body relative to A is formally defined by the cross product

$$\vec{L} \equiv \vec{r} \times \vec{p}. \tag{12.1}$$

According to the properties of the cross product, $\vec{L}$ is perpendicular to the plane containing $\vec{r}$ and $\vec{p}$, and points in the direction determined by the right hand rule. (For a review of cross products, see Section 8.3.) The magnitude of $\vec{L}$ is given by $L = rp\sin\theta$, where θ is the angle between $\vec{r}$ and $\vec{p}$ ($\theta < 180°$). Alternatively, we may express the magnitude as $L = rp_{\perp}$, where $p_{\perp} = p\sin\theta$ is the component of $\vec{p}$ perpendicular to $\vec{r}$, or equivalently as $L = r_{\perp}p$ (see Figure 12.1b).

EXERCISE 12.1

Show that the angular momentum relative to point A of a body moving with no forces acting on it is a constant, that is, $\vec{L}(\vec{r}) = \vec{L}(\vec{r}\,')$.

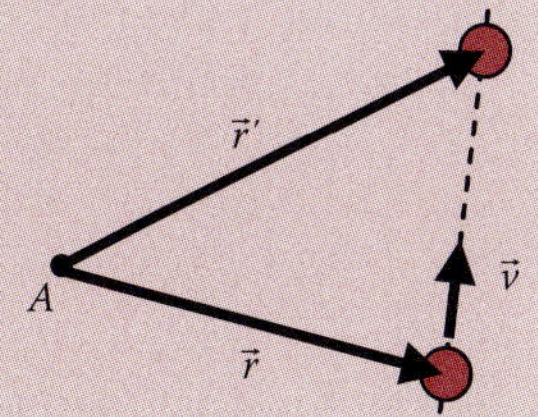

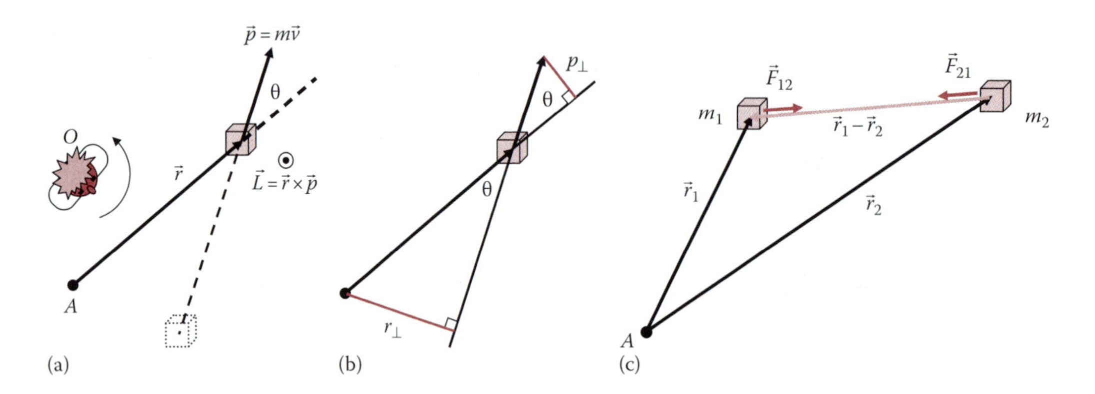

FIGURE 12.1 (a) The angular momentum $\vec{L}$ of a body relative to point A is defined as the cross product of the body's position $\vec{r}$ and momentum $\vec{p}$. If $\vec{r}$ and $\vec{p}$ are in the plane of the paper, then $\vec{L}$ points out of the paper. An observer O at A would need to turn his head to follow the body's motion. (b) The magnitude $L = rp\sin\theta = r_\perp p = rp_\perp$. (c) The forces of interaction between bodies m_1 and m_2, assuming they are *central* forces.

According to Equation 12.1, a body moving in a straight line can possess angular momentum. What is the connection between linear motion and spin? Referring to Figure 12.1a, imagine you are watching the body pass by. As it moves, you will need to turn your head (i.e., rotate) in order to follow its motion. But if the momentum $\vec{p}$ were parallel to $\vec{r}$, then the body would be moving radially away from or toward you, $\vec{r} \times \vec{p} = 0$, and you would not need to turn in order to follow the motion. In the first case, the body has angular momentum relative to you. In the second case it does not.

For a system of N particles $m_1, m_2, \ldots, m_N$ moving with momenta $\vec{p}_1, \vec{p}_2, \ldots, \vec{p}_N$, the system's total angular momentum is $\vec{L}_{tot} = \sum_{i=1}^{N} \vec{r}_i \times \vec{p}_i$. In this expression, all of the position vectors $\vec{r}_i$ must be measured from the same point A, so $\vec{L}_{tot}$ is the total angular momentum relative to point A.

If a different reference point A' were chosen for the position vectors, $\vec{L}_{tot}$ would have a different value.

EXERCISE 12.2

Three small bodies of equal mass m are located at positions that are equidistant from the origin at point A ($r_1 = r_2 = r_3 \equiv r$). Each body has a speed v with the direction shown in the figure.

(a) What is the magnitude of $\vec{L}_{tot}$ relative to point A?
(b) What is the direction of $\vec{L}_{tot}$?
(c) What is $\vec{L}_{tot}$ relative to point A'?

Answer (a) $1/2\ mrv$ (b) into the paper (c) mrv into the paper

Let's show that our earlier definition of spin angular momentum $\vec{L}_{spin} = I\vec{\omega}$ is consistent with the broader definition of $\vec{L}$ given in Equation 12.1. Consider the flat plate shown in Figure 12.2. It is rotating in a counterclockwise sense with angular speed ω about a vertical axis. According to the right hand rule for spin,

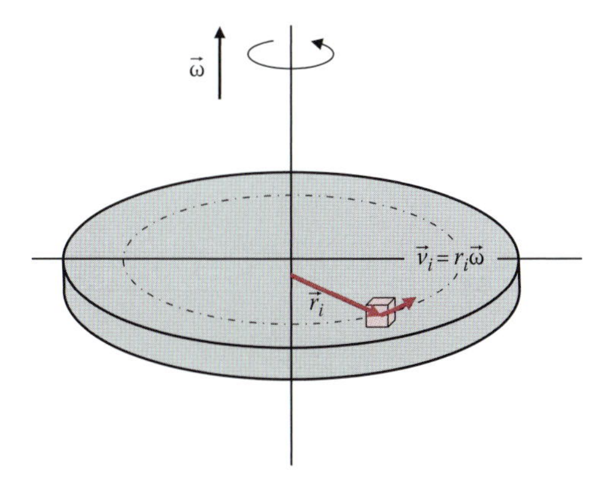

FIGURE 12.2 A disk rotating with angular velocity $\vec{\omega}$ can be divided into N tiny pieces. The ith piece is undergoing uniform circular motion with speed ωr_i, where r_i is its distance from the axis of rotation. The cross product $\vec{r}_i \times \vec{v}_i$ points in the direction of $\vec{\omega}$.

the vector $\vec{\omega}$ is pointing upward along the rotation axis. To find the total angular momentum of the plate, divide it into N tiny pieces. The ith piece has mass Δm_i and position $\vec{r}_i$ relative to point A, where the axis of rotation pierces the plate. Δm_i is in circular motion with velocity $\vec{v}_i \perp \vec{r}_i$ and speed $v_i = r_i\omega$. By the right hand rule, the cross product $\vec{r}_i \times \vec{p}_i = \Delta m_i \vec{r}_i \times \vec{v}_i$ points upward in the direction of $\vec{\omega}$. (Check this!) The total angular momentum of the plate, relative to point A, is

$$\vec{L}_{tot} = \sum_{i=1}^{N} \Delta m_i \vec{r}_i \times \vec{v}_i = \left(\sum_{i=1}^{N} \Delta m_i r_i^2 \right) \vec{\omega} = I\vec{\omega},$$

in agreement with Section 11.5. Be warned, however, that this equality is not generally valid! It holds only for thin plates or for bodies that are rotating about an axis of symmetry. Happily, it is true for all of the situations illustrated in Figure 11.3. A case where it is not true is presented in the following exercise.

EXERCISE 12.3

Two identical balls are attached to the ends of a massless rigid rod. The rod is pivoted about a frictionless bearing passing through the center of mass of the system. The rod tilted out of the horizontal plane by an angle θ. The direction of rotation is along the axis of the bearing and is indicated in the figure by $\vec{\omega}$. Show that, even though the axis of rotation passes through the *CM*, $\vec{L}_{tot}$ is *not* parallel to $\vec{\omega}$. Will the direction of $\vec{L}_{tot}$ remain constant with time?

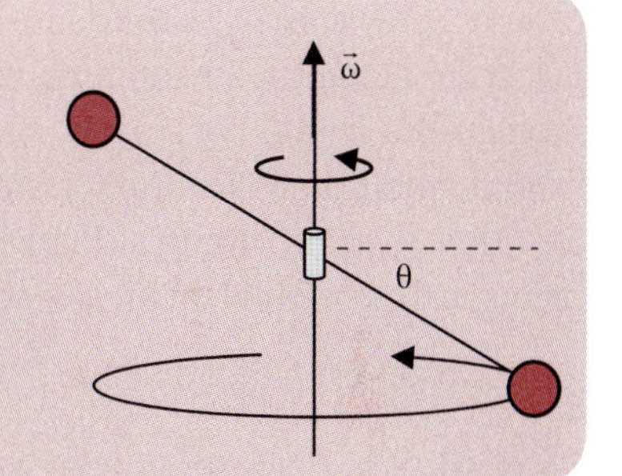

12.3 CONSERVATION OF ANGULAR MOMENTUM

If the system of particles is *isolated*, so that the particles interact with one another but not with any external bodies, the *law of angular momentum conservation* states that

$$\vec{L}_{tot} = \sum_{i=1}^{N} \vec{r}_i \times \vec{p}_i = \text{constant}. \tag{12.2}$$

No matter how often or how strongly the bodies within the system interact with one other, their total angular momentum is conserved. This is a fundamental principle of physics, like the conservation of momentum and energy, and just like those two conservation laws, it cannot be proven. (However, see Box 12.2.)

In certain cases it is possible to show that conservation of $\vec{L}_{tot}$ follows from Newton's laws. Consider, for example, an isolated system of two bodies, m_1 and m_2. Let $\vec{F}_{12}$ be the force on m_1 due to m_2, and let $\vec{F}_{21}$ be the reaction force on m_2 (see Figure 12.1c). Newton's second law tells us that $d\vec{p}_1/dt = \vec{F}_{12}$ and $d\vec{p}_2/dt = \vec{F}_{21}$. The total angular momentum of the system is $\vec{L}_{tot} = \vec{r}_1 \times \vec{p}_1 + \vec{r}_2 \times \vec{p}_2$. Taking the derivative of $\vec{L}_{tot}$ with respect to time,

$$\frac{d\vec{L}_{tot}}{dt} = \vec{r}_1 \times \frac{d\vec{p}_1}{dt} + \vec{r}_2 \times \frac{d\vec{p}_2}{dt} = \vec{r}_1 \times \vec{F}_{12} + \vec{r}_2 \times \vec{F}_{21}.$$

EXERCISE 12.4

What happened to the other terms of the derivative, $(d\vec{r}_1/dt) \times \vec{p}_1$ and $(d\vec{r}_2/dt) \times \vec{p}_2$?

But Newton's third law requires $\vec{F}_{21} = -\vec{F}_{12}$, so

$$\frac{d\vec{L}_{tot}}{dt} = (\vec{r}_1 - \vec{r}_2) \times \vec{F}_{12}.$$

If we make the additional assumption that $\vec{F}_{12}$ and $\vec{F}_{21}$ point directly from one body to the other (like Newtonian gravitational forces), then $\vec{F}_{12}$ and $\vec{F}_{21}$ are parallel (or antiparallel) to $\vec{r}_1 - \vec{r}_2$. Then the cross product above is exactly zero, and $\vec{L}_{tot} = \text{constant}$. But not all forces behave like this. *Noncentral* forces, where $\vec{F}_{12}$ and $\vec{F}_{21}$ do not point along the line joining m_1 and m_2, often arise in electromagnetic phenomena. Nevertheless, angular momentum is still conserved, so we must regard the principle of angular momentum conservation as more fundamental than Newton's laws. Whenever we find an apparent violation of the conservation law, we search for other "hidden" entities that carry angular momentum.* Conservation of angular momentum is a powerful principle that is valid on all size scales, from subatomic to galactic, everywhere in the universe.

EXERCISE 12.5

A particle of mass m moving with speed v strikes and sticks to the rim of a stationary horizontal disk at the position shown. The disk has mass $M = 6m$ and radius R. After the collision, the disk and particle rotate together about a vertical axis through the center of the disk with angular speed ω. Assume angular momentum is conserved.

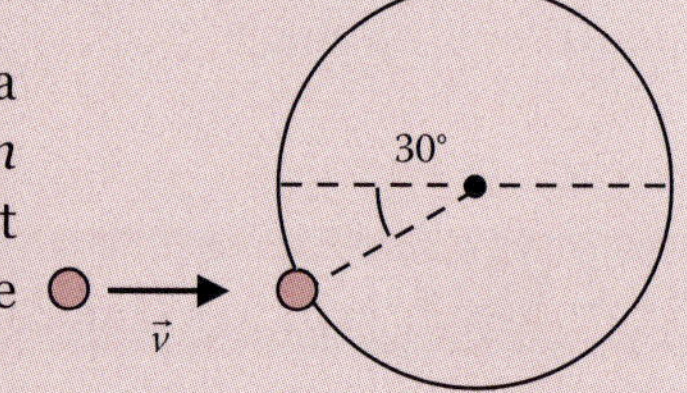

(a) What is the total angular momentum of the particle and disk?
(b) What is the total moment of inertia of the disc and particle after the collision (in terms of m and R).
(c) What is the angular velocity ω (in terms of v and R)?

Answer (a) $\frac{1}{2}mvR$ (b) $4mR^2$ (c) $\omega = v/8R$

* For example, electromagnetic fields (photons) carry both momentum and angular momentum.

12.4 APPLICATION: FINDING THE SWEET SPOT OF A BASEBALL BAT

Anyone who has played baseball (or tennis, or cricket) knows the satisfaction of hitting the ball "just right," striking it squarely without feeling the impact. When this happens, the ball has struck the bat at its "sweet spot," or center of percussion. Let's use conservation principles to study this phenomenon.

Imagine a stationary baseball bat of mass M and length ℓ that is struck by a baseball of mass m. Before the collision, the ball is moving with velocity $\vec{v}$ perpendicular to the axis of the bat (see Figure 12.3). The impact occurs at a distance s below the bat's center of mass. It sets the bat into translational motion with CM velocity $\vec{V}_{CM}$, and also induces rotation about the CM with angular velocity $\vec{\omega} = \omega\hat{k}$. Ignoring gravity,* the ball and bat form an isolated system, so their total momentum and total angular momentum are conserved in the collision. Just before impact, the ball's momentum is $\vec{p}_{ball} = mv\hat{i}$ and its angular momentum *relative to the bat's CM* is $\vec{L}_{ball} = r_{\perp}p\hat{k} = mvs\hat{k}$, where the unit vector $\hat{k}$ points out of the page in Figure 12.3. The bat's initial momentum and angular momentum are both equal to zero. Applying our conservation laws,

$$\vec{L}_{tot} = mvs\hat{k} = I\omega\hat{k} + mv's\hat{k}$$

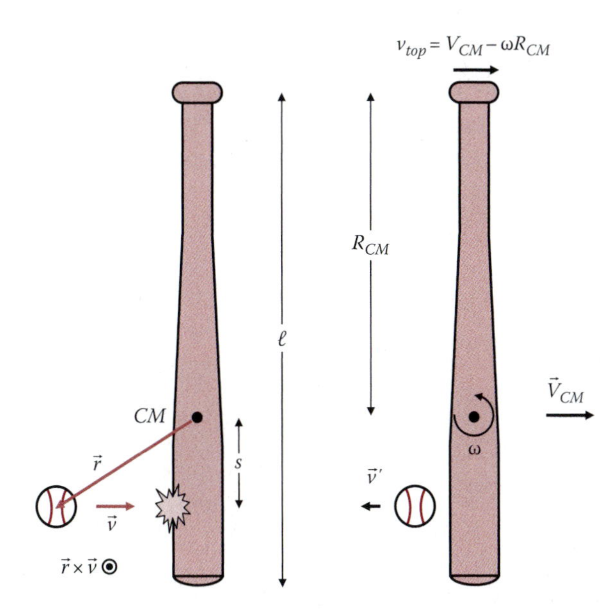

FIGURE 12.3 A baseball strikes a bat at a point that is a vertical distance s from the bat's CM. After the collision, the bat is moving in two ways: its CM is moving with velocity $\vec{V}_{CM}$ (to conserve momentum) and it is rotating with angular velocity $\vec{\omega}$ about its CM (to conserve angular momentum).

* For the purposes of a classroom demonstration, hang the narrow end of the bat from a long vertical string.

and

$$\vec{p}_{tot} = mv\hat{i} = M\vec{V}_{CM}\hat{i} + mv'\hat{i},$$

where, I is the bat's moment of inertia relative to its CM and $\vec{v}'$ is the velocity of the ball after the collision (We assume that the bat and ball both move parallel to the $\hat{i}$ direction after the impact.)

Eliminating the unit vectors, we obtain

$$MV_{CM} = m(v - v')$$

and

$$I\omega = m(v - v')s,$$

which can be combined to eliminate v and v':

$$V_{CM} = \frac{I\omega}{Ms}.$$

Immediately after the collision, the top end of the bat (held by the batter) moves by a combination of translation and rotation: $v_{top} = V_{CM} - \omega R_{CM}$, where R_{CM} is the distance from the top of the bat to its CM (Figure 12.3). The "sweet spot" is the value of s for which $v_{top} = 0$, that is, $V_{CM} = \omega R_{CM}$. Substituting for V_{CM} in the above equation, we find

$$\omega R_{CM} = \frac{I\omega}{Ms}$$

or

$$s = \frac{I}{MR_{CM}}.$$

If we approximate the bat as a long thin rod, then $R_{CM} = \ell/2$ and $I = \frac{1}{12}M\ell^2$, so $s = \ell/6$. The sweet spot is located roughly $\frac{1}{6}\ell + \frac{1}{2}\ell = \frac{2}{3}\ell$ from the batter's end of the bat. (You might test this result using a long stick or a real bat.)

What's so "sweet" about this spot? If $s > \frac{1}{6}\ell$, then $\omega > V_{CM}/R_{CM}$, and the batter's end of the bat will lurch forward (toward the pitcher) unless it is restrained by the batter's hands. On the other hand, if $s < \frac{1}{6}\ell$, then

$\omega < V_{CM}/R_{CM}$ and the upper end will recoil backward (toward the catcher) unless restrained by the batter. In either case, the batter must exert a force to hold the bat, so by Newton's third law, he or she will feel a reaction force from the bat. But when the ball strikes the bat's sweet spot, the reaction force on the batter is zero.

EXERCISE 12.6

What happens if the ball strikes the bat exactly at its *CM* ($s = 0$)?

The above problem was solved using conservation principles exclusively. The range of problems that can be solved in this way is limited, however, and we will more often need to employ Newton's laws plus a new concept, called *torque*, to tackle problems successfully that are more challenging. Torque and rigid body dynamics will be discussed in depth in Chapter 13.

12.5 KEPLER'S SECOND LAW REVISITED

Kepler's second law of planetary motion (Section 8.4) states that a line extending from the Sun to an orbiting planet sweeps out equal areas in equal times. The area swept out in time Δt (see Figure 12.4a) is given by $\Delta A = \frac{1}{2} rv \sin\theta \Delta t$, where θ is the angle between the planet's position vector $\vec{r}$ and its velocity $\vec{v}$, both measured relative to the Sun. This result can be expressed as a cross product:

$$\frac{\Delta A}{\Delta t} = \frac{1}{2} rv \sin\theta = \frac{1}{2} |\vec{r} \times \vec{v}| = \text{constant.}$$

Since $\vec{r}$ and $\vec{v}$ always lie in the plane of the orbit, $\vec{r} \times \vec{v}$ is always perpendicular to the orbital plane, so its direction is constant. Kepler's second law therefore implies that the cross product $\vec{r} \times \vec{v} = \text{constant}$. Ignoring interactions with other planets, the Sun and the chosen planet form an isolated system, so their total angular momentum must be constant. If the solar mass M_s is much greater than the planetary mass m, then the Sun is immobile and $\vec{L}_{tot} = \vec{L}_p = m\vec{r} \times \vec{v} = \text{const}$. For the planets of our solar system, $M_S \gg m$, and Kepler's second law is simply an expression of angular momentum conservation.

What if the two bodies are of comparable size, as in a binary star system or (perhaps) a gas-giant planet of another solar system? Is Kepler's law still obeyed? Once again, treat the two bodies as an isolated system. In their *CM* reference frame, the following three relations are true:

$$M_s \vec{v}_s + m\vec{v}_p = 0$$
$$M_s \vec{r}_s + m\vec{r}_p = 0$$
$$M_s \vec{r}_s \times \vec{v}_s + m\vec{r}_p \times \vec{v}_p = \vec{L}_{tot} = \text{constant.}$$

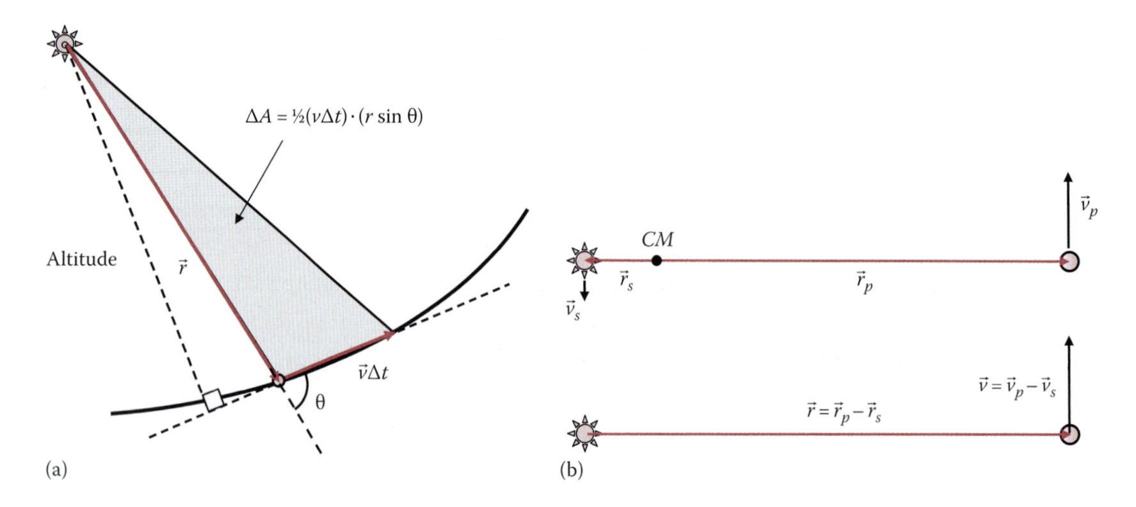

FIGURE 12.4 (a) A review of Kepler's second law. The area ΔA swept out by the planet's position vector in time Δt is shown as the shaded triangle: $\Delta A = \frac{1}{2}\,\text{base} \times \text{altitude} = \frac{1}{2} v\Delta t \times r \sin\theta$. (b) (top) The motion of a planet and the Sun viewed from the *CM* of the system. (bottom) The same motion when viewed from the position of the Sun.

The first equation states that the total momentum equals zero in the *CM* frame (Section 5.4). The second locates the *CM* at the origin of coordinates,* and the third asserts that the total angular momentum of the two-body system is conserved. Using the first two expressions, $\vec{r}_s = -\vec{r}_p m/M_s$ and $\vec{v}_s = -\vec{v}_p m/M_s$ (see Figure 12.4b). Eliminating $\vec{r}_s$ and $\vec{v}_s$ from the third (angular momentum) equation, we obtain

$$\vec{L}_{tot} = M_s \frac{m^2}{M_s^2} \vec{r}_p \times \vec{v}_p + m\vec{r}_p \times \vec{v}_p = m\left(1 + \frac{m}{M_s}\right)\vec{r}_p \times \vec{v}_p.$$

The position and velocity of the planet *relative to the Sun* are

$$\vec{r} = \vec{r}_p - \vec{r}_s = \left(1 + \frac{m}{M_s}\right)\vec{r}_p$$

and

$$\vec{v} = \vec{v}_p - \vec{v}_s = \left(1 + \frac{m}{M_s}\right)\vec{v}_p.$$

* Recall from Chapter 5, $\vec{r}_{CM} = (m_1\vec{r}_1 + m_2\vec{r}_2)/(m_1 + m_2)$. In the CM frame, $\vec{r}_{CM} = 0$, so $m_1\vec{r}_1 + m_2\vec{r}_2 = 0$.

Using these relations to express $\vec{r}_p$ in terms of $\vec{r}$, and $\vec{v}_p$ in terms of $\vec{v}$, we find

$$\vec{L}_{tot} = m\left(1+\frac{m}{M_s}\right)^{-1}\vec{r}\times\vec{v} = \mu\vec{r}\times\vec{v}, \tag{12.3}$$

where $\mu = mM_s/(m + M_s)$ is the *reduced mass* that we met earlier when studying coupled oscillators (Section 7.11). The vectors $\vec{r}$ and $\vec{v}$ in Equation 12.3 are the position and velocity that appear in Kepler's second law. Since μ is a constant, the "equal area" law remains true, even when both bodies are in motion. Just as before, conservation of angular momentum leads directly to Kepler's second law.

EXERCISE 12.7

A planet of mass m is moving on an elliptical orbit about a star of mass $M_s \gg m$. The distance between the two bodies at aphelion is three times the distance at perihelion. The maximum speed of the planet is 30 km/s.

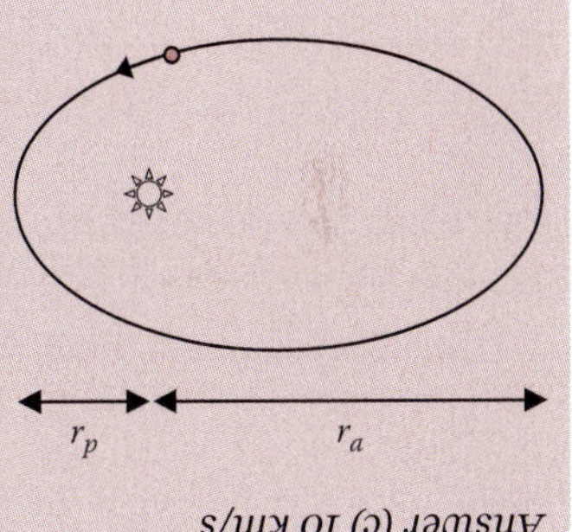

(a) Where does the planet reach its maximum speed?
(b) Where is its speed a minimum?
(c) What is the minimum speed of the planet?

Answer (c) 10 km/s

12.6 APPLICATION: FINDING THE SEMIMINOR AXIS OF AN ELLIPTICAL ORBIT

Kepler's greatest achievement was proving that planets moved in elliptical—as opposed to circular—orbits. Let's briefly review the features of elliptical orbits (see Figure 12.5a). An ellipse has a semimajor axis a, a semiminor axis b, and two focal points F_1 and F_2 that are displaced from the center of the ellipse by a distance εa, where the eccentricity ε is defined by $b = a\sqrt{1-\varepsilon^2}$. The Sun occupies one of these foci. Let P be a point on the ellipse, a distance r from F_1 and r' from F_2. For all points P, $r + r' = 2a$.

Now imagine a planet in orbit about the Sun. For simplicity, assume $M_s \gg m$. From Equation 10.13, the planet's energy is

$$E = \frac{1}{2}mv^2 - \frac{GM_s m}{r} = -\frac{GM_s m}{2a}. \tag{12.4}$$

If we know the planet's velocity $\vec{v}$ at any location $\vec{r}$ along its trajectory, we can calculate E and find the semimajor axis a of the orbit. Knowing $\vec{v}$ and $\vec{r}$, we can also calculate the angular momentum $\vec{L}$, and this will

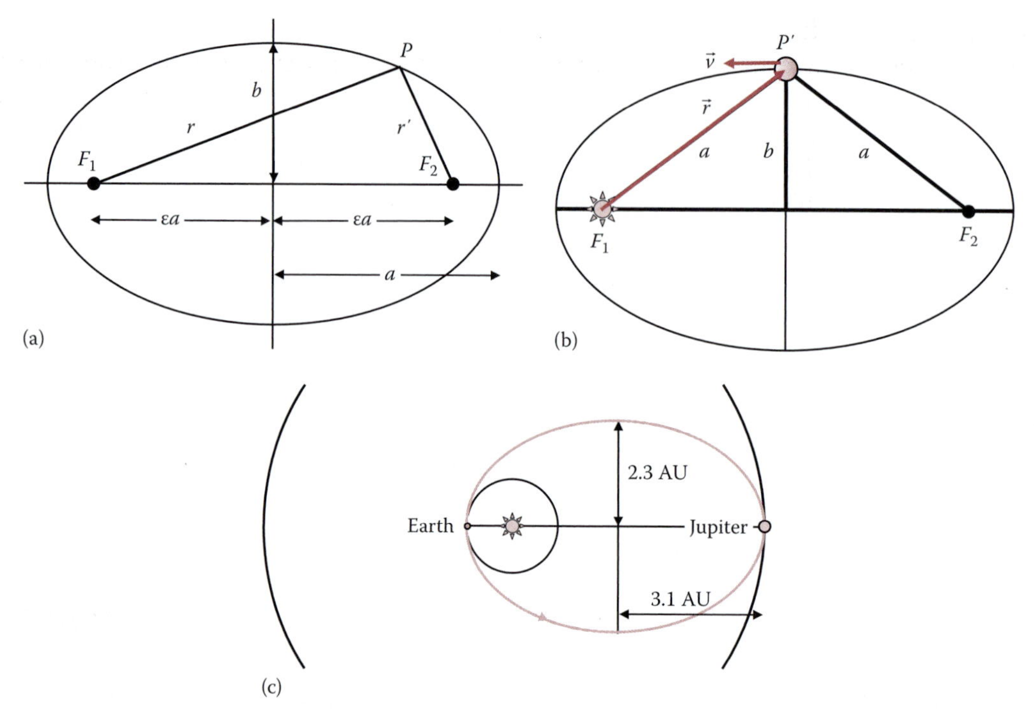

FIGURE 12.5 (a) A review of the properties of an ellipse. (b) When the orbiting body lies on the semiminor axis of the ellipse, it is a distance a from either focus, and is traveling parallel to the major axis of the ellipse. $L = m|\vec{r}\times\vec{v}| = mr_\perp v = mvb$. (c) The Hohmann transfer orbit between Earth and Jupiter.

allow us to determine the semiminor axis b. Suppose that the planet is located on the semiminor axis, point P' in Figure 12.5b. At this location, $r = r' = a$, and Equation 12.4 yields (remember that $E < 0$)

$$\frac{1}{2}mv^2 = \frac{GM_s m}{2a} = -E = |E|$$

or

$$v = \sqrt{\frac{-2E}{m}}.$$

At P', the planet is moving parallel to the major axis of the ellipse, so the magnitude of its angular momentum is

$$L = |\vec{L}| = m|\vec{r}\times\vec{v}| = mr_\perp v = mvb$$

or

$$b = \frac{L}{mv} = \frac{L}{\sqrt{-2mE}} = \frac{L/m}{\sqrt{GM_S/a}}. \tag{12.5}$$

While the semimajor axis depends exclusively on the orbital energy E, the semiminor axis is a function of both energy and angular momentum. For example, Figure 12.5c shows the Hohmann transfer orbit of a spacecraft traveling between Earth and Jupiter. Recall from Chapter 10 that this trajectory is tangential to Earth's orbit at perihelion, and tangential to Jupiter's orbit at aphelion. The radii of the two planetary orbits are 1.0 and 5.2 AU, respectively, so the semimajor axis of the Hohmann orbit is $a = \frac{1}{2}(1.0 + 5.2) = 3.1$ AU. Using Equation 12.4 to calculate the speed of the spacecraft at perihelion,

$$E = \frac{1}{2} m v_p^2 - \frac{GM_s m}{r_p} = -\frac{GM_s m}{2a}$$

or

$$v_p^2 = \frac{2GM_s}{r_p}\left(1 - \frac{r_p}{2a}\right).$$

EXERCISE 12.8

For $r_p = 1$ AU, $GM_s/r_p = 8.89 \times 10^8$ J/kg. Use this result to find v_p.

Answer 38.6 km/s

At perihelion, $\vec{v} \perp \vec{r}$, so $L = mv_p r_p$. But L is a constant throughout the orbit, so we can use its value calculated at r_p to determine b in Equation 12.5:

$$b = \frac{L/m}{\sqrt{GM_s/a}} = \frac{v_p r_p}{\sqrt{GM_s/a}}. \tag{12.6}$$

Plugging in the numbers, using the result of Exercise 12.8, we find $b = 2.28$ AU. The orbit eccentricity is $\varepsilon = \sqrt{1 - b^2/a^2} = \sqrt{1 - 2.28^2/3.10^2} = 0.46$.

EXERCISE 12.9

What is the speed of the spacecraft when it crosses the semiminor axis, for example, when it is at point P' in Figure 12.5b? Use angular momentum rather than energy to find the answer.

Answer 16.9 km/s

EXERCISE 12.10

Let's check these results for a circular orbit. Start from Newton's second law for uniform circular motion, $mv^2/r = GM_s m/r^2$. Show that Equation 12.6 correctly predicts that $b = a = r$.

12.7 FINDING THE ORBIT'S ORIENTATION

We now know how to calculate the orbit parameters a and b from the velocity $\vec{v}$ of the orbiting body at a single point $\vec{r}$. Furthermore, the vectors $\vec{r}$ and $\vec{v}$ define the plane of the orbit. However, we still don't know how the ellipse axes are oriented within the plane of motion. Figure 12.6a shows the position and velocity vectors of a planet at point P in its orbit about the Sun. The star occupies focal point F_1 of the ellipse, but until we locate the other (empty) focus F_2, we don't know the direction of the orbit's major axis (parallel to $\overline{F_1F_2}$) nor its minor axis (perpendicular to $\overline{F_1F_2}$). To determine the orientation of the orbit, we will use another feature of the ellipse called its *reflection property*.

The velocity vector is always tangent to the trajectory, so we know the angle θ_1 between $\overline{F_1P}(=r)$ and the trajectory at point P. The line $\overline{F_2P}(=r')$ extends from the second (empty) focus to the planet and makes an angle θ_2 with the trajectory at point P. *The reflection property states that* $\theta_2 = \theta_1$ (see Box 12.1 for a proof). Since we know r and a, we can easily calculate $r' = 2a - r$. This fixes the length of $\overline{F_2P}$, and the reflection property sets the direction of $\overline{F_2P}$, so we can locate F_2 unambiguously (Figure 12.6b). The major axis of the ellipse lies along $\overline{F_1F_2}$, and the minor axis coincides with the perpendicular bisector of $\overline{F_1F_2}$ (Figure 12.6c). *Now* we are able to draw the complete orbit.

Here is a numerical example. A new asteroid is discovered. It is presently located 2.0 AU from the Sun and is moving with speed $v = 21$ km/s (relative to the Sun). Its present velocity $\vec{v}$ makes an angle $\theta_1 = 30°$ with its position vector $\vec{r}$ (so the asteroid is moving *away* from the Sun.) See Figure 12.7a. To determine its orbit, start by calculating the total energy:

$$E = \frac{1}{2}mv^2 - \frac{GM_S m}{r} = \frac{1}{2}m(21\times10^3\ \text{m/s})^2 - 8.893\times10^8\frac{m}{r\,(\text{AU})}$$
$$= -2.24\times10^8\, m\,(\text{J}),$$

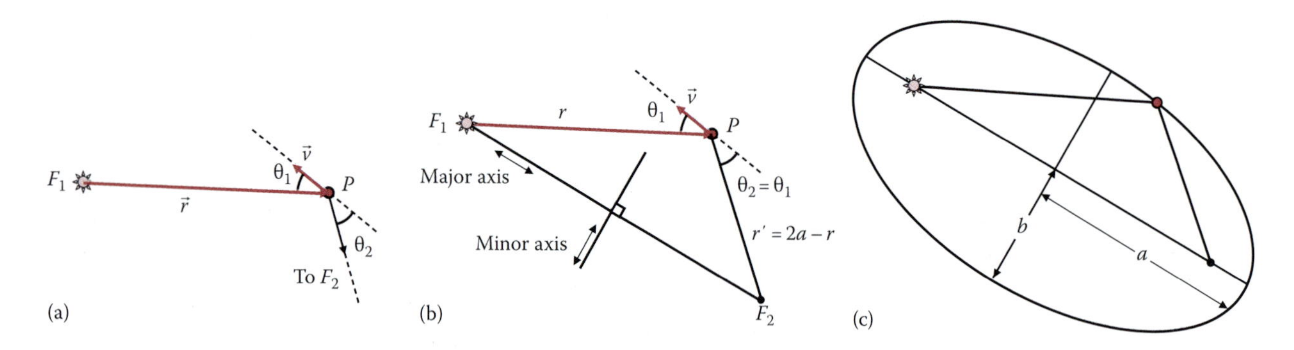

FIGURE 12.6 (a) A planet P is in orbit about the Sun at F_1. Its velocity vector $\vec{v}$ is tangent to the trajectory at point P and makes an angle θ_1 with the line $\overline{F_1P}$. (b) According to the reflection property of ellipses, angle $\theta_2 = \theta_1$, indicating the direction from P to the second focal point F_2. The minor axis is along the perpendicular bisector of the line $\overline{F_1F_2}$. (c) Once F_2 is located, the entire ellipse can be drawn.

BOX 12.1 THE REFLECTION PROPERTY OF AN ELLIPSE

Let P be a point on an ellipse, and F_1 and F_2 be the ellipse's two focal points. Draw lines $\overline{F_1P}$ and $\overline{F_2P}$ from the foci to P. For an ellipse, $\overline{F_1P} + \overline{F_2P} = 2a$, where a is the semimajor axis. Now draw line L tangent to the ellipse at P (see Figure B12.1c). Except for P, all points P' on L are external to the ellipse, so $\overline{F_1P'} + \overline{F_2P'} \equiv \ell_1 + \ell_2 \geq 2a$. Therefore, P is the location along line L that minimizes $\ell_1 + \ell_2$. We will use this fact to prove the reflection property for ellipses.

Next construct a new figure. Points F_1 and F_2 are joined by the line $\overline{F_1F_2}$ having length d and lying in the x–y plane. Let point P lie along a second line L parallel to $\overline{F_1F_2}$. Let ℓ_1 be the length of line $\overline{F_1P}$, and ℓ_2 be the length of $\overline{F_2P}$. We wish to find the location of P that minimizes the sum $\ell_1 + \ell_2$. To do this, draw two lines perpendicular to L, one passing through F_1 and the other through F_2. Referring to Figure B12.1a, $\ell_1 = \sqrt{x_1^2 + y_1^2}$ and $\ell_2 = \sqrt{x_2^2 + y_2^2}$, where $x_2 = d - x_1$ and $y_1 = y_2$ (since $\overline{F_1F_2}$ and L are parallel). To find the location of P, take the derivative of $\ell_1 + \ell_2$ with respect to x_1, and set it equal to zero:

$$\frac{d}{dx_1}(\ell_1 + \ell_2) = \frac{x_1}{\sqrt{x_1^2 + y_1^2}} - \frac{(d - x_1)}{\sqrt{(d - x_1)^2 + y_2^2}}$$

$$= \frac{x_1}{\sqrt{x_1^2 + y_1^2}} - \frac{x_2}{\sqrt{x_2^2 + y_2^2}} = 0. \qquad \text{(B12.1)}$$

The first term is just $\cos\theta_1$ and the second is $\cos\theta_2$, where θ_1 and θ_2 are the angles $\overline{F_1P}$ and $\overline{F_2P}$ make with L. Equation B12.1 simply says that the point P where $\ell_1 + \ell_2$ is a minimum is where $\theta_1 = \theta_2$.

This result is intuitively obvious. But what if L is not parallel to $\overline{F_1F_2}$? In this case, draw lines perpendicular to L passing through F_1 and F_2, as shown in Figure B12.1b. Once again, $x_2 = d - x_1$. In fact, the only difference from the above derivation is that now $y_1 \neq y_2$. But since y_1 and y_2 are constants, independent of the location of P, the differentiation to minimize $\ell_1 + \ell_2$ proceeds exactly as in Equation B12.1, and again we find that the point P where $\ell_1 + \ell_2$ is minimized is where $\theta_1 = \theta_2$.

Finally, compare Figure B12.1b and c. The result $\theta_1 = \theta_2$ does not depend on the slope of L, so we could choose L to be the line tangent to any point of the ellipse. Since P is the location on L that minimizes $\ell_1 + \ell_2$, lines $\overline{F_1P}$ and $\overline{F_2P}$ make equal angles ($\theta_1 = \theta_2$) with the tangent line L at any point of the ellipse. This is the reflection property of ellipses.

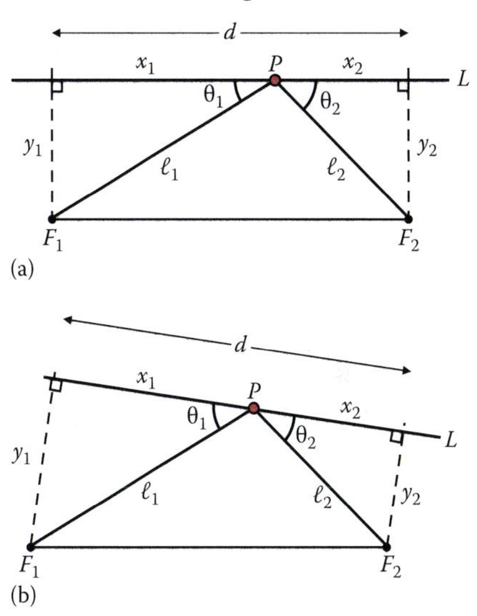

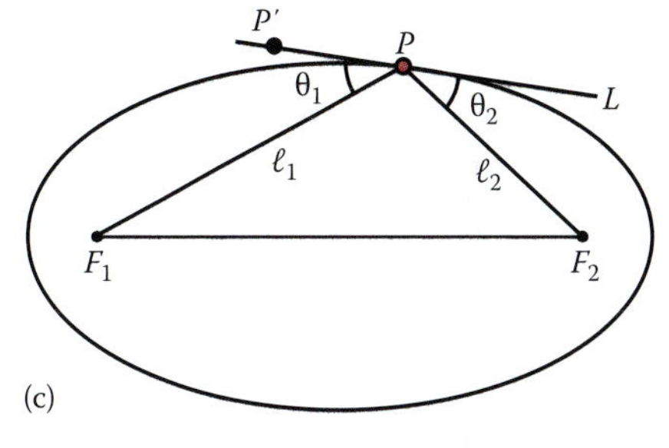

FIGURE B12.1 (a) When L is parallel to $\overline{F_1F_2}$, the point P that minimizes the sum $\ell_1 + \ell_2$ is easily found by differentiation to be where $\theta_1 = \theta_2$. (b) When L is not parallel to $\overline{F_1F_2}$, the same mathematics once again locates P where $\theta_1 = \theta_2$. (c) An ellipse is constructed with focal points F_1 and F_2, tangent to line L. All points on L lie outside the ellipse except for the tangent point P. Therefore, P is the point on L that minimizes $\ell_1 + \ell_2$, and so $\theta_1 = \theta_2$.

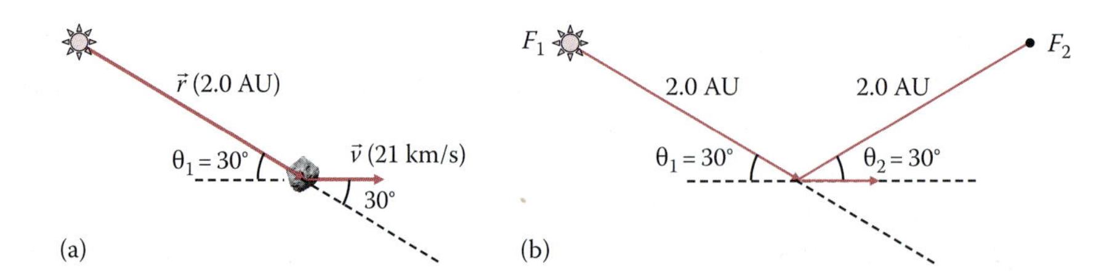

FIGURE 12.7 (a) An asteroid is observed at position $\vec{r}$, moving at velocity $\vec{v}$ in an elliptical orbit about the Sun (at F_1). (b) Using the reflection property, we can locate the second focal point F_2 and trace out the entire orbit.

where we have used the numerical information in Exercise 12.7. Since $E < 0$, we know that the asteroid is in a closed orbit. Next, use Equation 12.4 to find the semimajor axis:

$$\frac{E}{m} = -2.24 \times 10^8 \text{ J/kg} = -\frac{GM_S}{2a} = -\frac{8.893 \times 10^8}{2a\,(\text{AU})},$$

where a is measured in AU. Solving, $a = 2.0$ AU.

The angular momentum magnitude $L = mrv\sin 30° = 2(1.5 \times 10^{11})(21 \times 10^3)\sin 30° m$, or $L = 3.15 \times 10^{15} m \text{ kg m}^2/\text{s}$. Using Equation 12.5,

$$b = \frac{L}{\sqrt{-2mE}} = \frac{3.15 \times 10^{15} m}{\sqrt{4.48 \times 10^8 m^2}} = 1.5 \times 10^{11} \text{ m} = 1.0 \text{ AU}$$

Finally, use the reflection property to determine the orbit's orientation (Figure 12.7b). The angles between $\vec{v}$ and the lines joining the asteroid to the focal points F_1 and F_2 are equal: $\theta_1 = \theta_2 = 30°$. Since $r = 2.0$ AU, the distance from the asteroid to F_2 is $r' = 2a - r = 2.0$ AU. Now locate F_2, draw the semimajor and semiminor axes, and then sketch in the ellipse. We'll leave this for you to do as an exercise.

EXERCISE 12.11

Copy Figure 12.7b. (a) Using the calculated values of a and b, sketch the asteroid's elliptical orbit to scale on your figure. (b) Find the distance to the Sun at perihelion. (c) The asteroid in Figure 12.7 is at a special place in the orbit. Where is it?

Answer (b) 0.27 AU

12.8 APPLICATION: THE SKY IS FALLING!

Each day, more than 100 tons of interplanetary matter rain onto Earth. Most of it arrives as small particles that burn up harmlessly in the atmosphere, but about once every hundred years, a much larger object slams into Earth with devastating consequences. An impact with an asteroid >50 m in diameter releases more energy than a dozen nuclear warheads. Because of this threat, NASA's Near Earth Object (NEO) Program searches for comets and asteroids with trajectories that are converging on the planet. The program's official mission is to provide advance warning of an imminent collision, so that the probable impact area can be evacuated. Aside from their danger, NEOs are relics of the early solar system. If astronomers had advance notice of an approaching asteroid, they could observe the object closely and possibly obtain valuable clues about how the solar system formed 4.6 billion years ago. For either purpose, it is essential to determine the object's trajectory well in advance of its perigee (point of closest approach to Earth).

If the position and velocity of the asteroid are known when it is far from Earth, we can predict whether or not it will strike the planet. For simplicity, assume that the asteroid is close enough so that its gravitational attraction to Earth is much greater than its attraction to the Sun. Then we can ignore the Sun and treat the planet plus asteroid as an isolated system. This allows us to take advantage of conservation laws to predict the asteroid's trajectory. In particular, we want to know its perigee: Will it strike the Earth? In the planet's reference frame, let $\vec{r}_o$ and $\vec{v}_o$ be the asteroid's position and velocity measured at a point far from Earth ($r_o \gg R_E$); likewise, let $\vec{r}_p$ and $\vec{v}_p$ be its position and velocity at perigee. Conservation of energy and angular momentum require that

$$\frac{1}{2}mv_o^2 - \frac{GM_E m}{r_o} = \frac{1}{2}mv_p^2 - \frac{GM_E m}{r_p}$$

and

$$mv_o s = mv_p r_p,$$

where, M_E and m are the masses of the Earth and asteroid and $s \equiv r_o \sin\theta$ is the asteroid's *impact parameter* (defined in Figure 12.8a).

From the second equation, $v_p = v_o(s/r_p)$. Substituting this into the energy expression,

$$v_p^2 - v_o^2 = v_o^2\left(\frac{s^2}{r_p^2} - 1\right) = 2GM_E\left(\frac{1}{r_p} - \frac{1}{r_o}\right).$$

For $r_o \gg r_p$, we can safely ignore the gravitational potential energy at r_o, to obtain

$$v_o^2\left(\frac{s^2}{r_p^2} - 1\right) \simeq \frac{2GM_E}{r_p} \quad \text{or} \quad \frac{s^2}{r_p^2} = 1 + \frac{2GM_E}{r_p v_o^2}. \tag{12.7}$$

Equation 12.7 can be solved to find the perigee of the asteroid as a function of its impact parameter s and its far-away speed v_o. Multiplying through by r_p^2, we obtain $s^2 = r_p^2 + 2GM_E r_p / v_o^2$. Clearly, s increases monotonically with r_p. The converse must also be true: r_p must increase monotonically with s. If the asteroid is to avoid colliding with the planet, $r_p > R_E$, or

$$\frac{s^2}{R_E^2} > 1 + \frac{2G_E M}{R_E v_o^2}.$$

obtaining the first-ever images of a comet's surface, and confirming that it was indeed a "dirty snowball," interstellar dust grains glued together by frozen H_2O, CO, and CO_2. Comet research is only in its infancy. It will be interesting to see if the above hypothesis—that comets brought life to Earth—holds up in the coming decades. In August 2014, the ESA spacecraft *Rosetta** intercepted comet Churyumov–Gerasimenko and on November 12, dropped a small probe onto its surface. The probe, called *Philae*, is now riding piggyback on the comet to record the processes that occur as its host body circumnavigates the Sun. Stay tuned!

* Just as the Rosetta Stone unlocked the secrets of ancient civilizations on Earth, the aptly named *Rosetta* spacecraft is expected to unlock the mysteries of the early solar system, before planets were formed.

12.9 APPLICATION: THE "SLINGSHOT EFFECT" REVISITED

In Chapter 5, we studied the "slingshot effect," by which a spacecraft's speed relative to the Sun is boosted when the spacecraft passes close to a planet. That discussion was incomplete because we *specified* an angle of deflection without asking if it were even possible. The larger the deflection, the closer the spacecraft must approach the planet. Can it be deflected by the desired amount without crashing into the planet? Now, with angular momentum as a tool, we can complete the discussion. As in the previous section, we will ignore the Sun's influence and treat the planet plus spacecraft as an isolated system. Our task is to predict the spacecraft's deflection using its position and velocity when it is far from the planet.

In the planet's reference frame, the planet M is stationary while the spacecraft m is approaching it with far-away velocity $\vec{v}_0$. Since the spacecraft is not gravitationally bound to the planet, the total mechanical energy is *positive*:

$$E = \frac{1}{2}mv_0^2 - \frac{GMm}{r_0} > 0,$$

where $r_0 \gg R$ and R is the radius of the planet. From Section 12.5, we know that if $E < 0$, m would execute a closed elliptical orbit with semimajor and semiminor axes a and b given by Equations 12.4 and 12.5,

$$E = -\frac{GMm}{2a} \quad \text{and} \quad b = \frac{L/m}{\sqrt{GM/a}}. \tag{12.9}$$

If we retain these expressions for $E > 0$, a will become a negative number and b will be *imaginary*.† Let $a' = -a$ and $b' = ib$, where a' and b' are positive real numbers. Then, using $a'^2 = a^2$ and $b'^2 = -b^2$, the defining equation for an ellipse (see Box 8.1) is transformed (for $E > 0$) into

$$\frac{x^2}{a'^2} - \frac{y^2}{b'^2} = 1. \tag{12.10}$$

Equation 12.10 describes a *hyperbola* such as the solid red line shown in Figure 12.10. When $y = 0$, $x = a'$. For large values of x and y, the "1" on the right side of the equation is negligible, and the orbit approaches

† The imaginary number i is defined as $i = \sqrt{-1}$. Therefore, $(ib)^2 = -b^2$. If $a < 0$, then $b \propto \sqrt{a}$ is imaginary.

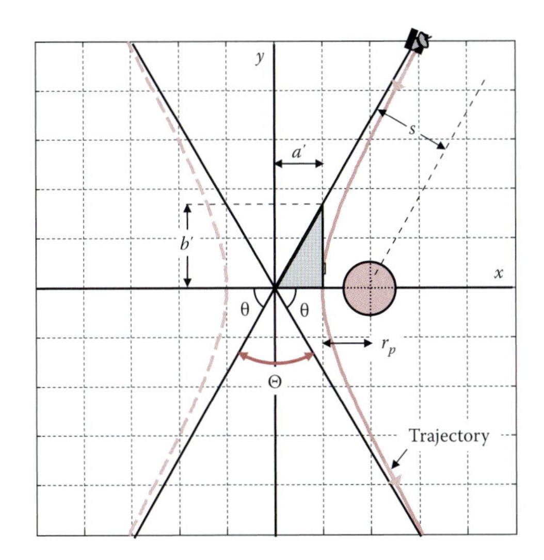

FIGURE 12.10 (solid red line) The hyperbolic orbit of a spacecraft as it executes a close flyby of a planet. The trajectory is plotted in the reference frame of the planet. Note $\tan\theta = b'/a'$, and $\Theta = 180° - 2\theta$. (The dashed line to the left of the origin is a second solution to Equation 12.8, and is unimportant.)

the two asymptotes $y = \pm(b'/a')x$. Each asymptote makes an angle θ with respect to the x-axis, with $\tan\theta = b'/a'$. For large r_0, assume that the spacecraft's kinetic energy is much larger than the potential energy, so $E \simeq \frac{1}{2}mv_0^2$, and Equations 12.9 become

$$E \simeq \frac{1}{2}mv_0^2 = \frac{GMm}{2a'}, \quad \text{or} \quad a' = \frac{GM}{v_0^2}$$

and

$$b' = \frac{v_0 s}{\sqrt{GM/a'}} = \frac{v_0 s}{\sqrt{v_0^2}} = s.$$

The asymptote angle θ therefore satisfies the equation

$$\tan\theta = \frac{b'}{a'} = \frac{v_0^2 s}{GM}.$$

Figure 12.10 shows the relationship between θ and the deflection angle Θ: $\Theta = \pi - 2\theta$. From basic trigonometry, we know $\cot(\pi/2 - \varphi) = \tan\varphi$, so $\cot(\Theta/2) = \cot(\pi/2 - \theta) = \tan\theta$, and

$$\tan\frac{\Theta}{2} = \frac{1}{\cot(\Theta/2)} = \frac{1}{\tan\theta} = \frac{GM}{v_0^2 s} = \frac{v_{esc}^2}{v_0^2}\frac{R}{2s}, \tag{12.11}$$

where we have again used Equation 10.7 for the escape velocity: $v_{esc} = \sqrt{2GM/R}$. Knowing the spacecraft's initial speed v_0, its impact parameter s, and the mass and radius of the planet M, we can now find the deflection angle Θ from Equation 12.11. Try the following exercise.

EXERCISE 12.13

A spacecraft is approaching Mars with speed 2.5 km/s relative to the planet. As it passes the planet, it changes direction by 90° in the planet's reference frame. The escape velocity from Mars is 5.0 km/s.

(a) Identify the deflection angle Θ.
(b) What is the value of the asymptote angle θ?
(c) Find the impact parameter s as a multiple of the planet radius R.

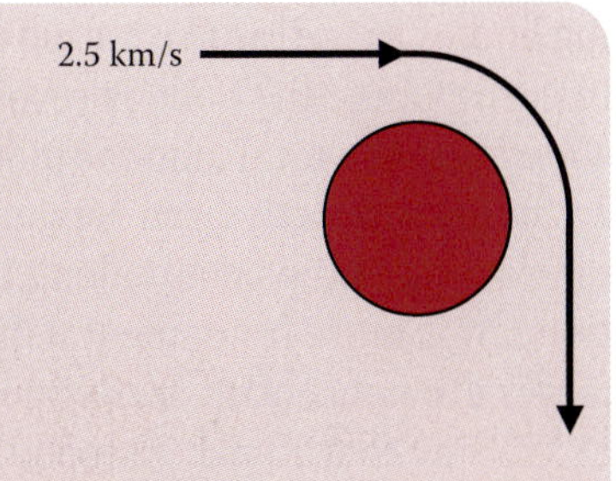

Answer (a) 90° (b) 45° (c) 2R

Finally, let's put some limits on the deflection angle. For a fixed value of v_0, Equation 12.11 shows that the deflection angle Θ increases as the impact parameter s decreases (as one would expect). The maximum possible deflection occurs for the minimum impact parameter s_{min}, that is, where the spacecraft grazes the surface of the planet at its closest approach. The value of s_{min} is found from Equation 12.8. Combining that result with Equation 12.11, we obtain

$$\tan\frac{\Theta_{max}}{2} = \frac{1}{2}\frac{(v_{esc}/v_0)^2}{\sqrt{1+(v_{esc}/v_0)^2}}. \tag{12.12}$$

To illustrate, return to Example 12.11, where $v_{esc}/v_0 = 2$. The maximum deflection angle is found from

$$\tan\frac{\Theta_{max}}{2} = \frac{1}{2}\frac{2^2}{\sqrt{1+2^2}} = 0.894$$

or $\Theta_{max} = 84°$. So the deflection specified in the exercise is just a little unrealistic.

EXERCISE 12.14

Find Θ_{max} in the limits (a) $v_0 \to \infty$ and (b) $v_0 \to 0$. Do these make sense to you? (c) Return to the one-dimensional flyby of Jupiter presented in Exercise 5.8, where for simplicity we chose $\Theta = 180°$. In reality, what would Θ_{max} be for the velocities given in the example? (The escape velocity from Jupiter is 60.6 km/s.)

Answer: (a) 0 (b) 180° (c) 94°

12.10 SUMMARY

We began this chapter with a broader definition of angular momentum, $\vec{L} = \vec{r} \times \vec{p}$, which applies to nonrotating bodies as well as to rotating ones. The new definition is consistent with the earlier one for spin, $\vec{L}_{spin} = I\vec{\omega}$, in cases where the body is revolving about an axis of symmetry. The third conservation law of mechanics states that the total angular momentum $\vec{L}_{tot} = \sum_i \vec{r}_i \times \vec{p}_i$ of an isolated system is conserved. Conservation of $\vec{L}_{tot}$ follows directly from Newton's laws in the special case of central forces, where the forces between two interacting bodies point along the line between them. But just like momentum and energy, the conservation law is more fundamental than Newton's laws, and is valid in all cases, on all size scales, even when the forces are noncentral or may not be well-known (as in interactions between subatomic particles).

We saw that Kepler's second (equal area) law is just angular momentum conservation in disguise, and then employed $\vec{L}$ to find the semi*minor* axis of an elliptical orbit. Using $\vec{L} = \text{const}$, the entire shape of a body's orbit can be found from its position and velocity at a single point. We then studied the trajectories of

BOX 12.2 NOETHER'S THEOREM: CONSERVATION AND SYMMETRY

In the words of Albert Einstein, "Emmy Noether (1882–1935) was the most significant creative mathematical genius thus far produced since the higher education of women began." It was her misfortune to pursue an education and an academic career at a time when German universities did not admit female students, and certainly did not hire female faculty. Assisted by the eminent mathematicians David Hilbert and Felix Klein, she nonetheless secured a position at the University of Göttingen, where for two decades she published prolifically in the field of abstract algebra. In 1915, shortly after the appearance of Einstein's theory of general relativity, she proved a theorem that inextricably linked the conservation laws of physics to fundamental symmetries found in nature. Noether's theorem is as important as it is beautiful, and is one of the bedrock tenets of theoretical physics. In the following paragraphs, we offer a simplified version–a faint shadow–of Noether's work. We hope it is sufficient to convey the gist of her insight, and will help you appreciate its profound significance.

Consider an infinite flat sheet lying in the x–y plane. The sheet is everywhere the same, and it induces a force on a particle moving in the space above it (see Figure B12.2b). The nature of this force is not important: it may be gravitational, electrical, or simply a contact force that is exerted only when the particle touches the sheet. Clearly, if we slide the sheet in the x- or y-direction, or move the particle in the x- or y-direction, the force on the particle does not change. The motion of the particle is invariant with respect to *x–y translation* of the sheet and/or the particle, relative to one another. Furthermore, if we *rotate* the sheet about a vertical axis perpendicular to the sheet, we come to the same conclusion: the force on the particle does not change, and again its motion is unaltered.

Now assume the force is conservative, so a potential energy function $U(x,y,z)$ can be defined. Because of translational invariance, U cannot be a function of x or y: $U = U(z)$. If the particle moves parallel to the sheet (z = const), U remains constant, so the particle's kinetic energy must also remain constant. Therefore, there is no acceleration in the x- or y-direction, and $F_x = F_y = 0$. (This result can be expressed concisely

(a)

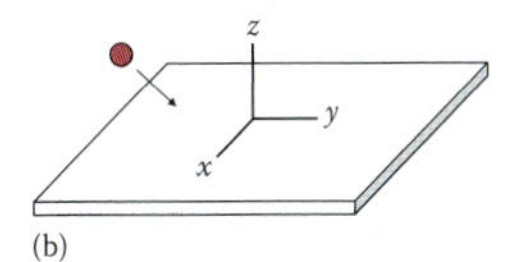

(b)

FIGURE B12.2 (a) Photograph of Emmy Noether, probably taken at Bryn Mawr. (Courtesy of Archives of the Mathematisches Forschungsinstitut Oberwolfach, © Universitäts-Archiv Göttingen, Göttingen, Germany.) (b) A point particle interacts with an infinite uniform sheet lying in the x–y plane. Because the sheet's properties are everywhere the same, the potential energy cannot depend on either x or y.

using partial derivatives: $F_x = -\partial U/\partial x = 0$, and likewise for F_y.) Since $\vec{F} = d\vec{p}/dt$, we conclude that p_x and p_y are constant, that is, they are conserved. Regardless of the nature of the force between the particle and the sheet, x–y translational invariance *guarantees* momentum conservation in these two directions.

Rotational invariance adds another conservation condition. Since $F_x = F_y = 0$, then $\vec{F} = F_z\hat{k}$ and the torque on the particle $\vec{\tau} = \vec{r} \times \vec{F}$ is perpendicular to the z-direction. Therefore, $\tau_z = 0$ and $L_z = \text{const}$. Regardless of nature of the force, rotational invariance about the z-axis guarantees conservation of L_z.

Exercise B12.1: An infinitely long cylindrical rod lies along the x-axis. The properties of the rod are the same at all points along its length, and it exerts a force on a nearby particle m. Use symmetry arguments to determine which components of $\vec{P}$ and $\vec{L}$ are conserved.

Answer: p_x, L_x

What if the sheet's properties are changing with time? For example, suppose the particle and the sheet are interacting gravitationally, and the magnitude of the gravitational constant G is slowly increasing.* As long as the properties change uniformly over the sheet, the conservation arguments discussed above are not affected: $F_x = F_y = \tau_z = 0$, but now F_z is a function of time t as well as z. Since force is a derivative of potential energy, U must also be a function of t and z. The total mechanical energy may be written as $E = \frac{1}{2}mv^2 + U(z,t)$, and its rate of change is

$$\frac{dE}{dt} = \frac{m}{2}\frac{d}{dt}\left(v_x^2 + v_y^2 + v_z^2\right) + \frac{\partial U}{\partial z}\frac{dz}{dt} + \frac{\partial U}{\partial t}.$$

Since $a_x = a_y = 0$, v_x and v_y are constant, and

$$\begin{aligned}\frac{dE}{dt} &= \frac{m}{2}\frac{d}{dv_z}\left(v_z^2\right)\frac{dv_z}{dt} + \frac{\partial U}{\partial z}\frac{dz}{dt} + \frac{\partial U}{\partial t} \\ &= mv_z a_z - F_z v_z + \frac{\partial U}{\partial t},\end{aligned}$$

where we have used the chain rule for differentiation. The derivatives of potential energy require some comment. $\partial U/\partial z$ represents the rate of change of the potential energy due solely to the change in the particle's distance z from the sheet. For a conservative force, we know from Chapter 9 (see Equation 9.13) that the force is $F_z = -\partial U/\partial z$. But even if z remains constant, the potential energy can change if the properties of the system are changing (e.g., $dG/dt \neq 0$). The term $\partial U/\partial t$ reflects this change. Since $ma_z = F_z$, the first two terms in the above equation cancel, so $dE/dt = \partial U/\partial t$. Energy is conserved only if the system properties are invariant with time ($\partial U/\partial t = 0$). Time invariance guarantees conservation of energy.

How is this related to *symmetry*? Symmetry means that an object or system appears the same after something has been done to it. Consider the infinitely long cylinder described in Exercise B12.1. If it were rotated about its central axis, it would look exactly the same to you, so we say that the rod has rotational

* This is not as bizarre as it sounds. A variation of G since the big bang would have a profound effect on our cosmological models. Experimental evidence indicates that G changes no faster than 1 part in 10^{12} per year.

(or cylindrical) symmetry about that axis. Likewise, you could not tell if the rod were slid parallel to its long axis, so it also has translational symmetry along its central axis. Now imagine that you and your lab partner perform an experiment using the particle and sheet described above. You launch the particle from position (x, y, z) and carefully record its motion. Then, while you briefly leave the laboratory, your classmate slides and/or rotates the sheet in the horizontal x–y plane. When you return, you repeat the experiment. Is there any way for you to determine if or how the sheet has been moved? (You can't ask your partner, and this is an infinite sheet, so technically there are no edges.) No! The particle will behave exactly as it did before, no matter how the sheet has been moved. The system therefore has translational symmetry in the x–y plane and rotational symmetry about a vertical (z) axis. You leave the room once again, and while you are away, your lab partner records the motion of the particle in two identical trials (without moving the sheet). When you return, you are given the experimental results. Is there any way for you to determine which trial was performed first? If the properties of the apparatus did not change with time ($\partial U/\partial t = 0$), the answer is again no. This is what is meant by the symmetry of time.

Noether's theorem goes well beyond the above discussion to consider systems of many interacting particles. But the conclusions are the same: Whenever a physical property of a system is conserved, an underlying symmetry of nature is at work, and vice versa. Conservation of momentum implies that three-dimensional space is homogeneous (having translational symmetry). If an experiment is carried out on an isolated system, it does not matter where in the universe it is conducted; the results will be the same. Similarly, conservation of angular momentum implies that space is isotropic (having rotational symmetry). Rotation of the experimental apparatus will not change the outcome. Conservation of energy implies that the laws of physics are not changing with time. Repeating the experiment at a different time will make no difference to the results. In order for our experiments to agree with these conclusions, *the laws of physics must be invariant with respect to translations in space and time, and rotations in space.*

Exercise B12.2: An isolated system contains two bodies that interact while conserving their total momentum. Which of the following force laws are possible?

(a) $\vec{F}_{12} = -\vec{F}_{21} = -k(\vec{r}_1 - \vec{r}_2)$, (b) $\vec{F}_{12} = -\vec{F}_{21} = k(\vec{r}_1 + \vec{r}_2)$, and (c) $\vec{F}_{12} = -\vec{F}_{21} = \hat{i}\left(x_1^2 - x_2^2\right)$

Answer (a)

The quest for symmetry is nowhere more apparent than in the world of particle physics. The "standard model" of the fundamental particles and forces (quarks, leptons, and their force carriers) is based on symmetries associated with their physical properties (e.g., mass, spin, charge) rather than on the symmetries of space and time. Symmetry was the driving force behind the 50 year, multibillion dollar search for the Higgs boson, and is the inspiration for new schemes, such as *supersymmetry,* which seek to probe the fundamental components of the universe at ever deeper levels. Although the symmetries of particle physics are more esoteric than the ones we have discussed above, Noether's theorem is still the guiding principle: conservation is linked to symmetry.

Emmy Noether's story has a tragic ending. Because she was a Jewish pacifist, she was expelled from Göttingen University in 1933 as the Nazis rose to power. She emigrated to the United States, and with the help of Albert Einstein and the distinguished mathematician Hermann Weyl, obtained a position at Bryn Mawr College near Philadelphia. She lectured frequently at the Institute for Advanced Study in Princeton. In 1935, she underwent surgery for cancer and died unexpectedly 4 days later. She was 53 years old. At her funeral, Hermann Weyl called her "a great woman mathematician, ... the greatest that history has known," and characterized her as "maternal, with a childlike warmheartedness," and a "heart that knew no malice." While hers is hardly a household name, her intellectual passion and self-effacing demeanor should be a model for all scientists—both men and women—everywhere.*

* See N. Angier, The Mighty Mathematician You've Never Heard Of, *The New York Times,* March 26, 2012.

asteroids and comets, and determined the conditions for impact with Earth. Finally, we reexamined gravity assisted trajectories, wherein a spacecraft's speed is boosted during a close approach to a planet. In the vicinity of the planet, the spacecraft's path is approximately hyperbolic and the angle of deflection (between the initial and final velocity vectors) is determined by the spacecraft's speed and impact parameter.

The year 1758 was an exciting and historic one for physics and astronomy. Seventy-five years earlier, the British astronomer Edmond Halley, perhaps the first person to recognize the importance and scope of Isaac Newton's work, had asserted that the comet of 1682 had been seen at least twice before, in 1532 and 1607. Using Newton's just-published theory of universal gravitation (in the *Principia*, 1687), Halley had calculated an elliptical orbit for the comet and predicted its return "about the end of the year 1758 or the beginning of the next [year]." Night after night throughout the year, astronomers in Europe and elsewhere eagerly scanned the skies in search of Halley's comet, each hoping to be the first to herald its return.

Nowhere was the event anticipated more keenly than in France. In 1757, the French astronomer Joseph-Nicholas Delisle speculated that the comet would first be visible 25–35 days before it reached perihelion (its closest point to the Sun). He used Halley's calculated orbit to predict where in the sky the comet would first be visible, and assigned comet-sighting duties to his assistant, Charles Messier.* By the end of 1758, Messier had been searching for comet Halley for more than one year.

Meanwhile, the French mathematician Clairaut had been working to refine Halley's prediction for the comet's return date. To pinpoint the date of perihelion, Clairaut included the gravitational effects of Jupiter and Saturn on the comet's orbital period. Since time was short, he employed two young assistants, Lalande and Lepaute, to carry out the laborious calculations (done entirely by hand!). The team worked assiduously, often from morning through dinnertime, to complete their work before the comet reappeared. They finished in late autumn, 1758, and—with no time to spare—presented their results in a paper read to the Academy of Sciences in Paris on November 14, 1758. Just 6 weeks later, the comet was sighted—but not by Messier!

On Christmas night, 1758, comet Halley was spotted by Johann Palitzsch, an affluent farmer and amateur astronomer in Germany. Hampered by cloudy skies over Paris, Messier could only manage a few scattered observations during November and December, and did not find the comet until January 21. When he and Delisle announced their discovery in April, culminating nearly 2 years of work, they were aghast to learn that they had been scooped by an amateur!

For Charles Messier, the story has a happy ending. Over the course of his career, he discovered 13 comets, earning him the admiration of King Louis XV and induction into the Academy of Sciences in Paris as well as

* Messier was hired by Delisle as a clerk. His job qualifications were his excellent handwriting and drafting skills.

the Royal Society of London. While searching for comets—his lifelong passion—he occasionally encountered fuzzy patches in the sky which masqueraded as comets. To avoid confusion between comets (which moved) and these fuzzy patches—or *nebulae* (which did not move)—he began to record the latter, eventually documenting the positions of over 100 nebulae in his catalog of *Messier objects*. We now know that many of these nebulae are galaxies.* Unwittingly, Charles Messier had set the stage for the 1920 "great debate" between Shapley and Curtis over the size of the universe, which motivated the pioneering work of Edwin Hubble.

* For example, the *Andromeda* galaxy is referred to as *M31*; that is, it is the 31st object listed in the Messier catalog of nebulae.

PROBLEMS

12.1 What is the total angular momentum about the origin of each of the systems shown? Each body has a mass m and a speed v. Do not forget to indicate direction!

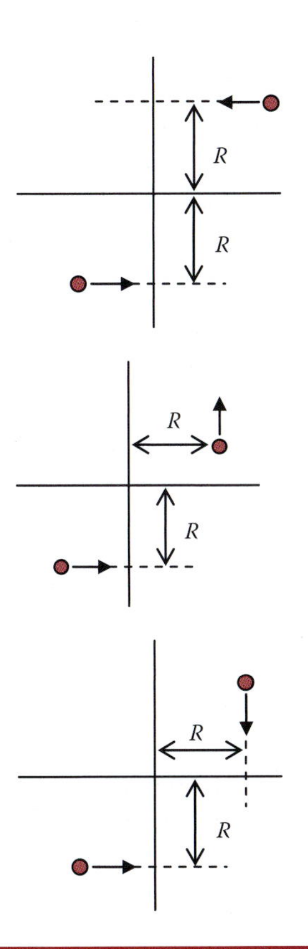

12.2 A circular track of mass $M = 2.0$ kg and radius $R = 0.5$ m is supported horizontally by lightweight spokes attached to a hub that can rotate without friction about a fixed vertical axle. (Neglect the mass of the spokes.) On the track, a battery-powered train of mass $m = 1.0$ kg starts from rest and reaches a constant speed of 2.0 m/s relative to the laboratory frame of reference.

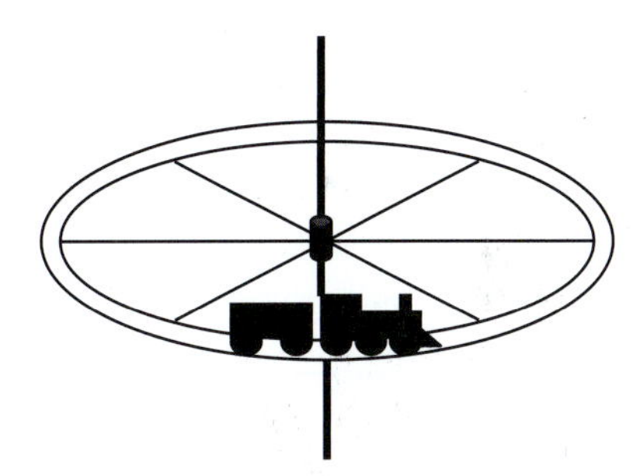

(a) What is the angular velocity of the wheel, assuming it is initially at rest? What is the direction of $\vec{\omega}$?

(b) The train's battery runs down and its motor stops. What is the final angular velocity of the wheel?

(c) Suppose the track is disconnected from the hub and laid flat on a frictionless horizontal surface. When the train is started from rest, discuss the motion of the system.

12.3 There are 1.1×10^8 automobiles in the United States, each with average mass of 2000 kg. One morning they all simultaneously start to move in an eastward direction, accelerating to a speed of 80 km/h (22.2 m/s). Assume they are at an average latitude of 40°, so that they are moving in a circle of radius $R_E \cos 40° = 4900$ km.

(a) What is the total angular momentum of these cars with respect to a N–S axis through the Earth's center?

(b) How much will the rate of rotation of the Earth change due to these cars? (The moment of inertia of Earth is 8.1×10^{37} kg m^2.)

12.4 Two automobiles of mass 1200 kg, traveling at 30 km/h (8.33 m/s), collide on an icy frictionless road. They were initially moving on parallel paths in opposite directions, offset from one another by 1.0 m. In the collision, the cars lock together, forming a single body of wreckage with a moment of inertia of 2.5×10^3 kg m^2 about the center of mass.

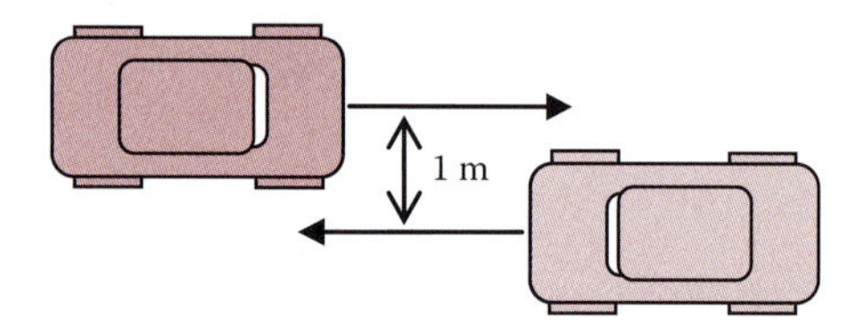

(a) What is the initial angular momentum about a point midway between the cars?

(b) What is the angular velocity of the wreck?

12.5 An atom of mass m, initially moving with a velocity $\vec{v} = v_0\hat{i}$, collides with a diatomic molecule of mass $2m$ that is initially at rest. The molecule's two atoms are identical, and the center-to-center distance between them is $2r$. The molecule's moment of inertia about its center of mass is $I = 2mr^2$. There is no mechanical energy lost in the collision.

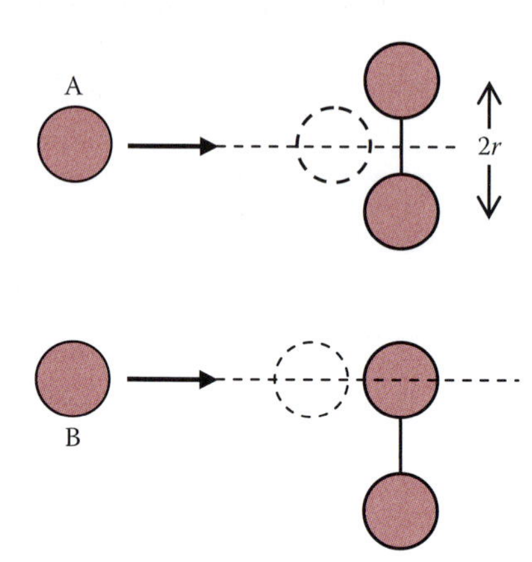

In the first case (A), the moving atom strikes the molecule midway between its two atoms.

(a) Calculate the center of mass velocity of the system and the velocities of both bodies relative to the *CM*, *before* they collide.

(b) Find the final velocities of the two bodies in the lab reference frame.

In the second case (B), the oncoming atom strikes one of the molecule's atoms head-on. After the collision, the atom is at rest.

(c) Describe the motion of the molecule after the collision. In what direction is it moving, and with what velocity? Indicate the direction of its rotational velocity vector $\vec{\omega}$.

(d) Find the magnitude of $\vec{\omega}$.

(e) For case B, show explicitly that momentum, angular momentum, and energy are conserved.

12.6 Two astronauts in deep space are approaching opposite ends of a long rod as shown in the figure. The mass of each astronaut is 75 kg, and they have equal and opposite velocities of magnitude 5 m/s. At the instant shown in the figure, they are each 50 m from the rod, and they will arrive at the rod ends simultaneously. The total length of the rod is 40 m.

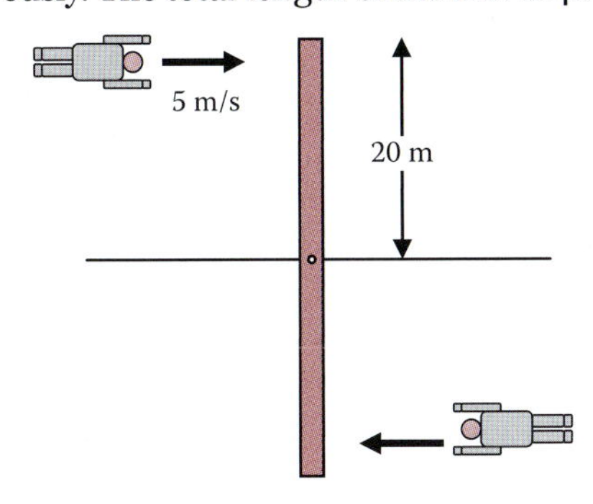

(a) What is the total angular momentum (magnitude and direction) of the astronauts relative to the center of the rod?

(b) The astronauts grab the rod at either end, and then pull themselves toward its center until their separation is 20 m. What is their speed at this time? Ignore the moment of inertia of the rod, and treat the astronauts as point masses.

(c) How much work did the astronauts perform while reducing their separation to 20 m?

(d) Now let's take the rod's moment of inertia into consideration. If the astronauts' rotational speed after seizing the ends of the rod is 3.0 m/s, what is the moment of inertia I of the rod?

12.7 Two astronauts in deep space are approaching opposite ends of a lightweight rod from opposite directions. The astronauts have identical masses, $M_1 = M_2 = 80$ kg but *unequal* approach speeds: $v_1 = 4$ m/s, $v_2 = 2$ m/s. They arrive at the bar at the same time and cling to it. The length of the bar is $\ell = 6$ m.

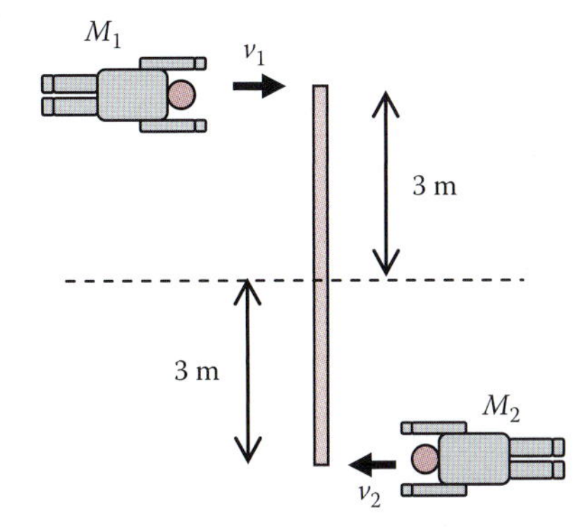

(a) Find the final velocity (magnitude and direction) of the center of mass (CM).

(b) Find the final angular velocity ω of the system. Treat the astronauts as point masses.

(c) Is kinetic energy conserved? If so, support your answer with an explicit calculation.

(d) By pulling on the bar, the astronauts decrease the distance between them to $\ell/3$, that is, each astronaut is now 1 m from the center of the bar. By what factor does their rotational kinetic energy increase?

(e) What was the average force each astronaut exerted to pull themselves toward the center of the rod?

12.8 Deep in space, far from any significant gravitational forces, a tiny asteroid of mass $m = 15.0$ kg collides with a scientific instrument. The instrument is a long rod of mass $M = 60.0$ kg and length $\ell = 40$ m. The asteroid is initially moving with velocity $\vec{v} = 20\hat{i}$ m/s relative to the rod, and strikes it at point P shown in the figure. After the collision, the asteroid's velocity is zero.

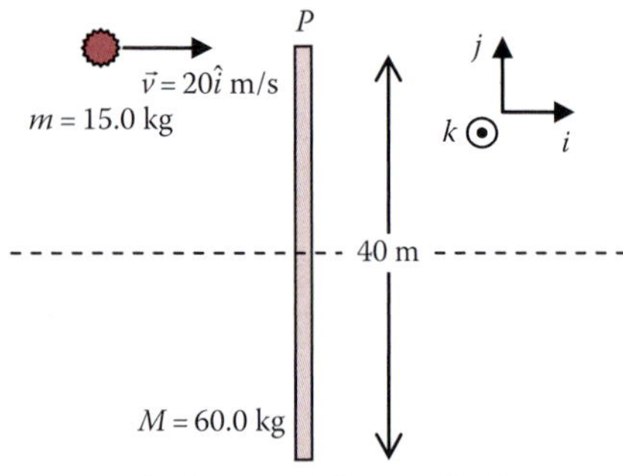

(a) Find the initial angular momentum of the asteroid *relative to the center of the stick*. Indicate its direction and include the proper units.

(b) Find the rotational velocity $\vec{\omega}$ of the rod after the collision. Be sure to indicate the direction of rotation.

(c) Find the velocity (magnitude and direction) of the rod's center of mass after the collision.

(d) Is this an elastic collision? Prove your answer by an explicit calculation.

12.9 An asteroid of mass $m = 4.0 \times 10^{18}$ kg collides with the Moon. The initial speed of the asteroid is $v = 2.0 \times 10^4$ m/s in the direction shown. After the collision, the asteroid recoils with velocity $\vec{v}'$ perpendicular to the surface. The Moon's radius $R = 1.7 \times 10^6$ m. Assume that it is not rotating initially.

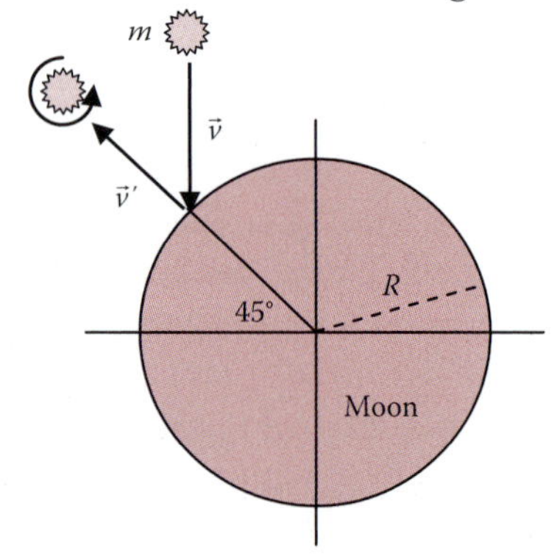

(a) Calculate the asteroid's angular momentum $\vec{L}$ relative to the center of the Moon, before the collision. What is the direction of $\vec{L}$?

(b) If the asteroid *does not rotate* after its collision, what is the Moon's angular momentum about its center of mass after the collision?

(c) The Moon's moment of inertia is roughly $2MR^2/5 = 8.6 \times 10^{34}$ kg m^2. After the collision, how long does it take for one complete rotation of the Moon?

(d) What is the direction of the Moon's final rotational velocity $\vec{\omega}$? Suppose the recoiling asteroid is rotating as shown in the figure. Does this increase, decrease, or have no effect on the magnitude of $\vec{\omega}$?

12.10 The *Dawn* spacecraft is presently closing in on asteroid Ceres after visiting its big sister Vesta. Shortly after launch from Earth, *Dawn* was spinning (for stabilization purposes) about an axis parallel to the rocket body. Wrapped around its perimeter were 2 cables of length $\ell = 12.2$ m, each tied to a small mass $m = 1.44$ kg. To reduce *Dawn's* spin rate, the two masses (called "yo-yo" weights) were unleashed, causing the cables to unwind. When fully extended and directed radially outward, as shown in the figure, the cables and yo-yo weights were jettisoned into space. This "yo-yo despin" technique can reduce the spacecraft's spin rate to zero or even reverse its spin direction.

In the following, assume that the spacecraft and the yo-yo weights were initially rotating together with angular velocity ω_o, and that after the weights were deployed and fully extended, the spacecraft was no longer spinning ($\omega = 0$) while the weights were each moving with speed v, as shown in the figure. Let I be the spacecraft's moment of inertia, and, for simplicity, treat *Dawn* as a solid cylinder of mass M and radius a.

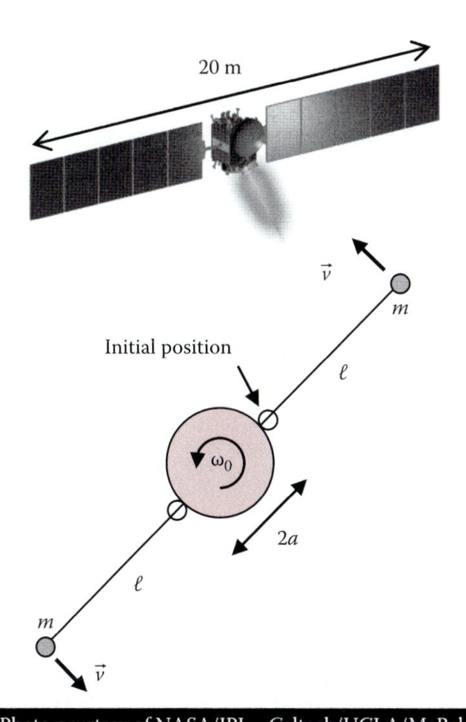

Photo courtesy of NASA/JPL—Caltech/UCLA/McRel.

(a) When the yo-yo weights were deployed, both energy and angular momentum were conserved. Find the cable length ℓ in terms of I, m and a. To simplify the math, assume $m \ll M$ and $\ell \gg a$.

(b) Using the values of m and ℓ given above, find *Dawn's* moment of inertia I. The mass of the spacecraft is 1200 kg. What is its effective radius a? (The real spacecraft is shaped more like a cube, with solar panels extending for about 20 m end to end. Check your answer for a against the figure.)

The actual cable length ℓ used was slightly greater than the value needed to stop the spacecraft's rotation, leaving *Dawn* spinning at 3 rpm in the *opposite* sense from its original spin direction. This reverse spin compensated for the compressed xenon gas (used for the ion propulsion fuel) that still spun within its tank at 39 rpm in the *original* spin direction. Friction between the gas and the walls of the fuel tank eventually reduced the rotation of the xenon and spacecraft body to near zero.

(c) The mass of the spacecraft body is 800 kg, while that of the xenon is 400 kg. Estimate the radius of the xenon storage tank, assuming it is cylindrical and centered within the spacecraft.

(d) A spinning ice skater *cannot* reduce her rotation to zero, or reverse her spin direction, merely by changing her moment of inertia. What is the important difference that makes this possible for the satellite? (*Hint*: read the second paragraph of the description carefully.)

12.11 The asteroid (90) Antiope, first discovered in 1866, was recently found to be a *binary asteroid*, consisting of two nearly identical bodies in circular orbit about their common center of mass. Each body has a radius of 44 km, and the center-to-center distance between them is 170 km. The orbital period of either body is known with high precision: 16.5051 h = 5.94184×10^4 s.

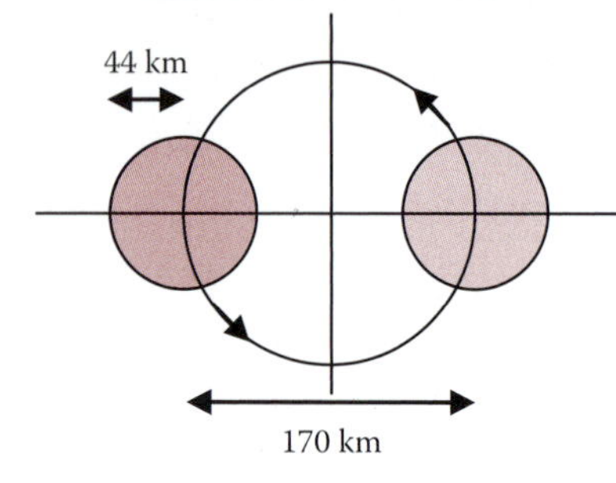

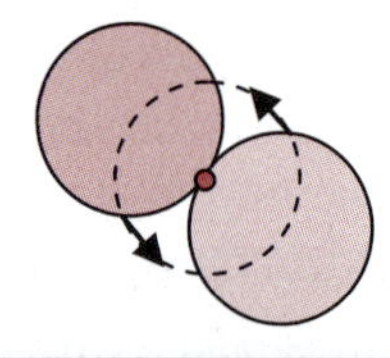

(Courtesy of William J. Merline, SwRI/W.M. Keck Observatories.)

(a) How fast are these bodies moving relative to their center of mass?

(b) Show that the mass m of either body is about 4.1×10^{17} kg.

It has been suggested that the two bodies were originally a single body. Imagine that they are touching one another and rotating about the point of contact, as shown in the figure.

(c) Show that their rotation period is ⅓ of the present orbital period, or 5.5 h.

(d) What is the minimum mass needed to hold the two bodies together?

(e) Imagine that the two bodies coalesce into a sphere of the same density as the two bodies. What would be the period of rotation? Would the single sphere cohere?

12.12 The moment of inertia of the Earth is $\frac{1}{3}M_E R_E^2$. If an asteroid of mass 5.0×10^{18} kg moving at 150 km/s struck and became embedded in the Earth's surface, by how long would the length of the day change? Assume the asteroid was traveling westward in the equatorial plane and struck the surface at an angle of 45°.

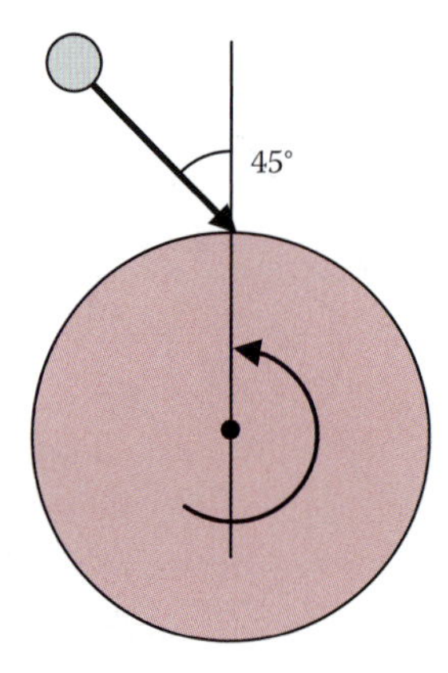

12.13 You wish to send a 500 kg spacecraft from Earth to Saturn on a minimum energy transfer orbit. Saturn's orbit has a radius of 9.5 AU. Neglect the gravity of Earth and Saturn in this problem.

(a) What is the total energy of the spacecraft in this orbit?

(b) What is its speed at perihelion?

(c) What is its angular momentum?

(d) Find the semimajor and semiminor axes of the orbit.

(e) What shape would an orbit have if its angular momentum is maximized while keeping its energy constant?

12.14 A small asteroid is spotted; it is presently moving at 15.3 km/s and is 2.54 AU from the Sun. As shown in the figure, the angle between its velocity and position vectors is 60°. From this information you can predict the asteroid's entire trajectory.

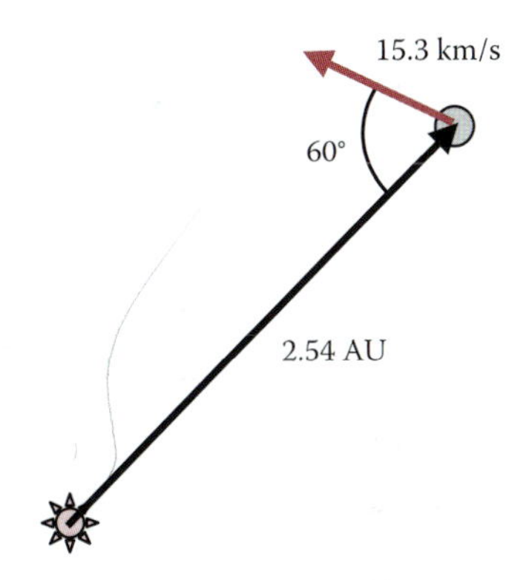

(a) Let m be the asteroid's mass. Calculate the value of E/m and find the semimajor axis a of the asteroid's orbit (in AU).

(b) What is the orbital period (years)?

(c) Calculate the value of L/m, and determine the semiminor axis b of the orbit (in AU).

(d) What is the speed of the asteroid at aphelion? At perihelion?

(e) What is the present distance to the second (empty) focal point of the ellipse? Copy the figure and locate the second focus. Then draw the ellipse's major and minor axes, and sketch the entire orbit to scale.

Your calculated orbital parameters should match those of asteroid 2012 UE, which passed within 0.005 AU of the Earth in October 2012. The NASA/JPL website http://neo.jpl.nasa.gov/ (Accessed on January 22, 2015.) lists the asteroid's orbital parameters and includes a figure of its orbit.

12.15 The *Odyssey* spacecraft reached Mars in late October, 2001. Its initial capture orbit was elliptical with a period of 17.0 h. The orbiter's mass was 725 kg, while the mass and radius of Mars are 0.642×10^{24} kg and 3.38×10^{6} m. At its closest approach (periapsis) to the planet, *Odyssey* was only 130 km above the surface. At this altitude, there was significant atmospheric drag on the spacecraft. Each time *Odyssey* passed through periapsis, it lost some of its velocity v_p, and the size and shape of its orbit changed. Suppose that each time it entered Mars' atmosphere, the spacecraft's velocity was reduced by 0.1 km/s while its altitude at periapsis remained unchanged.

(a) In the first column of the following table, the value of v_p is given after each *aerobraking* event. Fill in the table by finding the kinetic energy at periapsis K_p, the total energy E, the angular momentum L, and then the orbit parameters a, b, and the eccentricity ε. (This is an ideal spreadsheet exercise.) To help you get started, the initial values of these quantities are entered in the top row of the table.

(b) What is happening to the shape of the orbit?

(c) What is the period of the orbit when $v_p = 3.56$ km/s?

v_p (km/s)	K_p (J)	E (J)	L (J s)	a (m)	b (m)	ε
4.66	**7.87 × 10⁹**	**−9.73 × 10⁸**	**1.19 × 10¹³**	**1.60 × 10⁷**	**1.00 × 10⁷**	**0.780**
4.56						
4.46						
4.36						
4.26						
4.16						
4.06						
3.96						
3.86						
3.76						
3.66						
3.56						

12.16 Oh no! A giant asteroid is on a collision course with Earth! You and other members of an elite team are dispatched to intercept the asteroid and destroy it with a nuclear weapon. But you are thwarted by a maniacal group of doomsday believers who prevent you from deploying the weapon. The asteroid crashes into Earth, changing its direction of motion by 45° *but not changing its orbital speed* v (30 km/s) ($GM_S/R_{ES} = 8.89 \times 10^8$ J/kg).

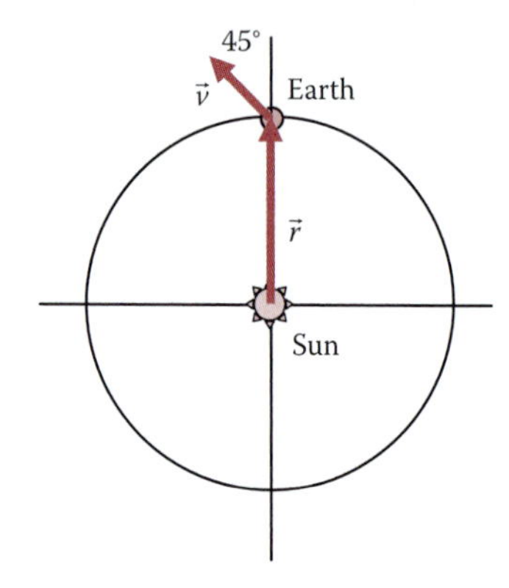

(a) What is the value of the semimajor axis of the ellipse after the collision (in AU)? What is the new period? (*Hint*: How has the energy changed?)

(b) Write an expression for the angular momentum of Earth, relative to the Sun, after the collision. Express L in terms of m_E, v_E, and r.

(c) Find the value (in AU) of the semiminor axis b of the Earth's new elliptical orbit.

(d) What is the eccentricity ε of the new orbit?

(e) Find the perihelion and aphelion distances (in AU).

(f) Using angular momentum conservation, find the orbital speed of the planet at

(i) perihelion, (ii) aphelion, and (iii) the point where Earth crosses the semiminor axis.

(g) How is the ellipse oriented? Sketch it on the figure above.

12.17 *Voyager 2* was launched from Cape Canaveral in August, 1977. The 700 kg spacecraft was the first to obtain close-up images of Jupiter, Saturn, Uranus, and Neptune, and used gravity assists from these giants to escape the solar system. It is presently more than 100 AU from the Sun, far beyond the orbit of Pluto, and remains in radio contact with Earth. In this exercise, we will examine *Voyager's* trajectory as it passed by Saturn in August, 1981. (The numbers below are approximate values.)

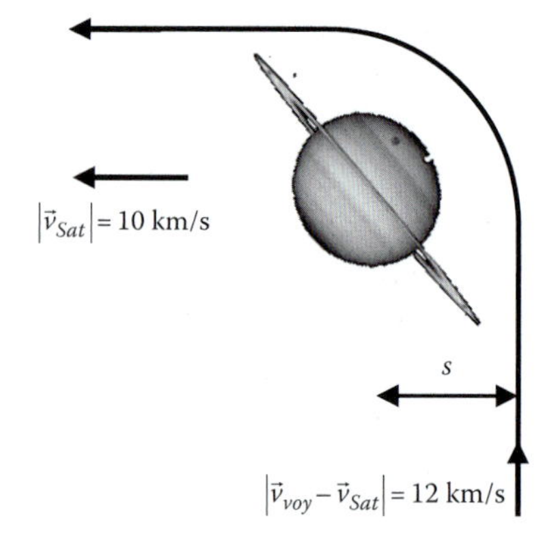

During the flyby, the spacecraft approached Saturn with relative velocity of 12 km/s and was deflected by 90° (in the planet's reference frame), with its final velocity parallel to Saturn's. This is shown in the figure. Saturn's orbital radius and velocity are 9.5 AU and 10 km/s, and the escape velocity from its surface is 37 km/s.

(a) Find the final speed of the spacecraft relative to the Sun. Is this speed sufficient to escape from the solar system? (Subsequently, *Voyager* received gravity assists from Uranus and Neptune.)

(b) Show that the spacecraft's impact parameter $s \simeq 4.8R$, where R is the radius of the planet.

(c) What was the minimum distance between the spacecraft and the center of Saturn? Express this distance in terms of R. Check your answer using Figure 12.8b.

(d) What was the maximum speed of the spacecraft relative to the planet?

(e) Find the *maximum deflection angle* possible for a spacecraft executing a close flyby of Saturn with an approach speed of 12 km/s.

12.18 A spacecraft having a mass m = 800 kg is making a close flyby of the planet Venus. Venus is in a near-circular orbit about the Sun with an orbital speed of 35 km/s. Initially, the spacecraft is traveling in the same direction as the planet. Its trajectory in the planet's reference frame is shown in the figure. Far from Venus, the craft approaches the planet with a relative speed of 10 km/s. It skims over the surface at point P and exits after being deflected by 90°.

The mass of Venus is 4.87×10^{24} kg, and its radius is 6.0×10^{6} m.

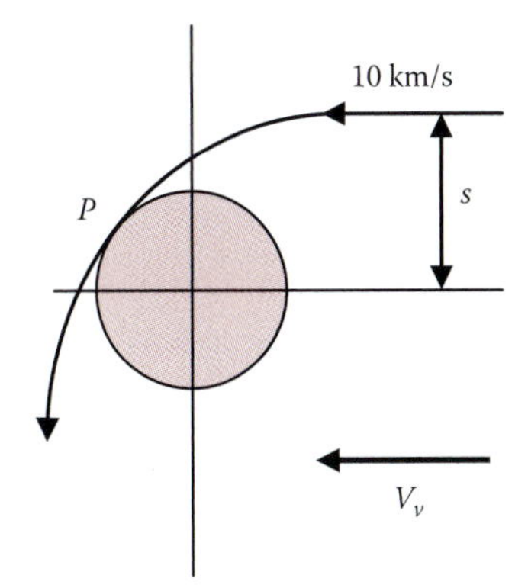

(a) Find the escape velocity from the surface of Venus.

(b) What is the smallest value of the impact parameter s such that the spacecraft does not collide with the planet?

(c) What is the spacecraft's angular momentum (magnitude and direction) relative to the center of the planet?

(d) What is the craft's speed *relative to Venus* at point P?

(e) What is the spacecraft's final vector velocity *relative to the Sun*?

12.19 You have detected a giant asteroid—the size of Texas—on a collision course with Earth. This is the big one! Based on your best estimates, the asteroid has a mass $m = 5 \times 10^{20}$ kg and will deliver a grazing blow to Earth at the equator, where its velocity *at impact* will be 20 km/s parallel to the Earth's surface (see the figure). After the collision, the asteroid will be imbedded in the Earth at its surface.

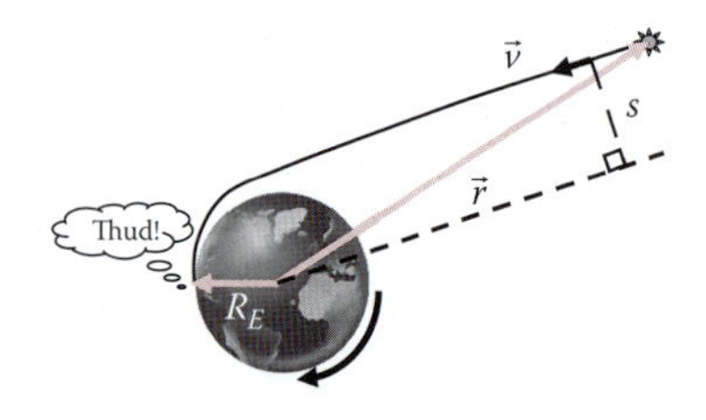

Earth's moment of inertia is

$$I_E = \frac{1}{3} M_E R_E^2 = 8.03 \times 10^{37} \text{ kg m}^2$$

(a) What is the asteroid's angular momentum relative to the Earth's center?

(b) By what fraction (or %) is the length of the day changed by the collision?

(c) What is the asteroid's speed when it is very far from Earth ($r \gg R_E$)? (Neglect the influence of the Sun and other planets.)

(d) Find the value of the impact parameter s for the glancing collision shown in the figure.

12.20 The bright comet that appeared in November 1680 terrified observers in Europe and America, but provided Isaac Newton with an ideal opportunity to test his law of universal gravitation. His contemporary Christiaan Huygens proposed an ingenious way to estimate the comet's distance from Earth. Huygen's method assumes that the comet's tail is always pointed directly away from the Sun and is finite in length. The maximum distance from Earth can be determined by measuring the comet's *elongation* (the angle between the comet and the Sun as viewed from Earth, shown as θ in the figure) and the *angle subtended* by the tail (as viewed from Earth, angle φ in the figure).

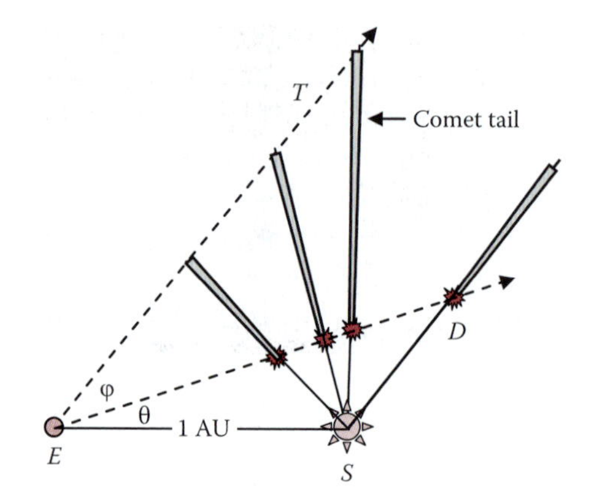

The figure shows four possible positions of the comet with the same θ and φ. In each case, the tail is drawn pointing radially away from the Sun. At position *D*, the tail would need to be infinite in length in order to subtend angle φ, so the comet cannot be farther from Earth than point *D*. On December 30, 1680, the comet's elongation θ was 26.5° and its tail subtended an angle φ ≈ 70°.

(a) Find the maximum distance between Earth and the comet on this day. (*Hint*: Line $\overline{SD}$ is parallel to line $\overline{ET}$.)

(b) Show that on December 30, the comet was closer than Venus to the Sun.

12.21 The Japanese spacecraft *Hayabusa* was launched in May 2003, and traveled to asteroid *Itokawa*, where it landed briefly in September 2005 and successfully scooped up a tiny amount of surface material. It returned to Earth in June 2010, becoming the first spacecraft to retrieve extraterrestrial matter from a body other than the Moon. Itokawa is in an elliptical orbit about the Sun, with a perihelion distance r_p = 0.953 AU and a perihelion speed v_p = 34.6 km/s. The mass of the asteroid is 3.5 × 10^{10} kg; very crudely, it is 500 m in length with a diameter of 250 m.

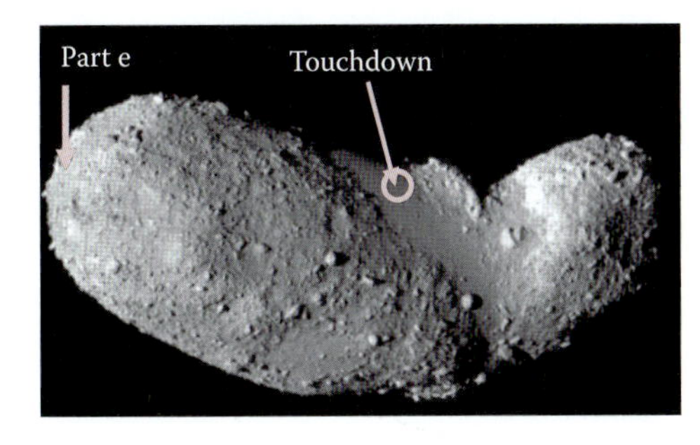

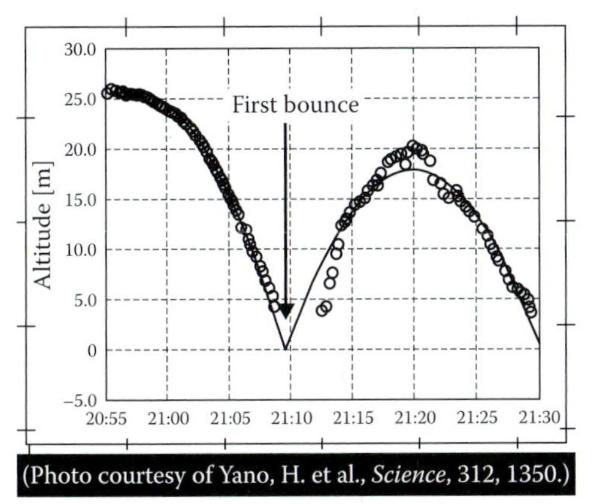

(Photo courtesy of Yano, H. et al., *Science*, 312, 1350.)

(a) What are the values of the semiminor and semimajor axes of the asteroid's orbit about the Sun?

(b) The spacecraft bounced several times before settling on the asteroid's surface. Its altitude (in m) vs. time (in min:s) is shown in the figure. What is the acceleration of gravity g at the landing site?

(c) With what speed did it hit the surface on the "first bounce?"

(d) What is the asteroid's orbital speed at aphelion? (*Hint*: Use angular momentum!)

(e) When *Hayabusa* reached the asteroid, it hovered about 7–20 km above the surface for about 6 weeks rather than entering a "parking orbit" about the asteroid. Why was a parking orbit impossible? (*Hint*: Compare the Sun's gravitational attraction to that of the asteroid.)

(f) Suppose the 500 kg *Hayabusa* had crashed into the asteroid at one end, as sketched in the figure. If the impact velocity had been 50 m/s, estimate the subsequent rotational period of the asteroid.

13

TORQUE, ANGULAR MOMENTUM, AND THE EARTH–MOON SYSTEM

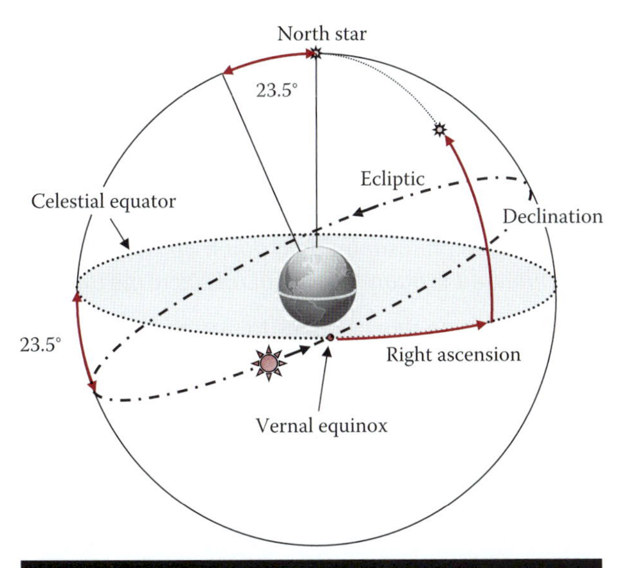

The Greeks' geocentric conception of the universe. The stars are embedded in a spherical shell that rotates once per day. The Sun's path (the ecliptic) is tilted from the celestial equator by 23.5°. The point where it crosses the celestial equator while heading northwards is called the vernal equinox.

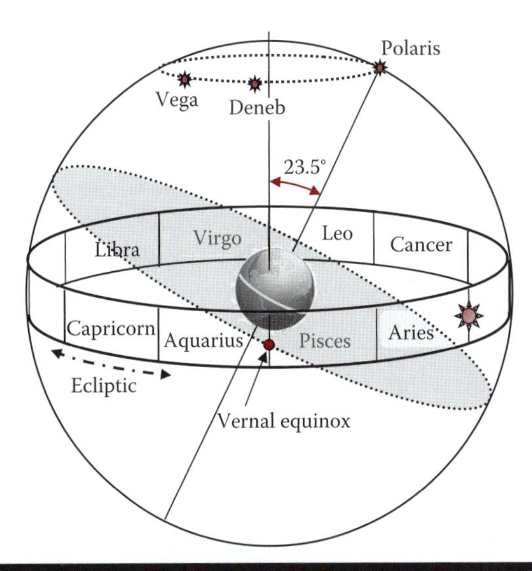

Each year, the Sun completes a full circuit of the heavens, passing through all of the constellations lying near the ecliptic—called the zodiac. Hipparchus noted that the position of the vernal equinox drifted by 2.1° over a span of 150 years.

When the sun come back
When the firs' quail call,
Then the time is come
Foller the drinkin' gourd.

Traditional American folk song

Polaris is an easy star to spot. It marks the end of the Little Dipper's handle, and lines up with two of the brightest stars of the Big Dipper. At the present time, it has the distinction of being Earth's "north" or "pole" star: the planet's axis of rotation points almost directly at Polaris, and as Earth turns, stars in the northern sky (including the Sun) appear to circle about it. For centuries, Polaris and other prominent stars functioned as a celestial GPS system, guiding mariners—and later, aviators—on their voyages and flights. In the mid-nineteenth century, fleeing slaves in America were counseled to make their escape in springtime ("when the sun comes back...") and to "follow the drinking gourd," code for using the Big Dipper to locate the north star and the direction to freedom. But Polaris' reign as the pole star is only temporary: in a mere 8000 years, Deneb (in the constellation Cygnus) will assume the role, and 2000 years later, Vega (in Lyra) will take over. 26,000 years from now, Polaris will regain its present status.

In the Ptolemaic universe, the stars were affixed to a vast celestial sphere that rotated westward once per day. The Sun's motion was a little more complicated: while it daily moved westward along with the stars, it also drifted slowly eastward relative to them, annually executing a complete circle around the celestial sphere. Throughout the year, this eastward drift traces out a path—called the ecliptic*—that is tilted by 23.5° from the celestial equator. The celestial equator is just the skyward extension of Earth's equator, so the two lie in the same plane. Just as we specify locations on Earth by latitude and longitude, a star's position on the celestial sphere is expressed by its* declination *and* right ascension. *Stars with zero declination lie on the equator. Right ascension is measured from the* vernal equinox, *the point where the Sun crosses the equator each year while moving northward along the ecliptic.**

* When the Sun crosses the celestial equator, Earth's N–S axis is perpendicular to the line joining the Earth and Sun, and daytime and nighttime have the same duration. The vernal equinox presently occurs on March 20. The autumnal equinox, when the Sun crosses the celestial equator going south, presently takes place on September 22.

In the second century BC, Hippachus compared his measurements of stellar positions to measurements taken 150 years earlier. He discovered that the stars had shifted eastward relative to the vernal equinox by 2.1°. At that rate, it would take about 26,000 years for the vernal equinox to rotate, or precess, one full turn (360°) around the celestial sphere. The universe would not be restored to its present state until that time. This slow precession of the equinoxes *spoiled the Greek ideal of perfect, unchanging, cosmic order.*

Nearly two millennia after the Greeks, a new kind of order was conceived, one where a single law governed the motion of all celestial bodies. Newton's law of universal gravitation successfully accounted for the motion of Galileo's moons, Kepler's planets, and Halley's comet. It accounted for the tides, the oblate shape of the Earth, and ultimately, the precession of the equinoxes. During the past 300 years, physics has advanced far beyond Newton's inverse square law, but the "rules of reasoning in philosophy" laid down by Newton in the Principia remain the nub of all scientific thought:

- *"admit no more causes to natural things than are true and sufficient to explain their appearances;"*
- *"to the same natural effects, we must, as far as possible, assign the same causes."*

"Nature does nothing in vain," Newton wrote, "and ... is pleased with simplicity." The precession of the equinoxes is a perfect example of what he meant. To understand the apparent motion of the celestial sphere, you need only understand the physics of a simple toy top.

13.1 INTRODUCTION

In Chapter 6, we introduced Newton's Laws and learned how to use free body diagrams to calculate acceleration. At that time, we asked *what* forces acted on the body, but did not ask *where* they were applied. In effect, we were treating the body as if it were a point particle and the acceleration we calculated was that of its

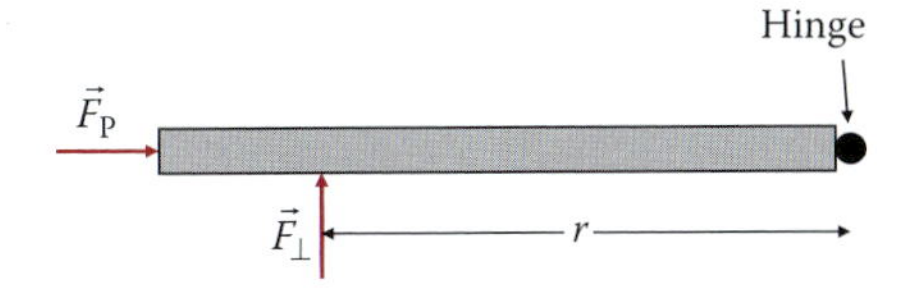

FIGURE 13.1 Top view of a door attached to its frame by vertical hinges. To open or shut the door, a force $F_\perp$ perpendicular to the door's face must be applied. The farther from the hinge the force is applied, the easier it is to move the door.

center of mass: $\vec{F}_{tot} = m\vec{a}_{CM}$. But *CM* motion isn't the whole story. Picture a stick lying flat on an icy surface. If you push on the center of the stick, it will simply accelerate in the direction of the force: $\vec{a}_{CM} = \vec{F}/m$. If you push off-center with the same force, the stick's *CM* acceleration is still the same, but the off-center force also causes the stick to spin. In addition to changing the stick's momentum, the force also changes the stick's spin angular momentum.

Suppose you want to close the door to your room. To do so, you must exert a force on the door. Where, and in what direction, should you apply the force? If you push parallel to the face of the door, straight toward the hinge, the door will not turn. It will turn only if the applied force has a component perpendicular to the face (see Figure 13.1). Moreover, the farther from the hinge r you apply the force, the easier it will be to close the door. Your ability to swing the door (changing its angular momentum) depends on $rF_\perp$, an expression you may recognize as the magnitude of the cross product $\vec{r} \times \vec{F}$.

In this chapter, we will formalize these ideas by introducing a new vector quantity called *torque*. Torque, denoted by the Greek letter $\vec{\tau}$ (tau), is related to angular momentum as force is to momentum: $\vec{\tau} = d\vec{L}/dt$ while $\vec{F} = d\vec{p}/dt$. Our goal is to derive equations that describe the complete motion of a rigid (solid) body—its rotation as well as its translation. Because we will be using cross products, the mathematics will be a little trickier than in earlier chapters. (That's why we saved this topic for last!) The equations and techniques that will be developed are essential for understanding the physics of everyday events on Earth as well as exotic phenomena in space. As always, we will introduce the physical concepts with familiar lab-scale examples, and then employ the same ideas to study the motion of the Earth, the Moon, and—finally—the precession of the equinoxes.

13.2 TORQUE AND ANGULAR MOMENTUM

Consider a point particle of mass m having an instantaneous position $\vec{r}$ and velocity $\vec{v}$ relative to the origin of coordinates at point O (see Figure 13.2a). The body has momentum $\vec{p} = m\vec{v}$ and angular momentum $\vec{L} = \vec{r} \times \vec{p} = m\vec{r} \times \vec{v}$. If there are no forces acting on the body, then it is isolated, and $\vec{p}$ and $\vec{L}$ are constant. When a force $\vec{F}$ is applied, the momentum changes according to Newton's second law, $d\vec{p}/dt = \vec{F}$, and the rate of change of the *angular* momentum is

$$\frac{d\vec{L}}{dt} = \frac{d}{dt}(m\vec{r} \times \vec{v}) = m(\vec{v} \times \vec{v}) + m(\vec{r} \times \vec{a}) = \vec{r} \times (m\vec{a}) \tag{13.1}$$

or

$$\frac{d\vec{L}}{dt} = \vec{r} \times \vec{F} \equiv \vec{\tau},$$

where $\vec{\tau}$ is called the *torque* acting on the body due to the force $\vec{F}$. Equation 13.1 both defines torque and equates it to the rate of change of angular momentum ($\vec{\tau} = d\vec{L}/dt$). Its magnitude can be expressed variously

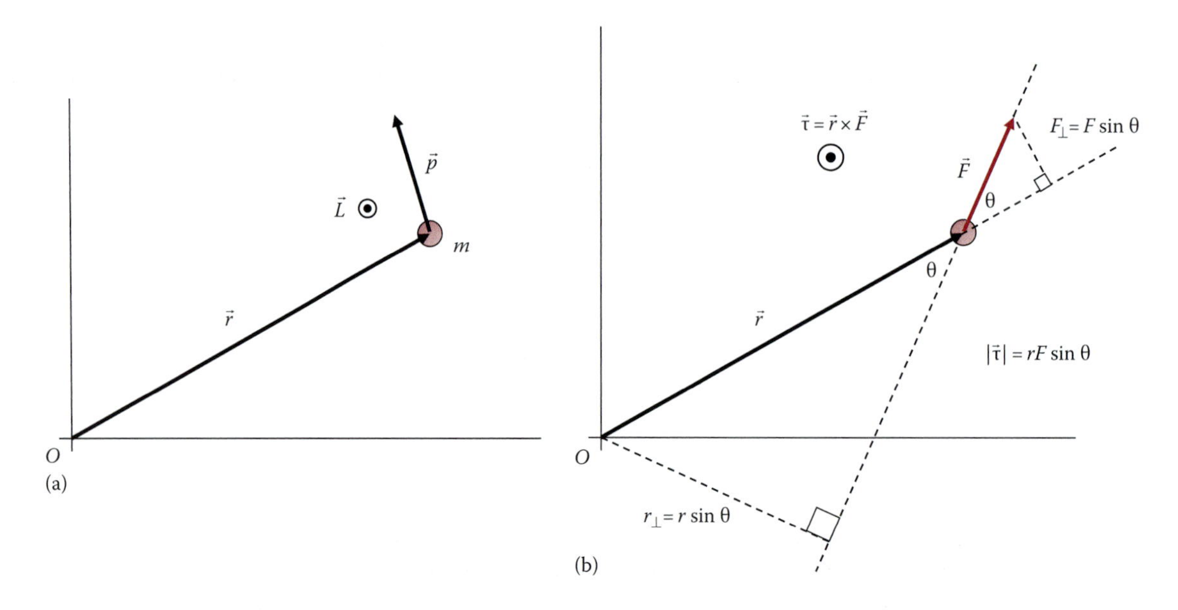

FIGURE 13.2 (a) The angular momentum of a point mass m, relative to the origin O at $\vec{r} = 0$, is given by $\vec{L} = \vec{r} \times \vec{p} = m\vec{r} \times \vec{v}$. The direction of $\vec{L}$ is out of the page. (b) Relative to the same point O, the torque acting on m due to force $\vec{F}$ is $\vec{\tau} = \vec{r} \times \vec{F}$ and is directed out of the page. The magnitude of $\vec{\tau}$ can be expressed in several ways: $\tau = rF\sin\theta = rF_{\perp} = r_{\perp}F$, where $F_{\perp}$ is the component of $\vec{F}$ perpendicular to $\vec{r}$, and $r_{\perp}$ is the component of $\vec{r}$ perpendicular to $\vec{F}$.

as $|\vec{\tau}| = rF\sin\theta = rF_{\perp} = r_{\perp}F$, where θ is the angle between $\vec{r}$ and $\vec{F}$ ($0 \leq \theta \leq 180°$) (Figure 13.2b). Note that $\vec{r}$ appears in the definitions of both $\vec{L}$ and $\vec{\tau}$. Equation 13.1 is valid only if $\vec{L}$ and $\vec{\tau}$ are calculated using the same $\vec{r}$, that is, only if they *are defined relative to the same point O*.

EXERCISE 13.1

A planet of mass m is in an elliptical orbit about a star of mass $M \gg m$.

(a) Prove that $\vec{\tau} = 0$ relative to the Sun at all points along the orbit.
(b) Now consider the torque relative to point O′ at the center of the ellipse. Identify points where $\vec{\tau} = 0$. Is angular momentum relative to point O′ constant throughout the orbit?

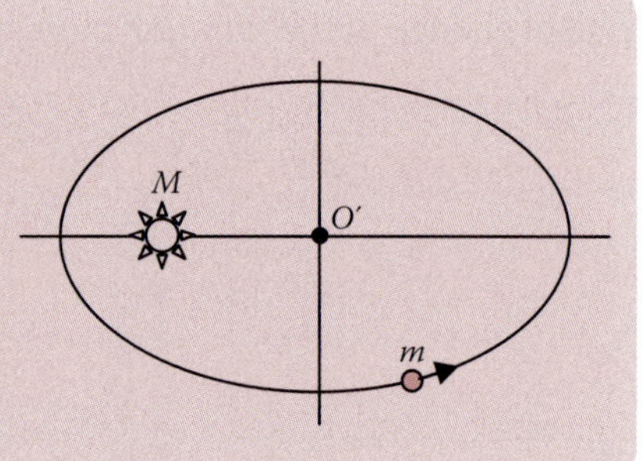

Answer (b) no

A real body may be represented as a collection of N interacting point particles m_i. The instantaneous location and momentum of the ith particle are $\vec{r}_i$ and $\vec{p}_i = m_i\vec{v}_i$, so the total angular momentum of the body is

$$\vec{L} = \sum_{i=1}^{N} \vec{r}_i \times \vec{p}_i.$$

The rate of change of the total angular momentum is

$$\begin{aligned}\frac{d\vec{L}}{dt} &= \sum_{i=1}^{N} \vec{r}_i \times \frac{d\vec{p}_i}{dt} + \sum_{i=1}^{N} \vec{v}_i \times \vec{p}_i \\ &= \sum_{i=1}^{N} \vec{r}_i \times \vec{F}_i\end{aligned}$$

where $\vec{F}_i$ is the total force acting on the ith particle. $\vec{F}_i$ may arise from internal interactions *between* the particles as well as from externally applied forces: $\vec{F}_i = \vec{F}_i^{int} + \vec{F}_i^{ext}$. In Section 12.3, we saw that for isolated systems, where only internal interactions are present, the total angular momentum is conserved. For example, the chemical bonds between neighboring molecules of a rigid body hold the body together and define its shape, but they cannot change its angular momentum (or its momentum). Therefore, $\sum_i \vec{r}_i \times \vec{F}_i^{int} = 0$ and the above equation becomes

$$\frac{d\vec{L}}{dt} = \sum_{i=1}^{N} \vec{r}_i \times \vec{F}_i^{ext} \equiv \vec{\tau}_{ext}. \quad (13.2)$$

A long thin stick is shown in Figure 13.3a. It has mass M and length ℓ and is free to pivot horizontally about a fixed pin through its center. It moves by rotation only, so its angular momentum is due entirely to its spin: $\vec{L} = \sum_i m_i \vec{r}_i \times \vec{v}_i = I\vec{\omega}$, where $I = \frac{1}{12} M\ell^2$ is the stick's moment of inertia about its *CM*. Let's apply a force $\vec{F}$ at position $\vec{r}$ measured from the center of the stick. The stick's *CM* cannot move, so $\vec{a}_{CM} = 0$, which implies that the pin must exert a force $\vec{F}_{pin} = -\vec{F}$ on the stick. Because $\vec{F}_{pin}$ is applied at the pivot point (where $\vec{r} = 0$), it does not produce a torque relative to the *CM*. ($\vec{F}_{pin}$ cannot cause the stick to rotate.) The only torque is due to the applied force $\vec{F}$: $\vec{\tau}_{ext} = \vec{r} \times \vec{F}$. By the right hand rule (RHR), $\vec{\tau}_{ext}$ points out of the page as shown in Figure 13.3a. If the stick is initially at rest, the torque will cause it to rotate CCW, producing an angular velocity $\vec{\omega}$ that also points out of the page. The time-dependent rotation of the stick is determined using Equation 13.2 and $\vec{L} = I\vec{\omega}$:

$$\frac{d\vec{L}}{dt} = I\frac{d\vec{\omega}}{dt} = \vec{r} \times \vec{F}.$$

The rate of change of $\vec{\omega}$ is called the *angular acceleration* $\vec{\alpha}$: $\vec{\alpha} \equiv d\vec{\omega}/dt$, and we may rewrite the above equation as $\vec{\tau} = I\vec{\alpha}$ in analogy to $\vec{F} = M\vec{a}$. If the force has a constant magnitude and is always perpendicular to

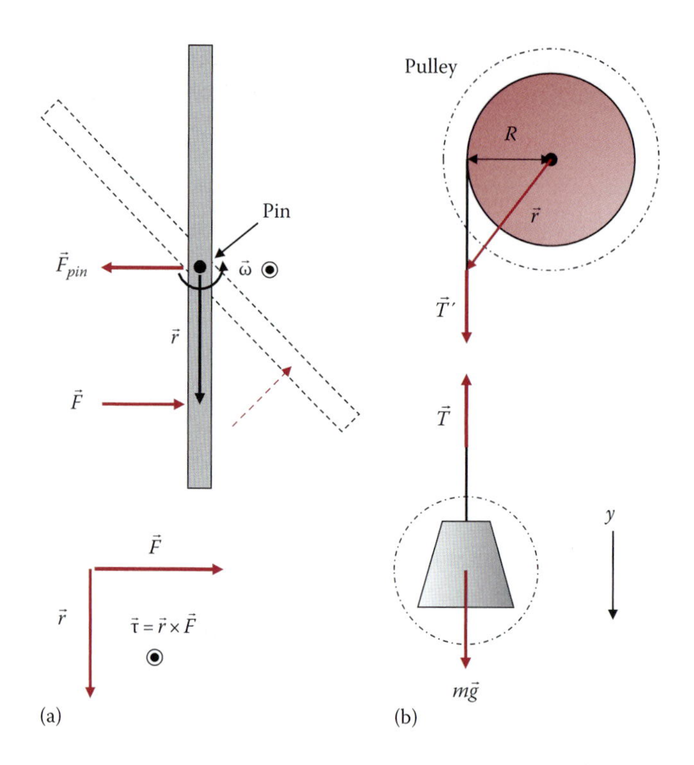

FIGURE 13.3 (a) A stick of length ℓ pivots about a fixed central pin. Relative to the center of the stick, only the applied force $\vec{F}$ contributes to the torque. By the RHR, $\vec{\tau}$ points out of the page. The stick's angular acceleration $\vec{\alpha} = d\vec{\omega}/dt$ also points out of the page. (b) A pulley of radius R and moment of inertia $I = \frac{1}{2} M_p R^2$ is attached to a hanging mass m by a lightweight string. The dotted circles separate the system into two free bodies for which equations of motion can be written.

the stick, then $\vec{\tau}$ and $\vec{\alpha}$ are both constant, and the angular velocity changes linearly with time: $\vec{\omega}(t) = \vec{\omega}_0 + \vec{\alpha}t$, analogous to the equation for constant linear acceleration $\vec{v}(t) = \vec{v}_0 + \vec{a}t$.

EXERCISE 13.2

A stick of length ℓ and mass M is mounted on a pin passing through its center. It is subjected to the three external forces shown in the figure. The forces all have the same magnitude F and are in the directions indicated. Each force rotates with the stick, and remains parallel or perpendicular to the stick as it turns.

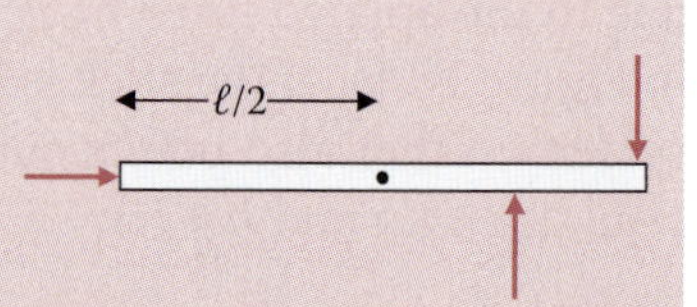

(a) In what direction does the stick turn?
(b) Find an expression for the angular acceleration α.
(c) What is the magnitude and direction of the force exerted by the pin on the stick?

Answer (a) CW (b) $\alpha = 3F/M\ell$ (c) $\leftarrow F$

A more challenging example is shown in Figure 13.3b. The hanging mass m is tied to a string which is wrapped around a pulley M_P. The pulley has radius R, moment of inertia $I = \frac{1}{2} M_P R^2$, and rotates freely about an axle through its center. When m is allowed to fall, its *CM* undergoes acceleration a. The pulley has no *CM* motion but rotates with angular acceleration α. To analyze the motion, begin by drawing two free body diagrams to identify the forces and torques acting on the bodies. Then write the relevant equation of motion for each body. For the hanging mass,

$$m\vec{a} = \vec{T} + m\vec{g}$$

while for the pulley,

$$I\vec{\alpha} = I\frac{d\vec{\omega}}{dt} = \vec{r} \times \vec{T}',$$

where, $\vec{T}$ and $\vec{T}'$ are the tension forces exerted by the string on m and M_P, respectively and $\vec{r}$ extends from the pulley axle to any point on the string beyond the free body boundary (so $\vec{r} \times \vec{T}'$ is an *external* torque).

If the string mass is negligible, then $\vec{T} = -\vec{T}'$. (Otherwise, the string acceleration would be infinite.) When the pulley rotates CCW through an angle $\Delta\theta$, the hanging mass drops by $\Delta y = R\Delta\theta$, so $v_y = dy/dt = R\omega$ and $a_y = R\alpha$. Define the positive direction of a_y to be downward and the positive direction of $\vec{\alpha}$ to be out of the page, in the same direction as $\vec{r} \times \vec{T}'$. The magnitude of the torque is $|\vec{r} \times \vec{T}'| = r_\perp T = RT$, so the vector equations of motion can be rewritten in scalar form as

$$ma_y = mg - T$$

and

$$I\alpha = I\frac{a_y}{R} = RT.$$

Solve the second equation for the tension: $T = Ia_y/R^2$, and then substitute this into the first equation to obtain

$$a_y = \frac{mg}{m + I/R^2} = \frac{m}{m + M_P/2}g.$$

The string tension reduces the acceleration of the hanging mass, and $a < g$.

EXERCISE 13.3

(a) Check the above expression for $M_P \to 0$ and $M_P \to \infty$. Does it give the correct result in these two limits? (b) Let $M_P = 4m$. Find the tension in the string in terms of mg. (c) Check that this tension provides the torque needed to rotate the pulley with angular acceleration $\alpha = a_y/R$.

Answer (b) $T = 2mg/3$

It is interesting to check the above result for conservation of energy. Each body starts from rest, so the initial kinetic energy equals 0. When m has fallen a distance Δy, $\Delta K = K = -\Delta U = mg\Delta y$:

$$\Delta K = \frac{1}{2}mv_y^2 + \frac{1}{2}I\omega^2 = mg\Delta y.$$

The pulley's moment of inertia is $I = ½ M_P R^2$, and $v_y = \omega R$, so $I\omega^2 = ½ M_P v_y^2$ and

$$v_y^2 = \frac{2mg\Delta y}{(m + M_P/2)}.$$

The equations of motion (Equation 6.4) of a body moving with constant acceleration is $\vec{r} = \vec{r}_0 + \vec{v}_0 t + ½\vec{a}t^2$ and $\vec{v} = \vec{v}_0 + \vec{a}t$. If the body starts from rest ($\vec{v}_0 = 0$), and falls vertically with constant acceleration a_y, then $v_y^2 = 2a_y\Delta y$. Comparing this with the previous equation for v_y^2, we obtain $a_y = mg/(m + M_P/2)$ in agreement with the previous result. Therefore, our solution conserves mechanical energy.

13.3 GRAVITATIONAL TORQUE AND STATICS

In many important applications, the gravitational force $M\vec{g}$ is the only external force acting on the body. In such cases, the external force acting on the ith particle of the body is $\vec{F}_i^{ext} = m_i\vec{g}$, so the total torque is

$$\vec{\tau}_{ext} = \sum_{i=1}^{N} \vec{r}_i \times m_i\vec{g} = \left[\sum_{i=1}^{N} m_i\vec{r}_i\right] \times \vec{g}.$$

From Equation 5.1, $\sum_i m_i\vec{r}_i = M\vec{R}_{CM}$, where M is the total mass of the body. Therefore,

$$\vec{\tau}_{ext} = \vec{R}_{CM} \times M\vec{g}, \tag{13.3}$$

exactly as if *the entire force of gravity* $\vec{F} = M\vec{g}$ *were applied at the body's CM*. Consequently, a uniform gravitational field (where $\vec{F}_i = m_i\vec{g}$) exerts no torque relative to the body's center of mass.

Equation 13.3 is especially useful for solving *statics* problems. A static body is one at rest. For it to remain at rest, the total force acting on it must be zero (so $\vec{a}_{CM} = 0$), and the total torque must also be zero (so $\vec{\alpha} = d\vec{\omega}/dt = 0$). For static bodies, if $\vec{\tau}_{ext} = 0$ relative to reference point O, it is also zero relative to any other reference point P.

To prove this, treat the body as a collection of N tiny particles. Let $\vec{r}_i^{\,O}$ and $\vec{r}_i^{\,P}$ denote the position of particle i relative to two reference points O and P separated by $\vec{R} = \vec{r}_i^{\,P} - \vec{r}_i^{\,O}$. Let $\sum_i \vec{F}_i = 0$ and $\vec{\tau}^{\,O} = \sum_i \vec{r}_i^{\,O} \times \vec{F}_i = 0$. Then,

$$\vec{\tau}^{\,P} = \sum_i \vec{r}_i^{\,P} \times \vec{F}_i = \sum (\vec{r}_i^{\,O} + \vec{R}) \times \vec{F}_i$$
$$= \sum_i \vec{r}_i^{\,O} \times \vec{F}_i + \vec{R} \times \sum_i \vec{F}_i = 0 + \vec{R} \times 0 = 0.$$

This is a powerful result. For *a static body*, the torque is zero about any reference point we select. We are free to pick the most convenient reference point to analyze the forces acting on the body. Use this to complete the following exercise.

EXERCISE 13.4

A 20 kg bar is hung from two cables located at the ends of the bar. (Use g = 10 m/s² to simplify your calculations.) It's easy to show that $T_1 + T_2 = Mg$ = 200 N. But this doesn't necessarily imply that $T_1 = T_2$.

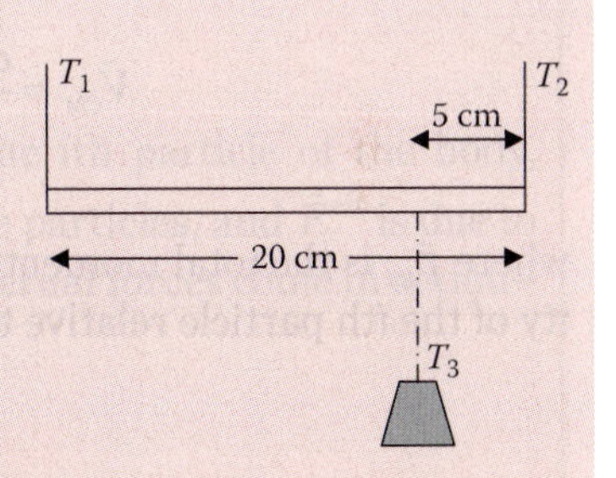

(a) Find the torques due to T_1, T_2, and Mg relative to the left end of the bar. What does this tell you about T_1 and T_2? Repeat for the right end of the bar.
(b) A 10 kg mass is now hung 5 cm from the right end of the bar. If the length of the bar is 20 cm, find the tensions T_1, T_2, and T_3.

Answer (b) T_1 = 125 N, T_2 = 175 N, T_3 = 100 N

13.4 RIGID BODY DYNAMICS

A rigid body moves by translation and rotation. We already know that Newton's second law governs the motion of the body's center of mass. Now we must derive an equation that describes the body's rotation. Recall that a body's kinetic energy splits neatly into two terms,

$$K = \frac{1}{2} M V_{CM}^2 + K',$$

where K' is the kinetic energy measured by an observer moving with the body's *CM*. For most cases of interest, a rotating solid body is spinning about an axis of symmetry passing through its *CM*. In these cases, $K' = \frac{1}{2} I\omega^2$, where I is the moment of inertia about the rotation axis. Like the kinetic energy, the angular momentum splits into two terms (see Box 13.1 for a proof):

$$\vec{L} = M\vec{R}_{CM} \times \vec{V}_{CM} + \vec{L}'. \tag{13.4}$$

Using the definition of the *CM* position, Equation 5.1, the second term simplifies to $M\vec{R}_{CM} \times \vec{V}_{CM}$, which is the angular momentum associated with the *CM* motion. Next, express the position in terms of *CM* variables, $\vec{r}_i = \vec{s}_i + \vec{R}_{CM}$, to obtain

$$\sum_i m_i \vec{r}_i \times \vec{u}_i = \sum_i m_i \vec{s}_i \times \vec{u}_i + \vec{R}_{CM} \times \left(\sum_i m_i \vec{u}_i \right)$$
$$= \sum_i m_i \vec{s}_i \times \vec{u}_i \equiv \vec{L}'.$$

where we have invoked $\sum_i m_i \vec{u}_i = 0$ for a third time. The remaining term $\vec{L}'$ is the angular momentum measured in the *CM* reference frame. For a rigid body, $\vec{L}'$ is the spin angular momentum. When the body's rotation axis coincides with an axis of symmetry, $\vec{L}' = I\vec{\omega}$. Collecting these results, the total angular momentum of the body is

$$\vec{L} = M\vec{R}_{CM} \times \vec{V}_{CM} + I\vec{\omega}. \tag{B13.3}$$

Using $\vec{L}' = I\vec{\omega}$ the rotational kinetic energy can be expressed as $K' = \frac{1}{2} I\omega^2 = L'^2/2I$.

Now we are ready to tackle torque. The total torque acting on the body is defined by Equation 13.2:

$$\vec{\tau}_{ext} = \sum_i \vec{r}_i \times \vec{F}_i^{ext}$$

which can be expressed using *CM* variables as

$$\vec{\tau}_{ext} = \sum_i (\vec{R}_{CM} + \vec{s}_i) \times \vec{F}_i^{ext} = \vec{R}_{CM} \times \vec{F}_{ext} + \sum_i \vec{s}_i \times \vec{F}_i^{ext}$$
$$= \vec{R}_{CM} \times \vec{F}_{ext} + \vec{\tau}',$$

where the second term is the torque measured by an observer located at the *CM*. Differentiating Equation B13.3 with respect to time,

$$\vec{\tau}_{ext} = \frac{d\vec{L}}{dt} = M\vec{R}_{CM} \times \vec{a}_{CM} + \frac{d\vec{L}'}{dt}.$$

Comparing these last two equations, and noting that $M\vec{R}_{CM} \times \vec{a}_{CM} = \vec{R}_{CM} \times \vec{F}_{ext}$, we are left with the equation of motion for the rotational motion of the body:

$$\vec{\tau}' = \frac{d\vec{L}'}{dt}. \tag{B13.4}$$

The rate of change of the rotational angular momentum of the body is equal to the torque measured relative to the body's *CM*. Summarizing all of the above results,

$$\begin{aligned}
&\text{(a)} \quad K = \frac{1}{2}MV_{CM}^2 + K' \\
&\text{(b)} \quad K' = \frac{1}{2}I\omega^2 = \frac{L'^2}{2I} \\
&\text{(c)} \quad \vec{F}_{ext} = M\vec{a}_{CM} \\
&\text{(d)} \quad \vec{L} = M\vec{R}_{CM} \times \vec{V}_{CM} + \vec{L}' \\
&\text{(e)} \quad \vec{\tau}' = \frac{d\vec{L}'}{dt}
\end{aligned} \tag{B13.5}$$

The first term is due to the *CM* motion and is what you would calculate if the body were a point mass. For a rigid body, the second term is the spin angular momentum $\vec{L}' = I\vec{\omega}$ measured relative to the *CM*.

EXERCISE 13.5

A ball of mass m and radius r rolls with speed v along a horizontal surface. The moment of inertia of the ball, relative to its *CM*, is $I = \frac{2}{5}mr^2$.

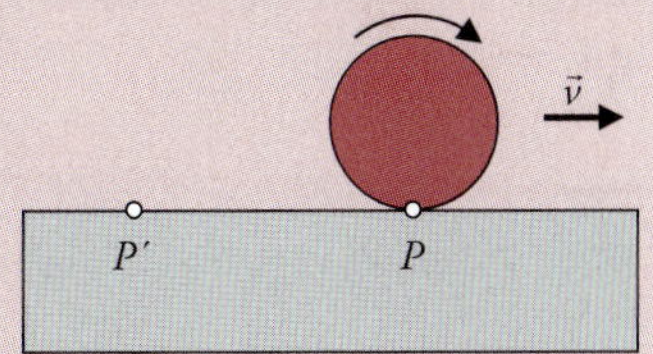

(a) Relative to the instantaneous point of contact P, what is the angular momentum associated with the *CM* motion?
(b) What is the angular momentum $\vec{L}'$ due to rotation?
(c) What is the total angular momentum relative to P?
(d) What is the total angular momentum relative to point P'?

Answer (a) mrv into page (b) $\frac{2}{5}$ *mrv into page (c)* $\frac{7}{5}$ *mrv (d) same as (c)*

Differentiating Equation 13.4,

$$\vec{\tau}_{ext} = \frac{d\vec{L}}{dt} = M\vec{R}_{CM} \times \vec{a}_{CM} + \frac{d\vec{L}'}{dt}. \tag{13.5}$$

In general, $\vec{\tau}_{ext}$ is due to one or more external forces $\vec{F}_n^{ext}$ applied to the body at various locations $\vec{r}_n$: $\vec{\tau}_{ext} = \sum_n \vec{r}_n \times \vec{F}_n^{ext}$ (see Figure 13.4). Switching to *CM* variables, $\vec{r}_n = \vec{R}_{CM} + \vec{s}_n$, where $\vec{s}_n$ is the vector from the *CM* to the point where $\vec{F}_n^{ext}$ is applied:

$$\begin{aligned} \vec{\tau}_{ext} = \sum_n \vec{r}_n \times \vec{F}_n^{ext} &= \sum_n \vec{R}_{CM} \times \vec{F}_n^{ext} + \sum_n \vec{s}_n \times \vec{F}_n^{ext} \\ &= \vec{R}_{CM} \times \vec{F}_{ext} + \sum_n \vec{s}_n \times \vec{F}_n^{ext} \\ &= M\vec{R}_{CM} \times \vec{a}_{CM} + \sum_n \vec{s}_n \times \vec{F}_n^{ext}, \end{aligned} \tag{13.6}$$

where $\vec{F}_{ext}$ is the total external force acting on the body, and the second term is the total torque $\vec{\tau}'$ *measured relative to the body's CM.* Comparing Equations 13.5 and 13.6,

$$\frac{d\vec{L}'}{dt} = \sum_n \vec{s}_n \times \vec{F}_n^{ext} = \vec{\tau}'. \tag{13.7}$$

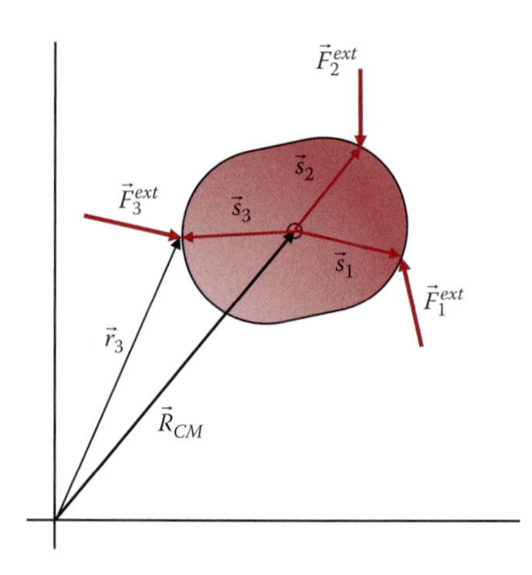

FIGURE 13.4 Three external forces are applied to a body at locations $\vec{r}_1 \ldots \vec{r}_3$. Relative to the body's *CM*, these forces are applied at locations $\vec{s}_1 \ldots \vec{s}_3$, and the torque $\vec{\tau}' = \vec{s}_1 \times \vec{F}_1^{ext} + \cdots + \vec{s}_3 \times \vec{F}_3^{ext}$.

The rate of change of the body's spin angular momentum is equal to the torque measured relative to its *CM*. Combined with Newton's second law, Equation 13.7 describes the full motion of a rigid body.

EXERCISE 13.6

Three forces of equal magnitude F act in the directions shown on a thin stick of length ℓ and mass m. One of the three forces acts at the center of the stick.

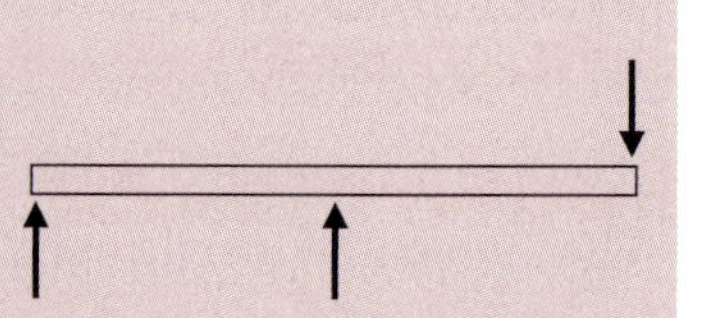

(a) In terms of m, ℓ, and F, describe the motion of the stick.
(b) Describe the stick's motion if the force on the right end were removed.

Answer (a) $a_{CM} = F/m \uparrow$, $\alpha = 12\ F/m\ell$ into page, rotating about the *CM*
(b) $a_{CM} = 2F/m \uparrow$, $\alpha = 6\ F/m\ell$ into page, rotating about the *CM*

When the only force acting on the body is $M\vec{g}$, $\vec{\tau}' = 0$ because the force effectively acts at the body's *CM*. In this case, $\vec{L}' = \text{const}$ and the rotation of the body about its *CM* is time-independent.

EXERCISE 13.7

A cheerleader's baton is tossed into the air, spinning while in flight. Describe its (a) *CM* trajectory and its (b) rotation.

Answer (a) *CM follows a parabolic path* (b) $\vec{\omega}$ is constant

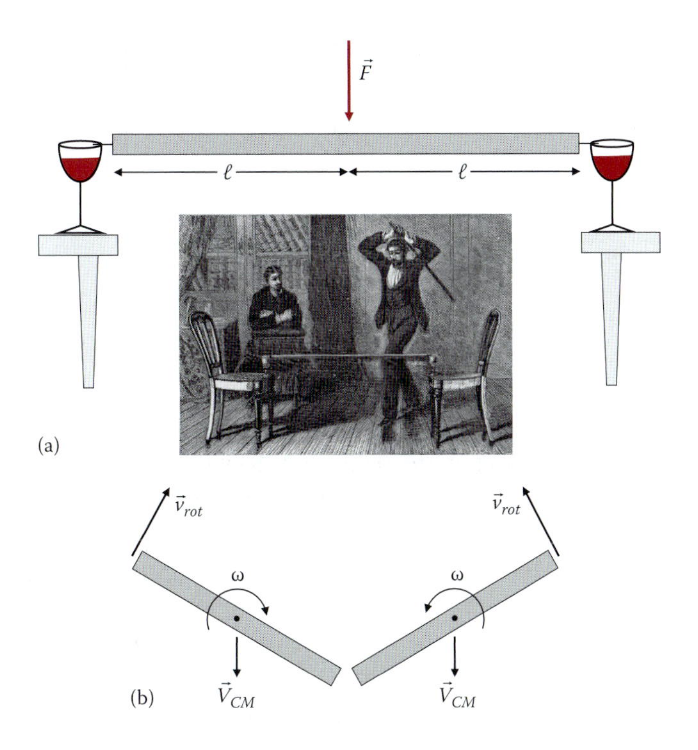

FIGURE 13.5 (a) In the "breaking broomstick" demonstration, a long thin stick is mounted on two wine glasses and given a sharp blow at its center, splitting it into two pieces. (b) The resulting motion of each piece is a superposition of translation and rotation about the *CM* of the piece. (From G. Tissandier, *Popular Scientific Recreations in Natural Philosophy*, 1881, public domain.)

The "breaking broomstick" demonstration is a terrific illustration of these ideas. It originated as a nineteenth century parlor trick, but today it is often performed by physics instructors to provide some relief from mathematical derivations. A straight pin is inserted into each end of a long stick, which is mounted horizontally by resting each pin on the rim of a delicate wine glass. See Figure 13.5a. (If your instructor is very brave, he or she may fill the wine glasses!) Using a second stick, the instructor forcefully strikes the first at its midpoint, snapping it into two equal pieces which crash to the floor. Amazingly, the wine glasses are undisturbed. How is this possible?*

The stick is struck by a strong force $\vec{F}$ of short duration Δt, splitting it into two pieces, each of length ℓ, mass m, and moment of inertia $I = \frac{1}{12} m\ell^2$ about its *CM*. Immediately after the blow, the *CM* of each piece is moving *downward* with velocity given by Newton's second law: $mV_{CM} = ma_{CM}\,\Delta t = F\Delta t/2$, where we have assumed that one half of the impulse is absorbed by each half of the stick† (Figure 13.5b). Each piece also receives an

* The demonstration is described well by K.C. Mamola and J.T. Pollack, in *The Physics Teacher* 31, 230–233 (1993). A complication is discussed by G. Vandergrift, in *American Journal of Physics* 65, 505–510 (1997).

† When a force $\vec{F}$ is exerted for a time Δt, the change in momentum is $\Delta\vec{p} = m\Delta\vec{v} = m\vec{a}\Delta t = \vec{F}\Delta t$. $\vec{F}\Delta t$ is called the *impulse* delivered by the force.

impulsive torque $\tau\Delta t = (F/2)(\ell/2)\Delta t$ that causes it to rotate about its *CM*. The left half rotates clockwise with angular momentum $\Delta L' = \tau\Delta t = I\omega$, or $I\omega = (F/2)(\ell/2)\Delta t$, yielding a value for the angular speed ω:

$$\omega = \frac{F\ell\Delta t/4}{m\ell^2/12} = \frac{3F\Delta t}{m\ell}.$$

The right half of the stick rotates CCW with the same angular speed ω. Therefore, the *rotational* velocity at the outer end of either stick (nearest the wineglass) is *upward*:

$$\begin{aligned} v_{rot} &= \frac{\omega\ell}{2} = \frac{3F\Delta t}{2m} \\ &= 3V_{cm}. \end{aligned}$$

The *total* velocity at either end of the stick, just after it snaps, is the vector sum $\vec{v} = \vec{V}_{CM} + \vec{v}_{rot}$, or $v = v_{rot} - V_{CM} = 2V_{CM}$ *upward*. Rather than smashing the wine glasses, the ends of the stick rotate harmlessly around them as the pieces fall to the floor! The wine glasses do not sense the blow and are undisturbed.

Torque provides a powerful new tool for solving dynamics problems. Consider the not-so-simple case of a ball rolling on an inclined plane (Figure 13.6a). The forces acting on the ball are gravity $\vec{F}_{grav} = m\vec{g}$, the normal force $\vec{N}$, and the friction force $\vec{F}_f$. Since we don't know $\vec{F}_f$, we can't use Newton's laws alone to determine the motion of the ball. But we can supplement them with torque and Equation 13.7 to solve for the motion. Relative to the ball's *CM*, $\vec{N}$ and $\vec{F}_{grav}$ produce no torque (Figure 13.6b). Only the friction force produces a torque about the *CM*: $\vec{\tau}' = \vec{s} \times \vec{F}_f = I\vec{\alpha}$, where $\vec{s}$ is the vector from the *CM* to point *P* where the ball contacts the ramp surface and $s = r$ is the radius of the ball. Let's guess that $\vec{F}_f$ points up the slope.* Then, $\vec{\tau}'$ and $\vec{\alpha}$ are directed into the page. Newton's second law and Equation 13.7 yield

$$ma_{CM} = mg\sin\theta - F_f$$

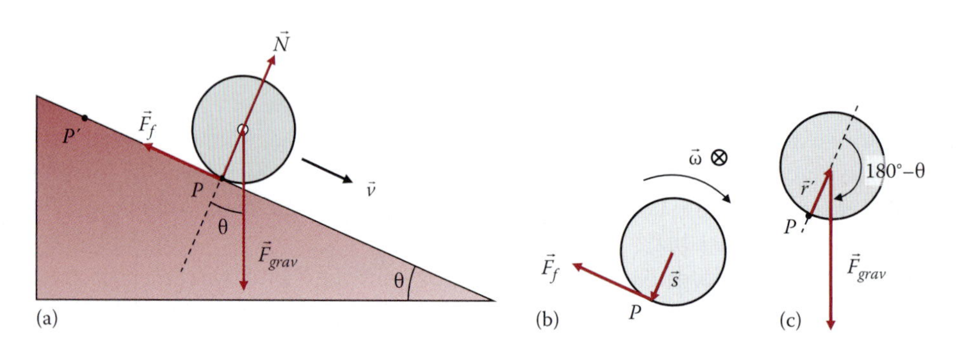

FIGURE 13.6 (a) A ball of radius *r* and mass *m* rolls down a ramp. Relative to the ball's *CM*, only $\vec{F}_f$ produces a torque on the ball. (b) This torque has a magnitude $\tau' = sF_f$ and, by the RHR, is directed into the page. (c) Relative to point *P*, only the gravitational force produces a torque: $\tau' = rmg\sin(\pi - \theta) = rmg\sin\theta$.

* If we guess the direction of $\vec{F}_f$ incorrectly, we will still arrive at the right answers. All that will happen is F_f will be a negative quantity.

and

$$I\alpha = sF_f$$

where $I = \frac{2}{5}mr^2$ is the moment of inertia of the ball about its *CM*. If the ball is rolling without slipping, $\alpha = a_{CM}/r$. Solving the above equations for a_{CM},

$$a_{CM} = \frac{g\sin\theta}{1 + I/mr^2} = \frac{g\sin\theta}{1 + \frac{2}{5}} = \frac{5}{7}g\sin\theta.$$

EXERCISE 13.8

(a) Find the magnitude of the friction force. (b) If we replaced the ball with a solid disk, what would be the magnitude of the friction force?

Answer (a) $F_f = \frac{2}{7}mg\sin\theta$ (b) $F_f = \frac{1}{3}mg\sin\theta$

We could have solved this problem by calculating the torque and angular momentum about a point other than the ball's *CM*. We might have chosen our reference point to be *P* or any other point *P'* on the surface of the ramp. Relative to *P*, the forces $\vec{N}$ and $\vec{F}_f$ produce no torque (Figure 13.6c). The only torque about *P* is due to the gravitational force: $\tau = |\vec{r}' \times m\vec{g}| = rmg\sin\theta$, where $\vec{r}'$ is the vector from *P* to the ball's *CM* ($\vec{r}' = -\vec{s}$). The ball's total angular momentum (relative to *P*) is given by Equation 13.4: $L = mrv + I\omega$ directed into the page. With $I = \frac{2}{5}mr^2$ and $v = \omega r$, $L = mrv + \frac{2}{5}mrv = \frac{7}{5}mrv$. Using $\vec{\tau} = d\vec{L}/dt$, $rmg\sin\theta = \frac{7}{5}mra_{CM}$, or $a_{CM} = \frac{5}{7}g\sin\theta$. The two approaches (*CM* frame and lab frame) yield the same answer, as they must. Try using the second strategy to solve the following problem.

EXERCISE 13.9

A bowling ball of radius r and mass m is launched parallel to the alley surface with speed v and negligible rotation. Friction with the surface causes the ball to skid and eventually roll without slipping on its way to the pins.

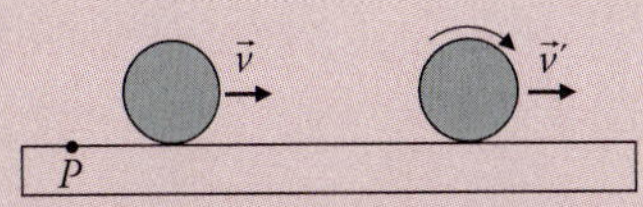

(a) Find the total torque acting on the ball relative to a point *P* on the floor surface.
(b) Find the angular momentum of the ball, relative to *P*, before it starts to spin.
(c) What is the final speed of the ball?
(d) Can you determine the friction force from the information given?

Answer (a) 0 (b) $L = mrv$ into paper (c) $v' = \frac{5}{7}v$, (d) no

13.5 APPLICATION: THE "PHYSICAL" PENDULUM

In Chapter 7, we studied the oscillation of a simple pendulum, which was represented by a point mass m hanging from a massless string of length ℓ. For small deflections, the pendulum's period was found to be

$$T = 2\pi\sqrt{\frac{\ell}{g}}. \tag{7.16}$$

Needless to say, real pendulum bobs are not pointlike and real strings are not massless. Equation 7.16 is only an approximate description of how a real pendulum behaves.

A pendulum whose mass is not concentrated at a single point, but is distributed in space, is commonly called a "physical" pendulum. Imagine a solid body of arbitrary shape that swings without friction about a fixed pivot A, as shown in Figure 13.7a. Denote the body's total mass by M, its moment of inertia relative

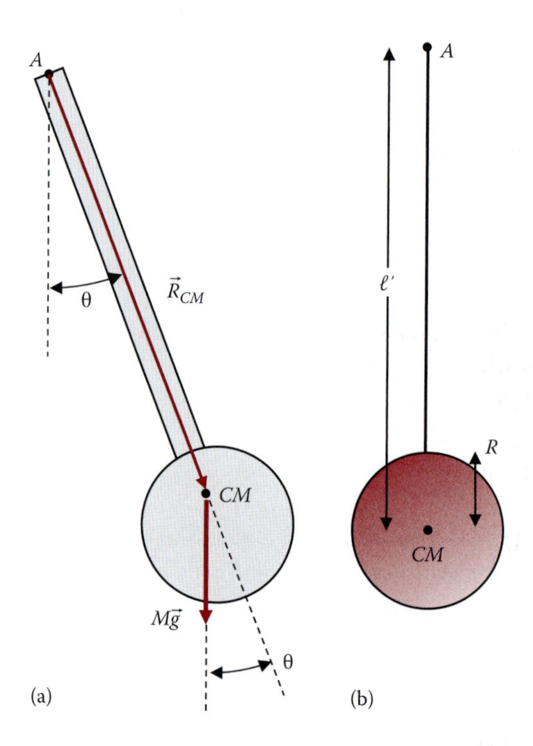

FIGURE 13.7 A physical pendulum pivots about point A. Relative to A, the torque on the pendulum is due to gravity, which acts effectively at the body's CM: $\vec{\tau} = \vec{R}_{CM} \times M\vec{g}$. For $\theta > 0$, $\vec{\tau}$ points into the page, and for $\theta < 0$, it points out of the page. (b) A physical pendulum made from a solid sphere of radius R suspended by a lightweight string. The sphere's CM is ℓ' from the top end of the string (point A).

to A by I_A, and let $\vec{R}_{CM}$ be the vector pointing from A to the body's CM. Relative to the pivot point, the only torque exerted on the body is due to gravity, which effectively acts at the CM. Thus,

$$\frac{d\vec{L}}{dt} = I_A\vec{\alpha} = \vec{\tau}_A = \vec{R}_{CM} \times M\vec{g}.$$

Both $\vec{L}$ and $\vec{\tau}_A$ are perpendicular to the plane of the figure. Let θ be the angle between $\vec{R}_{CM}$ and the vertical, and choose the positive direction of $\vec{L}$ and $\vec{\tau}_A$ to point out of the page in Figure 13.7. According to the RHR, when $\theta > 0$, $\vec{\tau}_A = \vec{R}_{CM} \times M\vec{g}$ points into the page, that is, in the negative direction. When $\theta < 0$, $\vec{\tau}_A$ points out of the page (positive direction). Consequently, the following scalar equation is valid for all values of θ:

$$\frac{dL}{dt} = I_A\alpha = -MgR_{CM}\sin\theta$$

or with $\alpha = d^2\theta/dt^2$ and $\theta \ll 1$,

$$\frac{d^2\theta}{dt^2} \simeq -\frac{MgR_{CM}}{I_A}\theta.$$

This is the equation for simple harmonic motion (see Equation 7.12), with solution $\theta(t) = \theta_0\cos(\omega t)$ and $\omega = \sqrt{MgR_{CM}/I_A}$. The oscillation period is

$$T = \frac{2\pi}{\omega} = 2\pi\sqrt{\frac{I_A}{MgR_{CM}}} \equiv 2\pi\sqrt{\frac{\ell_{eff}}{g}}, \tag{13.8}$$

where the "effective length" ℓ_{eff} is defined as the length of a simple (point mass) pendulum with the same period as the real physical pendulum.

For example, a simple pendulum having a string length $\ell = 1$ m has a period of 2.0 s. Suppose we wish to make a clock regulated by a physical pendulum that consists of a solid sphere of radius R hanging from a massless string. Let ℓ' be the distance from the sphere's CM to the string's support point. What length ℓ' should we choose so that our pendulum has a period of 2.0 s? (see Figure 13.7b). Using the parallel axis theorem, $I_A = I_{CM} + M\ell'^2 = \frac{2}{5}MR^2 + M\ell'^2$. The period is found from Equation 13.8, with $R_{CM} = \ell'$:

$$T = 2\pi\sqrt{\frac{\ell'}{g}\left(1 + \frac{2}{5}\frac{R^2}{\ell'^2}\right)} \equiv 2\pi\sqrt{\frac{\ell_{eff}}{g}}$$

where $\ell_{eff} = 1$ m is the length of a simple pendulum with period 2.0 s. Solving for ℓ',

$$\ell' = \frac{1}{1 + (2/5)(R^2/\ell'^2)}\text{(m)}.$$

If the sphere's radius is $R = 10$ cm, then $\ell' = 0.996\text{ m} = 99.6\text{ cm}$. This small correction (4 mm) is important, because without it, the pendulum's period would be 0.2% too long, and a clock regulated by this pendulum would lose nearly 3 min/day.

EXERCISE 13.10

(a) Derive an equation for the period of oscillation of a long thin stick of length ℓ that swings from a pivot point A at its top end.
(b) What is its effective length?
(c) What would be the stick's period if it were swung from an axle through its CM?

Answer: (a) $T = 2\pi\sqrt{2\ell/3g}$ (b) $\ell_{eff} = \frac{2}{3}\ell$

Pendula have played important roles throughout the development of physics and astronomy. In the early seventeenth century, Galileo studied pendula extensively and used them to time his experiments on accelerated motion. In the 1650s, Huygens developed the first pendulum clocks, improving the accuracy of timekeepers from minutes per day to seconds per day. In 1672, Richer traveled to French Guiana, near the Equator, to assist in a measurement of the distance to the Sun. (The Cassini–Richer experiment was discussed in Chapter 1). To his surprise, he discovered that a pendulum clock that kept accurate time in Paris lost 2½ min/day in French Guiana. This implied that the value of g at the Equator is smaller than it is in Paris (latitude 49°N). Later, Newton used Richer's measurements to prove that, because of the Earth's rotation, the planet is an "oblate ellipsoid," fatter at the Equator than along the poles by about 1 part in 300.

In the seventeenth and eighteenth centuries, pendula were used to measure g, but their accuracy was limited by the poor metallurgy of that age: the density of alloys could not be controlled precisely. Consequently, the location of a pendulum's center of mass—and its effective length ℓ_{eff}—had to be found by experiment, a difficult task given the high accuracy required. This problem was circumvented in 1818 by a British sea captain, Henry Kater, borrowing an idea originally proposed by Christiaan Huygens in 1673.

Imagine an arbitrarily shaped physical pendulum of mass M that pivots about point A with an oscillation period T_A. If the pendulum is turned upside down, a second pivot point B can be found, by trial and error, with the same period: $T_A = T_B$. Assume that two points A and B are collinear with the pendulum's CM, whose exact position is otherwise unknown (see Figure 13.8a). From Equation 13.8,

$$\ell_{eff} = \frac{I_A}{Ma} = \frac{I_B}{Mb} \tag{13.9}$$

where a and b are the (unknown) distances from the CM to the pivot points A and B. I_A and I_B are related to the pendulum's moment of inertia about its CM by the parallel axis theorem:

$$I_A = I_{CM} + Ma^2 \quad \text{and} \quad I_B = I_{CM} + Mb^2. \tag{13.10}$$

With a little algebra, Equations 13.9 and 13.10 can be combined to obtain $I_{CM} = Mab$ and $\ell_{eff} = a + b$.

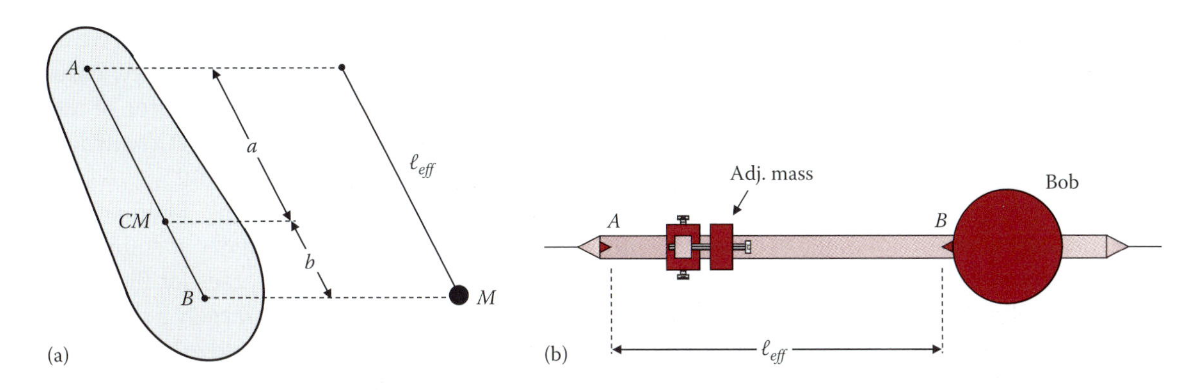

FIGURE 13.8 (a) An arbitrarily shaped physical pendulum. Its period of oscillation is the same when pivoted about points A and B, which are found by experiment. The distance between these two pivot points is the effective length of the pendulum: $\ell_{eff} = a + b$. (b) Kater's pendulum. Pivot points A and B are fixed, and the location of the CM is adjusted by sliding a small mass along the pendulum axis until $T_A = T_B$.

EXERCISE 13.11

(a) Prove these two results. (b) In Exercise 13.11, you found that the period of a thin stick pivoted at one end (point A) is $T = 2\pi\sqrt{2\ell/3g}$. What is ℓ_{eff} and how far is A from the stick's CM? Is there another pivot point on the stick (other than the opposite end) that will give you the same period of oscillation? If so, where is that point located?

Answer (b) *$\ell/6$ from the center of the stick*

The effective length of the pendulum is therefore equal to the distance between the two pivot points A and B, which can be measured with great precision. (For a "seconds" pendulum with $T = 2$ s, Kater measured $\ell_{eff} = 39.3708 \pm 0.0001$ in., and proposed this distance as a unit of length.) Kater's pendulum had the shape of a long slender bar pivoted on knife edges that were at fixed distance ($= \ell_{eff}$) apart. The location of the CM was finely adjusted (changing both a and b until $T_A = T_B$ precisely) by sliding a small mass along the bar (Figure 13.8b). A separate pendulum clock was calibrated to keep perfect time by synchronizing it with the motion of the stars overhead. Then the Kater pendulum (with precisely known ℓ_{eff}) was swung, and its period of oscillation measured. Using Equation 13.8, the local value of g could then be calculated.

Armed with accurate pendula, eighteenth and nineteenth century scientists and surveyors set out to measure g at locations around the globe to determine the exact shape of Earth.* The planet's shape was important for large-scale land surveys. Maps were made by triangulation. First, a large flat stretch of land was selected, and the straight-line distance between two widely separated points on this plateau was

* Pendula were also used in attempts to determine the density of Earth, for example, by lowering them into deep mine shafts and measuring the change in the period vs. depth.

measured precisely. From each end of this baseline, the angle to a distant third point was measured, allowing the surveyors to construct a triangle from the three points.* This triangle provided two new baselines from which new triangles were constructed, and the process was repeated until the entire area of interest was covered with interlocking triangles. For this strategy to work, the curvature of the Earth's surface had to be taken into account. This information was obtained by measuring g with a pendulum.

The Great Trigonometric Survey of India is perhaps the most famous of all triangulated surveys. It occupied British surveyors for much of the nineteenth century, and at the peak of activity, it employed over 700 persons. Four elephants were needed to transport the principal surveyors, 30 horses for the accompanying military personnel, and 42 camels for hauling supplies. Today, mapmaking is largely a desk activity, facilitated by imagery and data relayed to Earth from the global positioning system and other satellites.

* Do you remember that two angles and one side length, or two side lengths and one angle, are sufficient to construct a triangle?

13.6 EARTH–MOON SYSTEM

The Earth, Moon, and Sun interact strongly. But while the Sun controls the planet's orbit, it exerts zero average torque of the Earth–Moon system. (We are temporarily ignoring the 26,000 year precession of the Earth's rotation axis, to be discussed in Section 13.10.) The total angular momentum $\vec{L}_{tot}$ of the Earth and Moon is therefore constant. Relative to the center of the planet, $\vec{L}_{tot}$ has three contributions:

$$\vec{L}_{tot} = I_E\vec{\omega}_E + I_m\vec{\omega}_m + \vec{L}_{orb}. \tag{13.11}$$

The first term on the right hand side is the spin angular momentum of Earth, the second is the spin angular momentum of the Moon, and the last is the Moon's orbital angular momentum. For simplicity, neglect the tilt of the Earth's N–S axis, and assume that $\vec{\omega}_E$ and $\vec{\omega}_m$ are perpendicular to the plane of the Moon's orbit, that is, parallel to $\vec{L}_{orb}$. Then all of the vectors in Equation 13.11 are parallel, and we can dispense with vector notation. Figure 13.9 depicts the Moon in its orbit around Earth. Just as ocean tides on Earth are induced by the Moon's gravity (discussed in Chapter 8), the Moon is distorted by the planet's gravitational pull: tidal bulges arise on the near and far sides of the Moon. If the Moon were stationary and not spinning ($\vec{\omega}_m = 0$), the bulges would lie along line $\overline{EM}$ joining the centers of the Earth and Moon (Figure 13.9a). But due to the viscosity of the lunar mantle (imagine the Moon to be a ball of very thick goop), these bulges do not develop instantaneously; instead, they take shape over a time Δt. If the Moon were spinning but otherwise stationary (nonorbiting), the bulges would appear at an angle $\omega_m\Delta t$ from the line $\overline{EM}$ (Figure 13.9b). This creates a net torque on the Moon, which gradually slows its spin. Of course the real Moon is in a near-circular orbit with $T_m = 2\pi/\omega_{orb} = 27.3$ days. When $\omega_m = \omega_{orb}$, the bulge delay is matched to the orbital motion, so the tidal bulges remain aligned with $\overline{EM}$ as the Moon circles the Earth (Figure 13.9c). The tidal torque on the Moon then equals zero, so $\omega_m = \text{const} = \omega_{orb}$, and the same hemisphere of the Moon always faces the Earth. This is called *tidal locking* and is a common feature among the moons of the solar system.

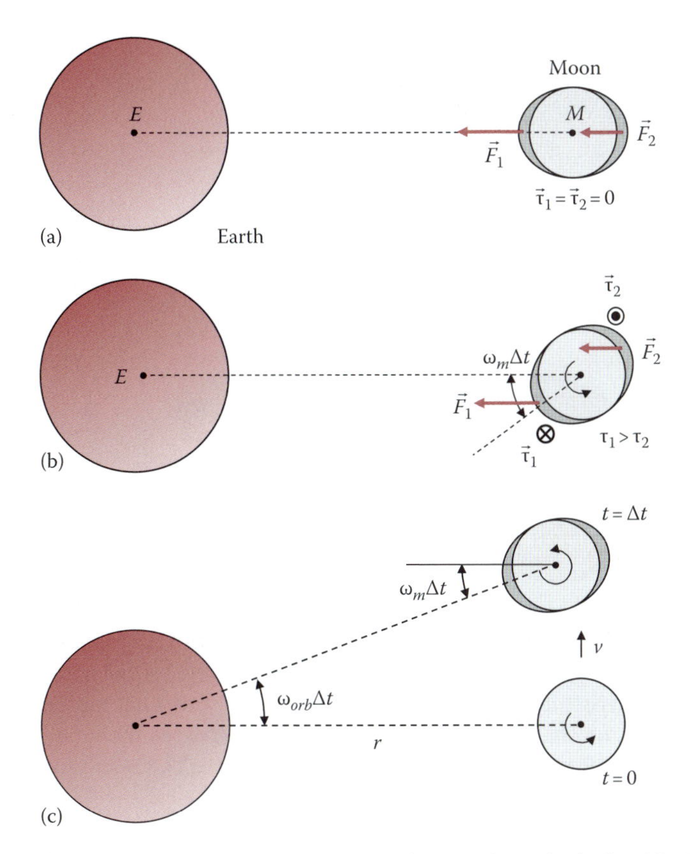

FIGURE 13.9 Tidal bulges slow the Moon's rotation rate until it is phase-locked with its orbital period. (a) For a stationary, nonrotating Moon, the two bulges are aligned with $\overline{EM}$ so the torque on the Moon is zero. (b) For a stationary, rotating Moon, the bulges appear at an angle $\omega_m \Delta t$ from line $\overline{EM}$ due to the viscosity of the Moon's mantle. This induces a torque ($F_1 > F_2$) which reduces the Moon's rotation rate. (c) An orbiting Moon: when $\omega_m = \omega_{orb}$, the tidal bulges are aligned with $\overline{EM}$ and the torque is zero.

EXERCISE 13.12

Show that tidal locking occurs whether $\omega_m > \omega_{orb}$ or $\omega_m < \omega_{orb}$ originally. (*Hint*: For each case, draw a picture like Figure 13.9c.)

Earth experiences similar tidal bulges because of the Moon's gravitational pull. Its tidal bulges are also delayed, giving rise to a torque on the planet which slows its spin: $d\omega_E/dt < 0$ (see Figure 13.10a). At the same time, the reaction forces on the Moon due to the Earth's tidal bulges give rise to a torque that increases the Moon's orbital angular momentum (Figure 13.10b). These two effects are linked by Equation 13.11. Since $I_m \ll I_E$, and $\omega_m \ll \omega_E$, we can safely ignore the second term on the right hand side of that equation. Let r and v be the radius and speed of the Moon's circular orbit, measured relative to the center of the planet, and

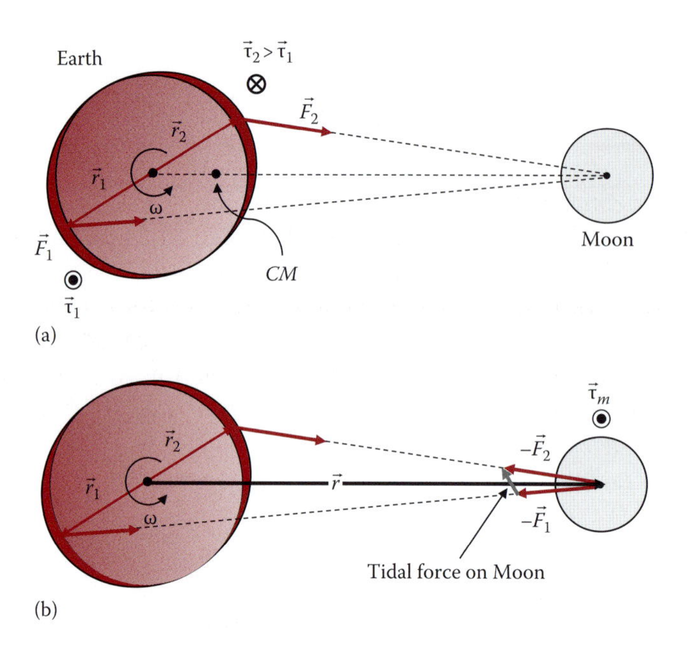

FIGURE 13.10 The Moon raises tidal bulges on Earth that are delayed by the viscosity of the planet's mantle. (a) Because the bulges are at different distances from the Moon, $F_2 > F_1$ and a net torque acts on the planet to slow its rotation. (b) The reaction forces on the Moon induce a torque that increases the lunar *orbital* angular momentum. The Earth–Moon distance $r \simeq 60R_E$ is not drawn to scale in the figure. The resultant force on the Moon is nearly tangential to its orbit.

let M_m be the Moon's mass. The Moon's orbital angular momentum is $L_{orb} = M_m vr$. From Newton's second law, $M_m v^2/r = GM_E M_m/r^2$, or

$$L_{orb} = \sqrt{M_m^2 v^2 r^2} = \sqrt{GM_E M_m^2 r} \equiv Ar^{1/2}, \tag{13.12}$$

where $A = \sqrt{GM_E M_m^2}$.

EXERCISE 13.13

Calculate L_{tot}, using $M_m = 7.35 \times 10^{22}$ kg and $r = 60R_E$, where R_E is the Earth's radius. The Earth's moment of inertia is $I_E = 0.33 M_E R_E^2$.

Answer 3.5×10^{34} J s

Invoking conservation of total angular momentum (Equation 13.11),

$$\frac{dL_{tot}}{dt} = I_E \frac{d\omega_E}{dt} + \frac{dL_{orb}}{dt} = 0$$

or, using Equation 13.12,

$$I_E \frac{d\omega_E}{dt} = -\frac{1}{2} A r^{-1/2} \frac{dr}{dt}. \tag{13.13}$$

In 1969 and 1972, Apollo astronauts deployed arrays of retroreflectors at four locations on the Moon's surface, enabling the Earth–Moon distance to be determined with about 1000 times more accuracy than ever before. The distance is found by *laser ranging*: a laser pulse is transmitted from Earth to the Moon, where its direction of propagation is reversed by one of the four retroreflector arrays. The reflected laser signal retraces its path back to Earth, arriving about 2.5 s after emission. The round-trip distance is found by multiplying the time delay by the speed of light. Measurements taken over the past four decades indicate that the Moon is currently receding from Earth at a rate dr/dt = 3.8 cm/year. From Equation 13.13, this proves that $d\omega_E/dt < 0$, as expected from tidal effects. The days (on Earth) *are* getting longer.

We can now derive a relation between the Earth's rotation rate ω_E and the Moon's orbital radius r. Multiply Equation 13.13 by dt and integrate both sides over time:

$$\begin{aligned}\omega_E(t_2) - \omega_E(t_1) &= \int_{t_1}^{t_2} \frac{d\omega_E}{dt} dt = -\frac{A}{2I_E} \int_{t_1}^{t_2} r^{-1/2} dr \\ &= \frac{A}{I_E}\left(\sqrt{r(t_1)} - \sqrt{r(t_2)}\right),\end{aligned}$$

where $\omega_E(t_i)$ and $r(t_i)$ are the Earth's rotation rate and distance from the Moon at time t_i. Let t_2 be the present time, so $r(t_2) = 60R_E$.* When the Moon was first formed, it was much closer to Earth, probably just beyond the Roche limit $R_{Roche} = 2.45R_E$ (see Chapter 8). How long was an Earth day at that time? From the above expression, with $\omega_E(t_i) = 2\pi/T_i$, where T_i is the Earth's spin rotation period at time t_i, we obtain

$$\frac{2\pi}{T_2} - \frac{2\pi}{T_1} \simeq \frac{AR_E^{1/2}}{I_E}\left(\sqrt{2.45} - \sqrt{60}\right).$$

Plugging in the numbers, a day on Earth was only about 5 h long during the early lifetime of the Moon!

Quite recently, scientists have questioned the assumption of constant Earth–Moon angular momentum as they strive to understand the origin of the Moon. When Apollo astronauts returned lunar rocks to Earth, geochemists discovered that they were igneous and depleted of volatile compounds such as water. This suggested that the Moon was created following a fiery collision with a giant asteroid, often called *Theia*. Incredibly, the lunar rocks were also found to be *isotopically identical* to material found on Earth. For example, the abundance ratios of oxygen isotopes $^{18}O:^{17}O:^{16}O$ and titanium isotopes $^{50}Ti:^{47}Ti$ are exactly the same for lunar rocks as for present-day Earth rocks. Computer simulations show that, using the present value of L_{tot} found in Exercise 13.13, Theia must have been the compositional twin of Earth—which is highly improbable. Otherwise, L_{tot} of the planet and asteroid before the collision must have been at least twice as

* This is the distance found by Hipparchus in 250 BCE. On the scale of billions of years, 250 BCE is just yesterday.

high as the present value. If this is true, where did the excess angular momentum go?* Whatever happened, angular momentum holds the key to the riddle of the Moon's origin.

* A recent theory suggests that the Sun drained it away in a complex exchange of angular momentum called an *evection resonance*. See A.N. Halliday, The origin of the moon, *Science* 338, 1040 (2012), and the papers cited by M. Ćuk and S.T. Stewart, and R.M. Canup, in that issue.

13.7 PHYSICS OF TOY TOPS

Everybody loves to spin a top. For thousands of years, children (and adults, too—see Figure 13.11a) have marveled at the gravity-defying antics of a spinning top. More than just a toy, it is a wonderful device for illustrating the counterintuitive physics of rigid body motion. Rotating bodies of all sizes are found

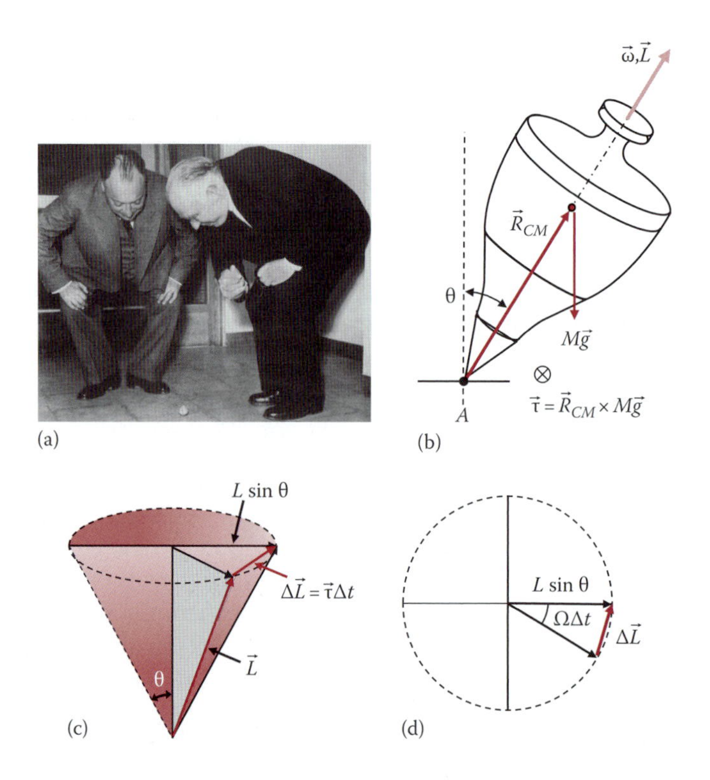

FIGURE 13.11 (a) Nobel laureates Wolfgang Pauli (left) and Niels Bohr spinning a "tippy top" in Lund, Sweden in 1954. The top flips upside down, apparently defying gravity, after it has spun for a few seconds. (b) The angular momentum $\vec{L}$ of a spinning top is directed along its central axis of symmetry. Relative to point A, where the top touches the floor, the torque is due solely to gravity and equals $\vec{\tau} = \vec{R}_{CM} \times M\vec{g}$. It is directed into the page. (c) Due to precession, the tip of vector $\vec{L}$ traces out a circle of radius $L\sin\theta$. (d) Top view of (c). In a short time Δt, $\vec{L}$ precesses through an angle $\Omega\Delta t$, and $\Delta L = L\sin\theta\Omega\Delta t$. (Photo courtesy of AIP Emilio Segre Visual Archives Margrethe Bohr Collection.)

throughout the universe, from nuclei and atoms to planets and stars, and their behavior has much in common with the motion of a simple toy top.

Consider the top shown in Figure 13.11b. Its *CM* lies along its axis of symmetry at a distance R_{CM} from point A where its tip touches the floor. The total angular momentum of the top, relative to A, is given by Equation 13.4: $\vec{L} = M\vec{R}_{CM} \times \vec{V}_{CM} + I\vec{\omega}$. If ω is large enough (to be discussed below), we can safely ignore the first term and set $\vec{L} \simeq I\vec{\omega}$. In this case, $\vec{L}$ is parallel to $\vec{R}_{CM}$ and points along the top's axis of symmetry. If the top is tilted from the vertical by angle θ, it experiences a torque about A due to the gravitational force acting (effectively) at its *CM*:

$$\vec{\tau} = \vec{R}_{CM} \times M\vec{g}. \tag{13.14}$$

At the instant shown in Figure 13.11b, $\vec{\tau}$ points into the page. From Equation 13.14, $\vec{\tau}$ is always perpendicular to $\vec{R}_{CM}$, and so it is also perpendicular to $\vec{L}$. Since it has no component parallel to $\vec{L}$, the magnitudes of $\vec{L}$ and $\vec{\omega}$ are constant. Moreover, because $\vec{\tau}$ is perpendicular to $M\vec{g}$, it is always horizontal ($\tau_z = 0$). Therefore, the vertical component of the angular momentum $L_z = I\omega\cos\theta$ cannot change, so the tilt angle θ remains constant. Finally, because the torque magnitude is $\tau = MgR_{CM}\sin\theta$, it too remains constant.

While its magnitude remains constant, the *direction* of $\vec{L}$ changes continually because $\vec{\tau} \neq 0$. In a short time interval Δt, $\Delta\vec{L} = \vec{\tau}\Delta t$, so $\Delta\vec{L}$ must point in the direction of $\vec{\tau}$ (into the page). The axis of the top turns into the page along with the vector $\vec{L}$. As time advances, $\vec{\tau}$ remains perpendicular to $\vec{L}$, implying that $\vec{L}$ and the top's axis of symmetry sweep out an inverted cone, as shown in Figure 13.11c. This motion is called *precession*. As the top precesses, the *tip* of vector $\vec{L}$ traces out a circle of radius $L\sin\theta$ (Figure 13.11d). Let T denote the time for $\vec{L}$ to rotate through 2π radians. The *precession rate* Ω is defined in the usual way: $\Omega \equiv 2\pi/T$. In an infinitesimal time interval Δt, $\vec{L}$ rotates through an angle $\Omega\Delta t$, so $\Delta L = L\sin\theta\Omega\Delta t$. But $\Delta L = \tau\Delta t$, so

$$L\sin\theta\Omega\Delta t = MgR_{CM}\sin\theta\Delta t \quad \text{or} \quad \Omega = \frac{MgR_{CM}}{I\omega}. \tag{13.15}$$

The precession rate Ω is independent of the tilt angle θ, and decreases as the spin rate ω increases. This justifies our neglect of the *CM* motion in the discussion above. The *CM* speed $V_{CM} = \Omega R_{CM}\sin\theta \propto 1/\omega$, so as ω increases, the angular momentum due to *CM* motion becomes smaller.

EXERCISE 13.14

(a) Show that, using Equation 13.15, the torque on the top can be written very concisely in vector form as $\vec{\tau} = d\vec{L}/dt = \vec{\Omega} \times \vec{L}$ where $\vec{\Omega}$ is the precession angular velocity (pointing vertically upward). (b) Suppose the top were spun in the opposite sense, so that $\vec{\omega}$ and $\vec{L}$ point along the symmetry axis *toward* point A. Prove that the direction of precession is reversed.

For smaller ω, the *CM* motion cannot be neglected, and the behavior of the top is more complicated—and more interesting! Instead of precessing smoothly, the top dips and rises as it turns, and the tip of vector $\vec{L}$ traces out a scalloped circular path like the one shown in Figure 13.12b. This behavior is called *nutation*.

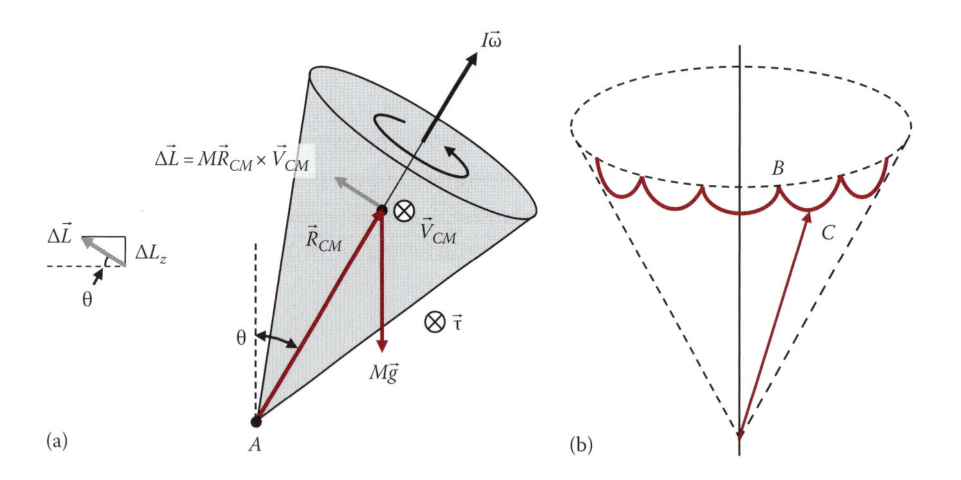

FIGURE 13.12 (a) When a top precesses, its *CM* motion makes a contribution $\Delta\vec{L} = M\vec{R}_{CM} \times \vec{V}_{CM}$ (gray arrow) to the total angular momentum. The *z*-component of $\vec{L}$ is $L_z = I\omega\cos\theta + MR_{CM}V_{CM}\sin\theta$. Since $\tau_z = 0$, L_z is constant, the top's tilt angle θ must increase to compensate for ΔL_z. (b) As the top precesses, the tilt angle θ bobs up and down, or *nutates*.

While a full mathematical treatment of nutation is quite challenging, its origin is not hard to understand. Imagine that you have set the top spinning at a tilt angle θ_0, but are holding it so it cannot precess. When you release the top, $\vec{V}_{CM}$ is initially zero and $L_z = I\omega\cos\theta_0$. Once precession is underway, the motion of the *CM* contributes to the total angular momentum. The *CM* contribution to the vertical angular momentum is $\Delta L_z = MR_{CM}V_{CM}\sin\theta$ (see Figure 13.12a). Since $\tau_z = 0$, the total L_z must be conserved: $I\omega\cos\theta_0 = I\omega\cos\theta(t) + \Delta L_z(t)$. The top's axis of rotation dips away from the vertical direction, decreasing the value of $\cos\theta$ as V_{CM} increases. Just like any oscillator, the top overshoots its "equilibrium" orientation, and bobs up and down as it precesses. (Alternatively, nutation can be understood using energy conservation. The kinetic energy needed for precession, $K = \frac{1}{2}MV_{CM}^2$, comes at the expense of the potential energy $U = MgR_{CM}\cos\theta$.) For ordinary tops, nutation typically dies out after a few oscillations. But bigger "tops" like the Earth have been precessing and nutating for billions of years. Earth's top-like behavior was explained by Isaac Newton in the *Principia*. The planet's precession (to be discussed in Section 13.10) was known since antiquity, while its tiny nutation ($\Delta\theta \approx 0.0025°$) was only discovered in 1748 by the British astronomer James Bradley.

13.8 GYROSCOPES

The gyroscope is an ingenious adaptation of a toy top. In its traditional configuration (Figure 13.13), it consists of a heavy flywheel cradled within two nested rings called gimbals. The flywheel spins on a pair of low-friction bearings (*A* and *A*′ in the figure) mounted on the inner ring. This ring swivels on another pair of bearings (*B* and *B*′) seated in the outer ring, which itself turns within vertical bearings *C* and *C*′ embedded in a rigid frame. The flywheel rotation axis $\overline{AA'}$ is perpendicular to the bearing axis $\overline{BB'}$, which in turn is perpendicular to the vertical axis $\overline{CC'}$. These three axes are aligned so they intersect at

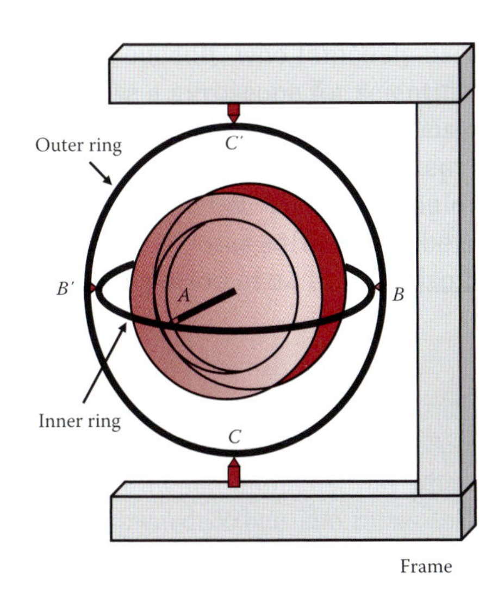

FIGURE 13.13 The classic gimbaled gyroscope. The outer ring turns on bearings C and C', the inner ring swivels on bearings B and B', and the flywheel (or rotor) spins on bearings A and A' (A' is hidden from view). The bearing axis $\overline{CC'}$ is perpendicular to axis $\overline{BB'}$, which in turn is perpendicular to axis $\overline{AA'}$. The three axes intersect at the rotor's center of mass.

the flywheel's *CM*. If the bearings were frictionless, the spinning rotor would be perfectly isolated from all external torques, including the torque due to gravity. The orientation of the rotor axis would then remain fixed in space, no matter what forces or torques were applied to the frame. Real bearings, of course, are not friction free. The small frictional torques transmitted by the bearings cause a slow precession of the rotor at a rate inversely proportional to its spin angular momentum $I\omega$, just like the top's motion in Equation 13.15. If $I\omega$ is large, this drift rate is very slow, so the spin axis keeps its orientation in space over long periods of time. This remarkable property makes the gyroscope a valuable instrument for navigation and science.

Let's see how it works. Denote the flywheel's spin angular momentum by $\vec{L} = I\vec{\omega}$, and consider only the flywheel and the inner ring. If bearings A and A' are frictionless, they offer no resistance to rotation about axis $\overline{AA'}$, so L and ω remain constant. In other words, the two bearings *exert no torque parallel to* $\overline{AA'}$.* Likewise, bearings B and B' cannot transmit a torque parallel to $\overline{BB'}$ nor can C and C' exert a torque in the direction of $\overline{CC'}$. If we cannot change the magnitude of $\vec{L}$, can we change its direction by turning the frame? The gyroscope shown in Figure 13.14 is positioned with $\overline{BB'}$ and $\overline{CC'}$ aligned with the y- and z-axes of the coordinate system. Let's apply a torque $\vec{\tau}$ to the frame and see how it affects $\vec{L}$. An applied torque $\vec{\tau} = \tau_z \hat{k}$ will rotate the frame about the z-axis. But the inner ring and flywheel are isolated from this rotation by bearings C and C': τ_z will turn the frame but not the outer ring. Therefore, L_z = const. An applied torque $\vec{\tau} = \tau_y \hat{j}$ will rotate the frame about the y-axis. However, bearings B and B' decouple the inner ring and flywheel from this rotation. Consequently, L_y = const. We already know that the magnitude $L = I\omega$ is a constant. Since $L^2 = L_x^2 + L_y^2 + L_z^2$, and L, L_y, and L_z cannot change, the remaining component L_x must also be constant, so the flywheel angular momentum $\vec{L}$ must be fixed in direction as well as magnitude. This result is true regardless of how we define our coordinate axes, and it remains true even if the gimbals are not perpendicular to each other. (Can you think of a gimbal configuration where it is not true?)

Now imagine a spinning gyroscope located on the Earth's equator, with its spin angular momentum $\vec{L}$ lying in the (local) horizontal plane pointing due east (see Figure 13.15a). As the Earth turns, $\vec{L}$ remains constant relative to the inertial reference frame of the stars. Six hours later, an observer stationed next to the gyroscope will find that $\vec{L}$ is pointing vertically and after another 6 h that it points westward, etc. Relative to this observer, the gyroscope appears to rotate 360° in 24 h, turning in the opposite sense of the Earth.

* It might help to think of it in another way: If the flywheel were initially stationary, you could not induce it to spin about $\overline{AA'}$ no matter how hard you pushed, pulled, or rotated the ring.

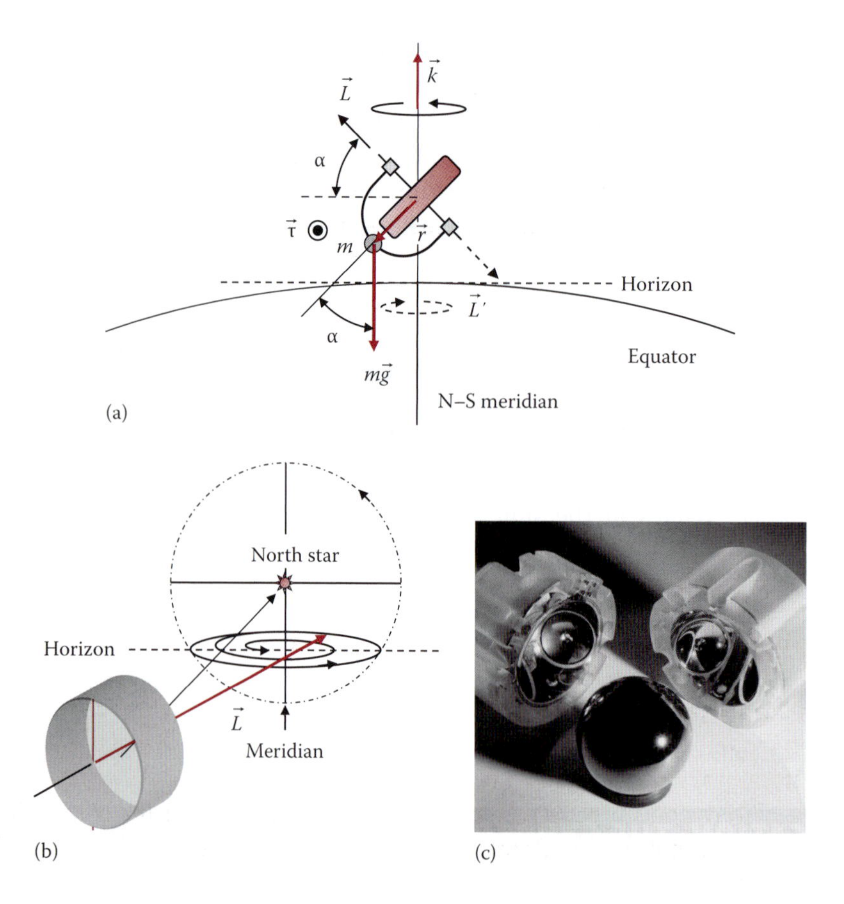

FIGURE 13.16 In a gyrocompass, the gyroscope is deliberately unbalanced by the addition of mass m, inducing a torque directed out of the page. (a) Whether the angular momentum is pointed above the horizon ($\vec{L}$) or below the horizon ($\vec{L}'$), the vector turns out of the page toward the north. (b) Without the added mass m, $\vec{L}$ would swing in a full circle once per day. With mass m, $\vec{L}$ moves in a highly eccentric ellipse centered on the N–S meridian and the plane of the horizon. Frictional damping reduces the size of the ellipse until $\vec{L}$ points steadily along the meridian. (c) The precision gyroscope used in the Gravity Probe B test of the general theory of relativity. The quartz sphere has a radius 1.9 cm, accurate to about ±20 atoms. (Photo courtesy of Gravity Probe B Media Archive, Stanford University, Stanford, CA. With permission.)

turn westerly, but now the Earth's rotation Ω *decreases* α, and both α and τ are equal to zero when $\vec{L}$ is facing due west. Subsequently, the Earth's spin drives $\vec{L}$ below the horizon ($\alpha < 0$) so the torque now reverses direction, driving $\vec{L}$ northward, crossing the meridian and halting briefly when it is once again facing east. This east–west oscillation of $\vec{L}$ across the N–S meridian would persist indefinitely, making the instrument useless as a compass, except that frictional *damping* is introduced to reduce the oscillation amplitude to a small fraction of a degree in a few oscillations.

Today, precision gyroscopes are used extensively for navigation of ships, submarines, aircraft, ballistic missiles, and even spacecraft (e.g., the Hubble Space Telescope). The most precise gyroscopes ever built were designed for the space-based Gravity Probe B experiment. Launched in 2004, the experiment was designed to test certain predictions of Einstein's general theory of relativity. According to the theory, a rotating body such as Earth drags space-time around with it. This so-called "inertial frame-dragging" effect is very small; nearly 30 million years are required for space-time to execute a complete revolution about the planet. If the gyroscopes carried by the GP-B spacecraft kept a fixed orientation relative to this slowly rotating reference frame, they would precess relative to the "fixed" stars at a rate 10^{-5} degree/year. The GP-B gyroscope rotors (Figure 13.16c) were near-perfect quartz spheres of radius $R = 1.9$ cm $\pm$ 8 nm (8 nm $\approx$ 20 atoms!). If left undisturbed in a zero-g orbital environment, they could remain spinning for more than 15,000 years. To test the frame-dragging hypothesis to an accuracy of 1%, the intrinsic drift of the rotors had to be $<10^{-7}$ degree/year. In spite of several glitches that reduced the accuracy of the experiment, the predictions of general relativity were found to be in good agreement with measurements.

13.10 PRECESSION OF THE EQUINOXES

We live on a gyroscope. There are no gimbals to support the spinning Earth, but the planet's near-circular orbit allows it to continually "fall" toward the Sun while maintaining a constant average distance from it. If Earth were a spherically symmetric body, the gravitational force due to the Sun would be the same as if all of the planet's mass were concentrated at its center. In this case, the torque due to the Sun, measured relative to the Earth's *CM*, would be exactly zero. The same would be true for the torque due to the Moon. Since these are the only astronomical bodies that exert significant force on Earth, the N–S rotation axis of a perfectly spherical Earth would maintain its orientation in space indefinitely, just like the ideal gyroscope discussed above. But because of its daily rotation, Earth's equatorial radius is about 21 km greater than its radius at the poles. This slight distortion, plus the tilt of the planet's N–S axis relative to the plane of its orbit, allow small torques to be exerted by the Sun and Moon. These torques induce a slow precession of the Earth's rotation axis, like the top in Figure 13.11, or the weighted gyroscope in Figure 13.16. As the axis changes direction, it turns away from Polaris, new stars assume the role of the "North Star," and the vernal and autumnal equinoxes—where the Sun crosses the celestial equator—drift slowly through the constellations of the zodiac.

With vectors and modern astronomical data at our disposal, we can derive an expression for the Earth's precession rate. For simplicity, model the ellipsoidal Earth as a sphere of mass M_E girded by a thin hoop of mass m wrapped around the Equator. See Figure 13.17a. From the shell theorem, we know that the gravitational torque on the *sphere* (relative to its *CM*), due to the Sun or Moon, is zero. Therefore, a calculation of the torque on the *ellipsoidal* planet reduces to finding the torque on the hoop m.

To find the torque, divide the hoop into four 90° sectors and treat each sector as a point particle of mass $m_i = m/4$ ($i = 1,\ldots,4$) located at its center. First consider just the torque due to the Moon. Let $\vec{F}_i$ be the

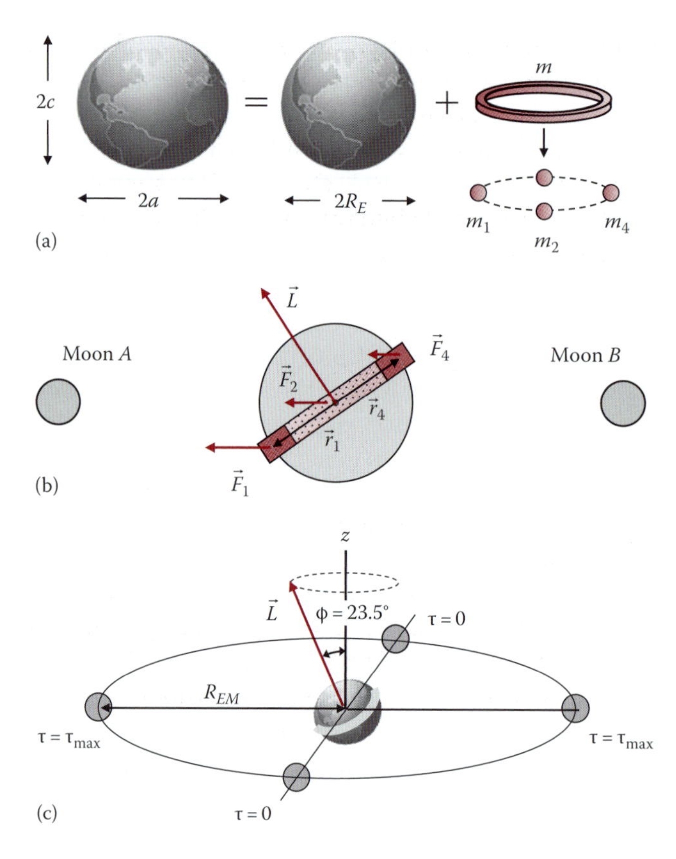

FIGURE 13.17 (a) Due to its rotation, Earth's equatorial radius a is greater than its polar radius c by about 1 part in 305 (21 km). The flattened planet can be modeled as a perfect sphere plus a close-fitting equatorial hoop. In the calculation of torque, the hoop is approximated by four identical, equally spaced masses. (b) The Earth's equator is tilted by 23.5° from the plane of the lunar orbit. When the Moon is at position A, $F_1 > F_4$, and the total torque on the planet is directed *into* the page. (c) The torque exerted on Earth is a maximum when the Moon is at positions A and B, and zero at the two points midway between A and B, so $\tau_{avg} = ½\tau_{max}$.

gravitational force on m_i due to the Moon and $\vec{r}_i$ be the vector from the Earth's core to m_i (Figure 13.17b). When the Moon is at position A, the force on each sector is

$$F_1 = \frac{GM_m m_1}{(R_{EM} - R_E \cos\phi)^2} \simeq \frac{GM_m m_1}{R_{EM}^2}\left(1 + 2\frac{R_E}{R_{EM}}\cos\phi\right),$$

$$F_4 = \frac{GM_m m_4}{(R_{EM} + R_E \cos\phi)^2} \simeq \frac{GM_m m_4}{R_{EM}^2}\left(1 - 2\frac{R_E}{R_{EM}}\cos\phi\right),$$

and $F_2 = F_3$ because m_2 and m_3 are equidistant from the Moon. In the above expressions, M_m is the mass of the Moon, R_E is the Earth's average radius, R_{EM} is the center-to-center distance between the two bodies, and $\phi = 23.5°$ is the tilt of the planet's rotation axis.

The torque acting on m_i is $\vec{\tau}_i = \vec{r}_i \times \vec{F}_i$. Since $\vec{r}_2 = -\vec{r}_3$, then $\vec{\tau}_2 = -\vec{\tau}_3$, or $\vec{\tau}_2 + \vec{\tau}_3 = 0$. The torque acting on m_1 is $\tau_1 = R_E F_1 \cos\phi$ pointing into the page, whereas $\tau_4 = R_E F_4 \cos\phi$ pointing out of the page. Since m_1 is closer to the Moon than m_4, $F_1 > F_4$ and $\vec{\tau} = \vec{\tau}_1 + \vec{\tau}_4$ points into the page of Figure 13.17b. (Verify this using the RHR.) The total torque on the planet, due to the Moon at position A, is, therefore,

$$\tau = \frac{GM_m m}{R_{EM}^3} R_E^2 \sin\phi \cos\phi,$$

where we have used $4m_i = m$. Because we have represented the hoop by four point masses, the above result is a slight underestimate. If we had done a proper integration about the tilted hoop, we would have arrived at the same result, but with a prefactor equal to 3/2.

When the Moon is located in front of or behind the planet, in line with m_2 and m_3, $\vec{\tau}_2 = \vec{\tau}_3 = 0$ and $\vec{\tau}_1 = -\vec{\tau}_4$, so the total torque on the hoop is zero. When the Moon is on the opposite side of Earth, at position B, $\vec{F}_4 > \vec{F}_1$, but once again $\vec{\tau}$ points into the page. (Check this using the RHR.) During one full orbit of the Moon, the torque oscillates between its maximum value $\tau_{\max}$ at positions A and B, and zero at the intermediate positions. In the time for the Earth's axis to precess by 1° (72 years), the Moon orbits the planet nearly 1000 times, so we are justified in treating the motion of Earth as a steady precession Ω due to a constant average torque $\tau_{avg} = ½\tau_{\max}$.

EXERCISE 13.16

(a) What is the torque on the Earth if its N–S axis is perpendicular to the plane of the Moon's orbit? (b) What is the torque if Earth's axis lies in the plane of the orbit? (c) At what angle ϕ is the torque a maximum?

Answer (c) $\pi/4$

The torque on Earth due to the Sun is evaluated in exactly the same manner. Since $M_m/R_m^3 \simeq 2M_S/R_S^3$, the average torque due to the Sun is about half as large as that due to the Moon.

As shown in Box 13.2, a good estimate of the hoop mass is $m = 0.0022M_E$. To find the precession rate Ω, use Equation 13.15, the expression derived earlier for the precession of a spinning top: $dL/dt = I\omega \sin\phi\, \Omega = \tau$. (Alternatively, use the vector expression $d\vec{L}/dt = \vec{\Omega} \times \vec{L} = \vec{\tau}$.) The Earth's moment of inertia about its rotation axis is $I = \frac{1}{3} M_E R_E^2$. Its rotation rate is $\omega = 2\pi/T_E$, where $T_E = 1$ day $= 8.64 \times 10^4$ s.

BOX 13.2 FINDING THE EQUIVALENT HOOP MASS m

In the *Principia*, Newton showed that the Earth's spin flattened the planet slightly, changing its shape from a sphere to an *oblate ellipsoid* whose equatorial radius a is greater than its polar radius c by about one part in 300, that is, $(a - c)/c \simeq 1/300$. See Figure 13.17a. The moment of inertia of an oblate ellipsoid of uniform density and total mass M_E about its N–S axis c can be found by integration:

$$I = \int r^2 dm = \frac{2}{5} M_E a^2,$$

the same equation as for a sphere of radius a. Because Earth's core has a higher density than its surface, its moment of inertia is slightly smaller:

$$I \simeq \frac{1}{3} M_E a^2.$$

In our calculations of torque, it is convenient to replace the ellipsoidal Earth by a spherical planet of radius $R_E = c$ plus a circular hoop of mass m wrapped around the sphere's Equator. The moment of inertia of the hoop is $I_h = mR_E^2$. To find the hoop mass m, set the combined moment of inertia of the sphere and the hoop equal to the moment of inertia of the real planet:

$$\frac{1}{3} M_E a^2 = \frac{1}{3} M_E R_E^2 + mR_E^2.$$

Rearranging,

$$\begin{aligned} mR_E^2 &= \frac{1}{3} M_E (a^2 - R_E^2) = \frac{1}{3} M_E (a + R_E)(a - R_E) \\ &\simeq \frac{2}{3} M_E R_E \Delta R_E = \frac{2}{3} M_E R_E^2 \frac{\Delta R_E}{R_E} \end{aligned}$$

or

$$m \simeq \frac{2}{3} M_E \frac{\Delta R_E}{R_E},$$

where we have used $(a + R_E) \simeq 2R_E$ and $\Delta R_E = a - R_E$ to simplify the result. From $\Delta R_E/R_E = (a - c)/c \simeq 1/300 = 0.0033$, we obtain $m \simeq 0.0022 M_E$.

From Equation 13.15, including the average torque due to the Sun $\left(\tau_{avg} = \tau_{Sun} + \tau_{moon} = \frac{3}{2}\tau_{moon}\right)$ plus the prefactor 3/2 mentioned above and the factor 1/2 due to averaging the torque over time,

$$\tau = \frac{3}{2}\cdot\frac{3}{4}\frac{GM_m m}{R_{EM}^3} R_E^2 \cos\phi \sin\phi = \frac{1}{3} M_E R_E^2 \frac{2\pi}{T_E} \sin\phi\Omega = \frac{dL}{dt}.$$

simplifying,

$$\Omega = \frac{27}{16\pi}\frac{GM_m}{R_{EM}^3}\cdot\frac{m}{M_E} T_E \cos\phi.$$

The mass of the moon is M_m = 7.36 × 10^{22} kg, and its average distance to Earth is R_{EM} = 3.80 × 10^8 m. Plugging in the numbers, we obtain Ω = 8.38 × 10^{-12} s^{-1}.

EXERCISE 13.17

Finish the calculation by showing that the precession period is approximately equal to 26,000 years, the time measured by Hipparchus in the second century BC.

Just as we did above, physicists often employ reasonable approximations to make difficult calculations easier. As you continue in your science studies, you will surely find this to be an important and useful skill.

13.11 SUMMARY

In previous chapters, we were concerned only with a body's *CM* motion. Newton's second law governs the translation of the *CM*: $\vec{F}_{ext} = d\vec{p}/dt = M\vec{a}_{CM}$ and treats the body as if it were a point particle. But real bodies are not point masses, and their motion is a superposition of translation and rotation. To describe the full motion of a solid body, we introduce a new physical concept, *torque*, which governs the change in the body's total angular momentum:

$$\vec{\tau}_{ext} = \frac{d\vec{L}}{dt}$$

where

$$\vec{\tau}_{ext} = \sum_i \vec{r}_i \times \vec{F}_i^{ext}$$

When calculating $\vec{a}_{CM}$, the spot $\vec{r}_i$ where the force is applied is unimportant; but when calculating $d\vec{L}/dt$ it is very important. Because of this distinction, torque and force provide complementary strategies for analyzing motion.

Like kinetic energy, a body's angular momentum splits neatly into two terms, one associated with *CM* motion and the other with motion relative to the *CM*:

$$\vec{L} = M\vec{R}_{CM} \times \vec{V}_{CM} + \vec{L}'.$$

If the body is rotating with angular velocity $\vec{\omega}$ about an axis of symmetry (which necessarily passes through the *CM*), then $\vec{L}' = I\vec{\omega}$. Only externally applied torques can change a body's angular momentum. The rate of change of angular momentum is

$$\vec{\tau}_{ext} = \frac{d\vec{L}}{dt} = \vec{R}_{CM} \times \vec{F}_{ext} + \frac{d\vec{L}'}{dt}.$$

The spin angular momentum $\vec{L}'$ obeys its own equation of motion:

$$\frac{d\vec{L}'}{dt} = I\frac{d\vec{\omega}}{dt} = \sum_i \vec{r}_i{}' \times \vec{F}_i^{ext} = \vec{\tau}',$$

where $\vec{\tau}'$ is the sum of the external torques *measured relative to the body's CM*. In the special case where the only externally applied force is gravity ($\vec{F}_i^{ext} = m_i\vec{g}$), $\vec{F}_{ext}$ acts as if it were applied at the *CM*, so $\vec{\tau}' = 0$ and $\vec{L}' = \text{const.}$

Three lab-scale applications were studied in this chapter: the pendulum, the top, and the gyroscope. Each of these devices has played a critical role in the progress of physics and astronomy. The mass of a real pendulum is not concentrated at one point, and the expression for its period of oscillation is a bit more complicated than the one derived in Chapter 7:

$$T = 2\pi\sqrt{\frac{I}{MgR_{CM}}} \equiv 2\pi\sqrt{\frac{\ell_{eff}}{g}}$$

where, I is the moment of inertia of the entire pendulum relative to the point of suspension and R_{CM} is the distance from the suspension point to the pendulum's *CM*.

Even in the case of a spherical bob hanging from a lightweight string of length ℓ, the difference between ℓ_{eff} and ℓ must be taken into account to predict the pendulum's period accurately, or to use the pendulum for accurate measurements of $\vec{g}$. The gimbals of a gyroscope isolate its rotor from external torques. Consequently, the orientation of the rotor's spin axis remains fixed in space, a property that can be exploited for a variety of purposes: for example, to make a nonmagnetic navigational compass (gyrocompass) or to test aspects of general relativity. Finally, the gravitational torque on a toy top is perpendicular to the top's spin angular momentum $\vec{L}' = I\vec{\omega}$. As a result, $\vec{L}'$ pirouettes, or precesses, about the vertical axis at a rate Ω set by L' and the applied torque $\tau = MgR_{CM}\sin\theta$:

$$\Omega = \frac{\tau}{I\omega\sin\theta}$$

where θ is the angle by which the top is tilted from the vertical. The same equation describes the 26,000 year precession of Earth's axis of rotation in response to the gravitational torques induced by the Moon and Sun. Trusting in the economy of nature, Isaac Newton admonished us to "assign the same causes... to the same natural events." In his gifted hands, the physics of the toy top revealed the solution to the ancient problem of the precession of the equinoxes.

Throughout the text, we have seen the importance of the pendulum to our understanding of nature. "The pendulum," wrote Léon Foucault,* is "one of physics' most precious instruments, one of Galileo's most beautiful conceptions." We have seen how it revolutionized the measurement of time, and how it was used to determine the shape of the Earth and reveal the Earth's rotation. Newton used pendula to conduct the collision experiments that ultimately led to his laws of motion, and later to test the equality (to within 1 part in 10^3) of inertial and gravitational mass: $m_{iner} = m_{grav}$. In 1785, Coulomb developed a *torsional* pendulum to prove that the electrostatic force between charges obeys an inverse-square law (like gravity). Soon afterward (1798), Cavendish constructed a colossal torsional pendulum to measure the gravitational constant G, or, in his words, to "weigh the Earth." In 1885, von Eötvös extended Newton's measurements by employing a torsional pendulum to show that $m_{iner} = m_{grav}$ to within 1 part in 10^8. In the twentieth century, Dicke improved this result to 1 part in 10^{11}, and the work continues even today: the Eöt-Wash group, led by E. Adelberger at the University of Washington,† has now established the equality of m_{iner} and m_{grav} to within 1 part in 10^{13} (Box 13.3).

The motivation for this exacting work is nothing less than the holy grail of physics—the "theory of everything." In the quantum mechanical "standard model" of particle physics, three of the four known fundamental forces of nature are conveyed by quanta; for example, the quantized carrier of the electromagnetic force is the photon. Gravity, the outlier, does not appear to have a quantum carrier, but is instead mediated by the curvature of spacetime. A "final theory" requires the unification of the two greatest achievements of modern physics: quantum theory and general relativity. String theorists propose that space has more than three dimensions, and the quantum aspects of gravity are entangled within these extra dimensions. If so, evidence might be found by searching for tiny violations of the principle of equivalence ($m_{iner} \neq m_{grav}$ exactly). In any case, the recent discoveries of dark energy and dark matter compel us to test our most cherished beliefs about the laws of nature to the utmost precision possible. Does it surprise you that, today, some of the most exquisite tests of fundamental physics are carried out using the most basic of instruments—the pendulum?

---

* In *Journal des Débats*, March 31, 1851. This quote was taken from a fascinating biography of Foucault, by W. Tobin, *The Life and Science of Léon Foucault: the Man Who Proved the Earth Rotates*, Cambridge University Press, Cambridge, U.K., 2003.

† See N. Jones, Tough science: Five experiments as hard as finding the Higgs, by *Nature* 481, 14–17 (2012).

BOX 13.3 FROM FOUCAULT'S PENDULUM TO THE GYROSCOPE IN YOUR POCKET

"You are invited to watch the Earth turn," wrote Léon Foucault to the French newspaper *Le Nationale*, and in 1851, the citizens of Paris flocked to the Panthéon, one of the city's tallest buildings, to watch Foucault's giant pendulum swing from the building's dome. The slow precession of the pendulum's motion offered visual proof that the Earth was turning, and spectators watched with rapt attention as the floor of the Panthéon visibly turned westward under the swinging bob.

The pendulum is a close cousin of the gyroscope: its plane of oscillation resists change, just as the gyroscope's axis of rotation resists change. If you own a smartphone (or a video game console), it contains a high-tech nanofabricated device that functions like a gyroscope, even though it has no spinning rotor. Surprisingly, the equation of motion governing your pocket "gyroscope" is the same as the equation describing the motion of the pendulum. In principle, your cell phone could be used like Foucault's pendulum—to reveal that the Earth turns.

Functionally, the nano-gyroscope is equivalent to the device illustrated in Figure B13.1a. It is a block of mass m connected to four identical springs of stiffness k aligned along the $\pm x$- and $\pm y$-directions. The block slides with negligible friction over a smooth supporting surface. Ignoring higher-order terms,* the equations of motion for the block are

$$F_x = ma_x = -2kx \quad \text{and} \quad F_y = ma_y = -2ky, \tag{B13.6}$$

where x and y are displacements from equilibrium in the horizontal and vertical directions. These two expressions can be combined into a single equation for the vector displacement $\vec{r} = x\hat{i} + y\hat{j}$. Multiply the first equation by the unit vector $\hat{i}$, the second by $\hat{j}$, and add the two together to obtain

$$\vec{F} = m\vec{a} = -2k\vec{r}. \tag{B13.7}$$

This is a remarkable result. Suppose we set the block into oscillation by displacing it along the $+x$-direction. The restoring force is $\vec{F} = -kx\hat{i} + 0\hat{j}$. Since $F_y(t) = 0$, the block will oscillate indefinitely in the $\pm x$-direction. What happens if we set the block into oscillation along a different direction, at an angle θ to the x-axis? Equation B13.7 predicts *the same motion*: the restoring force is always antiparallel to the displacement, that is, the force perpendicular to the displacement is zero, so the block oscillates indefinitely along the new line of motion. We might start the block oscillating in the original $\pm x$-direction and *rotate the device* by an angle θ while the block is moving (see Figure B3.1b). This is equivalent to the previous situation, and the initial $\pm x$-direction of motion is maintained. In other words, the block is insensitive to the rotation of the supporting springs: the block-and-spring device is like a gyroscope.

* If m is displaced horizontally by a small amount x, the vertical springs stretch by $x^2/2\ell$, where ℓ is the equilibrium length of the spring. The vertical springs therefore contribute a force F_x th-at is smaller than the force of the horizontal springs by a factor $x/2\ell$. For small displacements x, this contribution can be ignored.

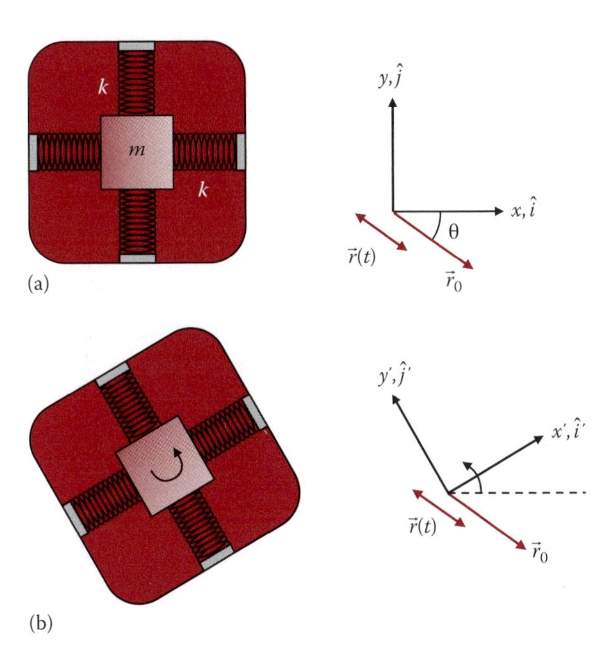

FIGURE B13.1 (a) A block m is cradled by four identical springs of stiffness k and slides without friction over a flat surface. When released from rest from $\vec{r}_0$, the block oscillates along a straight line between $\pm\vec{r}_0$, regardless of the direction of $\vec{r}_0$. (b) If the device is rotated while oscillating, the path of the block is unaffected.

Exercise B13.1: Suppose the spring constant of the vertical springs was different from the horizontal springs, that is, $k_x \neq k_y$. (a) Would the motion of the block remain unaffected by rotating the apparatus? (b) Could you write a vector equation like Equation B13.7?

In Chapter 7, we derived the equation of motion for a simple pendulum swinging in the $\pm x$-direction: for small displacements from equilibrium, $F_x = -mgx/\ell$, where ℓ is the length (more properly, the *effective* length) of the pendulum. If the bob is swinging in the $\pm y$-direction, $F_y = -mgy/\ell$. Combining these two expressions as before, we obtain

$$\vec{F} = F_x\hat{i} + F_y\hat{j} = -\frac{mg}{\ell}(x\hat{i} + y\hat{j}) = -\frac{mg}{\ell}\vec{r},$$

which has the same form as Equation B13.7 with mg/ℓ replacing k. Therefore, like the oscillating block, the pendulum is unaffected by rotation. If either of these systems were placed at the North Pole, its line of motion would remain fixed in space, oblivious to the Earth turning beneath it. To an observer on the surface of the revolving Earth, of course, it would appear that the trajectory of the block or pendulum is rotating in the opposite sense from the planet, making one full revolution about the N–S axis every 24 h.

In these two examples, the oscillating body was confined to a plane, and the axis of rotation was perpendicular to this plane. What happens if the oscillator is placed somewhere on the Earth's surface other than the North or South Pole, at a latitude $\lambda < 90°$? We can express the planet's rotation $\vec{\Omega}$ in terms of its horizontal and vertical components $\vec{\Omega} = \vec{\Omega}_H + \vec{\Omega}_V$, and treat the Earth's spin as the superposition of these two independent rotations (see Figure B13.2a). The block is the simpler system to analyze. It is constrained by gravity and the normal force $\vec{N}$ to remain in contact with the surface supporting it, so $\vec{\Omega}_H$ has no effect on its motion relative to this surface. On the other hand, $\vec{\Omega}_V$ is perpendicular to the oscillation, and, as we have seen, the block is insensitive to a rotation about the vertical axis. Accordingly, if you are standing near the apparatus on the rotating Earth, you will see the trajectory of the oscillating body precess in the $-\vec{\Omega}_V$ direction, making a full revolution with period

$$T = \frac{2\pi}{\Omega_V} = \frac{2\pi}{\Omega \sin\theta} = \frac{24\,\text{h}}{\sin\lambda}. \tag{B13.8}$$

The same physics describes the motion of a pendulum: the bob is constrained to swing in the horizontal plane by gravity and the string tension force. Its precession period is likewise given by Equation B13.8.

The latitude of Paris is 49° ($\sin 49° \simeq 3/4$), and Foucault's pendulum completed a full revolution in 32 h rather than 24 h. Although he proposed Equation B13.8 for the latitude-dependent precession period of the pendulum, he could not prove it. Despite the immense appeal of Foucault's pendulum demonstration,

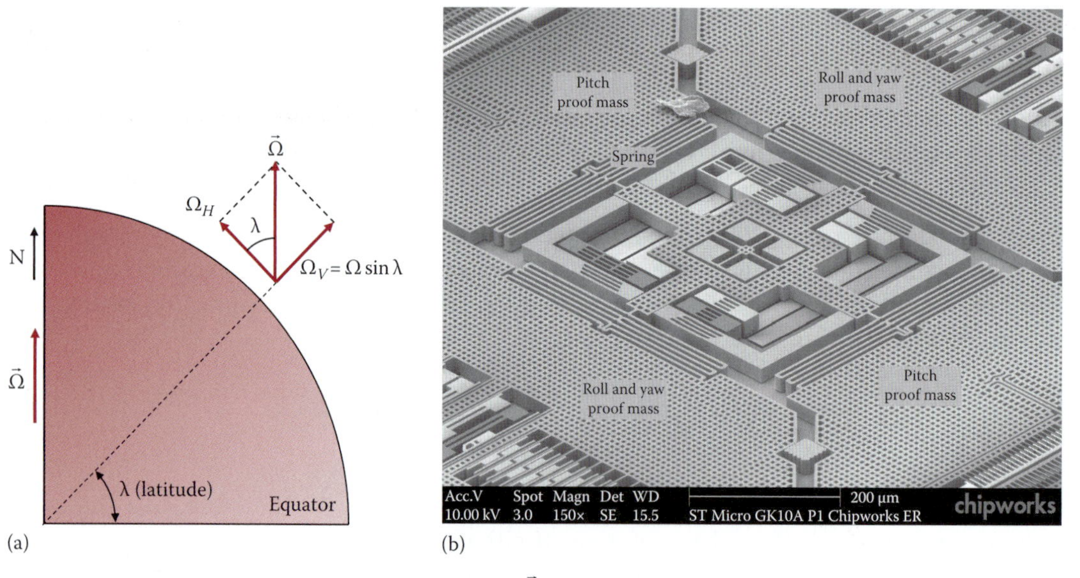

FIGURE B13.2 (a) At a latitude λ, Earth's rotation vector $\vec{\Omega}$ can be split into components parallel ($\Omega_H = \Omega\cos\lambda$) and perpendicular ($\Omega_V = \Omega\sin\lambda$) to the local horizon. The vibrating gyroscope of Figure B13.1 and Foucault's pendulum are sensitive only to Ω_V. (b) Scanning electron micrograph of a three-axis nanofabricated MEMS gyroscope. Note the springs and the central mass and the scale of the image. (Courtesy of Chipworks, Inc., Ottawa, Ontario, Canada.)

the sin θ factor was unsettling for him and the more thoughtful spectators at the Panthéon. That is why Foucault developed an independent demonstration using a gyroscope, which did precess with a 24 h period.

Exercise B13.2: Suppose the apparatus in Figure B13.1 were replaced by a block suspended from 6 identical springs extending along the $\pm x$, $\pm y$, and $\pm z$-directions, and the block set into horizontal vibration in an N–S direction. At a latitude $\lambda = 45°$, would an observer on Earth observe (a) precession of the line of motion? (b) If so, what would be the period of the precession?

Answer: (b) 24 h

The nanofabricated "gyroscope" in your smartphone is a marvel of modern nanofabrication. It has blocks and springs etched from a solid slab of pure silicon. A detail of a typical device is shown in Figure B13.2b. Note the scale of the image: the smallest features have one dimension on the order of 1 μm! The complete device fits within a thin electronic "chip" measuring about 5 mm × 5 mm, and costs just a few dollars. The gyroscope is one of the latest examples of microelectromechanical systems, or MEMS. Other MEMS devices include accelerometers (to deploy your car's airbags in the event of a collision), micromotors complete with gear trains and transmissions, and microfluidic "labs on a chip" for chemical or biological research and medical diagnosis. MEMS is twenty-first century manufacturing, and its potential to influence our lives positively seems limited only by our imagination. The next time you play an action-packed video game, be aware that you are using an incredibly sophisticated, state-of-the-art device made possible by spectacular advances in physics, chemistry, and engineering.

PROBLEMS

13.1 A square plate pivots in the horizontal plane about a pin through its center. Forces of 1.0 N are applied as shown in the figure.

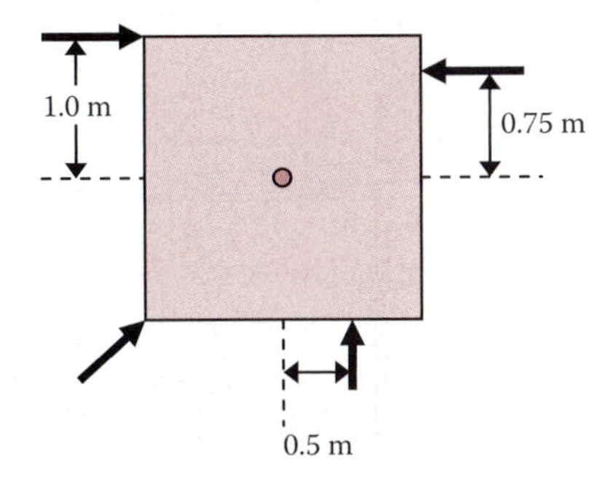

(a) What is the total torque relative to the pivot point? (Do not forget units!)

(b) What is its direction?

13.2 A bar of length $D = 1$ m and mass $m = 1$ kg is suspended by two strings, as shown. A mass $M = 2$ kg is hung from a point 75 cm from the left end of the bar.

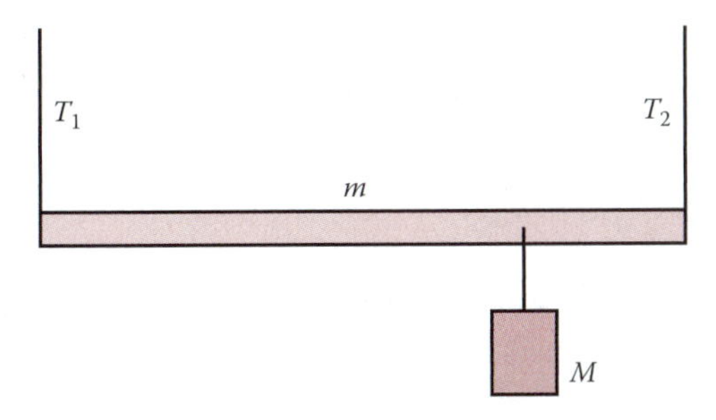

(a) Write an expression, in terms of M, m, D, T_1, T_2, and g, for the torque acting on the *bar*, relative to the *left end* of the bar.

(b) Find the values of T_1 and T_2.

(c) Similarly, write an expression for the torque acting on the bar, measured relative to the *center* of the bar. Once again, use this expression to find T_1 and T_2.

(d) Repeat by writing the torque relative to the right end of the bar, and solving for T_1 and T_2.

(e) The maximum string tension is 40 N. What is the maximum mass M that can be suspended by the two strings? Which string breaks first?

13.3 A spacecraft is in the shape of a wheel (hoop) of mass $m = 100$ kg and radius $r = 2$ m, plus 6 equally spaced "spokes," each of mass m', extending outward from the hub of the wheel to its rim. The craft is orbiting a distant planet, and initially it is not rotating.

(a) The total moment of inertia of the spacecraft about an axis through its hub perpendicular to the wheel is $I = 2mr^2$. Find the mass m' of a single spoke.

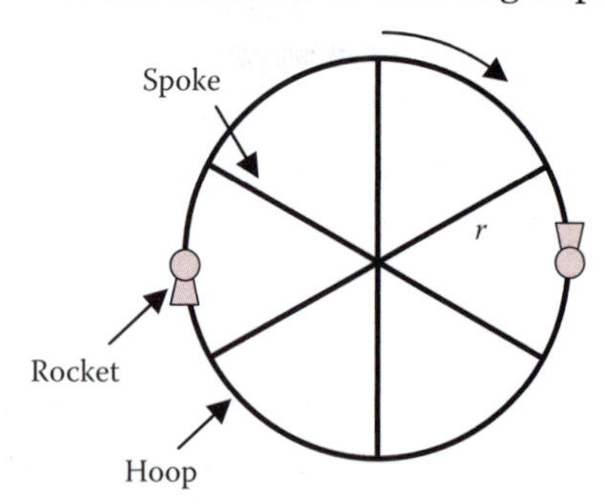

(b) Using two small, identical rockets (ignore their mass) located on the rim, the craft is "spun up," reaching an angular velocity $\omega = 10\ \text{s}^{-1}$ in a time $\Delta t = 50$ s. What is the thrust delivered by each rocket?

(c) The figure to the right shows a side view of the spinning spacecraft.

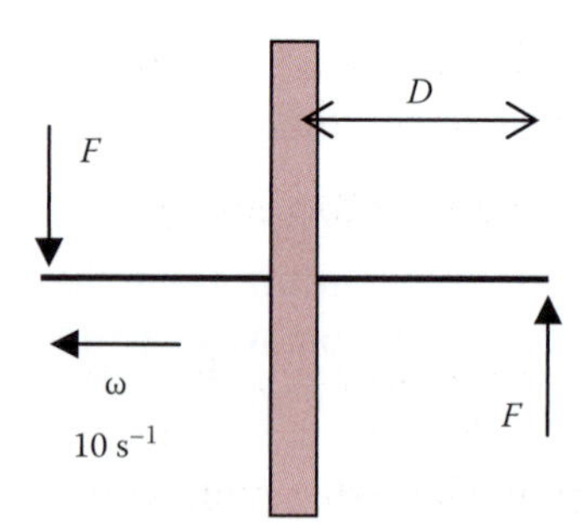

A lightweight *axle of length* 2D passes through the hub. Note the direction of $\vec{\omega}$. Equal and opposite forces F are applied perpendicular to the ends of the axle, as shown, and the craft *changes its orientation by 90°*. What is the new direction of $\vec{\omega}$?

(d) If $D = 1$ m, what forces F are necessary to turn the spacecraft by 90° ($\pi/2$ radians) in 100 s?

13.4 A thin uniform stick of mass M and length L stands vertically on a frictionless tabletop. When it is released, it teeters, slips, and falls.

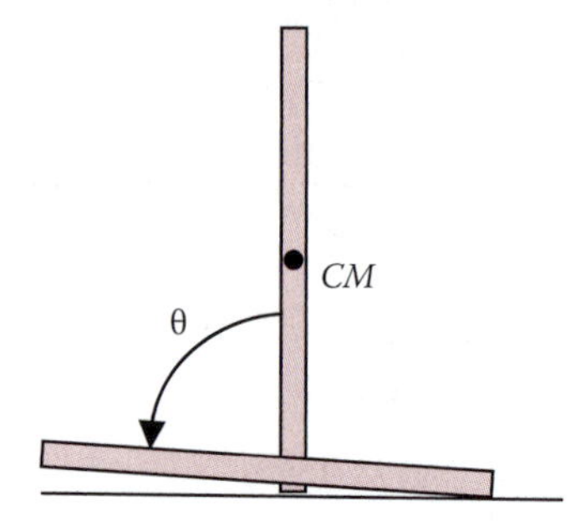

(a) As it falls, the stick rotates about its center of mass. What force provides the torque for this rotation?

(b) Does the *CM* fall to the left or right of its initial position, or does it fall straight down?

(c) The height of the *CM* is $y = L\cos\theta/2$. Use the chain rule to find a relation between the velocity V_{CM} and the two variables θ and $\omega = d\theta/dt$.

(d) Find the *CM* velocity just before it hits the table.

13.5 A solid cylindrical body of mass $M = 1.0$ kg has a radius $R = 5$ cm. The object sits on an inclined plane that makes an angle 30° with respect to the horizontal. The cylinder has a circular hub of radius $r = 3$ cm about which a string is wrapped that supports a hanging mass m. The cylinder and mass m do not move.

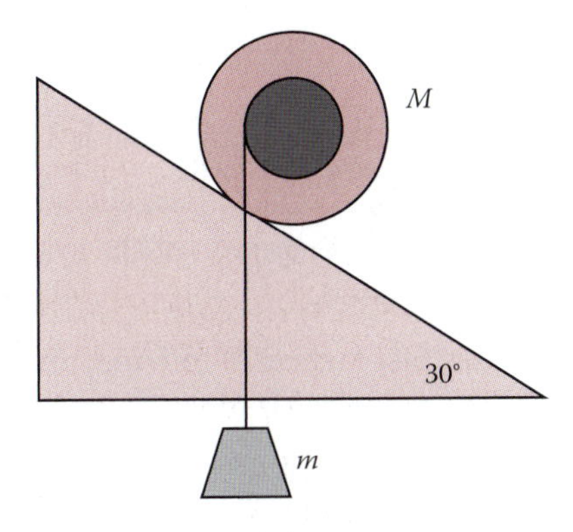

(a) Draw a free body diagram showing the forces that act on the cylinder.

(b) Write an equation for the total torque about the *CM*.

(c) Find the mass m.

(d) Find the friction force (magnitude and direction) acting on the cylinder.

(e) What is the minimum value of the static coefficient of friction μ_s?

13.6 In the final stage of its launch, the *Odyssey* spacecraft was spun about its cylindrical axis to achieve stability. Model the spacecraft as a cylinder of mass $M = 700$ kg with a radius $R = 1.1$ m and length $L = 2.2$ m. In addition to the cylinder, the craft has two "ballast" masses $m = 50$ kg attached to its "belly," as shown below.

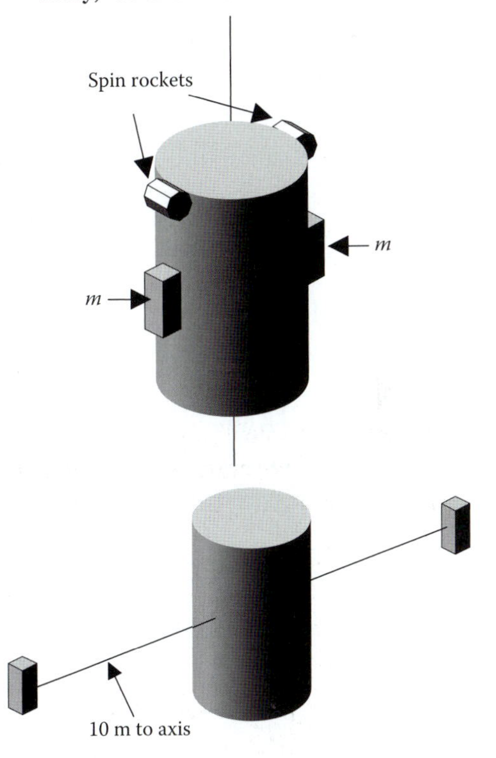

(a) Calculate the total moment of inertia of the spacecraft about the axis shown in the figure.

(b) Each rocket (located 1.1 m from the spin axis) is fired for 5 s with thrust 270 N. Assuming *Odyssey* had zero initial angular momentum, what is its final angular velocity ω?

Once the spacecraft is spinning, it fires its final boost rockets, sending it on its way to Mars. The next step is to despin the spacecraft. This is done by extending the 50 kg ballast masses from (massless) wires to a distance of 10 m from the axis of rotation.

(c) For simplicity, assume the ballast masses rotate with the same angular velocity as the satellite body. What is the value of ω after the masses are fully deployed?

(d) What is the tension in each wire after the ballast masses are deployed?

(e) The ballast masses are disconnected from the spacecraft. Does the angular velocity of the spacecraft (i) increase, (ii) decrease, or (iii) stay the same after they are jettisoned?

13.7 A spring of stiffness k = 40 N/m is connected by a string to a hanging mass m = 2.0 kg. The string passes over a pulley of mass M = 3.0 kg and radius R = 10 cm.

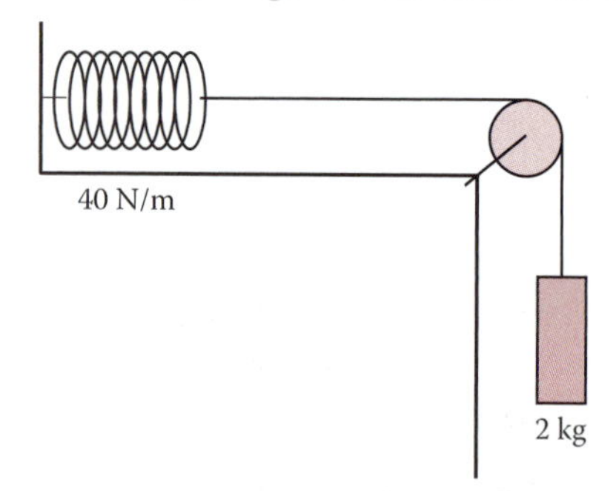

(a) By how much is the spring stretched at equilibrium?

(b) The mass is pulled down a distance 0.5 m from its equilibrium position. What is the *change* (from equilibrium) in the potential energy of the mass–pulley–spring system? Is this change positive or negative?

(c) The mass is then released. What is its maximum kinetic energy?

(d) Calculate the maximum velocity of the mass.

13.8 Imagine a thin stick of length ℓ and mass m, resting on a horizontal frictionless surface. A small puck of equal mass m, moving with speed v perpendicular to the stick, strikes the end of the stick and adheres to it.

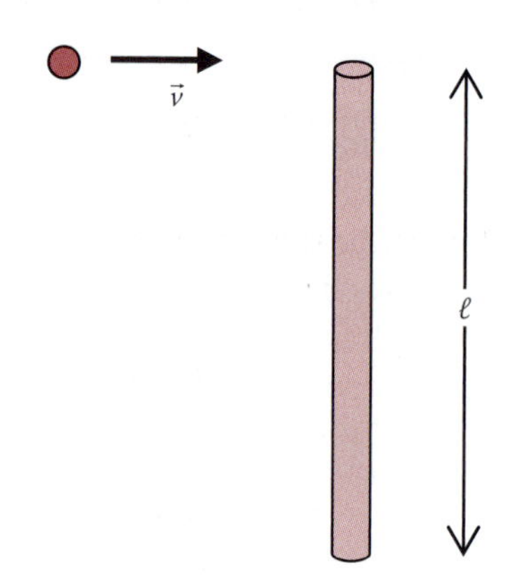

(a) After the collision, where is the *CM* of the system?

(b) Describe the motion of the system after the collision.

(c) How fast is the *CM* moving, and in what direction does it move?

(d) What is the angular momentum of the system with respect to the *CM*?

(e) What is the moment of inertia of the system relative to the *CM*?

(f) Find the angular velocity ω about the *CM*.

13.9 A man is trying to roll a barrel of mass M and radius R along a level street by pushing forward along its top rim. At the same time another man is pushing backward at the middle, with a force of equal magnitude. The barrel rolls without slipping.

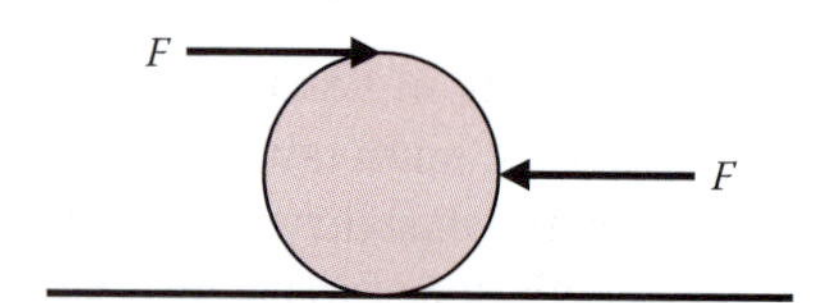

(a) Which way does it roll?

(b) What is the magnitude of the barrel's acceleration? Assume $I = \frac{1}{2}MR^2$.

(c) Find the magnitude and direction of the friction force acting on the barrel.

(d) Suppose the men push with different forces, so that the barrel does not roll. In this case, find the ratio F_{top}/F_{middle}.

13.10 A block of mass M = 0.4 kg is attached to a massless spring of stiffness k = 10 N/m. The block slides without friction over a smooth horizontal surface. Its maximum velocity is 1.0 m/s and its maximum displacement from equilibrium is 0.2 m.

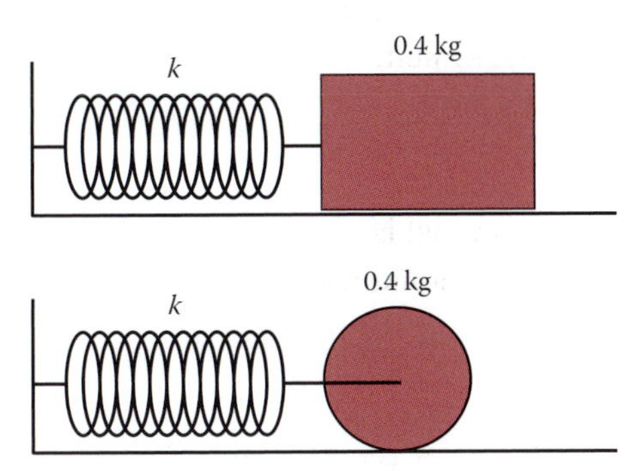

(a) Find the period of the oscillation.

(b) The block is replaced by a round object of radius R (see the figure) having the same mass M = 0.4 kg, which rolls without slipping over the (horizontal) surface. The maximum velocity of the object is 0.82 m/s, and its maximum displacement is 0.2 m. Identify the shape of the object (i.e., ball, disk, hoop, etc.).

(c) When the object is moving (rolling) to the right of its equilibrium and slowing down, what direction is the friction force acting on it? Discuss clearly.

13.11 A block of mass M = 2 kg rests on a frictionless inclined plane at a 30° angle with respect to the horizontal. The block is attached to two other blocks by a string passing over a pulley, as shown in the figure. In parts (a) and (b), assume the pulley's moment of inertia is negligible. The blocks do not move when released. But after the string between m_1 and m_2 is cut, M moves down the slope with acceleration $a = g/3$ m/s².

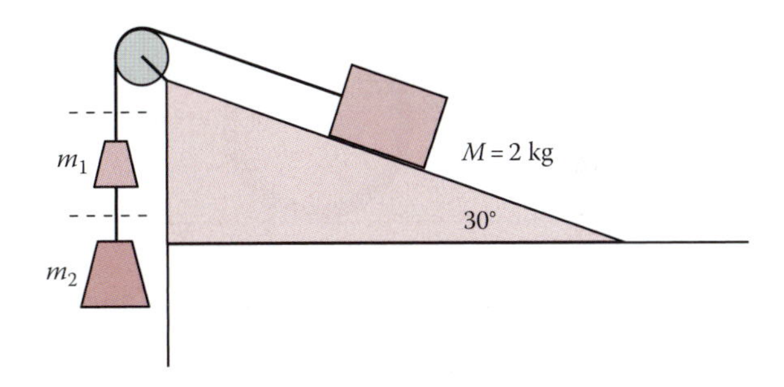

(a) What is the tension in the string attached to M after the string is cut? Draw a free body diagram showing all forces on M.

(b) Find the masses m_1 and m_2.

Now consider a real pulley with a moment of inertia I and radius R = 0.06 m. The string is wrapped tightly around the pulley for several turns. When the string is cut *above* m_1 (as opposed to between m_1 and m_2), M accelerates down the slope with acceleration $a = g/3$ (again).

(c) What is the torque acting on the pulley, measured relative to its axle, after the string is cut as described?

(d) Find an expression for the moment of inertia of the pulley I in terms of M and R.

13.12 A "hula hoop" of mass M and radius R is spinning with angular velocity ω_0 about a horizontal axis. It drops vertically onto a surface where the friction with the hoop sets it into rolling motion. Find an expression for the final angular velocity ω of the hoop once it is rolling without slipping. What is the speed of its CM? (*Hint*: There is a point about which the torque on the hoop is zero.)

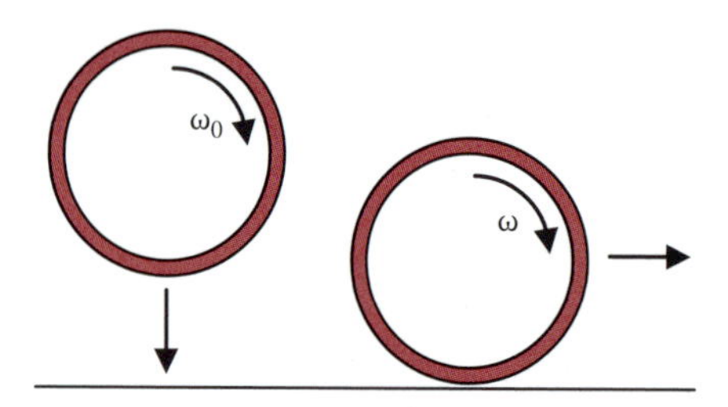

13.13 A constant horizontal force of 9 N is applied to the hub of a wheel of mass $M = 10$ kg and radius $R = 0.25$ m, as shown in the figure. The wheel rolls without slipping on a horizontal surface, and the acceleration of its center of mass is 0.50 m/s².

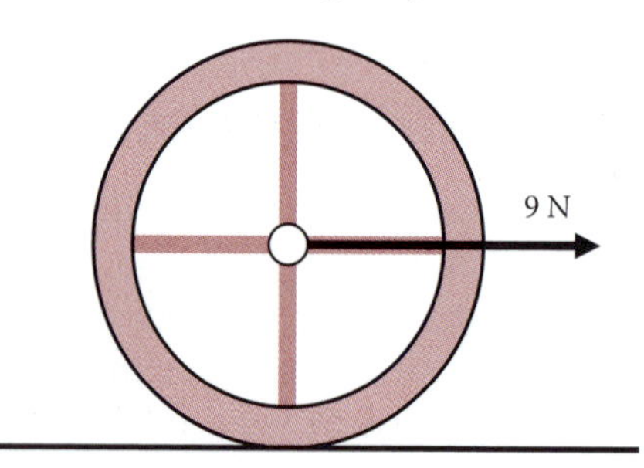

(a) What are the magnitude and direction of the frictional force on the wheel?

(b) Let $I = xMR^2$ be the moment of inertia of the wheel about its hub. Find x.

(c) How does your answer to (a) change if R is increased to 0.5 m while M and x stay the same?

13.14 Two unequal masses M and m hang by a lightweight string from a pulley. The pulley's moment of inertia is $I_p = \frac{1}{2} M_p R^2$ where M_p and R are the mass and radius of the pulley. The string does not slip with respect to the pulley.

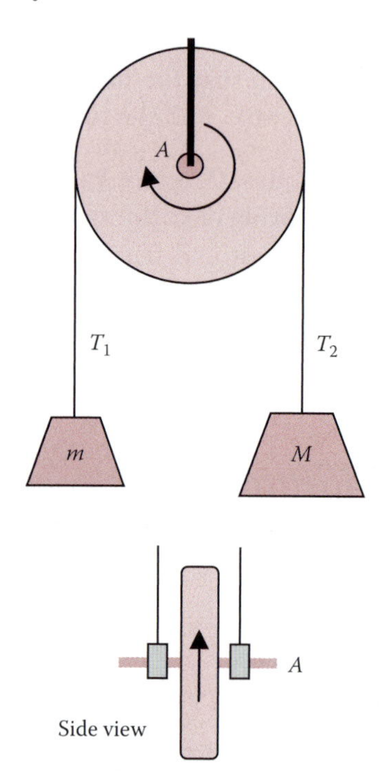

(a) Draw free body diagrams for the two masses and the pulley. Define what direction(s) will be positive on your diagram.

(b) Write an expression for the torque on the pulley as a function of the tensions on the string.

(c) Derive expressions for the acceleration of each mass and the angular acceleration of the pulley.

(d) Solve your three equations to find the acceleration of the mass m.

(e) The pulley is hung from the ceiling by ropes rather than being rigidly mounted. Point A is the end of the pulley's axle that is pointing out of the page. After the masses have started to move and the pulley is rotating clockwise, you give the right side of the pulley a push directed into the page (in the side view shown above). In response, point A moves:

(i) up, (ii) down, (iii) left, (iv) right

Justify your answer.

13.15 A uniform cube of mass M rests on a horizontal plane. The front edge of the cube is held in place by a hinge, so the cube can tip forward. A force $\vec{F}$ is applied by a horizontal string attached to the center of the top surface of the cube.

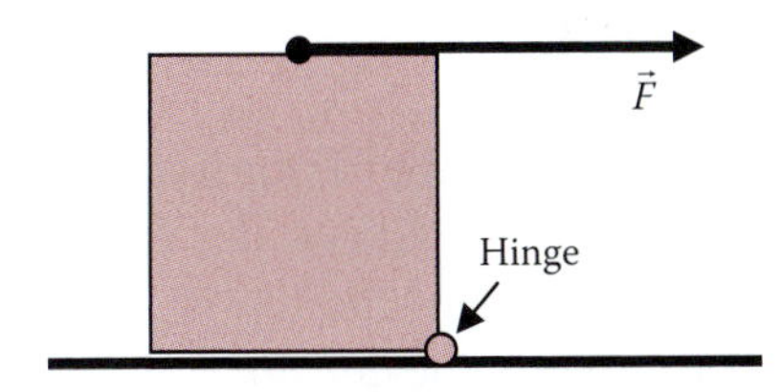

(a) At what value of $\vec{F}$ does the cube start to tip?

(b) What force is exerted on the cube by the hinge *just before* the cube starts to tip?

(c) Instead of being held by the hinge, the block is allowed to slide without friction over the surface. Is there a force $\vec{F}$ such that the cube tips over? If so, calculate its magnitude. (*Hint*: Use the center of the cube as your reference point for calculating torques.)

13.16 A spool of thread having mass M and radius R is unwound under a constant force $\vec{F}$. The spool is a uniform solid cylinder that rolls over a horizontal surface without slipping.

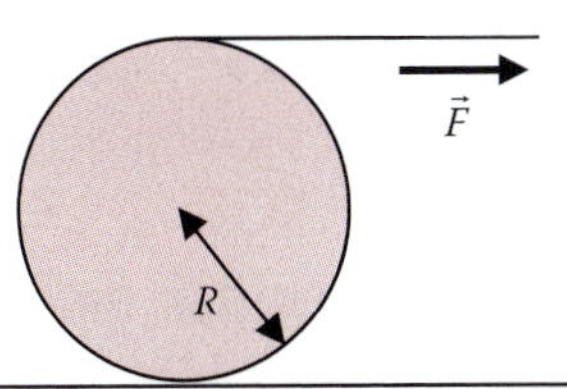

(a) Find the acceleration of its center of mass.

(b) Find the force of friction (magnitude and direction).

(c) If the cylinder starts from rest, what is the velocity of its *CM* after it has rolled a distance d? Assume that $\vec{F}$ is constant. Find the total kinetic energy. Does your answer violate the work–energy theorem? Explain clearly. (*Hint*: How much thread has unwound from the spool?)

13.17 A cylinder of mass M and radius R rolls without slipping on a horizontal surface. On each of its two ends there is a hub of smaller radius r. Thread is wound around each hub in the direction shown, and equal tensions $T_1 = T_2 = T$ are maintained in each thread.

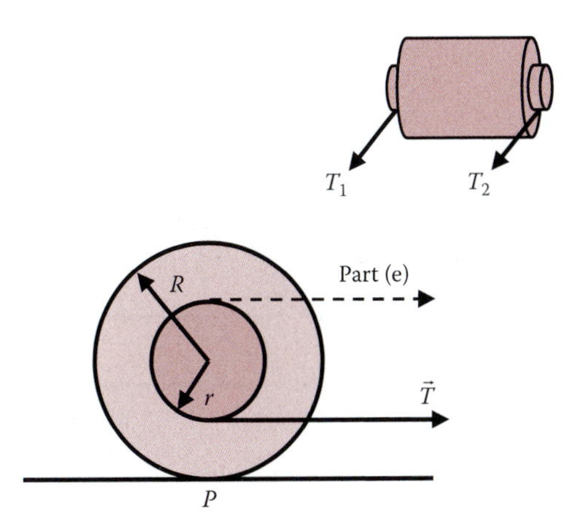

(a) Which way does the cylinder roll? (*Hint*: What is the torque relative to P?)

(b) Derive an expression for the acceleration of the cylinder as a function of M, R, r, and T.

(c) Find the friction force F_f in terms of the same variables as in (b). What is its direction?

(d) Do your answers make sense? Check them for the easy cases of $r = 0$ and $r = R$.

(e) Now suppose the threads are wound in the opposite sense about the two hubs, so that they come off the *tops* of the hubs. Calculate a_{CM} and F_f for the case $r = R/2$. What is the direction of F_f when $r > R/2$ and when $r < R/2$?

13.18 A yo-yo is resting on a horizontal table over which it is free to roll. If the string is pulled by a horizontal force $\vec{F}_1$, which way will the yo-yo roll? Describe what happens if the string is pulled vertically $(\vec{F}_3)$. What happens if the applied force $(\vec{F}_2)$ is directed along the line passing through the point of contact P of the yo-yo and the table?

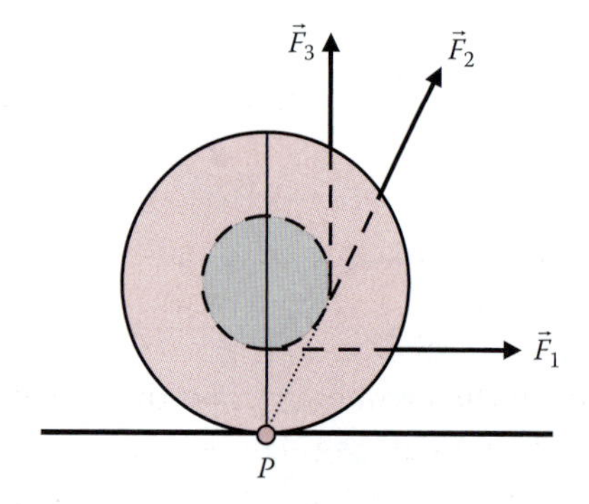

13.19 A simple conical toy top has height h and base radius R. Let z be the distance from its tip to a point along the axis of symmetry. The radius of the cone is a linear function of z: $r(z) = Rz/h$.

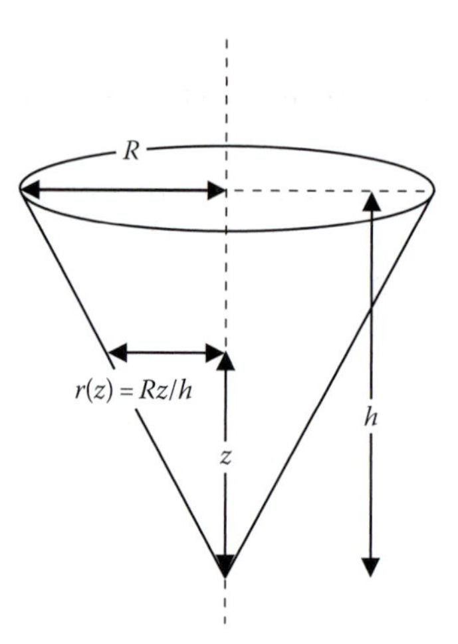

(a) By dividing the toy top into disks of infinitesimal thickness dz, prove that the total mass of the top is $M = \frac{1}{3}\pi\rho R^2 h$, where ρ is the density.

(b) By symmetry, the top's *CM* lies along the z-axis, so

$$Z_{CM} = \frac{1}{M}\int z\,dm.$$

Divide the top into thin disks of mass $dm = \rho\pi r^2 dz$ and integrate to show $Z_{CM} = \frac{3}{4}h$.

(c) Once again, divide the top into thin disks of mass dm. The moment of inertia of each disk about the axis of symmetry is $dI = \frac{1}{2}r^2dm$. Integrate over the entire top to find its moment of inertia about its axis of symmetry.

(d) The top is set spinning with angular speed ω. Find the top's precession rate Ω in terms of h, R, M, g, and ω.

13.20 A gyroscope is constructed from a wheel of mass M = 0.25 kg and radius R = 10 cm, mounted on a lightweight axle such that the center of mass of the wheel is 5 cm from the pivot point P. The wheel is spinning at a rate of ω = 100 s^{-1}, and the sense of the rotation is such that the angular momentum points in the direction shown in the figure at time t_0.

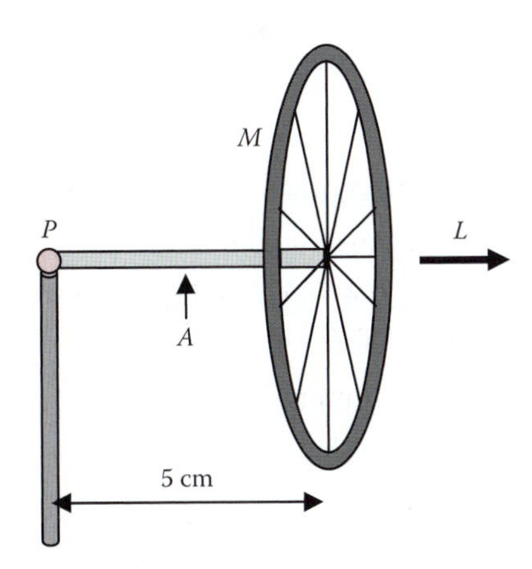

(a) Which way is the wheel spinning?

(b) Calculate the moment of inertia of the wheel and the magnitude of its angular momentum.

(c) Find the torque on the gyroscope relative to the pivot point. What is the direction of the torque?

(d) Let Ω equal the precession rate about the vertical axis. By equating the torque and the rate of change of angular momentum, write an equation relating the precession rate Ω to the angular velocity of the wheel ω.

(e) Using the information given, find Ω and the time needed for the gyroscope to precess one complete turn about the vertical axis.

(f) Now imagine that the wheel axis is tilted 30° above the vertical axis. Find the new precession rate Ω.

(g) What is the force exerted on the axle by the vertical support?

(h) Why doesn't the axle drop downward and fall off the pivot? (If it did, then what direction would $\Delta\vec{L}$ be in? Is there a torque in that direction?)

(i) Point A is located between the pivot point and the hub of the wheel. In what direction should you apply a force (vertical up/down, horizontal in/out) at A so that you stop the precession of the wheel?

13.21 Imagine a spacecraft situated in deep space, far from any gravitational influences. It carries a gyroscope that is mounted on a pivot point as shown in the figure. The gyroscope is made of a spinning wheel (hoop) with a mass M = 0.5 kg and radius R = 0.2 m. The center of the wheel is at a distance r = 0.3 m from the pivot point, and the wheel is spinning with an angular velocity ω = 200 s^{-1} at an angle of 45° relative to the vertical.

The spacecraft ignites its rockets and undergoes an upward acceleration $\vec{a}$ causing the gyroscope to precess as shown with a period of 10 s. Due to the acceleration, the gyroscope experiences a force $M\vec{a}$ applied by the pivot.

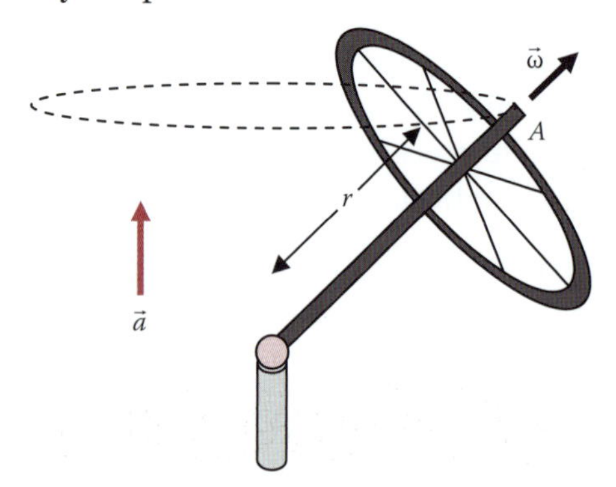

(a) In terms of a and other symbols, write an equation for the magnitude of the torque relative to the center of mass of the wheel.

(b) At the moment shown in the figure, what is the direction of motion of point A?

(c) Can the gyroscope's precession be used to determine the spacecraft's acceleration? If so calculate a numerical value for a.

(d) As the gyroscope precesses, the angular momentum vector changes. Find the magnitude and direction of the *change* in $\vec{L}$ over a 5 s time interval starting at the position shown in the figure. Indicate the direction of $\Delta\vec{L}$ with an arrow.

13.22 A physical pendulum is made from a long thin bar of mass M and length $L = 1$ m, plus a disk of mass m and radius $r \ll L$ attached to one end of the bar. Treat the disk as a point mass. The CM of the apparatus is found to be $2L/3$ from point A at the upper end of the bar.

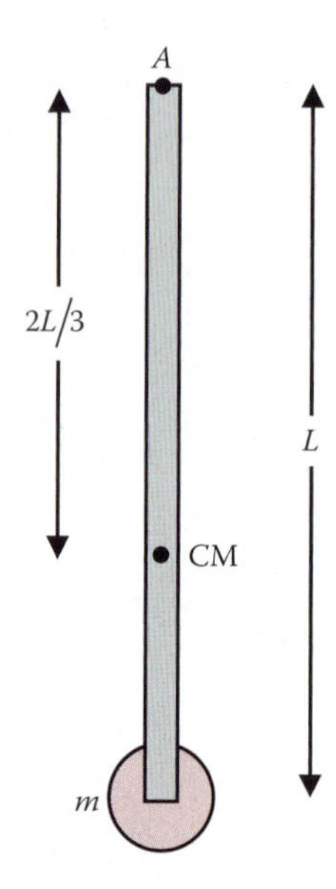

(a) Find the mass of the disc ($m/M = ?$).

(b) Derive expressions for the moment of inertia of (i) the bar and (ii) the disk about point A.

(c) Find the pendulum's period when it is swung from point A.

(d) A second pivot point B on the bar gives the same period as point A. What is its position?

13.23 Consider two gyroscopes like the spinning wheel shown in Problems 13.20 and 13.21. Both are spinning at the same rate ω.

(a) The two gyroscopes have the same size, but one is twice as heavy as the other. (The wheel is made from a higher density material.) Compare the precession rates Ω of the two devices.

(b) Suppose instead that the gyroscopes are made of the same material, but one is twice as large as the other in all aspects. (*Note*: The volume of the wheel increases by a factor $2^3 = 8$.) Compare the precession rates of the larger and smaller gyroscopes.

13.24 A simple top is made from a ball of mass 100 g and radius $r = 10$ cm, pierced by a thin needle of negligible mass that passes through the sphere's center. The pointed end of the needle, which is in contact with the floor, is 20 cm from the center of the sphere. When the top is spinning at 600 rpm, its precession period is 1.34 s.

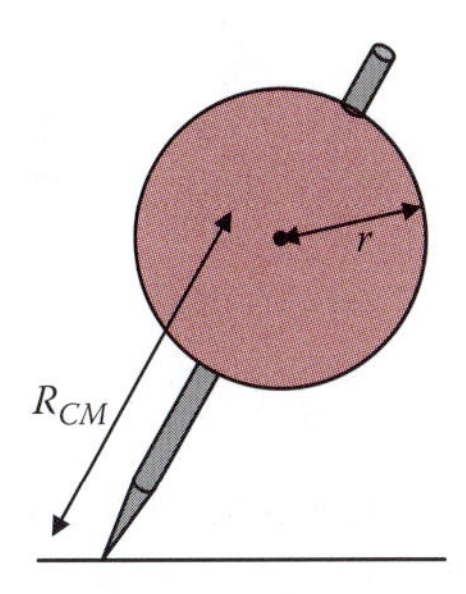

(a) Find the moment of inertia of the sphere.

(b) Is the ball solid or hollow? (See Figure 11.3.)

13.25 Light carries momentum as well as energy. When it is absorbed, reflected, or radiated by a body, it exerts a tiny force on the body. If the body is irregularly shaped, a tiny torque can result. Surprisingly, this tiny torque can change the spin rate of asteroids over the course of millions of years. This is called the YORP effect (Yarkovsky–O'Keefe–Radzievskii–Paddack effect). It was predicted in 1901 and has recently been observed for the near-Earth asteroid 2000 PH5, now called 54509 YORP in recognition of the discovery.

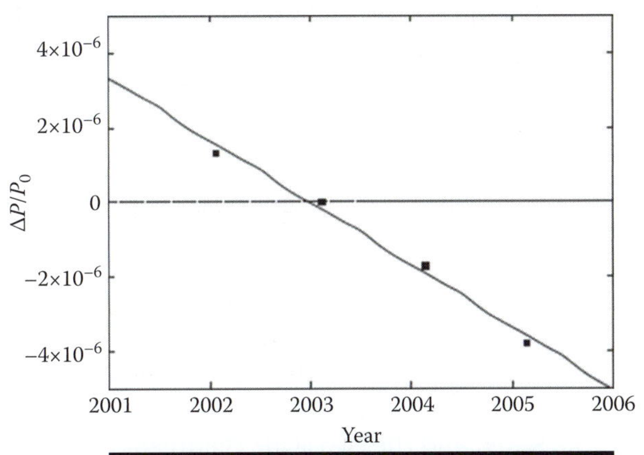

(From Lowry, S.C. et al., *Science*, 316, 272, 2007, Figure 2, p. 273. With permission.)

The asteroid currently has a spin rotation period $P_0 = 0.2029\,\text{h} = 730.4$ s. Over the course of 4 years of observation, the period has been growing shorter at a constant rate, as shown in the figure.

(a) Using the figure, show that the spin is increasing at a rate $d\omega/dt = 4.7 \times 10^{-16}$ rad/s² or $d\omega/dt = +2.0 \times 10^{-4}$ deg/day².

(b) The average radius of the asteroid is 57 m. Estimate its mass and moment of inertia, using $\rho = 3000$ kg/m³ as a "typical" density.

(c) Calculate the torque due to the interaction of the asteroid with light from the Sun.

(d) Many asteroids are rubble heaps, that is, loose aggregations of rock and ice held together by gravity. Can 54509 YORP be such a rubble heap? (*Hint*: See Problem 8.15.)

(e) Examine the importance of the YORP effect as a function of asteroid size.

If the density remains the same, the moment of inertia scales as $MR^2 \rightarrow R^3 \cdot R^2 = R^5$. The amount of light striking the asteroid, and thus the force exerted on the body, varies as R^2, so the torque scales as $R^2 \cdot R = R^3$. (Physicists often use scaling arguments to draw general conclusions about things that would be difficult or impossible to treat exactly.)

13.26 A regulation baseball bat is suspended by two supports positioned 15.8 cm (6 in.) and 63.4 cm (24 in.) from the knob end of the bat. Each support is mounted on a scale, and the two scale readings (in mass) are M_6 = 0.092 kg and M_{24} = 0.944 kg. The bat is 86.5 cm long.

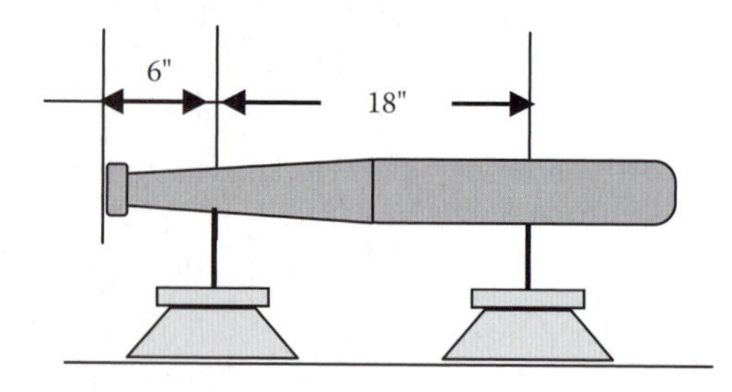

(a) What is the mass of the bat?

(b) How far from the knob end of the bat is its *CM*?

(c) The bat is hung vertically from a pivot located 15.8 cm from the knob (about where a batter's hands would be). Its period of oscillation is measured to be 1.42 s. What is the bat's moment of inertia relative to the pivot point? (The speed at which a ballplayer can swing a bat is related to its moment of inertia.)

(d) Is there another pivot point that yields the same period of oscillation? If so, locate it.

13.27 Tidal effects are slowing Earth's rotation. Far in the future, the planet's rotation will be phase-locked to the orbit of the Moon, and the same hemisphere of Earth will always face the satellite. At that time, the Moon's orbital angular momentum will be nearly equal to the total angular momentum of the Earth and Moon, $L_{tot} \approx L_{orb}$. ($L_{orb} \approx 0.8L_{tot}$ at the present time.) Use this approximation to answer the following questions:

(a) How far away will the Moon be at that time? Express your answer in Earth radii.

(b) What will be the length of an Earth day? What will be the period of the Moon's orbit? Express your answers in terms of a present Earth-day.

(c) Approximately when will this occur?

13.28 At the present time, the Moon is receding from Earth at a rate of 3.8 cm/year. Show that this implies that the length of an Earth day is increasing by 21 μs/year.

Section V
GOING BEYOND

14

DARK MATTER, DARK ENERGY, AND THE FATE OF THE UNIVERSE

My life has been an interesting voyage. I became an astronomer because I could not imagine living on Earth and not trying to understand how the universe works.

Vera Rubin*

* V.C. Rubin, An interesting voyage, *Annual Review of Astronomy and Astrophysics* **49**, 1–28 (2011).

One evening, when I was a child about ten years old, my mother told me that she knew something about astronomy that no one else knew. To this day, I remember thinking that this was extraordinary. What I learned later was that all scientists discover things that no one else knows (because this is the point of science), that some ideas are more interesting or more important than others, and that what my mom alone then knew was the beginning of the story of dark matter.

David Rubin, USGS†

† V.C. Rubin, An interesting voyage, *Annual Review of Astronomy and Astrophysics* **49**, 1–28 (2011).

Vera Rubin hated her high school physics class. According to her, it was a "big macho boys club," and the physics labs were a "nightmare." Her teacher was unaware of her passion for astronomy, and when she won a scholarship to Vassar College, he quipped, "As long as you stay away from science you should do okay." She graduated from Vassar in 1948, the college's only astronomy major that year, but was turned down for graduate study at Princeton because the school did not accept women into the astronomy program. (This remained Princeton's policy until 1975.) Instead, she earned a master's degree in astronomy from Cornell, and then a PhD from Georgetown in 1954. The next decade was devoted to family—four children, all of whom earned PhDs in science or mathematics. In 1965, she joined the Carnegie Institution's Department of Terrestrial Magnetism, where she teamed up with physicist and instrument builder Kent Ford to study the rotation of galaxies. Ford had designed an electronic "image tube" to intensify the light collected by telescopes, allowing the pair to collect and analyze light from the dim outskirts of galaxies much faster than previously possible. Focusing first on the Andromeda galaxy (aka M31), they used the Doppler effect to measure the rotational velocity of regions far from the galactic center, in order to determine the galaxy's total mass. They were in for a big surprise.

Most of the light emitted by a galaxy comes from its nucleus, so it was natural to assume that most of a galaxy's mass is concentrated within the nucleus. But by measuring velocities far from the center, Rubin discovered that about 90% of Andromeda's mass resided well outside its nucleus, in regions of space that emitted no light! She had uncovered vast troves of dark matter *within the Andromeda galaxy, a discovery that profoundly altered our understanding of the universe. Over the next few decades, she studied hundreds of other spiral galaxies and obtained similar results in every case: spiral galaxies contained about 10 times more dark matter than bright matter. Rubin's work established the existence of dark matter and showed scientists how shockingly little they knew about the universe. For more than 2000 years, they had been studying only a tiny fraction of its contents.*

14.1 INTRODUCTION

Physics was originally motivated by our ancestors' deep desire to explore the cosmos and know their place within it. Long before the Greeks, the sky watchers of ancient Babylon, Egypt, and the New World kept careful records of celestial events—lunar and solar eclipses, first sightings of stars—to schedule important religious celebrations and to regulate civic and agricultural activities. Greek astronomers continued this tradition of careful observation; more importantly, they employed the rigorous mathematical methods of Euclidean geometry to interpret what they saw. Among their many stunning achievements, they measured the Earth's radius, the Moon's diameter, and the Moon's distance from Earth. In addition, they concocted an ingenious geocentric model of the universe that described, with good accuracy, the apparent motions of the Moon, Sun, planets, and stars. But as astronomical instruments and observations improved, their geocentric model grew more cumbersome in the struggle to "save the appearances," that is, to resolve discrepancies between their observational data and the circular cycles and epicycles imposed by their cosmic model. At the close of the classical Greek era (ca. AD 100), the many brilliant achievements of Greek astronomy were compiled in Ptolemy's *Almagest*, the bible of astronomical thought that endured unchallenged for nearly 1400 years.

In the sixteenth and seventeenth centuries, the Ptolemaic world view was toppled by Copernicus, Kepler, and Galileo, who freed science forever from Aristotelian doctrine and laid the foundations for all of modern science. Copernicus demoted Earth to a mere planet orbiting the Sun; Kepler used the data of Tycho Brahe to disclose the true elliptical shape of planetary orbits; Galileo used a primitive telescope to observe the phases of Venus (disproving the geocentric model conclusively), measure the height of the Moon's craters, and discover Jupiter's "solar system" of moons.

In subsequent work, Galileo elevated *experiment* to its present status as the arbiter of scientific truth. With simple table-top experiments (using projectiles, rolling balls, and pendula), he overturned Aristotelian dogma and launched the quest for verifiable physical laws. Newton, who was born in the year of Galileo's death (1642), accredited him with the first two of his three laws of motion. Newton's inverse square law of gravitation successfully explained the elliptical orbits of the planets, the motion of comets, and much, much more. His crowning achievement was his demonstration that the inverse square law governed projectiles on Earth as well as celestial bodies in space; by inference—it held for *all* bodies.

The law of universal gravitation is a prime example of the underlying assumption of physics: the laws of motion that prevail here on Earth hold sway throughout the universe. The three conservation laws that we found by studying simple laboratory experiments apply to all bodies everywhere, from subnuclear particles in the Large Hadron Collider to galaxies at the edge of the observable universe. To use these laws productively, we introduced new mathematics: geometry, vectors, dot and cross products, and differential and integral calculus. The payoff has been prodigious. We can predict the trajectories of planets, moons, comets, and asteroids as well as the paths of projectiles here on Earth or anywhere else. We know how to navigate space by rocket propulsion and gravity assists and understand how a planet's size and temperature control its atmospheric composition. We understand the 12 hour rhythm of the tides, know why astronomical bodies are (or are not) spherically shaped, and know where to look for planetary moons and rings. We can determine the masses of planets, stars—even galaxies—and know how to find new worlds and new solar systems circling other stars. We understand how pulsars acquire their rapid spin, and can "see" invisible black holes by observing bodies in orbit about them. We know—in part—the history of the universe, and have glimpsed its ultimate fate. All of this is made possible by the three conservation laws and the auxiliary concepts of force and torque.

Yet the very things that Newton dismissed as self-evident have resurfaced to haunt us: time, space, and mass. Einstein's special and general theories of relativity have shown that these concepts are anything but obvious. Relativity tells us that time depends on speed and gravity; there is no "absolute" time that flows at the same rate for all observers. The same is true for "absolute" space. Moreover, it is the equality of inertial and gravitational mass—first appreciated by Newton—that inspired Einstein to reformulate gravity as a curvature of spacetime. In the twenty-first century, the cosmos holds as many mysteries for us as it did for Newton at the end of the seventeenth century.

In this, the final chapter of the text, we examine two contemporary cosmic mysteries that keep physicists and astronomers hard at work today. They are offered, not as additional topics to be mastered, but as examples of current scientific concerns that can be addressed using the mathematics and principles of mechanics you encountered earlier in this course. We hope that you will enjoy reading and thinking about them! If you are unable to discuss this material in class during the semester, we hope you will be inspired to read the chapter on your own—*after* the final exam. That would be truly wonderful!

14.2 THE CASE FOR DARK MATTER

Vera Rubin was not the first to conceive of dark matter. The idea originated in 1933 with theorist Fritz Zwicky, who argued that a galaxy's luminosity is not a reliable measure of its mass. To prove his point, Zwicky examined the Coma cluster, a group of perhaps 10,000 galaxies located about 100 Mpc from the Milky Way (Figure 14.1a). These galaxies are in random motion relative to the cluster's center of mass. To estimate the total mass of the cluster, Zwicky invoked the *virial theorem*, which states that for a gravitationally bound system, the average kinetic and potential energies are related in a simple way:

$$K_{avg} = -\frac{1}{2}U_{avg} \tag{14.1}$$

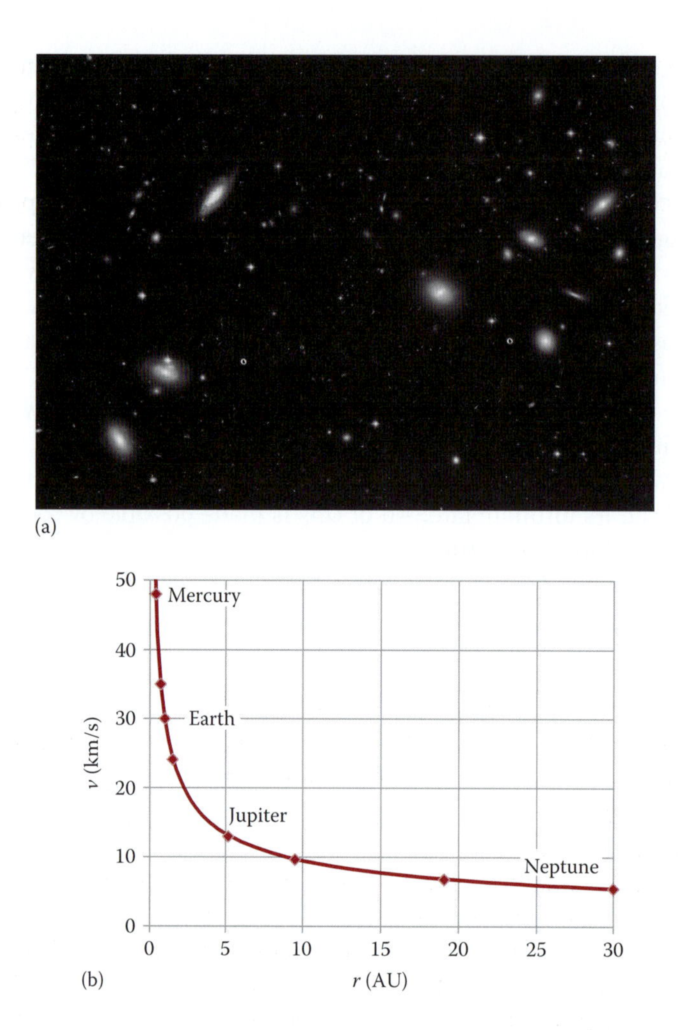

FIGURE 14.1 (a) Part of the Coma cluster of galaxies, imaged by the Hubble space telescope. The cluster center is about 100 Mpc from the Milky Way and is moving away from us at a speed of about 7000 km/s (in agreement with Hubble's law). There are roughly 10,000 galaxies within the cluster. (b) The "rotation curve" of the solar system. The orbital velocities of the eight planets obey the expected relationship $v \propto r^{-1/2}$. (Photo courtesy of NASA, ESA, and the Hubble heritage team (STScI/AURA).)

Equation 14.1 is certainly true for a body in circular orbit (see Equations 10.11 and 10.12). It is also true for one in an elliptical orbit. Rather than show this rigorously, let us simply note that the total energy of the body m in orbit about M is given by Equation 10.13: $E = -GMm/2a$, where a is the semimajor axis of the orbit, and its potential energy is $U = -GMm/r$. Therefore, the kinetic energy of the body at position r is

$$K = E - U = GMm\left(\frac{1}{r} - \frac{1}{2a}\right)$$

Assuming that the average distance between M and m is $r_{avg} = a$, then $K_{avg} = GMm/2a = -½U_{avg}$. The virial theorem, Equation 14.1, is simply an extension of these results to *multibody* bound systems.

The total kinetic energy of a galactic cluster can also be written as $K = ½Mv^2_{rms}$, where M is the total cluster mass and v_{rms} is the root-mean-square speed of a galaxy relative to the cluster CM (see Section 10.5). If the cluster can be approximated as a uniform density sphere of radius R, the gravitational potential energy can be shown to be

$$U \simeq -\frac{3}{5}\frac{GM^2}{R}$$

(see Problem 14.1). Inserting these expressions for K and U into Equation 14.1, we obtain an expression for the total galactic mass:

$$M \approx \frac{5}{3}\frac{v^2_{rms}R}{G} \tag{14.2}$$

Using modern redshift (Doppler) measurements of the Coma cluster, $v_{rms} \simeq 1500$ km/s and $R \simeq 1.5$ Mly $\simeq 4.6 \times 10^{19}$ km, so $M \approx 2.6 \times 10^{45}$ kg $\approx 1.3 \times 10^{15} M_{Sun}$. However, the total luminosity of the Coma cluster is only about 10^{13} times that of the Sun, so the cluster's mass is 130 times greater than what we would estimate from the amount of light it is emitting. Troubled by this discrepancy, Zwicky suggested that either the inverse square law of gravitation is not valid at the intergalactic distance scale or the cluster is held together by copious amounts of dark (nonradiant) matter. Most astronomers ignored Zwicky's paper, and assumed that the discrepancy would disappear when better data were available.

But the discrepancy did not disappear. Three decades later, Vera Rubin and Kent Ford began their study of galactic rotation. Rubin was interested in learning how a spiral galaxy's spin influences its structure and shape. Since most of the light emitted by the galaxy comes from its nucleus, it was reasonable to assume that most of the galaxy's mass is located there as well. Let's model the galaxy as a spherical nucleus of radius r_n and constant density ρ_n, surrounded by a thin disk of much lower density. Assuming that stars orbit the galactic center in circular orbits, it is straightforward to calculate what a star's velocity v should be as a function of its distance r from the galactic center. Within the nucleus ($r \leq r_n$), the shell theorem and Newton's second law combine to give

$$\frac{mv^2}{r} = \frac{GM(r)m}{r^2} \tag{14.3}$$

where, m is the mass of the orbiting star and $M(r) = \frac{4}{3}\pi r^3 \rho_n$ is the nuclear mass contained within a spherical shell of radius r.

Outside the nucleus, $\rho \ll \rho_n$, so that $M(r) \simeq M_n$, where M_n is the total mass of the nucleus. Solving for the velocity, we find

$$v = \sqrt{\frac{GM(r)}{r}} \propto \begin{vmatrix} r & \text{for } r \leq r_n \\ r^{-1/2} & \text{for } r > r_n \end{vmatrix}$$

so one would expect to observe a *rotation curve* $v(r)$ like the one shown in Figure 14.2a (and 14.1b). Instead, Rubin discovered that their rotation curves *leveled out* at large r, and $v(r)$ remained roughly constant as far from the galactic center as it was possible to gather data (Figure 14.2b). This was an

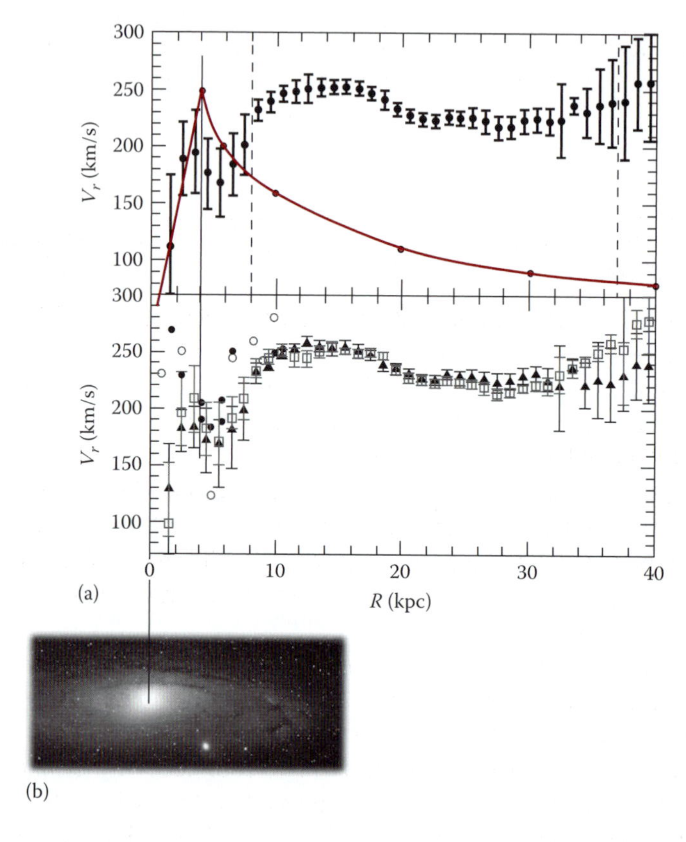

FIGURE 14.2 (a) Rotation curve for the Andromeda galaxy from Corbelli et al. (2010). The bottom panel shows data from both sides of the galactic nucleus. The top panel is the average of the two sides. Solid red curve is the theoretical curve for rotational velocity vs. distance assuming a constant density nucleus of radius 4 kpc. (b) Visible image of Andromeda, scaled to the same dimensions as the graphs above, scaled to the same dimensions as the graphs above, showing that visible matter is largely confined to a disk of radius $r < 10$ kpc. (Rotation curve of M31: Corbelli, E. et al., *A&A*, 511, A89, 2010; Image of Andromeda: Bill Schoening, Vanessa Harvey/REU program/NOAO/AURA/NSF. Reproduced with permission, copyright ESO.)

astounding discovery. From Equation 14.3, $v^2 = GM(r)/r$, so if $v(r) \approx$ const, $M(r) \propto r$. Since data was taken out to $r \approx 10\,r_n$, at least 90% of the galaxy's mass is located outside the nucleus, in regions where little or no light is emitted!

Rubin's discovery was compelling evidence for the existence of dark matter, and forced us to admit that, after more than two millennia of study, we still know nothing about most of the matter in the universe. In a lecture she delivered in 1995, she remarked: "In a spiral galaxy, the ratio of dark-to-light matter is about a factor of 10. That's probably a good number for the ratio of our ignorance to knowledge. We're out of kindergarten, but only in about the third grade."

> EXERCISE 14.1
>
> Rubin and Ford found that the velocity at $r = 24$ kpc $(= 7.4 \times 10^{20}$ m) was about 200 km/s. Find the mass of the galaxy. Express your answer in terms of solar masses ($M_{Sun} = 2 \times 10^{30}$ kg).
>
> Answer $M_A = 2.2 \times 10^{11} M_{Sun}$

14.3 EVIDENCE FOR DARK ENERGY

In 1929, Edwin Hubble discovered that light emitted by distant galaxies is redshifted by an amount proportional to the galactic distance: $\Delta\lambda/\lambda \propto r$. Hubble used his redshift data to show that the universe is expanding. He associated the redshift with the classical, Doppler effect: $\Delta\lambda/\lambda = v/c$, where v is the galaxy's radial speed, and concluded that $v = H_0 r$, where the proportionality constant H_0 is called the Hubble constant. In Chapter 3, we described the Hubble expansion as the colossal eruption of *space*—the big bang—that began nearly 14 billion years ago and continues today. Picture a two-dimensional space represented by the surface of a balloon, with spots painted on it. As you inflate the balloon, the separation of the spots increases. Similarly, as real space expands, it drags the galaxies along with it. The redshifts measured by astronomers occur—not because of the Doppler effect—because as the universe expands, the wavelengths of light propagating through space stretch with it.

Let's develop these ideas mathematically. For simplicity, imagine a one-dimensional space represented by a long, taut elastic band. Attach galaxies along the band at uniform intervals (see Figure 14.3a). When you grasp the ends of the band and pull, the band (space) will stretch and the distance between any two galaxies will increase. Let $r_1(t_0)$ be the distance between two neighboring galaxies at the present time t_0. (The subscript 0 indicates present time.) At a later time $t = t_0 + \Delta t$, the band has stretched and the intergalactic distance has changed by an amount $\Delta r_1(t)$. The change in the distance between two *next*-nearest galaxies is $\Delta r_2(t) = 2\Delta r_1(t)$, etc. The change in the distance between *any* two galaxies is therefore proportional to their separation: $\Delta r(t) \propto r(t_0)$. Dividing by Δt, this is mathematically equivalent to Hubble's law: $v = \Delta r/\Delta t \propto r$ or $v = H_0 r$.

More generally, the change in the distance between any two *points in space* is proportional to their separation. Suppose we pick two closely spaced points on the band that are presently separated by $\Delta r(t_0) = \Delta r_0$ (Figure 14.3b). As time advances, the separation increases, $\Delta r(t) = a(t)\Delta r_0$, where $a(t)$ is called the *scale*

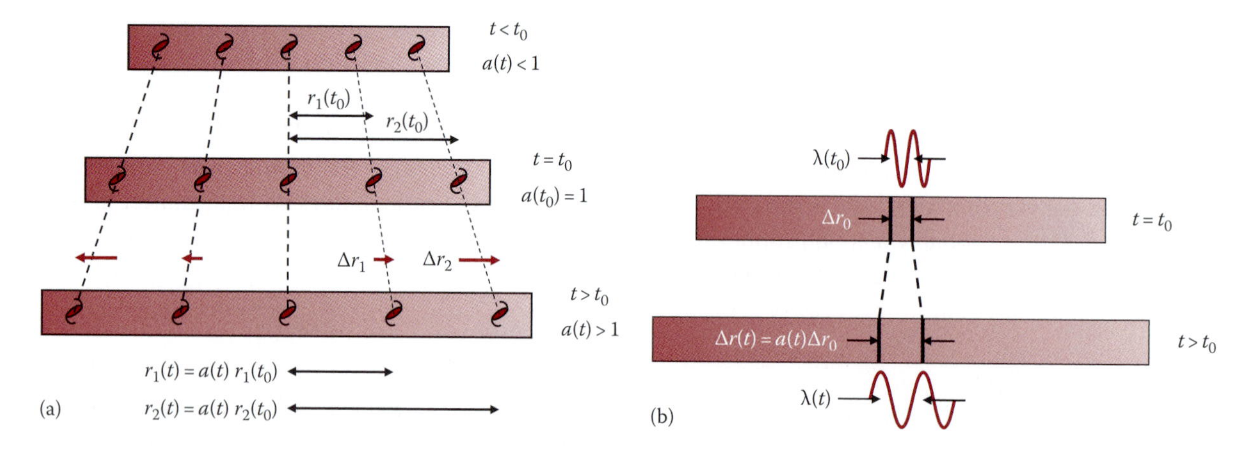

FIGURE 14.3 (a) A one-dimensional universe is represented by an elastic band with galaxies spaced uniformly along its length. When the band is stretched, the distance between two neighboring galaxies increases by Δr_1. The distance between next-nearest neighbors increases by $\Delta r_2 = 2\Delta r_1$, etc. The stretch can be expressed mathematically in terms of a scale factor $a(t)$ that increases with time: $r(t) = a(t)r_0$. (b) The separation of any two points increases when the band is stretched: $\Delta r(t) = a(t)\Delta r_0$. If $\Delta r_0 = \lambda_0$, where λ_0 is the wavelength of light at time t_0, then at a later time, $\Delta r(t) = \lambda(t) > \lambda_0$, that is, the light will be redshifted.

factor describing the expansion of space. Clearly, $a(t_0) = 1$, and $a(t) > 1$ for $t > t_0$. (Likewise, $a(t) < 1$ for $t < t_0$.) For two arbitrary times t_1 and t_2,

$$\frac{\Delta r(t_2)}{\Delta r(t_1)} = \frac{a(t_2)}{a(t_1)} \tag{14.4}$$

Suppose our two points mark the locations of successive crests of a light wave, that is, their separation is equal to the wavelength of the light: $\Delta r(t) = \lambda(t)$. Since $\Delta r(t)$ increases with time, $\lambda(t)$ also increases as the light propagates through space. This is the present interpretation of the Hubble redshift. Let t_e be the time when the light was emitted from a distant galaxy and t_0 be the present time—when the light is detected here on Earth. Then, from Equation 14.4, the fractional change in the wavelength between the times of emission and detection is

$$z \equiv \frac{\Delta\lambda}{\lambda_e} = \frac{\lambda_0 - \lambda_e}{\lambda_e} = \frac{a_0}{a_e} - 1$$

or (14.5)

$$z + 1 = \frac{a_0}{a_e} = \frac{1}{a_e},$$

where we have introduced the customary symbol z to indicate the redshift. λ_e is the wavelength at the time of emission; it is the *unshifted* wavelength that an observer would measure from a nearby source at rest relative to the observer. According to Equation 14.5, the measured redshift z is inversely related to the scale factor at the time of emission. Since $a_e < 1$, $z > 0$, or $\lambda_0 > \lambda_e$. Let's use these results to express Hubble's law in terms of the redshift z and the scale factor $a(t)$.

Hubble's law $v = H_0 r$ interprets the redshift as a Doppler effect: $z = \Delta\lambda/\lambda = v/c$. Combining these two equations,

$$cz = H_0 r$$

Now let's express H_0 in terms of $a(t)$.

EXERCISE 14.2

The value of the Hubble constant determined by Freedman et al. (Figure 3.7b), for galaxies within 400 Mpc, is $H_0 = 72 \pm 7$ km-s^{-1}/Mpc. What is the largest value of the redshift z included in this data?

Answer $z_{max} \approx 0.1$

For galaxies that are not too far away ($z \ll 1$), light will reach us in a time that is short compared to the age of the universe: $t_0 - t_e = \Delta t \ll t_0$. Consequently, the scale factor $a(t)$ does not change very much while the light is propagating, allowing us to expand $a(t)$ in a Taylor series (Section 9.9) about t_0. Keeping only the first order term: $f(x_0 - \Delta x) \simeq f(x_0) - f'(x)\big|_{x_0} \Delta x$, or

$$a_e = a_0 - \left.\frac{da}{dt}\right|_0 \Delta t$$

The notation $|_0$ indicates that the derivative is to be evaluated at the present time t_0. Using the above result and the binomial expansion $(1 + x)^n \simeq 1 + nx$, Equation 14.5 can be expressed as

$$1 + z = \frac{a_0}{a_e} = \frac{a_0}{a_0 - \left.\frac{da}{dt}\right|_0 \Delta t} \simeq 1 + \frac{1}{a_0}\left.\frac{da}{dt}\right|_0 \Delta t$$

or

$$z = \frac{1}{a_0}\left.\frac{da}{dt}\right|_0 \Delta t\,.$$

Finally, $\Delta t = r/c$, where r is the distance to the galaxy, so Hubble's law in terms of z and $a(t)$ is

$$cz = \frac{1}{a_0}\left.\frac{da}{dt}\right|_0 r \equiv H_0 r$$

and the Hubble constant is (14.6)

$$H_0 = \frac{1}{a_0}\left.\frac{da}{dt}\right|_0 .$$

EXERCISE 14.3

Let $a(t) = (t/t_0)^n$, where n is a constant. Show that $H_0 = n/t_0$.

Keep in mind that H_0 is an *experimentally* determined quantity equal to the *present* expansion rate of the universe. In general, the Hubble "constant" will change with time. The quest of cosmology is to trace the expansion $a(t)$ in the past and predict it for the future. The scale factor informs us about the birth and the ultimate fate of the universe.

Hubble's law is limited to galaxies with $z \ll 1$ (see Exercise 14.2). To learn about the expansion rate in the distant past, we need to observe galaxies that are farther away ($z \approx 1$). While light from these remote galaxies was *en route* to Earth, the universe expanded significantly. How is z related to the present distance to these galaxies? Consider a light wave from a very remote source, emitted at time t_e and reaching us here on Earth at time t_0. During a short time interval $(t, t + dt)$ while the light was propagating toward Earth, the wave advanced by $dr(t) = cdt$. By the time the wave is detected, this distance has stretched to $dr(t_0)$ due to the expansion of space. From Equation 14.4, with $a(t_0) = 1$, $dr(t_0) = dr(t)/a(t) = cdt/a(t)$. To find the present distance r_0 to the galaxy, we must sum—or integrate—over all of the incremental distances $dr(t_0)$ along the wave's path to Earth:

$$r_0 = r(t_0) = \int dr(t_0) = c\int_{t_e}^{t_0} \frac{dt}{a(t)} \tag{14.7}$$

We do not know $a(t)$ *a priori*, but if we have an independent means of measuring r_0, we can use it along with the measured redshift z to find $a(t)$. In particular, we can find out if the expansion is slowing down, speeding up, or proceeding at a steady rate.

Let's explore those three options. In Section 10.8, we examined three models of the universe. The simplest was the "empty universe" model, in which the gravitational attraction between galaxies is ignored. The second model was the "critical density" model, wherein the density of matter is just sufficient to halt the expansion in the distant future. The third model was one in which a mysterious repulsive force, dubbed "dark energy," dominates the expansion. Equations 10.17, 10.18, and 10.19 described the expansion as the changing radius $r(t)$ of a sphere containing a fixed amount of mass. The same three equations can be reexpressed in terms of the scale factor by substituting $a(t)/a_0 = r(t)/r_0$:

$$a(t) = \begin{vmatrix} t/t_0 & \text{empty universe} \\ (t/t_0)^{2/3} & \text{critical density} \\ e^{H_0(t-t_0)} & \text{dark energy only} \end{vmatrix} \tag{14.8}$$

In the first case, the universe is expanding at a steady rate; in the second, the expansion is slowing down; and in the third, it is accelerating. As you will see, there is now compelling evidence for the third option.

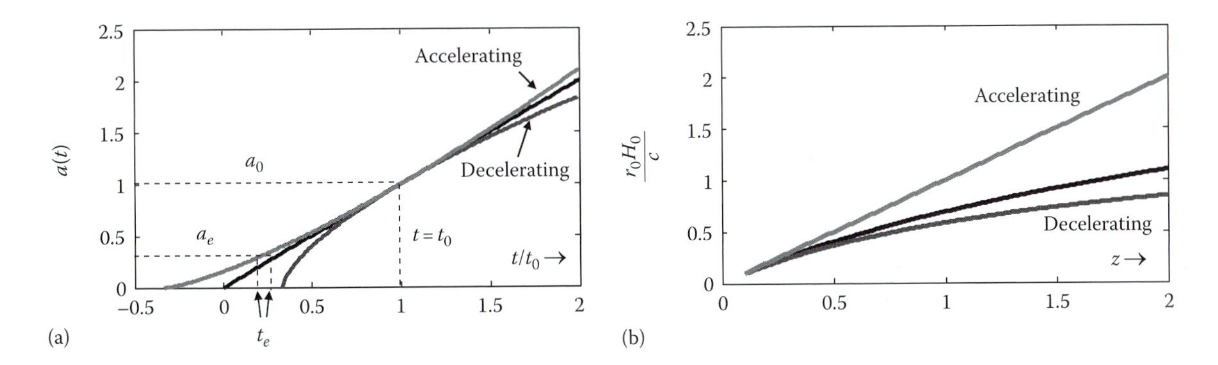

FIGURE 14.4 (a) The scale factor vs. time for the three models discussed in the text. For the same redshift $z = a_o/a_e - 1$, a_e has the same value for each model. This implies that t_e is earlier in the dark energy (accelerating) model than in the empty or critical density models. (b) The present distance r_o to a galaxy vs. its redshift z, where r_o is measured in units of c/H_o, the distance light emitted at the dawn of the universe would travel in the empty universe model. r_o is largest for the dark energy model, and least for the critical density model.

Figure 14.4a plots the scale factor for each of the above options. The three curves have been shifted horizontally so that they are tangent to one other at the present time t_o. This must be done because we have defined $a(t_o) = a_o = 1$ and because—no matter what model we choose—the derivative (slope) of $a(t)$ at the present time t_o must be equal to the presently measured value of the Hubble constant H_o (Equation 14.6). As a consequence, the age of the universe is different for each model. The time $t_o = 1/H_o$ is the age of the universe—the time since the big bang—only if the expansion rate has been constant, that is, only for the "empty universe" model.

EXERCISE 14.4

The present value of the Hubble constant is H_o. For the empty universe model, what was the value of the Hubble constant when the universe was half as old as it is now?

Answer $H(t_o/2) = 2H_o$

Using Equation 14.7, we can now calculate the present distance r_o in terms of the scale factor at times t_e and t_o for each of the three models. Then, we can rewrite r_o as a function of the redshift z using Equation 14.5. For the empty universe case,

$$r_o = ct_o \int_{t_e}^{t_0} \frac{dt}{t} = ct_o \ln\left(\frac{t_o}{t_e}\right)$$

or, with $a_0/a_e = t_0/t_e = 1 + z$

$$r_o = ct_o \ln(1+z) = \frac{c}{H_o} \ln(1+z). \quad \text{empty universe}$$

The results for the other two models are derived similarly:

$$r_0 = \begin{cases} \dfrac{2c}{H_0}\left[1-(1+z)^{-\frac{1}{2}}\right] & \text{critical density} \\ \dfrac{cz}{H_0}. & \text{dark energy} \end{cases}$$

These results are compared graphically in Figure 14.4b. For a given value of z, r_0 is greatest for the accelerating (dark energy) model and least for the decelerating (critical density) model. This makes sense, because if the expansion rate is increasing with time, then in the past it was slower than it is now. To experience the same redshift $z = a_0/a_e - 1$, the light wave must have traveled for a longer time than if the expansion had taken place at the present rate H_0. Therefore, the galactic source must be farther away. (See also Box 14.1.)

Earlier in this book (Section 1.8) we defined the *luminosity* L of a radiant body as the amount of energy emitted by the body per unit time. The *apparent brightness* b of the body—how bright it appears to an observer at a distance r—is inversely proportional to the square of that distance: $b = L/4\pi r_0^2$. If we knew the luminosity of the source, we could directly determine its distance by measuring its brightness. Edwin Hubble assumed that there was an upper limit to a star's luminosity, so by measuring the brightest stars of a galaxy, he could crudely estimate the distance to that galaxy. (His calculated distances were too low by a factor of 7.) In Section 3.7, we described a better "standard candle" called a *type 1a supernova*. This is the death throe of a white dwarf (a collapsed star that has run out of nuclear fuel) that has formed a close binary system with a large active star. Over millions of years, the white dwarf drains mass from its companion until it becomes gravitationally unstable and collapses, triggering a violent explosion that often gives birth to a pulsar. For a few weeks, the supernova radiates an enormous amount of energy ($L_{SN} \approx 10^{10} L_{Sun}$), often outshining the rest of its host galaxy. It has been found that the peak luminosity, the color spectrum and the light curve (luminosity vs. time) of all type 1a supernovae are nearly identical. They are ideal standard candles for determining distances.

Type 1a supervovae provide a way to probe the history of the Hubble expansion. The redshift is a measure of the expansion since the supernova exploded, and the observed brightness allows us to measure the present distance r_0 to the supernova site. The greater this distance, the farther back in time the event occurred. Therefore, by measuring the brightness vs. redshift for many supernovae over a wide range of z-values, we can obtain the functional relationship between r_0 and z, as in Figure 14.4b. It is then possible to extract the history of the expansion of the universe.

Figure 14.5a displays the calculated brightness vs. redshift for our three models. Note that the graph is a log-log plot, and that the brightness *decreases* in the upward direction.* Compare these three curves

* Decreasing brightness corresponds to increasing "magnitude" in the standard scheme used in astronomy, dating back to the writings of Hipparchus. The calculated quantity is $b \propto 1/r_0^2(1+z)^2$. The extra factor of $(1+z)^2$ arises from the quantum (photon) properties of light.

BOX 14.1 GALACTIC DISTANCES

In this box, we show that for a given redshift z, the present distance r_0 to a galaxy is greatest for an accelerating expansion. For simplicity, let $a(t) = (t/t_0)^n$. If $n = 1$, the universe is expanding at a steady rate, like the zero gravity "empty universe" model we considered in Section 10.8. If $n < 1$, the expansion is slowing down, as we would expect due to the gravitational attraction between galaxies. If $n > 1$, the expansion is accelerating due to some bizarre repulsive force, which we called "dark energy" in Chapter 8. For the simplest case ($n = 1$), we find

$$r_0 = ct_0 \int_{t_e}^{t_0} \frac{dt}{t} = ct_0 \ln\left(\frac{t_0}{t_e}\right) \tag{14.7}$$

For this case, Equation 14.5 tells us that $t_0/t_e = a_0/a_e = 1 + z$, so the solution of Equation 14.7 is

$$r_0 = ct_0 \ln(1+z)$$

For the more complicated cases where $n \neq 1$, a careful integration (and this is the only hard part of the derivation) yields

$$r_0 = \frac{ct'_0}{1-n}\left[1-\left(\frac{t_0}{t_e}\right)^{n-1}\right] = \frac{ct'_0}{1-n}\left[1-(1+z)^{\frac{n-1}{n}}\right]$$

Notice that we have denoted the present time as t'_0 to indicate that the age of the Universe may depend on the expansion rate. In fact, Equation 14.6 tells us that

$$H_0 = \frac{1}{a_0}\frac{da}{dt}\bigg|_0 = n\frac{t_0'^{n-1}}{t_0'^{n}} = \frac{n}{t'_0}$$

or $t'_0 = n/H_0$ for our three choices of n. Combining all of these results, the solution to Equation 14.7 is

$$r_0 = \begin{cases} \dfrac{c}{H_0}\ln(1+z) & \text{for } n = 1 \\ \dfrac{n}{1-n}\dfrac{c}{H_0}\left[1-(1+z)^{\frac{n-1}{n}}\right] & \text{for } n \neq 1 \end{cases}$$

Exercise B14.1: (a) Find r_0 for $z = 1$ for the three cases $n = 1$, $n = 1.5$ (an accelerating expansion), and $n = 0.5$ (a decelerating expansion). Express each answer as a multiple of c/H_0. (b) Show that for $z \ll 1$, the two expressions above yield the same result. This exercise shows that, for a fixed redshift z, the present distance to a galaxy is greatest for an accelerating expansion.

(*Hints:* $\ln(1 + x) \simeq x$ for $x \ll 1$. Simplify the second expression using the binomial theorem.)

Answer (a) $r_0 = 0.693c/H_0$, $0.780c/H_0$, $0.415c/H_0$

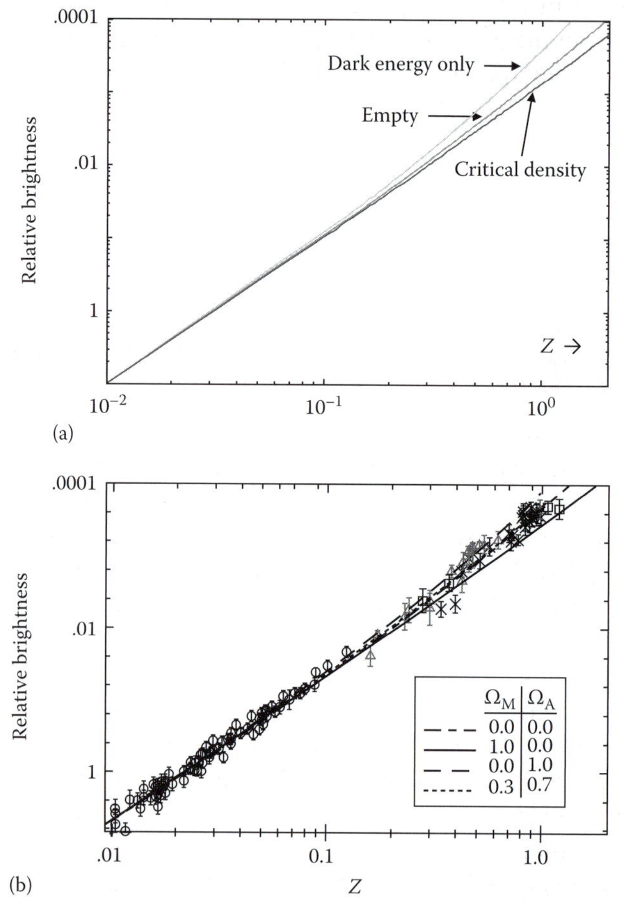

FIGURE 14.5 (a) The theoretical brightness of a distant galaxy vs. its redshift, calculated for the three models described in the text. Note that the three lines converge for $z \ll 1$, and also note the shape of the curves for higher z. For the same redshift, the source is farther away in the dark energy model, and so it is dimmer than for the other models. (b) Data collected by the high-z supernova search team led by Adam Riess and Brian Schmidt, as of 2006. The solid line corresponds to the critical density model, the upper dashed line to the pure dark energy model, and the middle line is for an empty universe. The high-z data lie above this middle line. (From A. Clocchiatti et al., *Astrophys. J.*, 642, 1, 2006. With permission.)

to the data* shown in Figure 14.5b. For any expansion governed by gravitational attraction between galaxies, the data should lie below the "empty universe" line. Instead, it lies *above* that line, between it and the curve representing the pure dark energy universe. The current "consensus" model—a "best fit" to the data—suggests that only 30% of the energy density of the universe is due to gravitationally interacting mass (calculated using $E = mc^2$, where m includes dark matter), while 70% is associated with dark energy. Scientists are now confident that they know the overall ingredients of the cosmos: 5% ordinary (baryonic matter), 25% dark matter, and 70% dark energy. We understand only about 1/20 of what the universe is made of.

Based on the data shown in Figure 14.5b, the future of the universe looks rather chilling. Space will continue to expand, drawing galaxies farther apart. Star formation will cease and existing stars will run out of nuclear fuel, collapsing to form black holes. One by one, the lights of the sky will be extinguished.

* A. Clocchiatti et al., *The Astrophysical Journal* **642**, 1–21 (2006).

14.4 SUMMARY

The topics presented in this chapter are just two of the surprises the universe holds in store for us. After more than 2000 years of study, we can only claim to understand a small fraction of its contents. What we do know has emerged from the cumulative work of countless men and women living on Planet Earth, in isolation from intelligent beings elsewhere. Their achievement is replete with creativity, brilliance, and heroic effort, yet also fraught with ordinary human frailty: pride, jealousy, and intellectual bias. It has been a thoroughly human activity, driven as much by ego as insatiable curiosity about our world and our role in it.

That role has been diminishing, in a cosmic sense, since the time of Copernicus. Nobel laureate Steven Weinberg, the author of the popular book *The First Three Minutes*, once quipped, "The more the universe seems comprehensible, the more it also seems pointless." He meant that, in the eyes of science, the universe is an impersonal place, and there is no cosmic drama in which we play the starring role. Weinberg added that if our only role is to make Planet Earth a "little island of warmth and love and science and art," then our existence is far from meaningless.†

At this moment (2014), *Voyager I* has left the solar system and entered interstellar space. It will soon be out of radio contact with us. Will it be detected by an alien civilization? Will the *Pioneer* spacecraft—with its plaque containing directions to Planet Earth—be intercepted by an intelligent civilization far beyond the solar system? Certainly not in our lifetime nor in the lifespan of our grandchildren or great grandchildren. Even if a successor to the *Kepler* Observatory were to detect

† Weinberg made these remarks in an interview for the series "Faith and Reason" aired by the Public Broadcasting System.

life on a distant exoplanet (which seems probable), we—and they—will remain in physical isolation for millennia, perhaps forever. No one is coming to rescue us from our own ignorance and misbehavior.

In 1990, as *Voyager I* journeyed beyond Saturn, astronomer Carl Sagan asked NASA scientists to reorient the spacecraft so that it could take one final picture of Earth and relay it back to us. That picture, the famous "pale blue dot" photograph shown in Figure 14.6, inspired Sagan to ponder the significance of human endeavor: "To my mind, there is perhaps no better demonstration of the folly of human conceits than this distant image of our tiny world. To me, it underscores our responsibility to deal more kindly and compassionately with one another and to preserve and cherish that pale blue dot, the only home we've ever known."*

* From C. Sagan, *Pale Blue Dot: A Vision of the Human Future in Space*, Random House, New York (1994).

FIGURE 14.6 The "Pale Blue Dot" (within the white circle) is the final image of Earth taken by *Voyager* as it passed beyond Saturn in 1990. (The vertical gray band is an artifact due to scattered light from the Sun. Image: NASA/JPL.)

PROBLEMS

14.1 Consider the change in potential energy when a small mass dM is carried from infinity to the surface of a larger spherical body of radius r and mass $M(r)$:

$$dU = -\frac{GM(r)dM}{r}$$

When dM is accreted by the larger body, it spreads out uniformly over the body's surface, increasing its radius by dr. As more mass is added to the body, its radius and mass increase until they reach their final values R and M. If the body has a uniform density ρ, then $M = \frac{4}{3}\pi R^3\rho$.

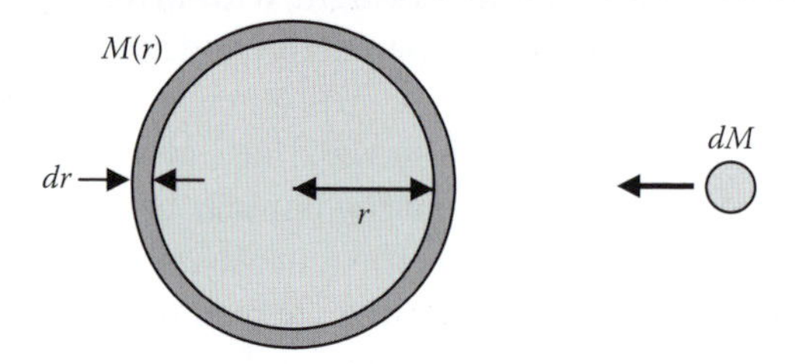

(a) Show that the change in radius dr when dM is added to the sphere is $dr = dM/4\pi r^2\rho$.

(b) Integrate dU to derive the total change in potential energy when the body is assembled from diffuse matter initially located at infinity. Express your answer in terms of the final mass M and radius R.

(This derivation is needed for the estimation of dark matter using the virial theorem.)

(c) A pulsar is a neutron star with mass $M \approx 1.4 M_{Sun}$ and radius $R \approx 10$ km. Use your result from part (b) to show that the pulsar's gravitational "self-energy" U is about 10% of its rest mass energy $E = Mc^2$. This means that the inertial mass of the pulsar is about 10% less than the total inertial mass of its ingredients before they are compressed together. (Einstein's equivalence principle states that the gravitational and inertial masses of a body are exactly equal. It has never been tested rigorously for astronomical bodies held together by strong self-gravity, such as a pulsar. In 2014, a triple star system containing a millisecond pulsar was discovered that may make such a test possible. If $m_{grav} \neq m_{iner}$ for such bodies, Einstein's version of general relativity will be disproved.)

14.2 The total *visible* mass of the Coma cluster is about $10^{13}\, M_{Sun}$. If the cluster galaxies are distributed uniformly within a spherical volume of radius $R = 1.5$ Mly and are moving with a root-mean-square velocity $v_{rms} \simeq 1500$ km/s, show that the cluster cannot be bound together gravitationally by the visible mass alone.

14.3 In the empty universe model, the time since the big bang, when $a(t) = 0$, is given by $t_0 = 1/H_0$. Consider instead the critical density universe, $a(t) = (t/t_0)^{2/3}$.

(a) How long ago was the big bang in this model? Express your answer in terms of H_0. (*Hint:* Use Equation 14.6. Also see Figure 14.4a.)

(b) In principle, the most distant objects we can see today are ones whose light, emitted at the time of the big bang, is just reaching us now. The present distance to these objects is called the *horizon distance* r_{hor}. Use Equation 14.7 to calculate r_{hor} for the critical density model. Express your answer in terms of H_0.

(c) In the "consensus model," the early universe was dominated by gravity (decelerating), whereas it is presently dominated by dark energy (accelerating). The theoretical horizon distance is $r_{hor} = 3.24c/H_0$. Calculate this distance in terms of megaparsecs (Mpc).

14.4 Galaxy NGC4258 is a spiral galaxy similar in size and luminosity to Andromeda. Doppler analysis shows that its disk gas is in nearly perfect Keplerian rotation ($v \propto 1/\sqrt{r}$) about a supermassive black hole at the galactic center. The center of the galaxy is moving away from us with speed 470 km/s.

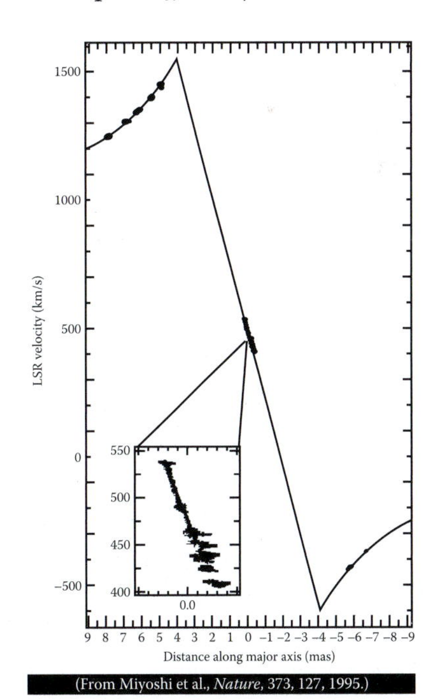

(From Miyoshi et al., *Nature*, 373, 127, 1995.)

(a) Find the distance to the galaxy, in Mpc.

(b) The figure shows the speed of the disk (relative to us) as a function of distance from the center. The x-axis (distance) is plotted in units of milliarcseconds (mas). Show that 1 mas is equivalent to a distance of 0.032 pc (1 pc = 3.09×10^{16} m).

(c) The solid line is the best fit to the data, assuming Keplerian rotation. From the high velocity data ($|x| > 4$ mas), find the mass of the black hole at the galactic center. Express your answer in terms of M_{Sun}.

14.5 An alternative explanation of the flat rotation curves measured by Rubin and Ford suggests that Newtonian mechanics must be modified for very small accelerations: $a < a_0 \approx 10^{-10}$. For these tiny accelerations, $F \simeq ma^2/a_0$.

(a) Rubin measured the rotational speed v of Andromeda to be 200 km/s at a distance $r = 24$ kpc (7.4×10^{20} m) from the galaxy's center. Find the centripetal acceleration of a star at this value of r.

(b) Let $F = GMm/r^2$ be the inverse-square gravitational force on a star (mass m) due to the galactic mass M. Show that the modified force law yields a flat rotation curve for large r.

The modified Newtonian dynamics (MOND) hypothesis was tested in the laboratory by Abramovici and Vagar (1986),* and more recently by Gundlach et al. (2007).† In the latter experiment, an oscillator with period 795 s was set in motion with an equivalent amplitude A in the range $0.5 < A < 600$ nm.

(c) Assuming sinusoidal motion, what is the maximum acceleration of the oscillator at an amplitude of 10 nm?

(d) The maximum acceleration of the oscillator occurs at the turning points, where $x = \pm A$ and $F = kA$. According to the MOND hypothesis, would the oscillation period increase, decrease, or stay the same as $A \rightarrow 0$? (No deviation from Newton's second law was observed in the experiment.)

* A. Abramovici and Z. Yager, Test of Newton's second law at small accelerations, *Physical Review D*, **34**, 3240 (1986).

† J.H. Gundlach et al., Laboratory test of Newton's second law for small accelerations, *Physical Review Letters*, **98**, 150801 (2007).

EPILOGUE

If I had to ask a question of the infallible oracle..., I think I should choose this: "Has the universe ever been at rest, or did the expansion start from the beginning?' But, I think, I would ask the oracle not to give the answer, in order that a subsequent generation would not be deprived of the pleasure of searching for and of finding the solution."

Georges Lemaître (1931)*

---

* G. Lemaître, The beginning of the world from the point of view of quantum theory, *Nature* **128**, 704–706 (1931).

APPENDIX A: PHYSICAL UNITS

A.1 METROLOGY

Each time you fill your car's gas tank, you use a fuel pump that measures the volume of gas dispensed and calculates its cost. If you look carefully at the pump, you will find a small inspection sticker which indicates that a government agent has checked the pump and has certified that the volume of gas you are charged for has been measured with an accuracy of ±0.3%. Imagine if this were not the case. If gas pumps were accurate to only ±10%, or if their accuracy were never verified, the advertised price per gallon would have little meaning, and you would be reluctant to frequent any but a few trusted service stations. This would be a nuisance for you and a nightmare for the gas industry. In order to thrive, the industry must have accurate and traceable standards for measuring volume: a gallon of gas in New York must be equal to a gallon in California.

Science has similar concerns. In 250 BCE the Greek astronomer Eratosthenes estimated the circumference of Earth to be 252,000 *stadia*. How accurate was his estimate? We don't really know, since we have only a rough idea of the length of a stade. Two-thousand two-hundred and fifty years after Eratosthenes (1999), the *Mars Climate Orbiter* spacecraft was lost or destroyed as it attempted to enter the Martian atmosphere, due to a simple mix-up of units: ground controllers sent instructions to the orbiter's thrusters in English units (pounds), but the spacecraft software interpreted them as metric units (Newtons, 1 lb = 4.4 N). *Metrology*—the science of measurement—seeks to avoid such difficulties by providing a unique system of precisely defined units that can be used by scientists everywhere, allowing them to conduct experiments, exchange results, and clearly communicate their conclusions.

Today, all scientists use the Système International d'Unités, or SI, for the bulk of their work. SI is the outgrowth of the standardized units for mass and length introduced by the French in the eighteenth century to promote trade and commerce throughout Europe. In 1889, the kilogram and meter were formally defined in terms of physical artifacts stored in Sèvres, France (near Paris). To make use of these *primary* standards, they were copied as faithfully as possible and the replicas (*secondary* standards) were distributed to governments of other countries.* The same artifacts were used for scientific purposes; even today, when you use a research-grade scale in the laboratory, its calibration is traceable to one of the replica kilograms. But science demands an ever-increasing level of precision. Are we sure that the standard kilogram is not changing with time (due to surface oxidation, for example)? If it is, we may need to correct our current measurements (of atomic mass, say) so that they are consistent with measurements made 50 years ago.†

---

* The United States owns 2 of the original 40 copies of the kilogram, and about once every 40 years, they are returned to France to be compared directly against the primary standard.

† In fact, the secondary standards are well known to have diverged from the primary standard kilogram. Over a period of 100 years, the secondary standards disagree with the primary by up to ±25 μg.

In 1889, the meter was defined as the distance between two scratch marks on a metal bar, so the accuracy of the length standard was limited by the roughness of the scoring (±0.1 μm, approximately). This accuracy was sufficient for nineteenth century science, but not for modern research. The ultimate goal of metrology is to provide a complete set of practical standards that satisfy the evolving needs of science, can be reproduced *locally*, and are guaranteed by the laws of physics to yield the same accuracy anywhere and anytime.

A.2 SI UNITS

In the SI, there are seven independent *base quantities* with which *all* physical quantities can be expressed. These are listed in Table A.1 along with their common symbols. For the study of classical mechanics, the most important of these are length, mass, and time, and their associated SI units are the meter, kilogram, and second. In addition to the base quantities and units of Table A.1, there are an endless number of *derived* quantities and units defined as products of the base quantities, each raised to some power (see Table A.2). For example, acceleration is defined as meters/second^2, or m/s^2. Some of these derived units are grouped together and given a special name, often in honor of the scientist who originated the associated concept.

TABLE A.1 Base Quantities and Units of the SI

Base Quantity (Property)	Base Unit	Symbol
Length	meter	m
Mass	kilogram	kg
Time	second	s
Electric current	ampere	A
Thermodynamic temperature	kelvin	K
Amount of substance	mole	mol
Luminous intensity	candela	cd

TABLE A.2 Some Derived SI Units

Derived Quantity	Symbol	Derived Unit	Symbol
Area	A, S	Square meter	m^2
Volume	V	Cubic meter	m^3
Speed or velocity	v	Meter per second	m/s
Acceleration	a	Meter per second squared	m/s^2
Mass density	ρ	Kilogram per cubic meter	kg/m^3
Momentum	p	Kilogram meter per second	kg m/s
Angular momentum	L	Kilogram meter-squared per second	kg m^2/s
Angular frequency, angular velocity	ω	Radians per second	rad/s
Angular acceleration	α	Radians per second squared	rad/s^2
Moment of inertia	I	Kilograms meter squared	kg m^2
Torque	τ	Force meter	N m

TABLE A.3 Derived Units with Special Names

Derived Quantity	Name	Symbol	Expression in Base Units
Angle	radian	rad	m/m = 1
Solid angle	steradian	sr	$m^2/m^2 = 1$
Frequency	hertz	Hz	s^{-1}
Force	newton	N	$kg\ m/s^2$
Pressure	pascal	Pa	$N/m^2 = kg\ m\ s^{-2}$
Energy, work, heat	joule	J	$N\ m = kg\ m^2\ s^{-2}$
Power	watt	W	$J/s = kg\ m^2\ s^{-3}$
Electric charge	coulomb	C	A s
Electric potential	volt	V	$J/C = kg\ m^2\ s^{-3}\ A^{-1}$
Magnetic field	tesla	T	$N\ s/C\ m = kg\ s^{-2}\ A^{-1}$

Not surprisingly, the derived unit of force ($kg\ m/s^2$) is the Newton, and the unit of energy ($kg\ m^2/s^2$) is the Joule. The most important groupings used in mechanics are listed in Table A.3.

You may be wondering how *all* physical quantities can be described in terms of just seven base quantities. For example, how is magnetic field associated with meters, kilograms, and seconds? The force on an electric charge q moving with velocity $\vec{v}$ through a magnetic field $\vec{B}$ has been found experimentally to be $\vec{F} = q\vec{v} \times \vec{B}$. The derived unit of charge is the coulomb or ampere second (A s), so from the force law,

$$1\,\text{N} = 1\,\text{kg m/s}^2 = 1\,(\text{A s})(\text{m/s}) \cdot B.$$

The units of B are, therefore, $N\ A^{-1}\ m^{-1}$, or in terms of SI base units, $kg\ A^{-1}\ s^{-2}$, a grouping that is called the *Tesla* in honor of the inventor and electrical engineer Nikola Tesla: $1\ T = 1\ kg\ A^{-1}\ s^{-2}$. In deriving this result, note that we treated the units just like numbers: they obey all of the rules of algebra. For example, the acceleration of a body is $\vec{a} = \vec{F}/m$, with units $N/kg = kg\ m\ s^{-2}/kg = m\ s^{-2}$.

A.3 WORKING WITH UNITS

The base SI units are often too small or too large to express physical properties in a convenient and easy-to-grasp way. For such cases, the SI offers a set of prefixes that can be appended to the units. These prefixes (such as micro- or mega-) are shorthand for multiplying the base unit by 10^n, and the most commonly used ones are listed in Table A.4. The radius of a hydrogen atom is 5.3×10^{-11} m, but it is easier to visualize and express as 0.053 nm, even though the two descriptions are equivalent.

Another strategy for treating a large or small quantity is to define an appropriately-scaled *non*-SI unit, such as the electron volt (eV) or the astronomical unit (AU), to describe the physical property. A short selection of these units is given in Table A.5. To employ them, you must know how to convert them to SI units, and conversely, how to express SI units in terms of them. This entails the use of *conversion factors* such as the ones listed in Table A.5. As you know, the value of a physical quantity must always be written as the

TABLE A.4 Common Prefixes Used in the SI

Factor	Name	Symbol
10^{15}	peta	P
10^{12}	tera	T
10^{9}	giga	G
10^{6}	mega	M
10^{3}	kilo	k
10^{2}	hecto	h
10^{1}	deca	da
10^{-1}	deci	d
10^{-2}	centi	c
10^{-3}	milli	m
10^{-6}	micro	μ
10^{-9}	nano	n
10^{-12}	pico	p
10^{-15}	femto	f

"product" of a number and a unit. *Both the number and the unit obey the rules of algebra.* For example, the value of the Hubble constant is usually expressed as $H_0 \simeq 72$ kg s^{-1} Mpc^{-1} (megaparsecs). Since 1 km = 10^3 m and 1 pc = 3.09×10^{16} m, we can convert H_0 to SI base units in the following way:

$$72\frac{\cancel{\text{km}}\ \text{s}^{-1}}{\cancel{\text{Mpc}}} \times \left(\frac{10^3\ \cancel{\text{m}}}{1\ \cancel{\text{km}}}\right) \times \left(\frac{1\ \cancel{\text{Mpc}}}{10^6\ \cancel{\text{pc}}}\right) \times \left(\frac{1\ \cancel{\text{pc}}}{3.09\times10^{16}\ \cancel{\text{m}}}\right) = \frac{72\times10^3}{3.09\times10^{22}}\ \text{s}^{-1} = 2.35\times10^{-18}\ \text{s}^{-1}.$$

The numerator of each bracketed term is equal to its denominator (e.g., 10^3 m = 1 km), so each term is equivalent to unity, and multiplying a quantity by 1 does not change its value. Identical units are "cancelled" just as if they were equal numerical factors.

Since km and pc are both units of length, the *dimensions* of H_0 are length time^{-1}/length = time^{-1}, or L T^{-1}/L = T^{-1}. Our final expression for H_0 must have the same dimension (T^{-1}) as the starting expression. This kind of dimensional consistency can be used to check your unit conversions, and more generally, all of your calculations. All terms within an equation, on both sides of the equal sign, must have the same dimensions. Moreover, *the arguments of trigonometric functions and exponential functions must be dimensionless*; otherwise they do not make sense. To illustrate, the position of a simple harmonic oscillator is $x(t) = A\cos(\omega t + \theta_0)$. Since radians are dimensionless, the units of angular speed ω are s^{-1}, and the quantities ωt (= $T^{-1}T$) and θ_0 (expressed in radians, not degrees) are both dimensionless. Likewise, if a quantity $y(t)$ decays exponentially as $y(t) = Ae^{-\alpha t}$, the constant α must have dimension T^{-1} in order for the exponent to be dimensionless.

TABLE A.5 Non-SI Units Used in Mechanics

Quantity	Unit	Symbol	Conversion to SI Units
Time	Minute	min	1 min = 60 s
	Hour	h	1 h = 3,600 s
	Day	d	1 d = 86,400 s
	Year	y	1 y = 3.156×10^{7} s
Length	Light year	ly	1 ly = 9.46×10^{15} m
	Parsec	Pc	1 pc = 3.09×10^{16} m
	Astronomical unit	AU	1 AU = 1.50×10^{11} m
	Mile	mi	1 mi = 1,609 m
	Yard	yd	1 yd = 0.9144 m
	Foot	ft	1 ft = 0.3048 m
	Inch	in	1 in = 0.0254 m
	Angstrom	Å	1 Å = 10^{-10} m
	Fermi	fm	1 fm = 10^{-15} m
Mass	Atomic mass unit	u	1 u = 1.660×10^{-27} kg
	Metric ton	T	1 T = 1,000 kg
Volume	Cubic centimeter	cm^3	1 $cm^3 = 10^{-6}$ m^3
	Liter	L	1 L = 10^{-3} m^3
	Gallon (U.S.)	gal	1 gal = 3.785×10^{-3} m^3
	Barrel (42 gal)	barrel	1 barrel = 0.159 m^3
	Fluid ounce	fl oz	1 fl oz = 2.957×10^{-5} m^3
Velocity	Kilometers per hour	km/h	1 km/h = 0.2778 m/s
	Miles per hour	mi/h	1 mi/h = 0.4470 m/s
Angular frequency, angular velocity	Revolution per second	rev/s	1 rev/s = 6.283 rad/s
Angle	Degree	°	1° = π/180 rad
	Arc second	as or $\hat{''}$	$1\hat{''}$ = 4.848×10^{-6} rad
	Revolution per minute	rpm	1 rpm = 0.1047 rad/s
Energy, work, heat	Electron volt	eV	1 eV ≅ 1.602×10^{-19} J
	Erg	erg	1 erg = 10^{-7} J
	Calorie	cal	1 cal = 4.184 J
	British thermal unit	Btu	1 Btu = 1.054×10^{3} J
	Kilowatt hour	kW h	1 kWh = 3.6×10^{6} J
	Megaton	Mt	1 Mt = 4.18×10^{15} J
Force	Dyne	dyn	1 dyn = 10^{-5} N
	Pound	lb	1 lb = 4.448 N
Power	Horsepower	hp	1 hp = 746 W
Pressure	Bar	bar	1 bar = 10^{5} Pa
	Millimeter of mercury	torr	1 torr = 133.3 Pa
	Atmosphere	atm	1 atm = 1.013×10^{5} Pa
	Pound per square inch	psi	1 psi = 6.895×10^{3} Pa
Magnetic field	Gauss	G	1 G = 10^{-4} T

A.4 THE FUTURE OF THE SI

The use of physical artifacts to define base units is far from ideal. The 1889 standard kilogram and meter are stored in secure vaults, making them inaccessible to the scientific community. Moreover, they are susceptible to damage whenever they are transported or used to calibrate other standards; their properties change with age, and their accuracy is fixed by their mechanical design. Similar problems plague the unit of time. The second was originally specified to be 1/86,400 of a mean (average) solar day, but this definition was soon revised to compensate for Earth's erratic, ever-slowing rotation rate. Better standards are required to meet the evolving needs of modern scientific work.

In the 1950s, it became clear that clocks based on atomic properties are capable of very high accuracy and stability. In an atomic clock, a beam of gaseous cesium (Cs) atoms is irradiated by microwaves of frequency f. Cs atoms absorb this radiation strongly if f is within the extremely narrow range 9,192,631,770 ± 20 Hz (1 Hz = 1 cycle/s). The absorption is monitored and used to control the frequency emitted by the microwave source. The frequency where the absorption is strongest is measured by an electronic *counter*, which counts the exact number N of wave periods in a time interval Δt: $f = N/\Delta t$. If the counting time Δt is 1000 s, the frequency uncertainty is $\Delta f \simeq \pm 1/\Delta t = \pm 0.001$ Hz, or about 1 part in 10^{13}. But f is an *invariant of nature*: it depends only on the atomic properties of Cs, which do not change with time. Rather than using the counter to monitor frequency, metrologists assign an *exact* value to f: $f(\text{Cs}) = 9{,}192{,}631{,}770$ Hz, which effectively defines the second to be

> the duration of 9,192,631,770 periods of the radiation corresponding to the transition between the two hyperfine levels of the ground state of the cesium 133 atom.*

Note the strategy used above: the second has been *indirectly* defined by assigning an exact value to an invariant of nature, in this case the frequency of radiation f that is most strongly absorbed by Cs atoms.

Using the same strategy, the meter is indirectly defined by assigning an exact value to the speed of light in a vacuum, another fundamental constant of nature: $c = 299{,}792{,}458$ m/s. This fixes the meter to be

> the length of the path traveled by light in vacuum during a time interval 1/299,792,458 s.

This definition frees the base unit of length from its former dependence on a physical artifact. Now, scientists everywhere have access to the standards of length and time if they have a Cs atomic clock (to measure time) and can determine how far light travels in a fixed amount of time. A high priority goal of metrology is to define the seven base units of the SI by assigning exact numerical values to seven independent constants of nature.

---

* A thorough and highly readable description of the SI can be found in the on-line brochure, a concise summary of the International System of Units, the SI, on the website of the *Bureau International des Poids et Mesures* (2006), http://www.bipm.org/en/measurement-units/. See also Gordon J. Aubrecht II, Changes Coming to the International System of Units, *The Physics Teacher* **50**, 338–342 (2012).

At the present time, the kilogram is the only SI base unit that is still defined by a physical artifact. No matter how the primary standard (called the international prototype of the kilogram, or IPK) is eroded by time, its mass is by definition exactly 1 kg. This situation is likely to change soon. The kilogram will be defined indirectly by assigning an exact value to another fundamental constant, *Planck's constant h*, which plays a central role in the microscopic world of atoms, nuclei, and elementary particles. The latest (2014) measured value of Planck's constant, determined by the National Institute of Standards and Technology (United States), is $h = 6.62606979 \pm (0.00000030)$ J s. In base SI units, the units of h are kg m^2/s. Since the units of time and length have already been defined, the kilogram will be defined when an exact numerical value is assigned to h. To achieve a more stable standard than the IPK, h must be measurable to better than 1 part in 10^8. The apparatus for achieving this precision, called a watt balance, is still under development. When it is fully operational, perhaps as early as 2015, the standard kilogram in Sèvres will no longer remain equal to 1 kg.

APPENDIX B: ASTROPHYSICAL DATA

TABLE B.1 Planet Physical Data

Body	Mass (10^{24} kg)[a]	Mass (M_{Earth})	Equatorial Radius (10^6 m)[a]	Equatorial Radius (R_{Earth})	Surface Temperature (K)	Density (kg/m^3)	Roche Limit[b] (10^6 m)
Sun	1.99×10^6	3.33×10^5	696	109.1	5778	1410	1209
Mercury	0.33	0.055	2.44	0.38	340	5430	6.65
Venus	4.87	0.82	6.05	0.95	730	5240	16.3
Earth	5.97	1.0	6.378	1.0	290	5520	17.5
Mars	0.642	0.11	3.39	0.53	227	3930	8.29
Jupiter	1900	318	71.5	11.2	165	1330	122
Saturn	568	95.2	60.3	9.45	134	690	82.3
Uranus	86.8	14.5	25.6	4.01	76	1270	43
Neptune	102	17.2	24.8	3.88	72	1640	45.3

[a] Masses are expressed in units of 10^{24} kg, for example, $M_{Sun} = 1.99 \times 10^6 \times (10^{24}\text{ kg}) = 1.99 \times 10^{30}$ kg. The same scheme is used to express radii: $R_{Sun} = 696 \times (10^6\text{ m}) = 6.96 \times 10^8$ m.

[b] Calculated using an average moon density of 4000 kg/m^3.

TABLE B.2 Orbital Data of the Sun, Planets, Dwarf Planets, and Their Main Moons							
		Semimajor Axis					
Body	Central Body	(10^9 m)[a]	AU	Eccentricity	Orbital Period	Orbital Speed (km/s)	Mass (10^{21} kg)[a]
Sun	Milky way	2.6×10^{11}			200 Million years	250	1.99×10^9
Mercury	Sun	57.9	0.39	0.2	88 Days	47.9	330
Venus	Sun	108.2	0.72	0.007	225 Days	35.0	4.87×10^3
Earth	Sun	150	1.0	0.017	365.3 Days	29.8	5.97×10^3
Moon	Earth	0.385	0.0026	0.055	27.3 Days	1.02	73.5
Mars	Sun	228	1.52	0.093	1.88 Years	24.1	642
Phobos	Mars	0.0094			0.32 Days	2.14	1.1×10^{-5}
Deimos	Mars	0.023			1.3 Days	1.35	1.5×10^{-6}
Jupiter	Sun	779	5.2	0.048	11.9 Years	13.1	1.9×10^6
Io	Jupiter	0.422			1.77 Days	17.3	89.3
Europa	Jupiter	0.671			3.55 Days	13.7	48.0
Ganymede	Jupiter	1.07			7.15 Days	10.9	148
Callisto	Jupiter	1.88			16.7 Days	8.20	107
Saturn	Sun	1.43×10^3	9.54	0.057	29.4 Years	9.65	5.68×10^5
Tethys	Saturn	0.294			1.89 Days	11.3	0.62
Dione	Saturn	0.377			2.74 Days	10.0	1.1
Rhea	Saturn	0.527			4.52 Days	8.48	2.3
Titan	Saturn	1.22			15.9 Days	5.57	135
Iapetus	Saturn	3.56			79.3 Days	3.26	1.8
Uranus	Sun	2.87×10^3	19.2	0.047	83.8 Years	6.8	8.68×10^4
Ariel	Uranus	0.19			2.52 Days	5.5	1.3
Umbriel	Uranus	0.27			4.14 Days	4.7	1.3
Titania	Uranus	0.44			8.8 Days	3.7	1.8
Oberon	Uranus	0.58			13.5 Days	3.1	1.7
Neptune	Sun	4.5×10^3	30.1	0.009	163.7 Years	5.43	1.02×10^5
Triton	Neptune	0.354			5.88 Days	4.4	13.4
Pluto	Sun	5.91×10^3	39.5	0.249	248 Years	4.74	15
Charon	Pluto	0.02			6.39 Days	0.2	1
Ceres	Sun	414	2.8	0.076	4.6 Years	17.9	0.94
Haumea	Sun	6.45×10^3	43.1	0.2	283 Years	4.5	4.0
Eris	Sun	1.01×10^4	68.0	0.44	560 Years	3.4	16.7
Sedna	Sun	7.8×10^4	532	0.86	11,400 Years	1	Unknown

[a] Semimajor axis length and mass are expressed in units of 10^9 m and 10^{21} kg, respectively. For example, $a_{Earth} = 150 \times (10^9 \text{ m}) = 1.50 \times 10^{11} \text{ m} = 1$ AU and $M_{Earth} = 5.97 \times 10^3 \times (10^{21} \text{ kg}) = 5.97 \times 10^{24}$ kg.

APPENDIX C: PHYSICAL CONSTANTS

Constant	Symbol	Three-Figure Value	Best Known Value[a]
Speed of light	c	3.00×10^{8} m/s	299,792,458 m/s (exact)
Elementary charge	e	1.60×10^{-19} C	$1.602176565(35) \times 10^{-19}$ C
Planck's constant	h	6.63×10^{-34} J s	$6.62606979(30) \times 10^{-34}$ J s
Gravitational constant	G	6.67×10^{-11} m^{3} kg^{-1} s^{-2}	$6.67384(80) \times 10^{-11}$ m^{3} kg^{-1} s^{-2}
Boltzmann's constant	k	1.38×10^{-23} J/K	$1.3806488(13) \times 10^{-23}$ J/K
Universal gas constant	R	8.31 J mol^{-1} K^{-1}	8.3144621(75) J mol^{-1} K^{-1}
Avogadro's number	N_A	6.02×10^{23} mol^{-1}	$6.02214129(27) \times 10^{23}$ mol^{-1}
Stefan–Boltzmann constant	σ	5.67×10^{-8} W m^{2} K^{-4}	$5.670373(21) \times 10^{-8}$ W m^{-2} K^{-4}
Bohr radius	a_0	5.29×10^{-11} m	$5.2917721092(17) \times 10^{-11}$ m
Electron mass	m_e	9.11×10^{-31} kg	$9.10938291(40) \times 10^{-31}$ kg
		0.511 MeV	$5.4857990946(22) \times 10^{-4}$ u
			0.510998928(11) MeV
Proton mass	m_p	1.67×10^{-27} kg	$1.672621777(74) \times 10^{-27}$ kg
		938 MeV	1.007276466812(90) u
			938.272046(21) MeV
Neutron mass	m_n	1.67×10^{-27} kg	$1.674927351(74) \times 10^{-27}$ kg
		940 MeV	1.00866491600(43) u
			939.565379(21) MeV
Hubble's constant	H_0	72 ± 7 km s^{-1} Mpc^{-1}	
Luminosity of the Sun	L	3.85×10^{26} W	3.846×10^{26} W

[a] The number appearing in parentheses is the uncertainty in the last two digits, so $1.602176565(35) \times 10^{-19}$ is a concise way of expressing $(1.602176565 \pm 0.000000035) \times 10^{-19}$. Numerical values are those recommended by the Committee on Data for Science and Technology (CODATA), compiled by the Physical Measurements Laboratory, National Institute of Standards and Technology, Gaithersburg, MD, and posted on http://physics.nist.gov/cuu/Constants/.

INDEX

C

F

G

O

P

Q

R

U

V

W

Y

Z